D1073905

THE TEACHING AND LEARNING OF MATHEMATICS AT UNIVERSITY LEVEL

New ICMI Study Series

VOLUME 7

Published under the auspices of the International Commission on Mathematical Instruction under the general editorship of

Hyman Bass, President   Bernard R. Hodgson, Secretary

The titles published in this series are listed at the end of this volume.

The Teaching and Learning of Mathematics at University Level

An ICMI Study

Edited by

DEREK HOLTON

University of Otago,
Dunedin, New Zealand

Michèle Artigue

Urs Kirchgräber

Joel Hillel

Mogens Niss

and

Alan Schoenfeld

(Section Editors)

KLUWER ACADEMIC PUBLISHERS
DORDRECHT / BOSTON / LONDON

FLORIDA GULF COAST
UNIVERSITY LIBRARY

A C.I.P. Catalogue record for this book is available from the Library of Congress.

ISBN 0-7923-7191-7

Published by Kluwer Academic Publishers,
P.O. Box 17, 3300 AA Dordrecht, The Netherlands.

Sold and distributed in North, Central and South America
by Kluwer Academic Publishers,
101 Philip Drive, Norwell, MA 02061, U.S.A.

In all other countries, sold and distributed
by Kluwer Academic Publishers,
P.O. Box 322, 3300 AH Dordrecht, The Netherlands.

Printed on acid-free paper

All Rights Reserved
© 2001 Kluwer Academic Publishers
No part of the material protected by this copyright notice may be reproduced or
utilized in any form or by any means, electronic or mechanical,
including photocopying, recording or by any information storage and
retrieval system, without written permission from the copyright owner.

Printed in the Netherlands.

FLORIDA GULF COAST
UNIVERSITY LIBRARY

TABLE OF CONTENTS

PREFACE

This book is the final report of the ICMI study on the Teaching and Learning of Mathematics at University Level. As such it is one of a number of such studies that ICMI has commissioned. The other Study Volumes cover assessment in mathematics education, gender equity, research in mathematics education, the teaching of geometry, and history in mathematics education.

All of these Study Volumes represent a statement of the state of the art in their respective areas. We hope that this is also the case for the current Study Volume.

The current study on university level mathematics was commissioned for essentially four reasons. First, universities world-wide are accepting a much larger and more diverse group of students than has been the case. Consequently, universities have begun to adopt a role more like that of the school system and less like the elite institutions of the past. As a result the educational and pedagogical issues facing universities have changed.

Second, although university student numbers have increased significantly, there has not been a corresponding increase in the number of mathematics majors. Hence mathematics departments have to be more aware of their students' needs in order to retain the students they have and to attract future students. As part of this awareness, departments of mathematics have to take the teaching and learning of mathematics more seriously than perhaps they have in the past.

As a consequence, university mathematicians are more likely to take an interest in mathematics education and what it has to offer. In the past the contact between mathematics educators and practising university teachers had been poor. Thus there is a need to bridge the gap that exists in many countries, between mathematics educators and university mathematicians.

Finally, university mathematicians tend to teach as they were themselves taught. Unless they have a particular interest in teaching they are unlikely to make changes in their teaching or to exchange views, experiences or knowledge with their colleagues at other institutions. Hence this Study was commissioned to provide a forum for discussing, disseminating and interchanging, educational and pedagogical ideas between and among, mathematicians and mathematics educators.

As in every study, an International Programme Committee was appointed by the ICMI Executive Committee to oversee our Study's development. The members of the IPC were

Derek Holton (Ed.), The Teaching and Learning of Mathematics at University Level: An ICMI Study, v—viii.

© 2001 Kluwer Academic Publishers. Printed in the Netherlands.

Nestor Aguilera, Argentina
Michèle Artigue, France
Frank Barrington, Australia
Mohamed E.A. El Tom, Quatar
Joel Hillel, Canada
Derek Holton, New Zealand
Urs Kirchgraber, Switzerland
Lee Peng Yee, Singapore
Mogens Niss, Denmark
Alan Schoenfeld, USA
Hans Wallin, Sweden
Ye Qi-xiao, PRC.

The progress of ICMI Studies takes the following pattern. Once the IPC is appointed they produce a Discussion Document that contains a discussion of the key issues of the Study. This is widely circulated along with a call for reactions by way of abstracts of papers, proposals, the raising of other issues, etc. The Discussion Document for this Study appeared in the ICMI Bulletin, No. 43, December 1997.

As a result of the submissions, participants were invited to attend the Study conference that took place in Singapore in December 1998. This working conference included plenary sessions, submitted papers, panel discussions and working groups. The conference and the ideas and material developed at the conference forms the basis for this Study Volume. Extra material has been assembled since the conference by a number of authors.

One publication related to this Study, which is not in the general pattern of ICMI Studies, was the publication in February, 2000, of a special issue of the International Journal of Mathematics Education in Science and Technology. Papers produced for this issue were expanded versions of papers given at the Singapore conference.

As I said above, the Study conference was a working conference. It consisted of Plenary Sessions, Panel Discussions and Working Groups. The Plenary Sessions were as follows:

Claudi Alsina: Why the Professor should be a stimulating teacher: Towards a new paradigm of teaching mathematics at university level.
Michèle Artigue: What can we learn from didactic research carried out at university level?
Hyman Bass: Research on university-level mathematics education: (Some of) what is needed and why
Bernard Hodgson: Teaching and learning mathematics at the university level: a personal perspective.
Lynn Arthur Steen: Redefining university mathematics: the stealth campaign.

There were three Panel Discussions. The titles of these and the panel members are listed below.

Secondary/Tertiary Transition
   Frank Barrington, Myriam Dechamps, Francine Gransard
Mass Education
   Garth Gaudry, Gilah Leder
Technology
   Ed Dubinsky, Celia Hoyles, Richard Noss

Finally there were eleven working groups. The Titles and Chairs of these Working Groups are listed below. As the titles alone do not necessarily give a clear view of the area covered we have added some explanation.

   Secondary-Tertiary Interface, Leigh Wood and Sol Garfunkel
   the interface between secondary and tertiary mathematics learning and teaching; interactions between secondary and tertiary teachers.
   Mathematics and Other Subjects, Jean-Pierre Bourguigon
   what mathematics is needed in other disciplines; which department should undertake this teaching?
   Preparation of University Teachers, Harvey Keynes
   what is the role of technology in mathematics education at the tertiary level; what should that role be; what programmes exist that use technology?
   Assessing Undergradute Mathematics Students, Ken Houston
   principles and purposes of assessment; methods of assessment; obstacles to change.
   Trends in Curriculum, Joel Hillel
   what topics are common to many curricula; what changes have occurred in the recent past; what changes are anticipated in the future?
   Practice of University Teaching, John Mason
   some principles of teaching; examples of innovative practice.
   Mass Education, Nestor Aguilera and Hans Wallin
   mathematics as a service course; what mathematics do students need; what is a good model for teaching students with a range of abilities and interests?
   Preparation of Primary and Secondary Mathematics Teachers, Honor Williams
   what is the current state of preparation; how might this change in the future; what is the role of academic mathematicians in teacher preparations?
   Policy Issues, Hyman Bass
   what are the different means of policy development? how do these affect practice? in what ways can policy be effected?
   The Future of Research in Tertiary Mathematics Education, Annie Selden and John Selden.
   what research is being and has been undertaken; how can this be translated into practice; what new directions should be explored?

I would like to thank the participants of the various working groups for their input to the Study. In particular, I would like to thank those who made contributions to the working group reports that appear in this volume. Unfortunately there has not been space in this book to mention them all individually.

As the result of the Study conference and reflecting on the issues raised in the working groups and in the more formal sessions, the Study seemed to naturally fall into seven parts, the seven sections of this book. These are an Introduction, Trends in Curriculum and Teaching Practice, Research, Mathematics and Other Disciplines, Technology, Assessment in Tertiary Mathematics Education, and Teacher Education. Each section has been edited by the people named at the start of that section.

Finally, I should like to thank the following people. First, there are the other members of the IPC. Without their considerable help the Study would never have reached the conference stage. They also provided an invaluable initial refereeing of papers for the special issue of the iJMEST

Second, I would like to thank Lee Peng Yee and his Local Organising Committee. They worked extremely hard to produce a conference that ran like clockwork but that still had a friendly personal touch.

Third, I would like to thank the conference participants and contributors to this Study Volume. It is their expertise that enabled us to produce a book that provides the latest thinking in a range of aspects of university-level mathematics education.

Then fourthly I am extremely grateful for the contribution of the editors of this Volume. Their knowledge and ability have carried this volume over a wide range of areas to present a thorough overview of the topic, and their individual knowledge and skills have enabled the volume to extend to great depths in all areas of the Study.

Next I would like to thank Leanne Kirk, Lenette Grant and Irene Goodwin for their considerable secretarial help throughout my period of engagement with this Study.

Sixth, I would like to thank the two people who were Executive Secretaries of ICMI during the period of the Study, Bernard Hodgson and Mogens Niss. Bernard shepherded through the Study to its final published form; Mogens was indispensable to me throughout and was always available with wise counsel from the beginning to the end of the project. So much that happened could not have happened without his support and guidance.

Finally I want to thank my wife Marilyn for supporting me through this and many other endeavours.

Derek Holton
University of Otago, Dunedin, New Zealand
dholton@maths.otago.ac.nz

SECTION 1

INTRODUCTION

Edited by Derek Holton and Mogens Niss

CLAUDI ALSINA

WHY THE PROFESSOR MUST BE A STIMULATING TEACHER

Towards a new paradigm of teaching mathematics at University level

1. INTRODUCTION

Mathematics at the University level is a complex field to explore. The diversity of institutions and social and cultural contexts, the variety of curricula and courses, the reforms taking place at present, etc., may induce us to believe that perhaps it makes no sense to talk about general or common aspects of our academic activities. But after many years of observing our own profession, of visiting so many places around the world and interacting with so many colleagues I have identified some problems and some challenges that may be of interest for mathematicians who love mathematics and love teaching. The aim of this presentation is to share some critical thoughts and to point out some constructive ideas on the educational goals of teaching mathematics at the university level.

2. SOME CRITICAL VIEWS ON EXISTING MYTHS AND PRACTICES IN UNIVERSITY TEACHING OF MATHEMATICS

In this section I would like to unmask some very general existing 'myths' (Kirwan, 1991) and practices in the teaching of mathematics at the undergraduate level that have a negative influence (Lewis, 1975) on the quality of mathematics teaching.

The researchers-always-make-good-teachers myth. This university myth says that 'researchers are *ipso facto* good teachers ... therefore the key criteria for selection and promotion must be high quality research'. Following Kline (1977) we quote the statement that:

> Hence appointment, promotion, tenure and salary are based entirely on status in research... but for most of the teaching that the universities are, or should be, offering, the research professor is useless.

3

Derek Holton (Ed.), The Teaching and Learning of Mathematics at University Level: An ICMI Study, 3—12.
© 2001 *Kluwer Academic Publishers. Printed in the Netherlands.*

This myth calls for a number of observations.

1. Sound knowledge does not necessarily mean active research;
2. The majority of mathematics courses do not include advanced results reached in recent decades;
3. Research takes place in thousands of different specialities, most of it in very narrow fields, and lines of research are often a matter of free choice and quite unrelated to teaching;
4. Unfortunately, research criteria are closely related to the Department's interests and rarely include research into mathematics education.

Let us remember here the critical words expressed in Kline (1977):

> The mania for research has produced an invidious system of academic promotion, perversion of undergraduate education, and contempt for and flight from teaching.

While for graduate, doctoral and post-doctoral teaching activities there is no doubt that only the most up-to-date and active researchers can introduce students to the latest results, techniques and trends, this does not hold true for most undergraduate programmes (see Carrier et al, 1962).

The self-made-teacher tradition. This is another standard mathematical myth and is based upon the claim that excellence in university teaching does not require any specific training - it is just a matter of accumulated experience, clear presentation skills and a sound knowledge of the subject. This approach leaves room for a lot of creative freedom but at the same time it can lead to quite a lot of anxiety, especially for inexperienced young teachers, who will in general try to reproduce the models that they have been exposed to during their own education. This myth does not make provision for students who are exposed to various styles of teaching simultaneously and it also avoids the issue of critical input from colleagues as well as the positive training that one would expect from the institutions involved.

Some classical references on this topic come from the 70s (e.g. CTUM, 1979, EBLE, 1974, Rogers, 1975, Rosenberg, 1972, Wilson, 1974).

Clearly, teaching may benefit from training and this must be a compulsory activity for those who want to teach.

Context-free universal content. This idea justifies the content of many courses as 'basic skills and results which must be learned by everyone taking the course'. This myth generated classic courses that were given to almost everyone entering science or technological university studies. It is taken for granted that some elements of linear algebra, calculus, differential equations, discrete mathematics, probability, statistics, etc., constitute the 'core' curriculum of university mathematics. In particular this myth justifies the concept that teaching is context-free, i.e., independent of personal interests, of specific professional training, of cultural environment, of social circumstances, and so on. While this situation makes for a more flexible teaching organization (anyone can teach anything), it sacrifices students' interest and kills interdisciplinary approaches. This led to wide and even universal sales for some textbooks. We, however, believe that contents must be related to interest, special needs, context, and the like (see COMAP, 1997, Howson, 1988, Pollack, 1988, Steen, 1989).

Deductive organization. In this case, 'teaching' is thought to be assimilated thanks to representations of deductive thinking. Topics are presented linearly, definitions-theorems-proofs are sequentially stated in their most general form. In particular this presentation leads to the need for constant proofs (the more formal the better) and leaves little room for discussion or historical remarks ... "How?" becomes more important than "Why?". (Freudenthal, 1991). Is deduction more important than induction? Is formal reasoning more important than plausible thinking? Clearly, deduction is only one component of mathematical thinking.

The top-down approach. This approach holds that by teaching mathematical topics in their most general form, students will be able to deal with any particular case, any example, any application. This gets rid of the problem of real data and the main elements of mathematics modelling. Learning is a bottom-up process, so teaching top-down is not an effective way of helping learners (see e.g. Begle, 1979).

The perfect-theory presentation. Mathematics courses present positive results, solved problems, bona fide models. Students become convinced that mathematics is almost complete, that theorem proving is just a deductive game, that errors, false trials, and zig-zag arguments, which play such a crucial role in human life, have no place in the mathematical world. Unfortunately, in some ways many textbooks have inherited the cold research-journal style. This style of presentation kidnaps the 'human nature' of mathematical discoveries, the mistakes that were made, the difficulties and the need for simplifications. In some cases (e.g. statistics) this gives the false idea that the 'real subject' is 'the mathematical model', when we know that mathematics may be a powerful tool but it needs to be used in combination with other disciplines or techniques. In addition, we are presented with the paradox that very often this perfect presentation implies only an instrumental understanding instead of a relational understanding. This perfect-theory presentation turns a living discipline into a dead garden.

The 'master class'/formal lecture paradigm. Teaching has frequently been oriented towards 'communicating' mathematical knowledge. Typically, a class for undergraduates would consist of a large group of students sitting, listening and writing in a classroom where a professor delivers several hours per week of spoken-written presentation before a blackboard, see Bligh (1972). After the lectures, students are supposed to study the delivered content by reading notes, the textbook and by solving *ad hoc* exercises proposed for each chapter-talk. This reduces 'teaching' to lecturing, and 'learning' to an individual after-class activity of assimilating results and practising techniques. In particular, as noted by Clements (1998), students spend a lot of time inefficiently or unproductively:

> ... a considerable part of the time is devoted to the transference from the notes of the
> lecturer to the notepads of the students of relatively straightforward factual material.

While 'master classes'/formal lectures are fine when truly 'masterly', they could
nevertheless be combined with other techniques of communicating and working.

The mature students myth. At the freshman level, this myth assumes that during the
few weeks between high school and university registration, students have grown in
such a way that their integration into the new university atmosphere does not require
any special attention. In particular, students going into scientific or technical courses
are assumed to be already motivated and aware of the relevance of mathematics to
their training, and students going into other studies are assumed to constitute a low-
interest class. The diversity of backgrounds is often ignored. The high school
curriculum may often be unknown. Clearly, the transition from secondary schools to
universities needs special attention.

The routine individual-written assessment. This presents the final test, or a written
examination mixing questions and exercises, as an ideal method of marking, i.e., of
gauging how well students master the content delivered in lectures. The method
focuses on individual preparation and rarely opens doors to project work, group
activities, open questions, etc. In its most rigorous form, this assessment is reduced
to a final exam to be marked and rarely integrates other activities or information
attained during the course into the student's progress. More flexible assessment
resources should be considered (see e.g. Dossey, 1998).

The non-emotional audience. This tries to present students enrolled in a course as an
audience at a movie show or a theatre. The main goal 'for all' is simply
mathematics. Individual problems, emotional difficulties, personality features do not
belong to the teaching and learning of mathematics. Tuition is for solving technical
doubts or clarifying previous lectures. Outside the classroom or the scheduled office
hours there is no place for further human interaction. The university walls keep
human nature out. To sum up, let me quote Krantz (1993):

> I don't think that it is healthy for a mathematics teacher to worry about math anxiety.
> Your job is to teach mathematics. Go do it.

That's a terrible mistake. The 'audience' is a group of people in which each
individual needs attention.

We, as mathematics educators working at university level, need to destroy the
above myths, practices and considerations by taking some positive steps towards
another way of teaching (see Howson, 1994).

3. TOWARDS A NEW PARADIGM OF TEACHING MATHEMATICS AT UNIVERSITY LEVEL

In this section we will identify some changes to be considered, some questions which need to be faced urgently and some goals for our future as mathematicians and mathematics educators.

There is a need to redefine mathematical research as a university activity, combining it with a soundly based teaching excellence. The critical pressure of research has evolved into a crazy rolling snowball: publishing as many papers as possible, going into citation and impact indices, attending an increasing number of congresses. It is time to sit down and think about what the main goals of universities today are. It is just possible that good teaching, fine multimedia and educational materials, virtual projects, community work, etc. are becoming more relevant to administrators and society than subscriptions to journals, abstract announcements and department reports. This does not mean a change from the research-realm to the teaching-paradigm. The 'either-research-or-teaching' polarity is false. With a little wisdom both activities can be (and should be) combined. Research also means writing expository papers, critiques of trends, historical perspectives, good texts, analyses of pedagogical materials, improvement of proofs, suggestions as to new approaches or interdisciplinary applications. Institutions and authorities should recognise and stimulate scholarship and research. And there is no need to say that the creation of exclusive research institutions is to be welcomed. But universities cannot close their eyes to their teaching ends. It is not just a question of achieving one annual award or medal for academic distinction but rather it is a matter of continuously controlling and stimulating the quality of education. Good teaching is according to a classic definition: "building understanding, communicating, engaging, problem solving, nurturing and organizing for learning", a complete task that merits special attention and preparation (see Krantz, 1993).

Research into mathematics education at tertiary level may be itself an interesting field of research and may give rise to useful results for all teachers for application to their teaching. Research into mathematics education is a growing scientific discipline (see Niss, 1998, Thurston, 1990). Nowadays it involves many researchers focusing on a wide range of topics and levels. However, there is clearly still a rich agenda for research on teaching and learning problems at university level. It would be marvellous if in the years to come this university research attracted well qualified mathematics specialists. If institutions wish or need to pay more attention to their educational goals, then mathematics education may – or indeed is certain to – play an increasingly important role in people's *vitae*. Though non-educational research has been a priority in people's careers until now, it could well be healthier if future mathematics specialists combined research with more educational aims. Moreover, research into mathematics education gives rise to useful results which should be disseminated and used, so that all mathematics teaching staff may benefit from an up-to-date knowledge of this field (see Niss, 1998).

All mathematics teachers can benefit from efficient training. While the training of teachers for non-university levels is a well-acknowledged necessity, much has yet to be done at tertiary level. There is much to explore here. While young people entering Departments may benefit greatly from the most experienced instructors, this training needs to be institutionalised and rewarded. Possibly, distance learning training courses would help to solve this problem, and they could include such materials as interactive CD-ROMS with case studies or videotaped classes, audio-visual resources, examples of model teaching procedures, alternative skills in teaching, etc. The development of specific societies or associations (or working groups within the existing ones) would also facilitate this training.

Mathematical content of courses needs continuous reform, renewal and a close link between what students learn and what they will need in their near future. In short, context is important for what we teach, motivation is essential, modes of thought are key concepts. Contents are not ends in themselves but means to ensure efficient knowledge. Examples, applications, historical backgrounds, modelling processes, the results that have been achieved in recent years... all these are or must be the *spiritus movens* of the curricula. Let us remember here some words of Sol Garfunkel on the "Principles and Practice of Mathematics" in COMAP (1997):

> The central importance of mathematics in our technologically complex world is undeniable, and the possibilities for new applications are almost endless. But at the undergraduate level, little of this excitement is being conveyed to our students. Currently, attention is being focused on reforming calculus, the traditional gateway course into the undergraduate curriculum. No one is questioning the importance and beauty of continuous mathematics. However, reformed or not, calculus is one branch (and a highly technical one) of a very rich subject. We know the breadth and richness of our subject; how, then, do we expect the students who are starting their study to gain these insights?

> Our proposed new start or gateway into the college mathematics curriculum is only a revolutionary idea for our discipline; other disciplines have had such courses in place for years...

> We asked ourselves a simple question: in designing the first undergraduate course for math and science majors, what should such a course look like?...

> The course content stresses the breadth of mathematics, discrete and continuous, probabilistic as well as deterministic, algorithmic and conceptual. We emphasize applications that are both real and immediate. And the text includes topics from modern mathematics that are currently homeless in the undergraduate curriculum.

Innovative teaching approaches will have a positive effect on learning processes and may affect the mathematical contents, the dynamics of teaching and some basic educational problems such as developing sound assessment methods. We would like to distinguish three levels of innovative teaching, corresponding to tools to be used, new pedagogical strategies and the issue of assessment.

1. *Innovative technological tools.* Beyond chalk we have at hand software packages (graphical, numerical, symbolical...), videotapes and audiocassettes, CD-ROMs, CD-I... and a wide variety of technological devices: computers, networks, televisions, radios, phone-fax-modems, overhead/computer projectors, etc. These tools can be incorporated into multimedia classrooms, computer labs and virtual campuses, and allow us to combine synchronic and asynchronic communications, presentations and interaction (see Reznick, 1988). Visualizing statistics data with Minitab; analyzing polynomial approximations or splines by means of Mathematica, Derive, Maple, MATLAB or whatever; visiting the polyhedra sites on the World Wide Web; showing COMAP videos on real mathematical modelling processes; working on historical mathematics on Internet data bases; simulating random processes on a work station; navigating on an interactive mathematical CD-ROM. All these examples are non-traditional tools and can be applied perfectly well to university topics.

2. *Innovative pedagogical strategies.* Problem solving, Internet forums, on-line tutorials, homework on the net, guided reading, simulation experiences, workshops, *ateliers*, group work, cooperative learning, a rich collection of dynamic learning strategies may be combined. Some of them assign an unconventional role to teachers but all of them contribute to stimulating motivation, interest and active participation. A surprising fact, proved in research into mathematics education, is that people learn more by talking and writing than by listening. Some of these pedagogical strategies come either from high school level or from distance/open learning environments, but they have received scant attention at the university level and far more development is needed.

3. *Innovative modes of assessment.* Open problems, project work, group projects, portfolios, critical reading, these, and other examples may lead to a more interesting approach towards a positive assessment method in which students can show what they have achieved, their capacities, their success as learners, etc. Continuous assessment is more demanding than final grading and requires small groups, but it is certainly an excellent way to produce assessment tasks that help the students to realise their real level of achievement. This means assessment must be considered as an informative and positive tool for individuals, and not as the traditional final diagnosis (see Steen, 1989, and 1992).

Mathematics courses need a complete review and clarification of what kind of professionals are trained. Mathematics as a service subject may be controversial and may need special consensus between mathematics specialists and other professionals. But the present undergraduate mathematics courses in many places are completely out of this world. These courses train 'mathematicians', i.e., people who know 'mathematics', but they are often undefined in terms of the professionals they wish to offer to the market. The usual paradox is that mathematics majors may be quite well prepared for entry into pure or applied research where few jobs are available. But most people go into teaching jobs, commercial divisions and computer companies, positions for which they have not been trained. It would be desirable for mathematics courses to attract more brilliant people (many nowadays go into other technical studies) and for some elements of specialisation to be introduced: for teaching, finance, engineering. The 'pure or applied' dichotomy is just a false opposition. Perhaps it is time to break mathematics courses into several separate fields of studies with specific goals. The departmental structure needs a review.

Mathematics as a service subject must teach students a certain body of knowledge, techniques and principles which will be relevant to their training as professionals, and it must show students how to acquire more knowledge and how to use it. In Howson, Kahane, Langinie, and de Turckheim (1988), one can find interesting views on mathematics as a service subject. Let me quote here just two of the stronger statements that originated from the ICMI Study:

> It would be desirable that service mathematics courses enabled students to acquire a range of essential knowledge, skills and modes of mathematical thought; each professional activity demands a particular mathematical literacy so mathematics courses must include applications, examples, modelling processes, etc. which motivate students.

Mathematicians must teach these courses but they need to be ready to prepare themselves for the job. Very often, mathematics has played a selective role in undergraduate studies. This is not true anymore and time devoted to mathematics has suffered a dramatic cut. It is therefore essential to redefine service subject courses to include the most basic training in useful mathematics. Let me quote here the old claim of M. Kline:

> ... the only generally understood fact about the subject is that it is understandable professors are content to offer courses that reflect their own values at the expense of students' needs and interests.

If we do not want to be seen this way, then an interdisciplinary approach is called for (see e.g. Heyneman, 1990, Gaskell-Klamkin, 1974).

Technology is a keystone for innovative global teaching of mathematics at universities in a world-wide situation. We mentioned earlier the impact of technology on the content and dynamics of teaching standard mathematics courses (calculus, algebra, discrete mathematics, statistics, analysis, etc.). Here we would like to point out an area awaiting exploration: technology for distance learning or open programmes. We have a fascinating agenda in front of us: what will the 'classroom' be like in the near future? How will we use interactive materials and on-line tuition and e-mail homework? How will we be teaching students from different continents without moving from where we are? In a global world the walls of our classroom do not exist. Distance education was once the reserve of a limited number of institutions but now all universities are entering the area of high-tech developments to offer distance learning courses. Within a few years, all of us will be competitors on a world-wide scale. What consequences will this world-wide competition have? Virtual campuses, satellite courses, non-synchronic teaching, all this demands deep research and innovative approaches.

Learning is not only preparation for the near future but a lifetime project: this new demand opens major questions for mathematics education in universities. There are more and more people going into higher education. But we also need to face the question of what we are going to offer in continuous education. Distance learning materials, tele-teaching, short courses: what kind of content are we going to offer and how? The increasing changes of jobs and technologies in professional life open up new demands (see Riley, 1998). This is a challenging problem to be solved. Short specific courses based upon the new technologies as delivery systems may be a good place to start. Here again, the appropriate mathematics teaching for adults will be necessary.

Teaching mathematics at university level should be an enjoyable human experience in which professors share with students the discovery of a new mathematical world as well as their development as person. Teachers and students are, before all else, human beings with brains and hearts. They have the opportunity to share experiences and enthusiasm, to learn from each other. Motivation, attitudes, feelings, guidance, discovery... these are the words that must fill our teaching job. We have the opportunity to transmit our passion for mathematics and this is something that all students may remember. We do not need to be actors or actresses in a mathematical play but rather the partners in a rewarding journey.

4. TIME FOR ACTION?

It is time for action. It is time to promote the teaching vocation of mathematicians. It is time to redefine the equilibrium between research and teaching. I am convinced that it is better to undertake this reform from inside than to wait to have decisions imposed from outside. Today's universities cannot afford to have ivory towers. But we know from past experience that all educational changes need time, effort and consensus. It will take a lot of work to do away with the old traditions and to gradually build up a new approach. But the clock is running. New social demands from the job market, new needs in training, life-long learning, demographic explosions, technological developments, and above all our love for mathematics and our passion for sharing this enthusiasm with others... all this must shape our drive to find our challenges and our answers for the future. Their future is ours!

CLAUDI ALSINA

REFERENCES

Begle, E.G. (1979). *Critical variables in mathematics education.* Washington, D.C.: Mathematical Association of America.

Bligh, D. (1972). *What's the use of lectures?* Harmondsworth: Penguin.

Carrier, G.F. et al. (1962). Applied Mathematics: What is Needed in Research and Education. *SIAM Review*, 4, 297-320.

Clements, R.R. (1988). *Teaching Mathemaitcs to Engineering Students utilising innovative teaching methods,* in (Howson, H.G. et al., 1988) 45-57.

Clements, R.R., Wright, J.G. (1983). The use of guided reading in an Engineering Mathematics Degree course. *Int. J. Math. Educ. Sci. Technol.,* 14, 95.

COMAP (1997). *Principles and Practice of Mathematics.* New York: Springer.

Committee on the Teaching of Undergraduate Mathematics (1979). *College Mathematics: Suggestions on How to Teach It.* Washington, D.C.: Mathematics Association of America.

Dossey, J.A. (Ed.). (1998). *Confronting the Core Curriculum. Considering Change in the Undergraduate Mathematics Major.* MAA Notes 45, Washington D.C.: Mathematics Association of America.

Eble, K.E. (1974). *Professors As Teachers.* San Francisco: Jossey-Blass Publishers.

Freudenthal, H. (1991). *Revising Mathematics Education.* Dordrecht: Kluwer Academic Publishers.

Gaskell, R.E. and Klamkin, M. (1974). The industrial mathematician views his profession: a report of the committee on corporate members. *Amer. Math. Monthly*, 81, 699.

Heyneman, S.P. (1990). Education on the world market. *American School Board Journal*, 28-30.

Howson, G. (1994). Teachers of mathematics. *Proceedings of the 7th International Congress on Mathematical Education.* Sainte-Fey, Québec, Les Presses de l'Université Laval, 9-26.

Howson, A.G., Kahane, J.P., Langinie, P. and de Turckheim, E., (Eds.) (1988) *Mathematics as a Service Subject.* Cambridge: Cambridge University Press.

Kirwan, W. et al. (1991). *Moving Beyond Myths.* National Research Council. Washington, D.C.: The National Academy of Sciences.

Kline, M. (1977). *Why the professor can't teach mathematics and the dilemma of university education.* New York: St Martin's Press.

Krantz, S.G. (1993). *How to teach mathematics, a personal perspective.* Washington, D.C.: American Mathematical Society.

Lewis, L.S. (1975). *Scaling the Ivory Tower.* Baltimore: The Johns Hopkins University Press.

Monsley, J. and Sullivan, P. (1998*). Learning About Teaching.* Reston, VA: NCTM.

Niss, M. (1998). Mathematics and Mathematics Education Research, *Proceedings ICMU98.* Berlin: University of Berlin Press.

Pollak, H.O. (1988). *Mathematics as a Service Subject. Why?* In (Howson, A.G. et al., 1988) 28-34.

Reznick, B. (1988). *Chalking it Up.* Boston: Random House/Birkhauser.

Riley, R.W. (1998). The State of Mathematics Education: Building a Strong Foundation for the 21st Century. *Notices AMS*, 45, 487-490.

Rogers, H. Jr. (1975). The Future of the University in Mathematics Education. *The American Mathematical Monthly*, 82, 211-218.

Rosenberg, A. et al. (1972*). Suggestions on the Teaching of College Mathematics.* Report of the Committee on the Undergraduate Program in Mathematics. Washington, D.C.: Mathematics Association of America.

Steen, L.A. (ed.) (1989). *Reshaping College Mathematics*, MAA Notes 13, Washington D.C.

Steen, L.A. (ed.) (1992). *Heeding the Call for Change: Suggestions for Curricular Activity.* MAA Notes 22, Washington D.C.: Mathematics Association of America.

Thurston, W. (1990). Mathematical Education. *Notices of the A.M.S.* 37, 844-850.

Wilson, R. C., et. al. (1974*). College Professors and Their Impact on Students.* New York: John Wiley and Sons.

Claudi Alsina
Universitat Politècnica de Catalunya
alsina@ea.upc.es

ROBYN ZEVENBERGEN

CHANGING CONTEXTS IN TERTIARY MATHEMATICS: IMPLICATIONS FOR DIVERSITY AND EQUITY

1. INTRODUCTION

The international phenomenon of expansion of the higher education sector has resulted in greater diversity in the intake of students. No longer is higher education the domain of the elite, but now more students can access it than in any previous times. Mathematics is not apart from this new intake. In previous times, mathematics had been able to take the elite school leavers who were well prepared for their study of mathematics. However, such backgrounds can no longer be assumed. Indeed as some authors (Kagesten, 1999; Kitchen, 1999) have recognized, the differing backgrounds of students entering tertiary mathematics have a significant impact on the teaching and content that can be delivered. Students who, in earlier times, would not have gained access to (or even considered enrolling in) tertiary mathematics, are now coming to classes. These students have very different needs and expectations of their study and are likely to encounter difficulties with aspects of mathematics and mathematics teaching so it is imperative that educators become aware of equity issues in their teaching. Significantly more women are taking classes in mathematics and the patriarchal modes of teaching are no longer suitable for a range of mixed gender classes (Forgasz, 1996). Similarly, students from various racial, ethnic and social backgrounds are enrolling in mathematics in greater numbers than in the past. Their backgrounds position them very differently from their white, middle-class peers of the past, so it is critical that educators become aware of how the backgrounds of their students impact on learning and, as a result, ensure that the practices that are used in their classrooms ensure inclusion rather than exclusion.

2. EQUITY AND MATHEMATICS

There are significant numbers of studies that investigate aspects of equity in primary and secondary mathematics, but there are relatively few in tertiary mathematics. Indeed, in preparing the literature review for this paper, very few published studies exist - empirical or theoretical - that address issues of equity in tertiary mathematics. In a sense, this is hardly surprising since mathematics has been one of the last bastions of an hegemonic curriculum and so has been immune to considering issues of inclusion. Only the elite had previously undertaken tertiary

13

Derek Holton (Ed.), The Teaching and Learning of Mathematics at University Level: An ICMI Study, 13—26.
© 2001 *Kluwer Academic Publishers. Printed in the Netherlands.*

mathematics, but this is now changing. The few studies that do exist have tended to focus on gender issues or on bridging mathematics rather than on tertiary mathematics itself. Hence this paper is timely in raising issues associated with the teaching of tertiary mathematics as a whole. Drawing on the few studies that exist, we must extrapolate from what is known from other contexts and apply this to tertiary mathematics in order to develop an issues' theme.

In considering equity, it is important to define how the word is to be used, since it is often confused with equality. Equity refers to the unequal treatment of students (or people more generally) in order to produce more equal outcomes. In contrast, equality means the equal treatment of students with the potential of unequal outcomes. For example, if a target group, such as indigenous students, were to be considered within an equity programme it would be recognized that the students entered the programme with very different and often restricted backgrounds in mathematics and the language of mathematics. Such students would be offered extra support in an attempt to enable them to 'catch up' with their peers in order that they may participate and compete on a more equal footing. It is hoped that with the extra support or changed conditions (which does not mean a reduction in programme), that there will be greater parity in the outcomes for all students. In this context, outcomes are considered very broadly to include cognition, the affective domain, retention and continuation. Whatever measures are used, they will provide some indication of how successful the programme, the teachers, and the students have been in providing a better learning environment for the students. In contrast, a programme based on equality has as its fundamental tenets that all students should be given the same programmes and chances of success, that is, all students will receive equal treatment. There will be some students who take greater or lesser opportunity to exploit what is offered, but all students will be given the same opportunities to succeed. In considering the two approaches, the equity programmes are designed to be more proactive and seek to redress differences in prior experiences. In contrast, equality programmes are somewhat more conservative in orientation and seek to treat all students equally with the assumption that some will choose to make the most of the opportunities offered to them.

This paper supports the former orientation as it is argued that the context of tertiary mathematics in contemporary times finds educators dealing with very different groups of students than have been the case in the past. There is greater diversity in relation to background knowledge, experiences and needs for mathematics. These variables are critical for subsequent success and enjoyment of mathematics. If students are to continue with their studies in the area, then learning environments need to be developed that will support and enhance such opportunities. For ethical as well as economic reasons, it is not possible to justify excluding students from effective participation and success in mathematics due to their prior experiences.

In the following sections, I take issue with some of these aspects of mathematics and analyse practices of mathematics that can be seen to be excluding some groups of students. This potential exclusion is insidious in nature and often remains at the level of the unrecognizable. It is part of the culture of mathematics and as such is often invisible and taken-for-granted. My purpose in the subsequent sections is to highlight aspects of mathematics than can be culturally biased and hence potentially exclusory for some groups of students.

Language and Mathematics. Until the relatively recent past, mathematics was seen to be the academic discipline that was least influenced by language. Indeed, it has been a common perception that students whose first language is not English, or speak a restricted form of English, would still be able to undertake successful study in mathematics. In part, this was due to the abstract symbols used in mathematics. Such symbols function as a cue to the operations to be undertaken. In reading textbooks where the language is not English, it is still possible to see the decontextualised mathematical tasks and to be able to work through them. However, mathematics is not taught in a linguistic vacuum. The on-campus teaching or the off-campus materials will be linguistically based so students must be able to work within that language in order to make sense of the spoken or written words.

In writing this section on language, I use English language as the exemplar. Many of the grammatical or lexical issues that I raise in this paper could be applied to other languages so that they would be applicable beyond English. The issue of English is important also since many countries cannot support the development of mathematical texts in their language. This means that the language of instruction in tertiary mathematics is often English. Hence, the issues that I raise here should also alert educators to the difficulties that many students in these contexts could experience when compelled to learn mathematics in and through English where it is a second or other language.

A number of different aspects of language are be considered. For students whose language background is congruous with that of formal contexts of mathematics, then passage through subjects and courses is made somewhat more easily. In contrast, for those students whose language background is not that of the formal learning environments (social or textual), then progress can be impeded. In this context, language background becomes a tool through which success can be enhanced or hindered. Bourdieu (1983) uses the notion of *habitus* to refer to this embodiment of aspects of culture, in this case language, which can be influential in the rites of passage through tertiary education. Accordingly, the linguistic habitus of the students becomes a form of capital that can be exchanged for success in mathematics. In other words, having access to the formal language of instruction and text, students' progress is enhanced or impeded depending on their levels of familiarity and competence in the language of instruction.

2.1 *Mathematics as a Register*

Halliday has used the term *register* when referring to the functional variations in language. He argues that language varies according to the "functions it is being made to serve... who the people are that are taking part in whatever is going on ... and what exactly the language is achieving" (Halliday, 1988, p.44). These three aspects of language, the field, tenor and mode, are what he sees as the function, or register, of the language. As such, mathematics becomes a particular register in that it serves a particular function - is spoken by particular people and achieves very particular goals.

2.2   Relationships of Signification and Lexical Ambiguity

In the first instance, mathematical words have very particularized meanings within the context of mathematics as opposed to non-mathematical contexts so that often the words are ambiguous. The relationships of signification between words and meanings are highly contextualized and as such are determined by the field – in this case, mathematics. For example, words such as integral, differentiate, matrix, volume and mass have very different meanings inside and outside of mathematics. In order to participate effectively in mathematics, students must be able to recognize the field and hence identify the specificity of the signifiers (words) and their mathematical signifieds (meanings). Where a student [mis]recognizes a word and uses this to then identify the context for that word, there is potential for very different interpretations of the task and hence very different outcomes. Oates (1999) cited the example of a student for whom English was his second or third language and who was undertaking a calculus exam where dictionaries were not permitted. One exam question related to the calculation of volume. The student could not decipher the meaning of the question as he was interpreting 'volume' to mean the sound that was emitted from his stereo system. So the exam question appeared nonsensical. Subsequent discussions revealed that the student was able to complete the mathematics so that it was the field of the question that was problematic rather the mathematics. For the student, the signifier 'volume' served as a cue for a very different context than that demanded by the exam.

2.3   Mathematics, Grammar and Lexical Density

As noted in the previous section, mathematics is a highly specialized register with many words unique to it or words in it used in highly specific ways. There are many words used during instruction that can be problematic for students. While the more obvious words like those in the previous section can be easily identified, there are many words that are more problematic. For example, (McGregor, 1991) cites the difficulty with the use of prepositions such as by, to, from, when considering a statement of temperatures where the temperature fell *by... to... from*. These 'filler words' are small but impact significantly on the meaning of the task and the subsequent interpretations and calculations. Where students do not have a strong command of the language, these prepositions can cause difficulties in interpreting the tasks to be undertaken.

The structure of mathematical language is another aspect of language that needs to be considered. Mathematics is very concise and precise in its expression. Students with restricted English competency may not be able to access the deep meaning of statements and so rely on key words to act as cues. For example, words such as more or less may be assumed to equate with addition and subtraction. However, this is not always the case.

A further factor compounding access to language is the concise and precise expression found in mathematics. This makes for very few redundant words, which means that almost every word in a mathematical expression conveys meaning. Unlike other forms of expression, particularly oral language, the written language of mathematics is very precise. Halliday (1988) refers to this aspect of language as *lexical density*. Lexical density is a tool for considering the readability of a text. One way of calculating lexical density is to find the number of lexical items, which is then divided by the number of clauses to give the lexical density. Lexical items are "the words that carry the content of the text. They are the words that are found in dictionaries as they can be easily defined" (Hammond, 1990, pp. 35-36). Where the lexical density is high (or dense), students may need more assistance in deconstructing the text for the information contained within it. In some cases, the lexical density can be low so that the information is sparsely spread throughout the text making it difficult to keep track of. However, within this context, it should be noted that some lexical items have a greater familiarity than others do (for example, 'circle' and 'two' are more familiar than 'radius' and 'parallel') which also impacts on readability. Where the lexical items are more familiar, then there is a greater chance of readability than where the items are not as familiar. For example, consider the following extract taken from a newspaper.

> *(Two crows with a serious rubber fetish are infuriating commuters at Coomera railway station), (particularly when it rains). (The crows are swooping daily on cars in the two station car parks, and ripping out the rubber blades from the windscreen wipers). (Stationmaster David Hobbs estimates) (more than 100 cars have had their wipers removed in the past six months). (Gold Coast Bulletin, July 12 2000, p. 3)*

Example 1: Extract from a daily newspaper

In this example, the lexical items have been underlined while the clauses have been indicated by the brackets. Full names (David Hobbs) are defined as being one lexical item as they refer to the one item. The lexical items are all common to everyday experiences (other than the place name - Coomera), so readability at this level is easy. Here there are 34 lexical items and 5 clauses, making for a lexical density of 6.8. The text of written passages is far denser than oral or spoken text, where language in the latter is commonly filled with many grammatical items. One only has to consider how oral language is peppered with words that make the text easier to process. For example, "now, when you come to think about the problem in such a way, like there is really not much of a difference between the first example and the second one that I have just given you" would be a common statement. There are fewer lexical items and a greater number of grammatical items, thereby reducing the ratio of lexical: grammatical items to begin with. This is then further reduced by the number of clauses. Examples of these types of interactions can often have lexical density in the region of 3 and below, indicating that they are low insofar as measures of lexical density are concerned. This can be seen to be one of the reasons why spoken language is often more accessible than written language. Students must come to terms with the differences between the two forms and become conversant and fluent with both forms, but the written text should be recognised as a more complex form. However, mathematics is a very specific and precise language where the words are often complex, have unique meanings within a mathematics discourse and the grammar is very tight. Example 2 below, shows a common representation of mathematics text.

Find the solution of the differential equation:

$$\frac{d^2 y}{dx^2} - 4\frac{dy}{dx} + 4y = 8(x + e^{-2x})$$

which satisfies the initial conditions $y = -1$ *and* $\dfrac{dy}{dx} = 1$ *when* $x = 0$.

Example 2: Example of a mathematical task

In this text, there are no clauses while there are seven lexical items if one does not consider the mathematical symbols as being lexical items. In considering lexical density as a measure of complexity in language, the symbolism used in the discipline needs to be considered. The abstract symbolism (see the following section) contained with these types of equations carry significant meanings that would be complex if they were to be written in words. By definition, each of these items (e.g. dx^2, $y = -1$) could be defined as a lexical item as each conveys a specific meaning, thus increasing the lexical density of the text. The lexical density of these types of questions is far greater than for everyday written texts, such as newspapers. Indeed, the wording is very tight, and made even tighter by the complexity of the equation. However, without considering the equation, the lexical density indicates a high degree of succinctness which students must be able to unpack in order to make sense of the task. There are very few 'hints' as to what the students would need to do in order to complete the task.

Another common task found in tertiary mathematics can be seen in the following example where the question is far more text-based.

(A $30 000 mortgage was taken out for a 25-year term at an annual interest rate) (which is assumed to be fixed at 12%.) The mortgage is paid off in equal instalments of $p throughout the 25 years. The interest is calculated each year on the amount owing at the beginning of the year. (Derive a recursive relation to be satisfied by $x_n the amount owing after n years) and (find in terms of p the particular solution which satisfies the initial condition x_0 = 30 000). Then use the further condition $x_{25} = 0$ to calculate the required value of p.

Example 3: Applied Contexts and Lexical Density

Such questions are often used in contexts once referred to as applied mathematics where the mathematics is seen to be used in a real-life context.

In this question, there are 53 lexical items and 5 clauses making for a lexical density of 10.6, thus creating a dense passage. Students need to make considerable use of the available words to recognise the specific mathematical meaning of terms such as *p* and *n*, which are a specific mathematical code within this context, and also need to make sense of the overall passage. There are very few words in the passage that do not convey very specific meanings.

2.3.1 Symbols

In concert with lexical density is the related notion of **symbolization** that is so common in mathematics. In mathematics, language is progressively refined so that it becomes represented in a symbolic form. The lexical density of the written form progressively evolves into an abstract symbolization. Consider the evolution of the function. It moves from highly contextual examples based in real life, through to contracted equations and pure symbols. Synonymous with this move, there is an increase in lexical density and it may well be this shift is what causes difficulties for students for whom language is a potential barrier to learning. For example: *draw a sketch of the given inequality*: $3y - x + 6 \geq 0$, has a mixture of both written language and symbols. The symbols are abstract algebraic concepts along with highly specialized mathematical symbols including operations and inequalities. The mathematical symbols represent highly specialized constructs. The meaning that is conveyed within the sentence is very complex yet contained within minimal wording. To explain what is actually meant by the statement requires significantly more wording than is represented. In a number of instances the accompanying wording of tasks decreases so that there is minimal wording and significant symbolization. This can be seen in the example below.

$$\text{Find: } \int \frac{1}{(x-1)(x+2)} dx \text{ when } -2 < x < 1.$$

Example 4: Symbolization and mathematics

The lexical density in Example 4 is extremely high thus leaving very few linguistic cues to aid students. In this integral, the symbols function as complex meaning systems. In much the same way as discussed in the relationships of signification, the signifier in this case becomes a symbol rather than a word while the signified is often a complex set of mathematical ideas. In Example 4, one has to consider the individual symbols and what they mean, as well as a complete meaningful statement where all the individual signs come to construct a rather complex problem. Students must be able to decipher the individual signs and the overall statement in order to understand what is being asked. This process is similar to that of reading a sentence where the reader must analyse the individual words and then combine them to make sense of the overall sentence.

2.3.2 Language as a Form of Cultural Capital

In considering the concepts mentioned in the preceding subsections, the role of language in the access to mathematics becomes somewhat more apparent. For students for whom English is not a first language, the implications are most obvious. However, as has been noted by socio-linguists (Halliday, 1988 and Bernstein, 1990), the social background of students is also a critical variable in language acquisition and use. Bernstein has been particularly powerful in showing how there are substantial differences in the patterns of language use by middle-class and working-class students and that the formal school context values the middle-class register. Accordingly, even for students whose first language is English, there may be differences in the register of use so that those who are not as familiar with the middle-class register of the formal education context must learn the intricacies of this register in order to participate in linguistic interactions. In considering the use of English as the medium of instruction and writing, it is critical to recognize the differences in register within the English language. Students whose first language may not be the middle-class register of formal educational contexts are likely to encounter difficulties with the spoken texts during instruction as well as with the written texts of books and instructions. These students are often placed at a disadvantage as it is often assumed that their first language is English so that they are often perceived to have no problems with the language used in instructional contexts.

2.4 Communications, Interactions, Pedagogy

To be a competent communicator, one needs to be able to make sense of the knowledge being transmitted through a communication and of the communicative context itself. Both demand different skills and different cultural knowledge. As developed in the previous sections, a competence in the language of mathematics is essential to making sense of the mathematics within the language. However, there are also skills needed to unpack the communicative context of instruction. In formal teaching contexts, including the formal lecture situation as well as the less formal tutorial situation, there are particular unspoken rules about who speaks to whom, when and how. Where the lecturer stands in front of a large group of students and works through mathematical problems or examples, there tends to be very little interaction. However, the level of language is a mix between oral and written language, both of which the students must be able to decipher in order to understand the task and processes of calculations, including the rationale for such calculations. In order that students are able to make sense of the mathematics being taught, the students must be able to crack the code of the language of instruction. Such code cracking depends significantly on the level of command of English. However, and equally important, is the ability to crack the code of the cultural aspects of the lecture format. For many students, participation in the lecture, or any teaching situation, demands an understanding of the unspoken rules of participation. There has been a significant amount of research conducted at the level of classroom interactions through the field of ethnomethodology (Mehan, 1982 and Hester, 1985).

At the level of primary and secondary schooling, these studies have shown very ritualized patterns of interaction in classrooms (Zevenbergen, 2000). It would appear that similar studies have not been undertaken at the tertiary level, but such studies would be useful in highlighting the unspoken rules of interaction within the higher education sector. Such studies would indicate how students are expected to interact in classrooms and as such, demonstrate the unspoken rules of participation in higher education. Given that there is a progressive abstraction and formality in teaching contexts as students move through their school years, and given the formality and large teaching classes of many higher education forums, it could be expected that the issues identified in school studies can be extended, and perhaps exacerbated, in the higher education sector.

As more students from a wider range of backgrounds enter the higher education sector to undertake the study of mathematics, it becomes necessary to consider whether they are able to crack the code of the unspoken rules of interaction both within the context of formal learning environments and within the more social interactions with teaching staff. This issue has been raised in the context of interactions where Forgasz (1996) has found that women in tertiary mathematics classrooms have noted very gendered patterns of interactions that they see as sexist. This finding indicates that the women in this study found the styles of teaching and interactions to be very masculinist and hence exclusory of women. Similarly, there are very unique styles of interaction among the different groups of indigenous peoples. Malin (1990) for example, has noted the different expectations and roles within Indigenous Australian interactions and their incongruency with expectations of the formal school contexts. The use of electronic forms of communication also pose potential problems for communication. For example, the genre of writing for the Internet is very different from that of the genre for thesis writing or even writing a letter, yet they are all written texts. Students need to be able to crack the code for these new forms of writing. In her study of women using the Internet, Spender (1995) has noted that women are at risk of being marginalised by the forms of sexism that can be displayed through masculinist writing styles. If this is the case for gender, then it is equally likely to be the case for other disadvantaged groups. Whatever the group, the differences in interactional styles need to be identified and monitored across the diverse range of students who are now undertaking the study of mathematics, if there are to be more inclusive practices adopted for the diversity of students enrolled in mathematics courses. By way of concluding this section on language, I have drawn out some of the major language considerations that will need to be addressed if mathematics is to be more inclusive of the diversity of students who now enrol in courses. However, I would take heed of the comments made by Thomas (1995) when she argues that not only must language background be considered. In her work with non-English speaking background students, she notes the vast differences within language groups and across language groups. Within any language group, the social background of the student impacts significantly on the students' potential to participate effectively. It cannot be assumed that the language spoken by students will support or hinder their access to mathematics but other aspects of language must also be considered – namely the social background (since mathematics is often associated with the middle-class register) and the competency in the first language of the student (Thomas, 1995).

2.5 Contexts, Problems and Mathematics.

There has been a progressive movement in mathematics to apply mathematical concepts to situations that may be encountered by students. This has been on the grounds that it demonstrates the relevance of mathematics to the real world. Boaler (1993a; 1993b) has argued that such contextualizing is in itself highly problematic for both mathematics and students. Often the context used to embed the mathematics is purported to give a sense of meaning and relevance to students, yet this is often not achieved. For example, embedding a vector task in a scuba diving context may appear to provide a real life scenario of the application of vectors but two key considerations need to be raised. The first is whether participants would actually conceptualize scuba diving as a vector task, and the second is whether students would have equal access to the scuba diving context. Research in the area of situated learning (Carraher, 1988, Lave, 1988 and Lave and Wenger, 1991) raises questions as to whether a scuba diver would resolve a real-life experience using vectors. Furthermore Boaler (1993a; 1993b) and Dowling (1998) have argued that the last point has been particularly problematic for many students. They contend that often the contextualizing of problems raises serious questions about access for whom. Students who have no experience with scuba diving may be placed at a disadvantage when using such examples. In their analysis of national testing schedules in the United Kingdom, Cooper and Dunne (1999) have found the embedding of problems into purportedly real contexts is highly problematic for some students, whereby students have greater or lesser access to the problems. Whereas both working-class and middle-class students were able to perform equally as well on tasks where there was 'pure mathematics', middle-class students significantly out-performed their working-class peers on tasks that have been contextualised. These results raise serious concerns about who has access to, and success with, different types of mathematical problems.

Reconsider the example (Example 3) given previously where the students needed to calculate interest. In this example, it has already been noted that the readability is difficult due to its high lexical density, but it is further compounded by the application of the problem to a real world context. Many middle-class and Western values are embedded in this type of problem. It is assumed that students will purchase homes, take out mortgages, have identified savings programmes for deposits and expect to repay the loan within particular time frames. Such contexts may not be as relevant, and hence, accessible to all students, thus making the task more or less accessible depending on familiarity with the context.

While there appears to be no research conducted at the level of tertiary mathematics, Cooper and Dunne's (1999) research with secondary students has shown quite powerfully that embedding mathematics into contexts can serve as a distracter for some students – most notably students from working-class backgrounds. Using Bernstein's (1990) linguistic theories, they proposed that because of a greater familiarity with the range of registers, middle-class students are more likely to recognise a contextualised task as a mathematical task, thereby reducing the potential to misinterpret the context. In contrast, working-class students, who are less likely to be familiar with the change of registers when posing contextualised tasks, are at greater risk of interpreting and answering the task as if it were a 'real life' situation and thus, miss the mathematics of the task. Such findings indicate a need for mathematics educators to be critically aware of the potential difficulty caused by embedding mathematics in word problems and to question the results of assessment where there is such embedding of problems. It may be that some students may misinterpret the context of the task rather than the mathematics.

3. TECHNOLOGY

Within education in general, and higher education in particular, the use of technology has been purported to be the key to the future. Technology varies in definition from the rudimentary tools of rulers and blocks through to graphic calculators and high tech computers. Arguably, technology is also a key component of the contemporary workplace so that if courses are to adequately prepare their students for the market place, then there is strong justification for inclusion of technology-rich classrooms. However, there is an equally strong argument for technology as a pedagogical tool. Technology has been seen to enhance learning through two complementary features. In the first place it reduces the tedium of boring and mindless calculations and second (but not unrelated) is that it allows students to explore mathematics. Most mathematics educators would not advocate using technology to replace early developments in understanding – such as plotting equations. Once this understanding has been mastered, then it can be argued that there is little to gain in undertaking timely calculations and plottings when these can be undertaken by technological means. Indeed, there is ample research to show that students can gain substantially by reducing mindless calculations and replacing these with more explorations relevant to the topic. The potential that has been realized through calculators, spreadsheets and graphic calculators has been well documented (see, for example, Coutis, Farrell and Pettet, 1999). However, there is a need for caution. While the value of these new tools can be shown, there is need to ask who has access to such technology and with what consequences. In a study at Auckland University, the authors found that only 12% of their class had access to graphic calculators (Barton and Oakes, 1998). Similar results have been found with access to graphic calculators at the Year 12 level in Victorian schools (Routitsky and Tobin, 1998 and Routitsky, Tobin, and Stephens, 1998), where the ownership of graphic calculators rarely reached beyond 40%. While the figures varied depending on the sector, they indicated that not all students have access to the benefits of graphic calculators. Consequently, serious questions need to be asked as to whom has access to this technology and what might be the effect on learning outcomes and the impact of this for tertiary mathematics.

A further issue that needs to be tackled from a systemic perspective where there is an increased focus on computer-based learning environments, including Web-based courses and email tutorials, is the access that students have to these facilities. Clearly economic capital will have restricted some students' access to this forum for learning. For these students, familiarity with the equipment and software may be restricted so that the expectation that certain background knowledge can be assumed is inappropriate. Students' access to technology-rich experiences prior to entering the higher education sector may be restricted along social and economic lines. It cannot be assumed that all students entering tertiary mathematics classrooms will have basic skills and familiarity with the pedagogical tools to such as level that they can participate fully in classroom activities. In some senses, the provision of common access laboratories, often with extended hours of operation, are seen to be a means through which these differences can be addressed. However, the logic of the argument must be challenged. In many cases, economically disadvantaged students may need to be engaged in substantive employment in order to remain in study. In such cases their potential to access these facilities is very restricted. Similarly, many students entering the higher education sector are mature-age students who have other family commitments so that the possibility of organizing child care during the extended hours of access is impossible. Kagesten (1999) has noted similar difficulties in the Swedish context where mature-age students have difficulty attending some classes due to family commitments.

4. CONCLUSION

In the contemporary context of tertiary mathematics the changing nature of higher education has meant that more diversity is encouraged in the sector through the opening up of the sector. Mathematics, once the terrain of the elite, is now open to a wider range of students and teachers of mathematics must develop skills to deal with this diversity. Students are no longer taking mathematics at the same rate or with the same purpose as had been the case in past decades. There are greater numbers of students who see mathematics as part of the smorgasbord of university subjects, rather than a major to be undertaken as a coherent and sustained field of study. They may take one subject as part of the requirements of another programme. They may enter the classroom without the background knowledge that was once taken as assumed. They do not see themselves as a part of the mathematics community but rather as a passing participant who dips a toe into the pool of knowledge. These factors result in the mathematics to be taught, and the ways in which it is to be taught, as being in need of serious reconsideration.

In this paper, I have sought to raise questions about which and how students are being excluded from participation and success in tertiary mathematics. While my focus on language has been with English as a language, the linguistic issues raised here should be of relevance to other languages where aspects of their particular grammatical structures will pose similar problems for inclusion and exclusion. In raising issues such as those posed in this paper, it has been my intention to demonstrate some of the ways in which the practices of mathematics may seem to be apparently innocuous but are, in fact, politically laden and hence engender the exclusion of some students. This exclusion is not random, but has the potential to be particularly focused on some groups of students more than others. This is not to claim that all students from particular groups are at risk, but to say that some students from some socially disadvantaged groups may have more difficulties in accessing and succeeding in mathematics than their advantaged peers. In most cases, the issues that I have discussed are at the level of culture and as such, are often invisible to participants (both teachers and students).

If mathematics is to be made accessible to the more diverse range of students who are now undertaking studies in the discipline, then there is a need to critically examine the ways in which mathematics is being taught in order to make explicit, and redress, the inequities inherent in such practices. Given the changing enrolment patterns of students and the greater diversity in contemporary tertiary mathematics, issues of equity, access and success become critical touchstones for good practice in higher education mathematics teaching and learning. The dearth of literature in this area suggests that there is a strong need to consider the issues raised in this paper and for research to be conducted in these areas if there is to be a more inclusive tertiary mathematics education.

REFERENCES

Barton, B. and Oates, G. (1998). *Students access to technology.* Unpublished report. Auckland University, New Zealand.

Bernstein, B. (1990). *The structuring of pedagogic discourse: Vol 4. Class, codes and control.* London: Routledge.

Boaler, J. (1993a). Encouraging the transfer of 'school' mathematics to the 'real world' through the integration process and content, context and culture. *Educational Studies in Mathematics,* 25, 341-375.

Boaler, J. (1993b). The role of contexts in the mathematics classroom: Do they make mathematics more 'real'? *For the Learning of Mathematics,* 13(2), 12-17.

Bourdieu, P. (1983). The forms of capital. In J. G. Richardson (Ed.), *Handbook of theory and research for the sociology of education,* pp. 241-258. New York: Greenwood Press.

Cooper, B. and Dunne, M. (1999). *Assessing children's' mathematical knowledge: Social class, sex and problem solving.* London: Open University Press.

Coutis, P. F., Farrell, T. W. and Pettet, G. J. (1999). Improving engineering mathematics education at Queensland University of Technology. In W. Spunde, P. Cretchley, and R. Hubbard (Eds.), *The challenge of diversity: proceedings of the Delta '99 symposium of undergraduate mathematics,* pp. 69-74. Rockhampton: University of Central Queensland.

Dowling, P. (1998). *The sociology of mathematics education: Mathematical myths/pedagogical texts.* (Vol. 7). London: The Falmer Press.

Forgasz, H. (1996). Gender issues in tertiary mathematics education. In P. Clarkson (Ed.), *Technology in mathematics education,* pp. 194-199. Melbourne: Mathematics Education Research Group of Australasia.

Halliday, M. A. K. (1988). *Spoken and written language.* Geelong: Deakin University Press.

Hester, S. (1985). Ethnomethodology and the study of deviance in schools. In R. G. Burgess (Ed.), *Strategies of educational research: Qualitative methods,* 1989 ed., pp. 243-263. London: The Falmer Press.

Kagesten, O. (1999). The consequences of an expanding university system on mathematical teaching. In W. Spunde, P. Cretchley and R. Hubbard (Eds.), *The challenge of diversity: Proceedings of the Delta '99 symposium on undergraduate mathematics,* pp. 114-117. Rockhampton: University of Central Queensland.

Kitchen, A. (1999). The changing profile of entrants to mathematics at A level and to mathematical subjects in higher education. *British Educational Research Journal,* 25(1), 57-74.

Malin, M. (1990). The visibility and invisibility of Aboriginal students in an urban classroom. *Australian Journal of Education*, 34(3), 321-329.

McGregor, M. (1991). Language, culture and mathematics learning. In M. McGregor and R. Moore (Eds.), *Teaching mathematics in the multicultural classroom: A resource for teachers and teacher educators*, pp. 5-25. Melbourne: University of Melbourne, School of Mathematics and Science Education.

Mehan, H. (1982). The structure of classroom events and their consequences for student performance. In P. G. and A. A. Glatthorn (Eds.), *Children in and out of school: Ethnography and education*, pp. 59-87. Washington: Center for Applied Linguistics.

Oates, G. (1999). *Language difficulties on a calculus exam for English-as-a-second-language speakers.* Personal communication, Nov. 1999.

Routitsky, A. and Tobin, P. (1998). A Survey of Graphics Calculator Use in Victorian Secondary Schools. In C. Kanes, M. Goos, and E. Warren (Eds.), *Teaching Mathematics in New Times*, Vol. 2, pp. 484 - 491. Gold Coast: Mathematics Education Research Group of Australasia.

Routitsky, A., Tobin, P. and Stephens, M. (1998). Access to and use of graphic calculators in Victorian secondary school. In J. Gough and J. Mousley (Eds.), *Mathematics: Exploring All Angles*, pp. 383 - 389. Brunswick: The Mathematical Association of Victoria.

Spender, D. (1995). *Nattering on the net: Women, power and cyberspace*. North Melbourne: Spinifex Press.

Thomas, J. (1995). Bilingual students and their participation in tertiary mathematics. In R. P. Hunting, G. E. FitzSimons, P. C. Clarkson and A. J. Bishop (Eds.), *Regional collaboration in mathematics education*, pp. 703-712. Melbourne: Monash University.

Willis, P. (1977). *Learning to labour: How working class kids get working class jobs*. Aldershot: Gower.

Wilson, T. (2000). Rubber fetish fury. *Gold Coast Bulletin*, July 12, 2000, p. 3.

Zevenbergen, R. (2000). Mathematics, social class and linguistic capital: An analysis of a mathematics classroom. In B. Atweh and H. Forgasz (Eds.), *Socio-cultural aspects of mathematics education: An international perspective*, pp. 201-215. Mahwah, NJ: Lawrence Erlbaum and Assoc.

Robyn Zevenbergen
Griffith University, Queensland, Australia
r.zevenbergen@mailbox.gu.edu.au

JAN THOMAS

POLICY ISSUES

1. INTRODUCTION

Policy issues have tended to be incidental to those involved in the mathematical sciences. Occasional forays have been made into issues of school curricula, or concern expressed about inadequate support for research, but involvement has been sporadic and *ad hoc* at best. However mathematics does not live in a policy, social or cultural vacuum and, as bigger political agendas have become part of a globalised economy, mathematics has started to pay the price for neglect of the political milieu in which it operates. The extent of this was apparent to the group who participated in the policy discussions at the ICMI Study conference in Singapore in 1998. Now it is even more apparent that many countries, and the discipline of mathematics, can be identified as paying a very heavy price indeed for this neglect. But is not mathematics an international discipline, something all young people should have access to, and is this not therefore an international problem? And what of the discipline itself if it becomes the prerogative of an elite few?

A culture without mathematical ideas has not been found (Bishop, 1988). In the modern technological world the importance of mathematics, both culturally and as a tool for participation in such a world, increases. It is this dual role of mathematics as a discipline with inherent cultural value and worthy of study in its own right, and as a necessary tool in the modern world, that is the focus of much of the discussion reported here. This discussion is based on the deliberations of the policy group at the Singapore meeting but extends that discussion to include recent events.

An overarching theme of the Singapore discussion concerned the cultural and economic value of mathematics, including its importance in participation in democratic processes which increasingly requires understanding of data presented in a multitude of forms. A major consideration was the need for mathematics as a discipline to be visible and accessible to all young people as access to this fundamental tool affects both economic potential and participation in society.

As the perceived impact of recent policy and political actions in various nations were shared in the working group, concern began to be articulated for the discipline itself. This was particularly so in regard to cultural aspects. In some regards this is quite remarkable as only some ten years before, at the ICME conference in Budapest, there was still heated debate about mathematics as culture and in cultures.

However, even though more widely accepted in the mathematical community, the pivotal cultural role of mathematics—as well as its economic value—in a modern technological society, is still not well understood in the wider community. In May 2000 a guest speaker at a Mathematical Sciences Education Board meeting

Derek Holton (Ed.), The Teaching and Learning of Mathematics at University Level: An ICMI Study, 27—36.
© 2001 *Kluwer Academic Publishers. Printed in the Netherlands.*

in Washington spoke of mathematics in conjunction with Latin and Classical Greek. The implication was that the serious study of mathematics was destined to become a minority activity for an educated elite and that this did not matter. That such a statement could be made at a meeting such as this clearly demonstrates the misconceptions that still surround mathematics and its role in society. At various stages in the history of the world, Greek and Latin were important languages but they never achieved the pervasiveness of mathematical ideas across social and cultural groups. Over time their study became the prerogative of an educated elite but, as culture and language are explicitly linked, their cultural importance has diminished as use of these languages has diminished. The opposite is true of the role of mathematics in the modern world where mathematical ideas underpin economic potential and societal participation and its use has increased.

Thus as a discipline we will fail both our young people and our nations if we allow mathematics to become an activity for an educated elite. Mathematics will indeed become the gate-keeper to wealth and privilege in unprecedented ways. If this is to be stopped then understanding, and active involvement in, political processes becomes a necessary precursor to the findings of this ICMI study being considered seriously. The serious study of mathematics being perceived as a modern version of elite studies of Latin or Classical Greek for the few, and not for all, is a serious challenge for the discipline when access to quality mathematics education for all is becoming increasingly problematic even in wealthy nations. The challenges facing mathematics are global, not regional or national, and they need to be faced by the international community of mathematical scientists.

2. A PERSONAL PERSPECTIVE

I have just completed a report on the mathematical sciences in Australia (Thomas, 2000a). It is a bleak document showing a discipline in disarray. Yet Australia is a wealthy nation, the economy is considered sound, and those in the mathematical sciences have been more politically active and organised than in most parts of the world. For some years there has been excellent press coverage of both the emerging problems of mathematical sciences in Australia, and their positive contributions. Politically we have so far failed dismally to have any significant impact. The key issues identified as underpinning the problems in Australia will be analysed in more detail in the context of the deliberations of the working group. However, first let me share some perceptions of some of the challenges facing other nations. Many of these concern the mobility of mathematical skills and a world-wide shortage of those skills. The depth of what is rapidly emerging as an international crisis in the mathematical sciences needs to be understood, as well as some of the underlining problems. The brief national snippets below come from many sources and are illustrative only.

Israel: Israel has done well in the technology stakes in recent years and demand for mathematicians is high. However there would already be a serious shortage of secondary teachers if it had not been for recent immigration, especially from Russia. This supply of mathematical expertise is drying up. Israel is not the only country to have benefited from the excellent mathematics base that used to exist in Russia. The tragedy is that mathematics in Russia had been allowed to wither and many excellent Russian mathematicians now struggle to find a place in the world. Instead of doing mathematics, many are trying to establish themselves learning another language and finding a place in a new culture.

United States of America: As the economy has boomed so has demand for skills. This is reflected in the number of visas available for skilled people being greatly increased. Mathematics in the universities seems healthy compared with many other parts of the world but the amount of money available for research may be a two-edged sword. It enables too many to ignore some of the teaching and learning issues, many of which are very problematic. For example, teachers are caught between increasing demands for accountability through often poorly conceived high-stakes testing regimes and the revised Standards of the National Council of Teachers of Mathematics (2000) which are often at odds with expectations of their principals and school districts. At least the problems in relation to the teaching of mathematics are openly discussed and solutions being sought (Glenn, 2000). However the challenges ahead remain huge and while the USA economy is strong, and America fails to produce enough of its own people with mathematical skills, these will continue to be bought elsewhere. That produces problems for other nations.

England: Recent developments indicate that England is a very good model for what not to do if the mathematical sciences are to flourish. Teacher shortages are acute and generous financial incentives for people to go into teaching have not produced applicants in sufficient numbers. At most the incentives seem to have stopped the decline in the number of applicants. Much of this seems to stem from reactions against a system of government control that means that teachers have little control over their work. This is exemplified by a mandated national curriculum, testing, league tables of schools, inspectors who may not necessarily have any educational expertise and have immense power — the list goes on. Intelligent people do not want to go into a profession in which they have little autonomy over their professional lives. Some fairly intrusive and time-consuming accountability is also part of the university system. This includes a research assessment exercise which is widely perceived as damaging cross-disciplinary activity including that between mathematicians and mathematics educators.

Malaysia: Some of the newer universities do not have separate mathematics departments and the mathematicians employed in them are mainly there to teach service courses. They feel that mathematics is invisible and they struggle to find an identity or establish the infrastructure for research.

Japan: There are tensions surrounding a reduced school day that has resulted in a disproportionate loss of time for mathematics. The mathematicians feel they have been largely ignored in recent curriculum decisions.

I could continue and I have only considered countries where it would be expected that the mathematical sciences would be strong. The point I am making is that none of us in the policy group were comfortable about what we perceived as happening in policies affecting mathematics. The examples above give substance to our concerns which, I think, the policy group struggled to articulate. Especially in many Western nations, I believe we have been unprepared for the kinds of political interference which now occurs at all levels of education, sometimes with astounding rapidity.

3. THE ISSUES AND CHALLENGES

Three major issues emerged in the policy discussions in Singapore. Coincidentally three major issues were identified in the recent paper on mathematical sciences in Australia. Superficially there is little overlap but at a deeper level there emerges international commonality of concerns. This illustrates again that mathematics is an international discipline and the challenges to be faced are global, not regional or national.

In the following sections the issues identified in Singapore are described. The three Australian issues are then discussed. Finally links are made between the two in an attempt to demonstrate that the mathematical sciences have global issues that need to be addressed. These are intimately related to policy and political matters which will determine the shape of mathematics in the future.

3.1 Issue 1

Issue 1 concerned 'mathematical literacy' (ML) which has other labels including 'quantitative literacy' and 'numeracy'. The working group identified world-wide tensions between mathematics for all that underpins notions of ML, which is a form of practical numeracy, and the serious study of mathematics as a discipline. While ML is important, policy makers appear not to appreciate how much has changed in the mathematical needs of society.

Thus a key challenge was seen as the need to raise the level of ML and resist the danger that a low level form of numeracy would become national policy and the goal for the majority. The danger of a check-list outcomes approach to ML, which was tending to produce very prescriptive curricula in some countries, was noted. Also noted was the low levels being set for ML in some countries. An example are the 'numeracy benchmarks' in Australia which have been set as something that 80% of students should be able to achieve easily based on existing curricula and tests. Thus there has been little expectation that Australian students could, or should, do better or that existing curricula may be inappropriate. In England additional tests— 'world-class tests'—are being produced as well as the national assessments, presumably because these are seen as too low a level.

Curriculum instability and fads in education were identified as international problems with a number of illustrative examples emerging in discussion that show the rapidity with which ideas spread. Thus 'civics' and 'enterprise education' were identified as recent curriculum focuses in a number of countries. A recent fad that has emerged and currently seems to be spreading is a move to quite prescriptive homework policy. Thus a homework policy in England took only a few weeks to surface in parts of Australia. Recently homework surfaced in the New York Times as a major issue (*New York Times*, 7 November 2000), especially in regard to students who receive little or no educational support within the home. One of the ever-present dangers of fads is that they tend to become policies which schools have to follow even though they may further disadvantage some students. Thus as the school day is crowded with more and more subjects, time lost in mathematics is meant to be recouped by mathematics homework. The students most likely to be disadvantaged by this are the ones already struggling within the school environment and least likely to be able to be supported at home.

3.2 Issue 2

Issue 2 concerned the relationship of mathematics to society and government and had two parts to it:

- public views of, and actions toward mathematics, and
- the roles of mathematicians in policy making and implementation.

The discussion of public views focussed on changes in public support for mathematics. With the recent advances in technology, this seems to relate to perceptions that less mathematics is needed, rather than different mathematics. The enabling role of mathematics is not well understood by those outside of mathematics and the crucial work of mathematicians in other disciplines is largely invisible. This was also perceived as a major flaw in many national curricula where mathematics is limited to performing service roles rather than a challenging discipline worthy of study in its own right.

The reactions of politicians and their short time-lines for educational issues, was seen as particularly harmful to mathematics which has a very long time horizon. Educational fads such as those identified above cause curriculum instability. They also result in a crowded curriculum and mathematics, with a substantive time allocation, is often the subject reduced to accommodate these changes. This is an attractive option when there is a shortage of mathematics teachers.

The discussion concerning the roles of mathematicians in policy making and implementation focused on why they are so often overlooked or even excluded from such forums. A great deal of self-criticism emerged. This included views that mathematicians have taken their subject and their jobs for granted and assumed its importance would continue to be recognised. It was suggested that they are often inept in their arguments which are too often based on self-interest and therefore fail to convince others. Further, that mathematicians have little experience of communicating with others about issues of substance and are politically inept because they just ask for more.

From a personal and Australian perspective I think the mathematicians have been far more politically astute than many in other discipline areas. However, I do not think that any of us in mathematics, or in mathematics education, were prepared for the rise of the bureaucracy in the last decade or so and the political impact of this on policy. I discuss this further later in the paper.

3.3 Issue 3

Issue 3 concerned mathematics and mathematics education in the third world with a particular focus on the way in which access to advanced level mathematics was being made so expensive, and so inaccessible, that there was a risk of creating an even greater divide. There are several reasons for this which also relate to access to mathematics education within nations. The international demand for mathematical skills means that people with these skills can attract high salaries and are very mobile. This makes attracting, and keeping, well qualified mathematics teachers in poorer areas or nations very difficult. While this is a critical factor in poorer countries it is exacerbated by wealthy countries who have failed to produce enough teachers and now recruit world-wide.

The acceptance of quite low levels of ML by wealthier countries as an alternative to quality for the majority of their students could also contribute to the problems of poorer nations. That much of their education funding is influenced by international bodies such as the World Bank and UNESCO contributes to this. These bodies are much more likely to look at, and be influenced by, what is happening in developed countries than at poorer countries. As a result standards are set at what is acceptable for the majority in wealthier nations and the curricula and tests that are often imposed take little account of culture or local need.

The curricula also imply access to technology that may only be accessible to a tiny minority in some countries. But teaching mathematics well with limited resources to a larger number of students may be more valuable.

4. EMERGING THEMES

As the issues were discussed, there emerged a common theme that had as its focus the notion of mathematics as a cultural artefact which was going to have to fight to preserve its cultural identity in a hostile environment in a similar way that minority cultures have had to in order to preserve their culture within dominant societies. As with minority cultures, mathematics faces a choice between distinctiveness and assimilation. Assimilation focuses on the goals of the dominant group and mathematics is very easily assimilated. It is flexible and responsive and readily merges into sciences, finance, medicine, computing and many other areas. In doing so it becomes invisible and can lose its identity.

Those of us who live in multicultural societies know the cultural richness that emerges in an environment where difference is celebrated rather than suppressed. The vibrancy of the arts and music do not come by assimilation to a common norm that is seen as accessible to everyone but by encouraging access through diversity and celebrating creativity. It comes from knowledgeable teachers in schools and nurturing of the talented through various forms of interaction with established artists and musicians.

However, if the critical issues identified in Australia are considered, the dangers to a rich and vibrant mathematical culture in that country become apparent. The three main findings of a recent report on the state of mathematical sciences in Australia (Thomas, 2000a) were:

- The number of year 12 students studying advanced mathematical courses continues to decline, a consistent trend since 1990.
- Few mathematics graduates are choosing teaching as a career at a time when many experienced and well-qualified mathematics teachers are retiring. Many primary teachers need further studies in mathematics and it is estimated that about 40% of junior secondary students are taught mathematics by a teacher who has little or no background in mathematics and no pedagogical studies specific to mathematics.
- And in the universities a decline of some 25% in staff since 1995, a brain drain of both experienced and new researchers, marginalised or restructured departments, fewer applications for research grants, few if any new appointments, difficulties in making appointments in key areas such as financial mathematics and statistics, and some universities no longer offering a three year degree majoring in mathematics or statistics.

In many ways these epitomise the global policy issues that confront mathematics teaching and learning.

In Western cultures many students now look to other options in, for example, business and commerce rather than mathematics and science. Even in countries where the number of students studying mathematics at advanced level in schools has not declined, it does not appear to translate into more students studying advanced mathematics at university.

If we look at what is happening in school curricula, there is much greater emphasis on outcomes and testing but students are given little information about the role of mathematics in society and no reasons for studying the harder courses. Instead mathematics has become one of the major measuring tools for educational accountability and has condemned many students, teachers and schools to 'fail'. Crude test instruments now ignore gender, race and socioeconomic factors. If music, art or sport were treated in this way, they too would be damaged and students would cease to study them at advanced levels.

The damage done by these courses and testing regimes could be minimised in the hands of a well-qualified teaching force. Instead the profession of teaching is also damaged by over-prescription and control and teaching becomes a less attractive option leading to many under-qualified teachers.

Finally, no discipline can prosper if its talented people are not encouraged in a creative, supportive environment. This implies that there must be a strong tertiary base which must interact with the talented young still in schools.

The integrity and culture of mathematics in Australia is under serious threat. But so, I suggest, it is globally unless the political issues are identified and addressed. The use of mathematics as one of the major measuring tools for educational accountability is particularly damaging. Instead of being a subject of access, a tool to be used in creative and useful ways and accessible to everyone, it becomes a gatekeeper. This must be resisted.

5. THE PARADOX OF BUREAUCRACIES AND MATHEMATICS

The following is a personal communication from a mathematician in Canada:

> This was more or less the situation over the next 30 years until the present Conservative Government took office. Part of its mandate was to arrest what they saw as the deterioration of the education system, so they commissioned a reworking and, in their view rigorization ("back to the basics, no nonsense" of the curriculum, set up a College of Teachers to serve as a professional accreditation and disciplinary body (analogous to similar bodies for law and medicine), set up a body to oversee educational "quality" in the form of a testing program that affects pupils at the levels of Grade 3, 6, 9 and 11.

> The problem is that all of this is being brought in very quickly without much planning and time to work things out well, the ministry has an inadequate base to oversee what is going on (since it has been continually "downsized" over the last two decades) and the process has become very politicized and stressful, particularly since our tax cutting government has cut back on a lot of programs and since the performance of schools in the tests has become public information.

It could have come from most western nations. The paradox is that as ministries were 'downsized' it became much more difficult to influence what was happening in any kind of meaningful way. Ministries used to have significant groups of people with time to consult widely. They tended to be people with discipline specific knowledge who were comfortable with mathematicians, mathematics educators and teachers in schools. Curricula changed, but slowly, over several years rather than several months.

The smaller ministries now have few discipline specific people and much of the work that is now done is by tender or consultancy. Many of the people with the tenders or consultancies used to work in the now smaller ministries but they are basically accountable to no-one. The stability of the senior education bureaucrats in Australia is in stark contrast to the ever changing ministers of education or the situation in the actual ministries. Some of these senior bureaucrats have changed States, or moved between the federal and State bureaucracies, but they have survived. More and more they cross international boundaries and recently, from Victoria alone, one has gone to a major international education position in Paris, another to an agency in New York and a third has been advising the Indonesian government on their national curriculum.

This group of bureaucrats forms a circle of influence with a shared language. They are very good at telling politicians what they want to hear and words like outcomes, accountability, school improvement, standards, benchmarks and quality, roll off their collective tongues with ease. Mathematics is not a discipline with cultural significance but something that is utilitarian and readily measured. They have developed a form of pseudo-consultancy that is an anathema to the kind of intellectual discourse that occurs within the discipline of mathematics or the kind of consultancy that used to occur in education. Further, they have so much influence in educational circles that few challenge them in public. And so the mathematicians are side-lined because if they do challenge the bureaucrats they are always in a minority and easily dismissed as out-of-touch academics.

I believe the mathematicians in Singapore were unduly self-critical of their role in the political arena. Gardiner (1998) was the first to identify this bureaucratic force that we now have to deal with, a force which is visible yet invisible, consisting of a group of like-minded individuals who currently wield immense educational influence. It is not a force that is easily dealt with but mathematicians have, as many of them articulated in Singapore, often stopped the worst excesses of some recent curricula reforms. These included examples from Australia and America where mathematicians have had a profound effect on maintaining the mathematical integrity of some recent curriculum documents including, in America, the revised NCTM standards (National Council of Teachers of Mathematics, 2000)

However, mathematicians, mathematics educators and mathematics teachers must learn to work together or we will all continue to make mistakes. The excesses of 'New Mathematics' were driven by the mathematicians. The excesses of the current outcomes and testing approaches have been driven by bureaucrats and educators who often know little about mathematics or mathematics education. There can only be quality mathematics education when those who understand the discipline and the teaching of mathematics work together. This by itself will not be enough though.

I have suggested some actions elsewhere (Thomas, 2000b) but I want to stress one action that must be led by the people working in the discipline. In very few nations does there seem to be any concerted effort to give young people reasons for studying mathematics other than as a means to an end. More and more mathematics is being used as an accountability and achievement measure but less and less are young people being told of potential career options which will be both enhanced and more rewarding if they have appropriate mathematical skills.

Mathematics as a tool that should be accessible to everyone, and its societal importance, needs to be to conveyed to the community. Parents need to see that a system which condemns so many young people to be seen as failures is inherently unfair and damages their life chances. It is also destroying the discipline of mathematics.

6. CONCLUDING COMMENT

I became convinced while I was writing the report on the mathematical sciences in Australia that we must become story tellers. It was known that there was a brain drain in the mathematical sciences in Australia but collecting data had been difficult—no-one wanted to put numbers to it. So I started collecting names and the list started to grow, along with comments that told a story in ways that no numerical data could. Suddenly I had data that could be understood by anyone in profound ways because it was real—it put a human face on what was happening.

We will not win the war for mathematics unless we tell a story that puts a human face to participation in mathematics. But mathematics is a human activity that brings immense rewards to those who do participate in it. We need the community to understand why mathematics is important, and the how and the why of teaching it well. That has to come from us. Our challenge is to tell the story in new and creative ways that grab the attention of students and the wider community.

ACKNOWLEDGEMENTS

Notes and comments from Professor Lynn Steen were gratefully received during the preparation of this paper.

REFERENCES

Bishop, A. (1988). *Mathematical Enculturation*. Dordrecht: Kluwer Academic Publishing.

Gardiner, T. (1998). The art of knowing. *Mathematical Gazette*, 82, 353-372.

Glenn, J. (Chair) (2000). *Before it's too late*: A report to the nation from the National Commission on mathematics and science teaching for the 21st century. Washington, DC: Department of Education.

National Council of Teachers of Mathematics (2000). *Principles and Standards for School Mathematics*. Reston, VA: NCTM.

Thomas, J. (2000a). *Mathematical Sciences in Australia: Looking for a Future*. Canberra: Federation of Australian Scientific and Technological Societies (URL: www.FAST.org).

Thomas, J. (2000b). Policy issues in the teaching and learning of the mathematical sciences at university level. *International Journal of Mathematical Education in Science and Technology*, 31(1), 133-142.

Jan Thomas,
Victoria University of Technology, Melbourne, Australia
Jan.Thomas@vu.edu.au

JEAN-LUC DORIER AND VIVIANE DURAND-GUERRIER

POLICY ISSUES CONCERNING TEACHING AT UNIVERSITY LEVEL IN FRANCE

1. ORGANIZATION OF THE TERTIARY LEVEL OF EDUCATION IN FRANCE

In order to understand the policy issues at stake in teaching at University level in France, it is necessary to have an overview of the overall organization of the tertiary education sector. This is because universities in France do not represent the only education options in post-secondary education.

We will start by a brief account of the organization of the general educational system in primary and secondary schools in France. Most schools are run by the state (the few private schools usually have a contract with the state and a similar organizational structure) and most teachers are civil servants with tenured positions. The programmes including curriculum, content, objectives, activities, time organization, etc., are national.

Primary school - 5 Years	Age 6-11
First level of secondary school (**collège**) - 4 Years	Age 11-15
Second level of secondary school (**lycée**) - 3 Years	Age 15-18

At the end of these 12 years, students sit a national examination, the *Baccalauréat* (Bac for short) with different specialties corresponding to the orientation taken by each student at the end of the first year of the lycée.

With any type of Bac a student can theoretically enter any type of university within his geographical district. Therefore the Bac is not only the final examination of secondary school but also the first examination of tertiary education.

Beside the universities, the tertiary education system offers various parallel institutions. There are mainly three different types[1]:

- Preparatory School (*Classes Préparatoires* abbreviated to CP in the rest of the text);
- Advanced Technician Training (*Brevet de Technicien Supérieur* - BTS);

---

[1] We will not describe here other systems that are numerically unimportant.

Derek Holton (Ed.), The Teaching and Learning of Mathematics at University Level: An ICMI Study, 37—44.
© 2001 *Kluwer Academic Publishers. Printed in the Netherlands.*

- University Institute of Technology (*Institut Universitaire de Technologie - IUT*).

Each structure has a different organization, especially regarding policy issues.

2. LES CLASSES PRÉPARATOIRES

The CP represents a cycle of two years of study. There are mostly two kinds of CP:

- 'scientific': preparing students for the competitive admission examinations to engineering schools (Grandes Écoles[2]); and
- 'business': preparing students for the competitive admission examinations to business schools.

The CP are located within the *lycées* (secondary schools) in every big city in the country. Classes have between 30 and 45 students. This system trains those students who are destined to become the élite of the country; it is therefore highly selective. A board of teachers and administrative managers at each lycée, selects the students who have applied using evaluations made during the last two years of secondary school. There is no geographical condition of admission and consequently there is competition between the different lycées. Globally the teachers teaching in CP are the 'best' of secondary school teachers. Most of them have been trained in one of the 'Écoles Normales Supérieures'. These teachers have no obligation to undertake research, and most of the time, are not connected with research. Like secondary schools, the programmes are national, are determined by the government, and reflect the content of the competitive admission examinations of the 'Grandes Écoles'. Each teacher decides on the type of assessment to be used during the year but it is usually very intensive. Admission to the second year of preparation can be refused if the board of teachers judges that the results are too weak, but that is quite rare. Some students may also give up after a few weeks. At the end of the second year, if a student does not pass any competitive admission examination, or does not pass the one she or he wanted, she or he can be authorized to repeat (but this can be done only once). Students failing in this system usually continue their studies at a university.

Most engineers in France have been trained in this system. One advantage of this is that it ensures an initial high level training in mathematics and physics/chemistry by highly qualified teachers. On the other hand, the training is purely academic and oriented toward selection. It therefore gives very little professional motivation or opening/experience.

---

[2] These range from Polytechnique, École Normale Supérieure, École Centrale, École des Mînes, etc., with an international reputation and a historical background dating from the French revolution, to more obscure highly specialised engineering schools.

After the CP, there are quite a variety of Grandes Écoles in engineering and business. Some are run by the state; some are (semi-)private. There is a wide range of levels of qualification on offer and the content of the teaching is very varied. Some schools are quite specialized and oriented toward a specific type of job, others offer a very general curriculum and, in some cases, the possibility of research activities in private or state organizations including universities. Most of the schools have an association of former students[3] that, as the essential organization providing jobs to students, can be very influential in the design of the curricula. Most of these schools have a curriculum organized over three years. Each school decides on the number of students it can take each year. On the other hand, creating or abolishing a classe préparatoire is a decision taken by the government, but can be influenced by local politicians in collaboration with the executive staff of a lycée.

Alongside this system there are a few engineering schools who enrol students for 4 or 5 years straight after the Bac.

3. BREVET DE TECHNICIEN SUPÉRIEUR

This is a specific two-year training for advanced technicians with various options in tertiary (selling, trilingual secretary, etc.) or secondary activity (building, electricity, etc.). Like the CP, this structure is located within the lycées of most big cities (some specializations are only offered in a very few places). Classes typically contain from 30 to 40 students. Admission is subject to a decision taken by a board of local teachers and administrative leaders, based on evaluations of the candidates essentially made during the last two years of secondary school. The teachers are secondary school teachers, who do only part of their teaching in these classes (unlike the teachers in CP who do all of their teaching in CP). The programmes are totally determined by the state. Creating or abolishing a class is a decision taken by the government. Assessment during the year is the responsibility of each teacher. Admission into second year is decided by the board of teachers. At the end of the two-years' training, students sit a national examination that is not competitive. The diploma they get entitles them to find a job as an advanced technician in their specialty. Nevertheless, some students continue studying, either in a university or, for very few of the best ones, in a Grande École. This system is very close in its organization to the secondary school system except that it has more connections with industry and has various professional partners. Although it is of less prestige than the Grandes Écoles, it offers students good employment opportunities and it also represents a smooth structural change with secondary school (small group classes, same institution) especially as compared to university (large lecture groups, tutorials, necessity to change city, etc.). Many students prefer this system to the university system, but admission in first year is subject to quotas and therefore is difficult to gain.

---

[3] These are essential to ensure jobs for graduating students.

4. INSTITUT UNIVERSITAIRE TECHNOLOGIQUE

This two-year training lies between secondary school and university. It is attached to a university, but in many ways it is independent of it. Teachers are recruited locally and are either secondary school teachers or university-type teachers (teaching only half as much as the former but being involved in research). Classes contain between 30 and 40 students. Admission is subject to a decision taken by a board of local teachers and administrative leaders essentially based on evaluations over the last two years of secondary school. The programmes are designed by each IUT according to nationally imposed schemes and is subject to four-yearly approval by the government. The schemes imposed by the state give the main guidelines in terms of programmes, curricula, content in each subject, number of hours, assessment, relation with industry, etc. Therefore local innovation is real but limited. The decision to create a new IUT is local but subject to national approval. Moreover, every four years, the contract with the government has to be renewed if the IUT wants to remain in existence, and if the national guidelines have changed during the last four years, that will influence the new proposal.

Student assessment over the two years of the course is organized according to what has been decided in the original project approved by the government. At the end of the two years, on the basis of the criteria of this assessment, the local board of teachers decides who will be given Diplomas. These Diplomas are national and entitle the student to find a job as an advanced technician in a specific field corresponding to her or his specialization. Some students continue to study, either in a university or, for very few of the best ones, in a Grande École.

5. UNIVERSITIES

The university system is organized in three cycles. The first cycle is two years long. At the end of the first cycle the students who pass the examinations are given a diploma called the DEUG (Diplôme d'Etudes Universitaires Générales). The second cycle is also two years long. The first year examination is the 'Licence', the second is the 'Maîtrise'. The third cycle starts with one year of initiation to research that leads to a diploma called the DEA (Diplôme d'Etudes Avancées). This is followed by the doctorate, which is theoretically two to three years long. Any student who has passed their Bac, can gain admission to the first year of a university within their geographical district. In theory there are no quotas. Some students enter a university outside of their geographical district, but this is exceptional and subject to a local decision. Admission to further levels is subject to decisions taken by the appropriate board of local teachers according to the results obtained in the preceding years or cycles. From the first year, a student has to choose his specialization (sciences, literature, social sciences, economics, etc.) with its limited range of options.[4] The specializations become narrower as they progress through the three cycles.

---

[4] This is different from a system in which a student can choose any combination of two or three different major or minor subjects. In the French system the combinations are already decided in fixed proportions.

There are national guidelines for each type of specialization[5] concerning programmes, curricula, content, hours in each discipline and balance between small group tutorial and big group lectures, assessment, etc. Nevertheless, each university is responsible for the precise design of its programme. The different proposals, initially discussed and designed within departments, have to be approved at different levels within the university. Finally, the university centralizes the different propositions in a global programme that has to fit the national guidelines. This programme is presented, in order to be approved at a national level, in a contract that each university signs with the National Ministry of Education and Research every four years. Globally the general designs of the curricula are quite similar from one university to another, though the differences tend to be more important in the second cycles and even more in the third cycles. Each department is responsible for the design of the part of the curriculum corresponding to its discipline even if usually the decisions are taken in collaboration with the other departments. Finally, each teacher or group of teachers is responsible for the specific choices of content, material and organization within the framework of the contract signed by the university and the state. The diplomas granted by the universities are national.

Generally, students choose their university for geographical reasons, therefore the differences in quality between the universities are not significant globally. Yet, it may be quite different in some specializations. On the other hand, Parisian universities traditionally attract the best teachers and students, whereas small provincial universities offer limited choices. France is divided into 25 'Académies'. Originally each Académie had one to three universities (depending on the separation between sciences, social sciences and literature) of variable sizes, located in the main city of the Académie. In the last 20 years, a national political decision has led to an increasing number of students having their Bac, therefore entering university. Among other effects, there has been local political pressure to create universities in smaller towns of the Académies (the buildings are partly financed by the local political authorities). Starting as entities depending on the main university, offering teaching only at the first cycle level, these universities became rapidly more or less independent universities offering more and more variety of teaching at different levels and developing research departments. This was made possible because of a general political decision taken by the French government in the early 1980s regarding the 'decentralization' of the administrative structure at various levels, including those regarding education. In particular, finances for the building of universities are now dependent on the Regions, not on the State.

Teachers in universities are mostly part-time researchers holding tenured positions as civil servants. There are two categories: *Maîtres de Conférences* (lecturer) and *Professeurs* (professors). Both categories have the same obligations in terms of teaching (192 hours per year). Every year, each university centralizes the wishes of each of its teaching and research departments and proposes to the government an ordered list of new teaching jobs it would like to have available for the following year. On the basis of the wishes of the different universities, the government decides to create a certain number of jobs on the list of each university.

---

[5] In particular the different types of specialization that can be offered are imposed by the government.

Once the government creates a job, the university is in charge of the employment of the teacher subject to specific rules. The potential candidates have to be qualified by a national commission made of teachers in the same field (2/3 are elected, 1/3 are designated by the government). Anybody having been qualified at the national level can apply for any job at the level of his qualification (lecturer or professor). Appointment decisions are then taken locally. Each university has specific recruiting commissions for each main subject. The decision to appoint is mostly based on research criteria.

Some teachers are full time teachers detached from the secondary school structure holding tenured positions as civil servants in a university. Their number increased quite rapidly in the 1980s, with the increasing number of students in the first cycles. This was the result of a national orientation taken against the wish of most teacher-researchers. Indeed, for a similar salary, a full time secondary teacher has to teach twice the number of hours of a teacher-researcher.[6] It is economical for the government but it takes up two researcher positions.

Some teachers are employed on the basis of short-time contracts. They are usually students preparing for their doctorate or just completing it.

Some teachers are employed on an hourly basis. Although globally this represents only a low percentage of teaching hours, it covers a great variety of cases and may represent a large proportion of hours in regard to secondary subjects in a specialization, like English for scientists, or mathematics in social sciences.

Connections between universities and industry were traditionally somewhat rare in France. Universities represent the academic world and usually have very few contacts with industry. Nevertheless, in recent years, many students with a Baccalauréat could not get into a BTS or an IUT and therefore entered a university as their last choice. Yet, the aim of a university for a general training did not meet their wish for a rapid professional training. This partly explains the fact that while the number of students in the first cycles of universities increased significantly in the last 15 years, the number of students in the following cycles hardly increased or even decreased in some academic subjects. In response to this new situation, the universities were led to introduce a new type of training with closer connections with industry, especially the IUP (*Institut Universitaire Professionalisant*). Moreover, the possibilities of entering an engineering school from a university, without going through the classes préparatoires have increased.

6. CONCLUSION

As in many areas, the French system is very centralized and standardized by government control. Even if universities in France may seem to have gained more autonomy in recent years, individual initiatives and originality have to fit national guidelines and are subject not only to local control but also to national control. This tends to unify everything.

---

[6] It is not rare for a secondary school teacher employed in a university to have, or be preparing for, a doctorate.

Yet, even if centralized, the system is under democratic control. Indeed, the process of decision at the top of the system is controlled by different counter-power organizations: the unions of course, but also professional associations in different subjects, which may influence decisions more specific to the content and curriculum.

For Mathematics, there are four quite powerful associations. These are:

- the Association of Mathematics Teachers (APMEP) which represents most of the secondary school teachers;
- the Association of Teachers in Classes Préparatoires (UPS);
- the French Mathematical Society (SMF); and
- the French Applied Mathematics Society (SMAI).

Moreover, the Federation of the IREM (Institutes of Research in Mathematics Education) has different commissions (including one on teaching at university level) whose work and whose members can be consulted by politicians when taking decisions relevant to this field.

This system can be seen as rigid and very complex. For instance, the system of Grandes Écoles is a heritage from the French Revolution and centralization is inherited from Napoleon. Nevertheless, this system, although archaic in some sense, is what has so far preserved the teaching, by specialists, of academic subjects like mathematics and physics in engineering and business schools. It is not rare to see a student after graduating from an engineering school starting a doctorate in fundamental research in physics, computer science or applied mathematics.

It is clear that the Classes Préparatoires are the places where the teaching of mathematics is of the best quality and the students the most able and most motivated. Mathematics teachers in these institutions have a very high level of qualification (even if they are not usually in contact with mathematics research). They also have the best salaries in the French educational system and are very respected. Through their official representatives in the UPS, they can influence changes in curricula and have a great liberty in their pedagogical choices so long as they ensure good results by their students in the national competitive examinations. The IUT and BTS are small structures in which teachers have direct impact on policy matters at the local level, even if they are tied to national guidelines that can be more or less restrictive. The selection of students on entrance ensures a good rate of success. Moreover, in these two institutions, mathematics is usually a secondary subject as the training is orientated directly towards a profession. Therefore the connection of mathematics with other subjects and with industry is an important issue that each mathematics teacher has to integrate in his pedagogical choices. It is certainly within the universities, that mathematics teachers encounter the most difficulty. Except in those few universities with the best reputations, when teaching at the highest level, they have to face large audiences of students lacking professional or academic motivation. In the first two years especially, the rates of success are very low and the level of students very heterogeneous. Overall the teaching conditions are very poor. Further, the size of the structure does not favour individual initiative and most teachers' influence on policy issues is very limited. Moreover, a university teacher is generally also a researcher and is only evaluated on his research work. As a consequence, many university teachers spend more time and energy on their research (which represents their real career) than on their teaching, for which they do not get much satisfaction neither in terms of students' motivation, nor in terms of professional recognition from their institution.

Jean-Luc Dorier
Laboratoire Leibniz, Grenoble, France
jean-luc.dorier@imag.fr

Viviane Durand-Guerrier
IUFM, Lyon, France
durand-guerrier@lyon.iufm.fr

XIANG LONGWAN

MATHEMATICS EDUCATION IN CHINESE UNIVERSITIES

1. INTRODUCTION

In China, mathematics is considered to be one of the essential characteristics of the human race. It is considered not only to be a tool of science and technology but it is also believed to be the best way to train people in logical thinking and to help them to appreciate aesthetics. This tradition goes back to the time of Confucius (551 B.C. – 479 B.C.). Consequently mathematics is the only subject in China that is taken by all students at all levels from kindergarten to university.

The situation at Shanghai Jiao Tong University, one of the 'top 9' universities in China, is typical. There the total teaching hours for an undergraduate student in a 4-year degree is between 2400 and 2500. Most students majoring in sciences, engineering, economics and management must take at least 243 hours of mathematics, including calculus and ordinary differential equations (162 hours), linear algebra (36) and probability and statistics (45). It should be mentioned that many departments require their students to take more mathematics courses. For example, a physics major must take methods of mathematics-physics (90); a student majoring in economics or management must take operations research (72); and so on. A student majoring in computer science must take 505 hours (including mathematical analysis (216), linear algebra (54), probability and statistics (45), and discrete mathematics (108)). There are many other mathematics courses that can be selected. Among these, complex variable (36), partial differential equations (36 to 54), numerical methods (36 to 54), mathematical modelling (36) and experimental mathematics (18 to 36) are quite popular. It should be pointed out that even a student majoring in Humanities must take at least 108 hours of university level mathematics.

At most universities, the Mathematics Department offers a range of courses. The Education Administrative Office of the university determines the required courses. Then different departments choose from either the required courses or courses especially designed for their students.

Because of the importance of mathematics, the Chinese Ministry of Education pays a great deal of attention to mathematics education. As a result, they have set up three nation-wide teaching guidance committees. These are the Guidance Committee for Maths Majors (GCMM); the Guidance Committee for Science Majors (GCSM); and the Guidance Committee for Engineering Majors (GCEM).

The GCEM is the oldest, having been established in the 1962. The Teaching Guidance Committee for Humanity and Arts Majors (GCHM) has been established recently.

Derek Holton (Ed.), The Teaching and Learning of Mathematics at University Level: An ICMI Study, 45—48.
© 2001 *Kluwer Academic Publishers. Printed in the Netherlands.*

The members of the above committees are professors from universities all over China. The main tasks of those committees are to:

- determine the policy and direction of mathematics education reform;
- draw up projects of mathematics teaching and reform;
- exchange practice and experience from both inside China and from abroad;
- produce, and publish, teaching material;
- edit journals for mathematics teaching;
- organise national conferences.

So the guidance committees are responsible for leadership in matters of university level mathematics content and pedagogy. Although they design and publish various programmes, courses and modules, they do not have the right to demand what universities teach and how they teach it. The final decision is made by the mathematics department, though even then, individuals may decide to teach something else. However, the guidance committees suggestions do have a great deal of influence and form the basis for most university's curricula.

In addition to the guidance committees, there are two big professional mathematics organisations in China. These are the Chinese Mathematical Society (CMS) and the Chinese Society for Industrial and Applied Mathematics (CSIAM). Both of these organisations have education committees that are concerned with all aspects of mathematics education. The relationship between the two committees and the three guidance committees is very good. Some mathematicians are members of both an education committee and a guidance committee. Many activities are co-sponsored by both groups.

In recent years, innovation has been increasingly encouraged in China. Mathematics plays an even more important role than before as

- a solid base for many follow-up courses:
- a way to produce models of real world problems and to solve them;
- a basis and/or tool for programming and for software design;
- a guide for logical thinking; and
- a part of aesthetics education.

So, in China, an understanding of mathematics is considered to be absolutely necessary for a well-educated person in the 21st century.

The traditional way of teaching mathematics in Chinese universities did have some weak points that do not fit in with the new innovative spirit. For example, the content was too narrow. Too much attention was paid to calculus techniques and many practical and/or modern topics were neglected. In addition, teaching methodology was relatively out of date. One could easily find 'Teacher centred delivery', 'A piece of paper and a stick of chalk only' in most classrooms. And finally, talented students with unusual ideas were sometimes stifled and individual contributions overlooked.

Although the view that mathematics education must be reformed is commonly held, there are still some big obstacles. Firstly, mathematics is such a basic and important subject that any big change must be given careful consideration. We have had some bad experiences in the past. In the late 1950s and, especially during the 'Cultural Revolution' from 1966 to 1976, extreme ideas played a leading role in China. Some radical steps taken under the pretense of 'reform' actually retarded the development of mathematics education in China.

Secondly, most mathematics teachers are not prepared for educational reform. For various reasons many of them resist change. Thirdly, some current regulations are not advantageous for mathematics education reform. We pay too little attention to students' individual development. Further the uniform entrance examinations for undergraduate freshmen and masters degree candidates, may "encourage teaching towards the exam" and thus hamper initiative and reform.

Actually, the first step of mathematics education reform at university level in China took place outside the classroom. In 1988, the 'Mathematical Contest in Modelling' was introduced into China based on an initiative of Professor B.A. Fussaro and his colleagues in the United States. Now, 'Mathematical Modelling' has become a course, either required or selected, at many universities in China. More than 20 textbooks have been published in the area since 1987 and during that time some foreign monographs on mathematical modelling have also been translated into Chinese.

To promote mathematics education reform, the Ministry of Education set up a big project called 'Face the 21st Century'. About 50 universities have become involved in various subprojects of this initiative. These subprojects include 'what kind of mathematical knowledge is required for a humanity and social science major?', 'how can computers and modern teaching tools be used to improve mathematics education?' and 'mathematical experimentation at university level'. Now more and more mathematics educators have realized that different students have different levels of mathematical knowledge. However, it is felt that a well-educated person should have a basic knowledge of phenomena that are deterministic or random, continuous or discrete, and finite or infinite, in order to be able to deal with the range of problems that exist. Hence most undergraduate students take courses in four basic areas of mathematics:

- analysis (calculus, ordinary differential equations at least) for experience with continuous variables;
- algebra (linear algebra at least) for discrete variables;
- probability (probability and statistics at least) for random variables;
- experimental mathematics and modelling for experience in working with real world problems.

Many experts consider geometry (solid analytical geometry, differential geometry, and elementary topology) should be given more attention. In China, mixing algebra with geometry has been one of the new hot points of mathematics teaching. Since the 1980s, computer-aided instruction with multimedia has also been promoted.

Although reform is difficult and is still continuing, many results to date have excited interest in mathematics education circles. Every year, GCMM, GCMS and GCEM hold meetings or symposiums to exchange reform results. Bigger conferences are held every few years or so.

Some new developments in the 1990s, especially since 1995, should be mentioned. These are new text books reflecting reform results have been published; experimental mathematics has been given more attention; and multimedia courseware has been introduced as an important tool for mathematics teaching.

Over a dozen textbooks have been published recently with reform ideas and some well-known foreign textbooks have been translated into Chinese. Among the foreign books that have been translated are:

- Braun, (1980), Differential Equations and Its Applications;
- Bender, (1982), An Introduction to Mathematical Modelling;
- Strang, (1988), Linear Algebra and Its Applications;
- Lucas et al, (1996), Modules In Applied Mathematics;
- Burghes et al, (1997), Mathematical Modelling;
- Hughes-Hallet et al, (1997), Calculus;
- COMAP, (1998), Principles and Practice of Mathematics;
- Strang, (1999), Calculus.

Three textbooks with the title of 'Experimental Mathematics' have been published, each one with its own character. One book, containing 21 independent real problems, emphasises modelling and using appropriate methods. Another begins by introducing some useful numerical methods and providing a computer platform. It then asks students to do open experiments. The third book discusses how to calculate special constants (like π) and how to accelerate the convergence of the calculations by experiment, etc. At many universities, experimental mathematics has become a regular course although its delivery may vary. For instance, in some places it may be taken in a student's second academic year, it may be separated into different semesters or it may be linked with related courses in other disciplines.

In the early 1990s, the GCEM arranged for a number of universities to produce a bank of exercises and/or test problems. This bank has been used continually since then and has proved to be very valuable.

Finally, it should be noted that the conferences that are being organised by the guidance committees are becoming more and more significant. Now that educational reform is encouraged nationally, many teachers want to know in what directions others are going and what results they are getting. These conferences encourage dissemination of knowledge and form bridges between universities.

Xiang Longwan
Shanghai Jiao Tong University, China
jfxlwc@online.sh.cn

ANDERS TENGSTRAND

POLICY IN SWEDEN

1. DESCRIPTION OF THE UNIVERSITY SYSTEM IN SWEDEN

1.1 A short review of the university reforms 1967-97

In the middle of the 1960s there were five universities, two technical universities and a few special institutes of higher education in Sweden. The universities were autonomous and the education was classically academic. Besides the ministry of education there was a civil service department, UKÄ (The Office of the Chancellor of the Universities in Sweden), which had a national responsibility for higher education in Sweden. The student population increased rapidly and there was a need for decentralisation. The student revolution started a debate on the possibility for students to have an influence on their education. It also started a discussion on pedagogical methods in higher education. Difficulties for many students to get relevant jobs after completing their degrees initiated plans to reorganise higher education in order to adjust it to the needs of society. The following steps were taken:

- In 1967 four new units were created: The Affiliated Universities in Karlstad, Linköping, Växjö and Örebro. They were branches of the universities in Gothenburg, Stockholm, Lund, and Uppsala respectively. In 1970 the Affiliated University in Linköping became an independent university.
- In the 1970s there were administrative experiments with increased student influence - they now participated in the decision making process at several levels within the university organisation.
- An inquiry, UKAS, from the government suggested that higher education should be organised in study programmes. After protests from the student organisations the proposal was changed and a softer steering of education (PUKAS) was undertaken by the Swedish parliament.

A new inquiry, U68, was set up in 1968 and its task was to make proposals for the organisation of all tertiary education in Sweden. In 1977, this inquiry resulted in a radical change of the system of higher education. Twelve new independent university colleges were created including the three affiliated universities. The country was divided into six regions with a board for each region, which had to co-ordinate the education between the university and the university colleges in that region.

49

Derek Holton (Ed.), The Teaching and Learning of Mathematics at University Level: An ICMI Study, 49—56.
© 2001 *Kluwer Academic Publishers. Printed in the Netherlands.*

- Education was organised in study programmes. The location and the aim of a programme was decided by the government but the details and the content was left to the actual university or university college. Each study programme was managed by a committee with representatives from the academic staff, from the students, and from organisations outside the university with interest in the specific programme. There were also possibilities for the local universities to create study programmes of their own and for the students to follow shorter courses.
- The academic staff was organised in departments with one subject or several related disciplines. There was a board for every department.
- The students were represented on every level of decision making in the university or university college. On the board of the university or the university college there were representatives from the academic staff, from the students, and from working life. The board was appointed by the government. The vice-chancellor was the chairman of the board.
- Since the student revolution in 1968 the students have had formal opportunities to affect their education. As a rule they have a big influence on decisions but their term of office is in general only one year which creates lack of continuity.
- The civil service department changed its name to UHÄ (the Office of the Chancellor of the Universities and Colleges in Sweden).

1.2   The university system now

In 1993 there was another university reform. The aim was to make the universities and the university colleges more autonomous and to increase the freedom of choice for the students. The fundamental ideas from 1993 were still valid but a few changes have been made during the 1990s. The following now hold:

Central authority
- In 1993 the UHÄ was closed down and replaced by the National Agency of Higher Education. This authority carries out the decisions of parliament within higher education.

The organisation of the university
- The board of the university has the overall responsibility for the activity of the university. The government appoints most of the members of the board. The teachers and students have the right to be represented on the board.
- Under the board there is a vice chancellor who leads the activity of the university. The vice chancellor is a member of the board.
- The chairman of the board is appointed by the government. The chairman cannot be an employee of the university. There are four domains of research, one for humanities and social sciences, one for science, one for technology and

one for medicine. The university colleges can apply to the National Agency of Higher Education for a domain of research.

- If the university or a university college has a domain of research, it shall have at least one faculty board, which is responsible for research and postgraduate education. These boards are also responsible for undergraduate education if the university or university college does not establish a special board for this activity. The board of the university has the freedom to decide which faculty boards or special boards they want.

- The faculty boards or the special boards for undergraduate education consist of teachers and students at the university or university college. Most of them are academic teachers and most of these have scientific competence. The students have the right to have three members of the board.

- The students have formal opportunities to affect their education. As a rule they have real influence on decisions but their term of office is in general only one year which sometimes implies less knowledge of the university system and this in turn decreases their chances of affecting decisions.

Courses, study programmes and degrees

- All education is organised in courses. Several courses can be brought together to form a study programme. There is a syllabus for every course and this syllabus contains the number of points (a year corresponds to 40 points), the purpose, the content, the prerequisites, the forms of assessment, the literature, etc. There is a syllabus for every study programme, which describes the courses of the programme.

- The government decides which academic degrees can be obtained. The National Agency of Higher Education decides at which universities the degrees can be obtained. If a university wants to introduce a new degree it has to apply to the National Agency of Higher Education. The agency appoints a special committee that determines if the university has the necessary qualities. It also evaluates existing degrees and it has the possibility to demand necessary changes. The agency also has the possibility to withdraw the right of the university to give a degree.

- If a university has the right to give a degree (Bachelor of Science, Master of Science, Master of Engineering, etc.) it is free to compose study programmes that lead to that degree. The courses of the programme can be compulsory and/or optional. The construction of a programme is decided by a faculty board or by a special board for undergraduate education.

Economic questions

- Each year parliament decides how much money it will spend on higher education. Each university is given money in two parts, one part is dedicated to undergraduate education and one for research and graduate education. The board of the university can allocate the money without restrictions from national authorities but they cannot mix the two parts.

- Every university gives a budget-proposal to the government for a period of about three years. A committee of the vice chancellors of the universities and university colleges discusses common questions and this committee has an annual dialogue with the government.
- Most students finance their education by Government sponsored financial support. The support consists of two parts, a study grant and a loan element. The amounts and the proportions of the two parts are decided by parliament. If a student's progress is not satisfactory they will lose their support.

Within these rules the universities have freedom to organise the research and education. The central law does not, for example, prescribe anything about departments. Nevertheless, most of the universities organise their activity in departments and a lot of decisions are delegated to this level. But there is a great variation between the universities.

The regional boards do not exist any longer and there are no external representatives on the faculty boards or the special boards of undergraduate education. However some years ago there was an addition to the law of higher education: the university now by law has a responsibility to co-operate with society outside the university.

1.3 A classification of universities in Sweden

In Sweden there is a big diversity of institutions of higher learning. Chalmers University of Technology and Jönköping University are independent foundations. The others are public authorities. In the following we give some kind of classification.

Universities

There are six universities with more than three faculties - the universities in Uppsala, Lund, Stockholm, Gothenburg, Umeå and Linköping. The universities in Uppsala and Lund have long traditions and were founded in 1477 and 1669, respectively. The others were founded in the twentieth century.

There are three new universities - the universities in Karlstad, Växjö and Örebro. They became universities in 1999 and before that they were university colleges.

Universities have the right to decide in what disciplines they will give masters degrees and they will give postgraduate studies in at least one of the following areas: humanities and social science, science, technology and medicine. The university gets resources from parliament for postgraduate education in at least one of these areas.

Institutes of technology

There are three institutes of technology: Chalmers University of Technology, Royal Institute of Technology and Luleå University of Technology. There are also institutes of technology in Lund and Linköping but they are parts of the corresponding universities.

University colleges[1]

There are sixteen university colleges of various sizes. The biggest is Mitthögskolan, which consists of four campuses in the northern part of Sweden. The smallest is Gotland College in Higher Education. They have the right to give college degrees and bachelors degrees. In some disciplines they can give masters degrees after evaluation and approval by the National Agency of Higher Education. Some university colleges have the right to give postgraduate studies in science or technology.

Special institutes

There are some special institutes of higher education, for example, the Karolinska Institute (medicine), the Stockholm Institute of Education (teacher education), the Swedish Institute of Agricultural Education, and institutes for music, fine arts and drama.

1.4 Departments and other special organisations

The board of the university or university college has the overall responsibility and the vice chancellor leads the activities. The faculty boards have, by law, special responsibility for research, postgraduate education, and undergraduate education if the university board has not established a special unit for it. The operative units are mostly departments and are established by the board of the university. They can consist of one or more disciplines. Typical names are the Department of Mathematics, the Department of Mathematics, Statistics and Computer Science, the Department of Technology. The big universities have in most cases one department per discipline, the university colleges have often brought together related disciplines in one department.

In some universities there are other organisational forms. At Lund University there is, for example, a Centre for Mathematical Sciences with its own chairman. This centre consists of four departments: Mathematics within the Faculty of Science, Mathematics within the Faculty of Technology, Mathematical Statistics and Numerical Analysis. Each department has its own director of studies.

The operative units normally have boards, which consist of staff-members and students.

In every university faculty boards or special boards for undergraduate education have responsibility for undergraduate education. This includes entrance requirements, themes and contents of courses and programmes, literature, pedagogical questions, etc. However, many decisions are often delegated to the

---

[1] The official name is 'university' but in this text it is practical to use the word 'university college'.

departments. It is, of course, important that the departments support pedagogical reforms.

1.5 The National Agency of Higher Education

The National Agency for Higher Education (Högskoleverket) is a central authority for matters concerning institutions of higher education. Its tasks include evaluation and accreditation, carrying out quality audits, developing higher education, research and analysis, supervision, international questions and study information. It is also responsible for coordinating the Swedish university network (SUNET).

The Council of Renewal of Higher Education is a part of The National Agency of Higher Education. The purpose of this council is to promote and support efforts to develop the quality and renewal of undergraduate education. In particular, the Council awards grants for development activities and evaluates such activities.

1.6 Studies in Mathematics

Mathematics is a part of many study programmes such as engineering programmes, teacher-training programmes, programmes for computer scientists, etc. Mathematics is primarily a support to other subjects. There are a few special programmes where mathematics is the main subject and there is a masters degree in mathematics. To get a masters degree in mathematics the students have to pass courses totalling 160 points and this programme have to contain in-depth studies of mathematics corresponding at least 80 points including one or two papers worth 20 points.

The professional mathematicians are, as a rule, recruited from programmes where mathematics plays an important role such as programmes for mathematicians, scientists, engineers, computer scientists and future high school teachers. They have almost always studied subjects other than mathematics but the amount varies a lot. Mathematical research in Sweden is by tradition directed to analysis and this is reflected in the undergraduate and graduate studies. Many of the outstanding Swedish mathematicians such as Ivar Fredholm, Torsten Carleman, Arne Beurling, Lennart Carlesson, Lars Gårding and Lars Hörmander, have worked in the field of analysis.

The training of future primary and secondary mathematics teachers is undertaken at a university and consequently organised in programmes which are built up by courses in the same way as other programmes. The contents of the mathematical courses for primary school teachers are, of course, very elementary and many of them are devoted to mathematical didactics.

2. SOME EXAMPLES

As we mentioned before there are great variations in the organisations of the Swedish universities and university colleges. The Higher Education Act makes this possible. In the following we briefly describe how decisions are made in one university, one institute of technology and one university college. In order to be as concrete as possible, we start from the subject mathematics.

2.1 Uppsala University

Uppsala University was founded in 1477 and is the oldest university in the Nordic countries. In 1998 there were 36,000 students at the undergraduate level. In the autumn of 1998 there were 2,553 students active in graduate programmes.

Education and research at Uppsala University have their foundation in departments, at present about 80 in number after a series of mergers of smaller into larger ones. Departments are autonomous in their use of the funds allocated to them from internal and external sources. Departments are accountable regarding education and research to faculty boards, who are responsible for academic priorities and the quality content of education and research in their area. Faculty boards set up guidelines and targets for departments, distribute government allocations, and monitor results. There are a number of faculty boards and one of them is for Science and Technology.

The Department of Mathematics consists of two parts: Mathematics and Mathematical Statistics. The board of the Department consists of four members of the teaching staff, one postgraduate student, one undergraduate student and one representative of the technical and administrative staff. The head of the department is one of the members of the teaching staff and she/he is also the chairman of the board. The organisation of the teaching and educational development in mathematics is decided by the Department of Mathematics

The department serves a lot of programmes within Science and Technology with courses. The programmes are planned by four committees. There is one committee for Master of Science programmes in Mathematical and Computer Sciences, one committee for education of the civil engineers, one committee for shorter education of engineers and one for single subject courses. Under these committees there are advisory groups one for each programme. These groups are led by programme co-ordinators. The syllabus and the literature of the courses in Mathematics are decided by the advisory groups. A special board for undergraduate education has the overall responsibility for the programmes and its members are mainly the programme co-ordinators and students. This board follows and evaluates the activities.

The board of undergraduate education is established by the Faculty Board of Science and Technology. The members of the faculty board are the dean, the assistant dean, students, the chairman of the board for undergraduate education and the chairmen of the six sections (Mathematics and Computer Science, Technology, Physics, Chemistry, Biology, Geology) that are responsible for research and postgraduate studies. The dean of the faculty is vice rector of the university.

2.2 Chalmers University of Technology

Chalmers University of Technology offers Ph.D and Licentiate course programmes as well as MScEng, MArch, BScEng and nautical programmes. The University is situated in Gothenburg and was founded in 1829 following a donation by William Chalmers. Chalmers became an independent foundation on July 1, 1994. Around 10,200 people work and study at Chalmers, including more than 8,000 undergraduates.

Education and research takes place on a scientific basis in the University's 80 departments. The departments are normally led by a head of department. The departments are organised into eight schools, corresponding to the main areas of engineering. Some of the schools, for example the School of Mathematical and Computing Sciences, are run jointly with the Faculty of Science at Göteborg University. The dean and the deputy dean are responsible for the management of the school. There is a board of the school that consists of five representatives of the staff, two from working life, three from the students, two equal opportunities representatives and one representative chosen by the president, Chalmers.

The School of Mathematical and Computing Sciences is divided into two departments: the Department of Mathematics and the Department of Computing Sciences. The heads of department constitute together with the dean and the deputy dean the steering group of the school.

2.3 Mälardalen University

Mälardalen University is a twin university college, located in Eskilstuna and Västerås. Around 12 000 students study at the university college. It is organised around subjects. Thus staff not engaged in general duties work within one of the ten departments, the majority of which are engaged in activities both in Eskilstuna and Västerås.

There is a faculty board that has overall responsibility for the research and graduate education within the University College. There is also a board of undergraduate studies which has the overall responsibility for planning undergraduate education. It distributes the resources for undergraduate education to the departments and it establishes study programmes. Every programme has a host department that is determined by the board of undergraduate studies.

One of the ten departments is the Department of Mathematics and Physics. It provides the courses in mathematics and physics in all of the university's degree programmes in the natural sciences and engineering, and is also the department responsible for four of these programmes. The department determines the syllabus and the literature of each course. The board of the department consists of six members of the staff, one representative of the technical and administrative staff and two students.

Anders Tengstrand
Växjö University, Sweden
Anders.Tengstrand@msi.vxu.se

SECTION 2

PRACTICE

Edited by Joel Hillel and Urs Kirchgraber

JOEL HILLEL

TRENDS IN CURRICULUM

A Working Group Report

1. INTRODUCTION

The Working Group: Trends in Curriculum, examined the various forces which act on a mathematics curriculum, and on curriculum trends, both at local and national levels. 'Curriculum' was considered in its widest sense to mean "matters pertaining to the purposes, goals and content of mathematics education" (Discussion Document for this ICMI Study, 1997), as well as the means for achieving curricular goals. Hence, the discussions in the Working Group touched on undergraduate programmes[1], specific courses, mathematical content, degree of rigour, modes of delivery and interaction, and assessment schemes. Inevitably, the issues discussed in this Working Group overlapped substantially with the other Working Groups of the conference.

2. BACKGROUND

2.1 Who are mathematics students?

Curricular issues are inextricably tied to the question: a mathematics curriculum for whom? The teaching of mathematics at universities and colleges is quite diverse in its organization hence there is a wide range of students populating mathematics courses.

Among those enrolling in mathematics courses, there are students for whom mathematics is the primary subject of their undergraduate studies, possibly coupled with another discipline such as statistics, physics, computer science, or economics. We will refer to this group as '(maths) programme students' so as to distinguish them from 'client students'. The latter come from client departments, traditionally the physical sciences and engineering departments, though nowadays, computer science has become a prominent client, replacing physics in many countries. Other client students come from departments such as social sciences, commerce and

---

[1] Any international gathering immediately points to the different senses attributed to words such as 'programme', 'course', 'module', or 'paper'. In this document we have adhered to the North American usage of 'programme' and 'course' whereby a programme is made up of a collection of compulsory and optional courses, and a (one semester) course constitutes about 40 hours of instruction.

Derek Holton (Ed.), The Teaching and Learning of Mathematics at University Level: An ICMI Study, 59—69.
© 2001 *Kluwer Academic Publishers. Printed in the Netherlands.*

economics, and psychology, who are increasingly requiring their students to take some mathematics. Thus, a somewhat facile distinction between the two groups is that programme students want to study mathematics while client students have to. In any case, it is usually expected that client students terminate their mathematics studies after a year or two of their undergraduate studies.

Future school mathematics teachers, particularly at the secondary level can be considered as either programme or client students depending on national criteria for the training of teachers. For example, in certain European countries, prospective secondary teachers have to complete a full 5-year undergraduate mathematics curriculum and hence are, in our terms, programme students. On the other hand, future teachers in North America generally take only a certain number of mathematics courses rather than a full mathematics programme.

2.2   Organization of undergraduate teaching

Many departments of Mathematics are responsible for teaching all programme and client students. In fact, the teaching of client students (the 'service role') is often the bread-and-butter component of departments' teaching and it justifies having a large department of mathematics. Other institutions have 'mini-departments' of mathematics housed in engineering (Polytechniques), finance, economics, or education and teach exclusively the students of their discipline.

Mathematics departments who teach both programme and client students do so in different ways. Some require all their students to take the same courses, say, calculus, differential equations, or linear algebra, resulting in classes with a heterogeneous group of students whose background preparation, career ambitions, and interest in mathematics are quite varied. This 'one-curriculum-for-all' approach inevitably raises a range of issues as to the appropriate level and emphasis, and as to the type and depth of applications. Other departments offer a variety of courses that are specifically geared for one client group or another, viz. 'calculus for engineers', 'calculus for chemists', or 'algebra for teachers'.

Focusing on the mathematics curriculum specifically targeted for programme students, there are also wide variations depending on the traditions of the universities involved. These traditions have to do with: admission standards, the juncture at which a student can choose a mathematics option, the length of study, the number of courses required, course choices (both in mathematics and outside the discipline), and whether or not there is a requirement for studying mathematics together with a cognate discipline. The intended goal(s) of a mathematics programme (even if not explicitly articulated) are also dissimilar. To take but one example, in most Canadian universities, programme students complete either a major or honours in some field of mathematics (e.g. pure, applied, statistics). A major programme is comprised of a certain concentration of core and elective mathematics courses (which can amount to as little as a third of the total number of courses necessary to obtain a bachelor's degree). Except for the first year of the major (in a 4-year programme) where there are several compulsory courses in the other sciences and computer science, students have nearly complete freedom to choose courses complementary to the major. The honours programme, on the other hand, tends to be a more selective programme with a substantially greater number of advanced courses, and may take possibly an extra year to complete. The goal of the honours programme is to train highly qualified persons who can continue doing graduate studies and research or be employed in demanding mathematical fields. On the other hand, the goals for the major programme are more modest, namely to graduate students who are mathematically literate, and who can function comfortably in work situations requiring quantitative, analytic, and mathematical problem solving skills. (For more details on a major and honours programme, see Hillel, this volume, pp. 179-184.)

3. FACTORS INFLUENCING CURRICULUM

3.1 Changes within mathematics

The undergraduate mathematical landscape is always in some state of flux mirroring the organic nature of mathematics. New theories and mathematical tools, sometimes supported by powerful computers, are being developed within mathematics or as applications in cognate disciplines such as physics, computer science, and engineering. Certain new subjects become highly visible (e.g., dynamical systems, computer algebra), others experience a renaissance (e.g., geometry, number theory, numerical analysis), and yet others become more marginalized (e.g., category theory). In fact, Steen has written that "strong departments find that they replace or change significantly half of their courses approximately once a decade" and "as new mathematics is continually created, so mathematics courses must be continually renewed" (Steen, 1992). These on-going updates to the curriculum can be regarded, in a sense, as 'deterministic' aspects of curriculum change, ones that do not put into question the purpose, goals, and means of undergraduate education.

3.2 Changes in the pre-university math curriculum

Secondary school[2] mathematics curricula have undergone tremendous changes in the past 20 years. One most visible change in many countries is the reduction in the number of hours devoted to mathematics and science. For example, in France, up to 1994, secondary school (lycée) students had 15 hours of science teaching per week, of which 9 were in mathematics and physics. By 1999, there are only 8 hours a week in which to teach mathematics, physical sciences, biology and technology. Also in France, traditionally taught subjects like set theory and algebraic structures have been dropped, as well as the emphasis on definitions and proofs. Reports from other countries also allude to a de-emphasis on formal mathematics and on complicated

---

[2] Secondary school terminates after 11 years of schooling in some countries but lasts for up to 13 in others.

manipulations, the increased use of calculators with computer algebra systems capabilities, and the teaching of synthetic geometry, as well as a teaching approach which relies more on investigative project-oriented work. There are also increasing attempts to introduce quantitative, statistical and probabilistic reasoning in the secondary curriculum.

One should mention here that, quite often, changes in the secondary curriculum are brought about without coordination with the very universities and colleges that the students subsequently attend. Consequently, many university mathematicians are not always aware of the nature and extent of these changes nor of the pressures and constraints on the pre-university system that could explain, for example, why the number of hours devoted to mathematics is being reduced.

3.3 Changing clientele

Most countries have wisely abandoned the elitist view of university education in favour of a more open policy that makes university education accessible to a larger segment of the population. This policy has resulted in a great influx of students to universities (estimated to have increased 6-fold in the past 30 years, Steen, this volume, pp. 303-312), including a massive increase in the number of client students who are enrolled in agriculture, commerce, finance, social sciences, etc. These students tend to be heterogeneous in terms of their mathematical preparation, probably would rather not take mathematics at all if given the choice, and are not very interested in mathematical rigour and abstraction (nor even always convinced about the relevance of mathematics to their careers). At the same time, they constitute, in some universities, the main clientele of a mathematics department. Also, open immigration policies in some countries has resulted in an influx of students whose first language is not the language of instruction.

One general feature of incoming students is that they enter university having logged less hours of mathematics lessons because of the reduction of the number of hours devoted to mathematics at the secondary level. But even when a choice exists for taking more mathematics at the pre-university level, the trend is for students to forego this choice. For example, in England, there has been a significant drop in the number of students completing A-level mathematics. And, among those who do complete their A-level, the number of university mathematics candidates who have completed two A-levels in mathematics has dropped to about 1 in 10 in the 1990s whereas it was 1 in 3 in the 1960s (Simpson, 1998). A recent article indicates a drop of over 30% in students entering mathematics programmes in Germany (Jackson, 2000). Thus, the overall effect is that students' background preparation is not sufficient for meeting the rigours of traditional entry-level university courses in linear algebra and calculus, even for students who are relatively successful in their pre-university courses. (This point was also made by participants from Australia, Brazil, Canada, England, Japan, Malaysia, and the USA.)

Students' attitudes towards education, their study habits, and their expectations, are influenced by the traditions and values of the prevailing culture in which they live. There is also a sense that students nowadays are more career-oriented and thus interested in getting skills that lead directly to jobs. This explains why in some countries, the number of mathematics programme students has been dwindling dramatically and the ablest students are drawn to such fields as computer science, engineering, and finance, where career opportunities are more evident. Within mathematics, there is sharp increase in enrolment in actuarial mathematics, when this option is available (for example, in Australia, Canada - see Hillel, this volume, pp. 179-184, - and Switzerland - see Kirchgraber, this volume, pp. 185-190). There is also an increasing tendency for students to combine study and work, thus taking on part-time jobs to supplement their income.

Departments of mathematics are thus faced with the challenge of having to teach students whose background preparation, learning styles, study habits, and career ambitions are more and more at odds with the traditional lecture-style mathematical training with its Bourbaki-like curriculum, particularly, in pure mathematics. Furthermore, many departments are facing an increase in the number of client students and a decline in the number of programme students.

3.4 Resources

Certain countries have never had adequate resources for higher education; others are experiencing political and economic upheavals which greatly affect education, as well as all other aspects of life. Even more affluent and politically stable countries, are witnessing a continual erosion of the levels of government support for universities. Diminishing resources usually translate into less staff, larger classes, and pressures to be more efficient and financially accountable. In such instances, mathematics departments are finding themselves less able to offer specialized courses to a small number of students and so have redefined an appropriate core of an undergraduate mathematics programme.

3.5 Technology

Computers have impacted on the methods and results of several mathematical domains. Coupled with graphing software, computer algebra systems, dynamic geometry, or differential equations packages, computers and calculators pose interesting challenges to mathematics departments. They have led to questioning what mathematics content is central and what is redundant, as well as, how present-day learning, teaching and assessment practices, can be and ought to be changed.

3.6 External influences: Governments, Research Agencies and Business

There is a prevalent sentiment among mathematicians that they, as the professionals, are in the best position to define the undergraduate curriculum for their students. They view attempts to influence curricular choices by bodies external to the department as an unwarranted intrusion. However, governments who, in most cases, foot the universities' bills, have, in recent years, been much more vocal and less timid in exerting pressures on the universities. One often-heard demand is for universities to be 'accountable'. Accountability has a general sense of being fiscally responsible. But more specific to mathematics, accountability can be about: reversing the high failure rate in mathematics; gearing graduating students' training towards the job market; insisting that students become familiar with computer technology; and questioning the overproduction of Ph.D.s in pure mathematics in times of low demands for academic positions. Few university administrators can afford to remain oblivious to these demands. Thus, there is a certain ongoing tug-of-war between mathematicians and university officials or representatives of government and business, in order to establish sensible 'boundary conditions' between autonomy and outside interests.

Indirect external pressures which ultimately influence the curriculum are via the research granting agencies who can give preferences to, say, applied research, or fund new initiatives which emphasize university-industry linkages, cross-disciplinary projects, hardware and software development, etc. Similarly, high-profile employers have also been active in describing a set of skills they want from their future employees - these include specific mathematical skills, knowledge of programming languages, as well as communication and social skills.

3.7 Research in mathematics education

Research on the teaching and the learning of undergraduate mathematics is becoming more visible. Mathematicians who examine mathematics education research findings and reflect on their own pedagogy have likely accepted the idea that lack of understanding in mathematics is not simply a question of the 'weak' or 'unmotivated' student; that there are epistemological and pedagogical roots to some of the difficulties in mathematics; and that there are feasible curricular options (albeit more time consuming) to ameliorate things.

4. TRENDS IN CURRICULUM

A fairly accurate picture of undergraduate mathematics is that, by and large, it is still dominated by the 'chalk-and-talk' paradigm, a carefully selected linear ordering of course content, and assessment which is heavily based on a final examination. Even the highly publicized 'computer revolution' has not really made a sweeping impact on mathematics. Hence the recent statement that "technology has repeatedly promised to transform mathematics pedagogically ... That said, mathematics in 1999 looks a lot more like mathematics in 1939 than is the case with any of its sister sciences" (Borwein, 1999). The agenda is still basically defined by pure mathematics and one can reasonably claim that as long as the primary goal of mathematics education is conceived in terms of preparing future professional mathematicians, existing curricula function optimally if they just keep abreast of new developments within mathematics. Nevertheless, there are many calls from the general scientific community and professional associations of mathematicians and users of mathematics, to overhaul undergraduate mathematics education. For example, a recent panel of experts reporting about the state of the mathematical sciences in the U.S. wrote: "The curriculum in U.S. institutions for undergraduates needs to be strengthened, broadened, and designed for more active participation by students in discovery" and "[The academic mathematical community] must respond to a new trend in pedagogy that makes use of scientific problems to motivate mathematical theory and to increase the use of computers, especially with regard to visualization. Students need a sound understanding of mathematics as a basis for life-long learning. The dual nature of mathematics as a theoretical field of its own and as a discipline intrinsically linked to applications must be better reflected in the undergraduate curriculum" (NSF, 1998). Even a cursory reading of the above statement reveals that it touches on many dimensions of mathematics education, including: goals, epistemology, learning styles, motivational issues, technology, and breadth of training. It is rather hard to conceive how recommendations of such sweeping generality can be implemented.

In practice, it turns out that actual trends tend to be more modest and depend very much on the contexts and goals of the institutions involved. Changes are most discernible in departments that consider the goal of training future mathematicians as being too narrow, too expensive, or simply unrealistic in terms of who is actually enrolled in their programmes. Rather, they see their goals nowadays as being both academic and vocational. Among the trends that were discussed in the Working Group are the following.

4.1 Profiles for students

Some departments are becoming more explicit about their aims and objectives for courses and for programmes as well as in describing a desired 'profile' of a student completing each of their programmes. Examples of items which appear in several documents refer to students as having "an understanding of the principles, techniques and applications of selected areas of mathematics, statistics, operations research and computing"[3], "ability and confidence to construct and use mathematical models of systems and situations, employing a creative and critical approach", and "high level written and oral communication skills that facilitate professional interaction". This may look not unlike the 'Standards' of the NCTM. Not all mathematicians (nor all participants in the Working Group) like the idea of an explicit list of objectives. However, departments use such profiles as means to define components of their curricula. For example, insisting that students graduate from a mathematics programme possessing communication skills and ability to work in teams, is reflected by an increase in the weight given to students' written reports, solutions to assignments, oral presentations of an accessible mathematics article, or of the results of a team project. Assessments are then based not just on the mathematics content but on the organization, clarity and quality of the presentations. In the same vein, there is an increasing number of undergraduate courses with titles

---

[3] It is interesting to note that the European Union is currently trying to define a list of mathematical concepts and techniques appropriate to students at a particular age level.

such as 'communicating mathematically' or 'mathematical thinking' which specifically address issues of forms and styles of mathematical communications.

The issue of student profiles triggered some discussion in the Working Group regarding 'meta level' goals which can serve as a guide for curriculum design. Such meta level goals have to be in response to the question: "what is it exactly that one wants to impart to students?" Some of the suggested goals were: showing the way of life of a professional mathematician (including using logic, making generalization and abstractions, and proving theorems); unpacking or unfolding of mathematics so students can come to understand where conjectures and theorems come from; providing a feeling of what the subject is and that it is part of the culture. A connected question was whether curriculum can be organized less in terms of courses and prerequisites and more in terms of general themes and activities somewhat as exemplified in this diagram.

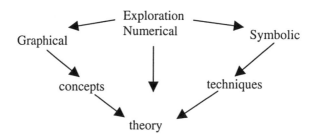

4.2   Reduced content and rigour

There is a general trend towards reducing the mathematical content of courses, both for programme and client students. This is, in part, due to client disciplines reducing the amount of mathematics required by their students. Coupled with this reduction of content there is also a trend towards undergraduate teaching which is less formal, more open-ended and which tries to build on students' intuitions, visualizations, and experimentation. This trend is perhaps best exemplified by the 'reform calculus' movement that offers an alternative approach to the teaching of the beginning calculus sequence. Though the approach is not without its detractors, it has been adopted by a good number of institutions in North America, South Africa, Australia, and other countries as well. At a slightly more advanced level, a course such as a "question-based approach to undergraduate analysis" (Alcock and Simpson, 1998), exemplifies the attempts to move away from the theorem-proof lecture format. There is also a renewed push towards a matrix-based approach to the introductory linear algebra sequence rather than the vector-space/linear-transformation approach (Carlson, Johnson, Lay and Porter, 1993).

4.3   Applications and modelling

There is an increased emphasis on applications and computer simulations both in main-stream mathematics courses and in courses targeted for client students.

Whereas, in the past, applications in courses such as linear algebra, calculus and differential equations were mostly in physics, they now tend to be in computer science/discrete mathematics, biology, financial mathematics, and actuarial studies. There are now in existence whole collections of modules (units) on applications appropriate at different levels. More generally, there is an emphasis on teaching mathematical modelling either as a specific course or by giving students some real problems to solve that either come from industry or are obtained from some data bank with real data. (This emphasis on modelling is also to be found in China, see Xiang, this volume, pp. 45-48.)

Related to modelling, part of the discussion in the Working Group centred on the importance of students being able to evaluate different models. The question was asked as to whether it is realistic to expect students to be able to evaluate models, given that in many cases, they are familiar neither with the specific context being modelled nor with the mathematics used.

4.4 Bridging courses

The transition problem from secondary to tertiary level has led to the appearance of bridging courses aiming to facilitate students' entry into university mathematics. Such courses either try to cover specific domains and concepts which incoming students lack (e.g. 3-dimensional coordinate and vector geometry, properties of real numbers, complex numbers, and polynomials, basic set theory) and/or introduce elementary logic, formal proof techniques and mathematical communication skills. An alternate and more systemic approach is the one in place in Québec where secondary school graduates enter a 2-year college system prior to starting university. These colleges are meant to provide the necessary 'buffer' between secondary schools and universities. (For more on this topic, see Wood, this volume, pp. 87-98.)

4.5 Technology

Though some uses of the computer in undergraduate mathematics can still be deemed to be 'cosmetic', individuals and departments have taken strong initiatives by substantially redesigning their courses in order to fully integrate the use of computer algebra systems, geometric, differential equations, and statistical software for the purpose of calculations, experimentation, visualization, and programming. This trend is more apparent in the first year calculus, differential equations and linear algebra courses. Similarly, departments are experimenting with using the Web for creating Web-based courses (particularly for education at a distance), (interactive) texts and lectures, and for setting up data banks of problems and applications.

4.6   Courses for clients

The one-maths-course-for-all model is giving way to customized courses for different clientele. Typical of such courses are discrete mathematics for computer science students, financial mathematics for commerce and economics students, applied calculus courses for life and social sciences, or specific engineering mathematics courses. Also, some courses for clients are now being team-taught by a multidisciplinary group which includes mathematicians as well as other professionals (physicists, chemists, engineers, etc.). Departments which are involved in teacher training are increasingly tailoring their courses to be centred around key issues of elementary number theory, synthetic geometry, or theory of one-variable real functions, and adding a strong emphasis on epistemological and cognitive aspects.

4.7   Assessment

Though assessment is still dominated by the end-of-year exams there is a move towards a more varied assessment based on projects, weekly tests, essays, report writing, and seminar presentation, and group projects. (See, for example, Houston, this volume, pp. 407-422.)

4.8   New degree options

Joint degrees, traditionally in mathematics and physics, have now given way to degrees such as mathematics and finance (Jackson, 2000), mathematics and ecology, mathematics and information technology. The new programmes are often designed along with colleagues from other disciplines. Also, co-operative programmes that combine mathematical studies with work experience are becoming more popular.

4.9   Research-based trends

Based on research findings, some undergraduate courses are being restructured, quite often by those involved in the research in the first place. The more visible of these efforts touch introductory courses in linear algebra, differential equations, and abstract algebra (see Artigue, 1994, Dubinsky and Leron, 1994 and Dorier, Pian, Robert and Rogalski, 1999).

All of the above trends were mentioned by various participants of the Working Group based either on the experience of their own institutions or their knowledge of the situation at large in their countries. There was also an informal international survey conducted by members of the Programme Committee prior to the conference in Singapore (Engelbrecht, 1998). One study worth mentioning here is the in-depth survey of the honours programme in England and Wales by Kahn and Hoyles, (1997). The study points to three main areas of change: a broadened range of mathematics courses including 'new' applications, less advanced content, and more emphasis on interim examinations for the purpose of assessment. The authors attribute the system of mass higher education and the changing profile of the mathematics undergraduate student as the main impetus for these changes.

REFERENCES

Alcock. L. and Simpson A. (1998). Definitions in university students' perceptions of structure, Proceedings of the International Conference on Mathematics Teaching, Samos, University of the Aegean, 19-21.

Artigue, M. (1994). Didactic engineering as a framework for the conception of teaching products. In R. Bielher, R.W. Scholz, R. Strasser and B. Winkleman (Eds.), *Didactics of Mathematics as a Scientific Discipline*, pp. 27-39. Dordrecht: Kluwer Academic Publishers.

Borwein, J. (1999), Talk given at Technology in Mathematics Education at the Secondary and Tertiary Level conference, Brock University, Canada, June 2-3.

Carlson, D., Johnson C., Lay D. and Porter, A. (1993). The Linear Algebra Curriculum Study Group recommendations for the first course in linear algebra. *College Mathematics Journal*, 24.1, 41-46.

Discussion Document (1997). ICMI Study on the Teaching and Learning of Mathematics at University Level. *Bull. ICMI*, No. 43, December, 3-13.

Dorier, J-L, Pian, J., Robert, A. and Rogalski, M. (1999). On a research program about the teaching and learning of linear algebra in first year of French science university, Pre-proceedings of this ICMI Study, 114-117.

Dubinsky E. and U. Leron (1994). *Learning Abstract Algebra with ISETL*. New York, Springer-Verlag.

Engelbrecht, J. (1998). What happened to undergraduate mathematics teaching over the past ten years? An international survey of undergraduate mathematics courses, conducted in conjunction with this ICMI Study. University of Pretoria, South Africa.

Jackson, A. (2000). Declining Students Numbers Worry German Mathematics Departments. *Notices of the American Mathematical Society*, Vol 47/3, 364-368.

Hillel, J. (2001). Departmental Profile, Concordia, Montreal, Canada, this volume pp.179-184.

Houston, K. (2001). Assessing Undergraduate Mathematics Students, this volume, pp.407-422.

Kahn, P.E. and Hoyles. C. (1997). The Changing Undergraduate Experience: A case study of single honours mathematics in England and Wales. *Studies in Higher Education*, Vol. 22/3, 349-362.

Kirchgraber, U. (2001). Departmental Profile, Eidgenössische Technische Hochschule (ETH), Zürich, Switzerland, the volume, pp.187-192.

NSF (March 1998): Report of the senior assessment panel of the international assessment of the U.S. mathematical sciences.

Simpson, A. (1998). The Impact of Tensions between Views of Mathematics and Students' Worldviews. Pre-proceedings of the ICMI Study Conference, 8-12 December 1998, Singapore, 234-237.

Steen, L. A. (1992). 20 questions that Deans should ask their mathematics department. *Bulletin of the American Association of Higher Education*, 44:9, 3-6.

Steen, L.A. (2001). Revolution by Stealth: Redefining University Mathematics, this volume pp.303-312

Wood, L. (2001). The Secondary-Tertiary Interface, this volume, pp.87-98.

Xiang Longwan, Mathematics Education in Chinese Universities, this volume, pp.45-48.

Joel Hillel
Concordia University
Montréal, Canada
jhillel@alcor.concordia.ca

JOHN MASON

MATHEMATICAL TEACHING PRACTICES AT TERTIARY LEVEL: WORKING GROUP REPORT

1. INTRODUCTION

My task in writing this chapter is to present the discussion of the working group while going beyond simply reporting what was said. The working group managed to bring to expression a variety of ways of (inter)acting with students and a variety of approaches to 'preparing to teach a topic'. I felt there was considerable overlap and commonality, but it is inevitable that as scribe I am tempted to perceive what was offered from my own perspective and to find my own unity. One manifestation of this is that I found myself connecting ideas as recorded with those I employ myself and even to subsume some of what others said under my own articulations.

2. AIM AND FOCUS OF GROUP

Before we met, it was suggested that the working group aim should be to build a collection of teaching acts which people have found useful and to consider associated pragmatic and philosophical issues. The practices were expected to range from general ways of preparing for sessions to specific things to do in the midst of a session.

The working group began by asking participants to describe specific forms of interactions that they use: particular devices that one might use in a mathematics session. This raised the question of why one would want to choose one device over another and for what purpose it might be most suitable. The second session was devoted to participants raising issues and describing problematic situations from their own setting. In the third session, seven participants each gave an example of a task they have used with students, as exemplary case studies addressing the question "How do you prepare to teach a topic?"

This chapter operates in more or less reverse order. It summarises the seven descriptions, using them as examples of Principles-In-Action, to draw out some inferred and explicit principles. It also addresses the question of the basis for some of the decisions, then provides some examples of devices or forms of interaction with students which have proved useful and which can be chosen because they promise to contribute to the achievement of goals. Finally, there is a summary of some of the issues raised by members of the working group, which need further elaboration and sharing of experience. There is, of course, insufficient room here to

Derek Holton (Ed.), The Teaching and Learning of Mathematics at University Level: An ICMI Study, 71—86.
© 2001 *Kluwer Academic Publishers. Printed in the Netherlands.*

do more than adumbrate the deep foundations of any set of principles, but reference is made to sources where further details can be found.

3. BACKGROUND: MAKING PRINCIPLED CHOICES

In a scientific age, one hopes that decisions made about how to teach a particular session are principled. Principles can comprise a value system (what one believes and values) and they can comprise assumptions or starting points from which one then derives appropriate acts or against which one evaluates the appropriateness of various possibilities. (For example, students need to be encouraged to be active and to assert rather than merely assent to what they are told). Of course, there are overlaps in the two forms of principles. Explicit discussion of principles is likely to draw participants into a wider discussion of pedagogic, psychological and socio-cultural assumptions and theories.

Exposing one's principles-beliefs-theories makes them available to questioning, critique and modification, whereas when they remain embedded below the surface of behaviour, they are not amenable to modification. Once exposed, they can provide a structure or framework for informing (literally in-forming) future actions, not by mechanical or automatic habit, but by increasing sensitivity to possible choices or options in the moment.

Teaching consists of a sequence of acts of teaching, while learning is something that has to be done by the student, for teaching takes place *in* time, while learning takes place *over* time (Griffin 1989). The acts of teaching require choices to be made both in advance and during the flow of the event. These choices are based on principles and on experience, whether explicitly or implicitly and they are influenced by the *weltanschauung* one has of the nature of mathematics, of how mathematics is learned and of the particular situation (students, purposes, context, etc.).

Many of these decisions have nevertheless been made years before and are frozen in habituated practices (James 1890, Dewey 1922). Habituated and automated practices may be effective, but some may need renewing and refreshing in order to release their full potential, while others may be unintentionally obstructing student learning. Other decisions are made locally in preparation of details, such as choice of examples, order of presentation and modes of interaction to be employed. Finally, there are choices made moment by moment as the event unfolds. These latter are often based on past experience of being taught.

All of these decisions are, of course, informed by the teacher's goals (usually that the students understand the concepts and achieve sufficient competence to pass the examination) and hopefully all are consonant with students' goals, or with goals students can agree to. Several presenters at the conference (including Legrand 1999, Simpson 1998, Mason 1999) advocated a revised didactic contract, seeking to balance developing competency with enculturation into mathematical thinking, rather than succumbing to student desire to minimise effort and simply be trained in requisite behaviour. Several other contributions to the conference decried growing pressures to focus on training in behaviour to the extent of displacing education of awareness. These pressures come at a time when students arrive at university with widely differing desires and aims, competencies and experiences. There seemed general agreement that a major aspect of the role of mathematics teaching is to engage students in the process of thinking mathematically, to show that scientific debate (Legrand 1999) offers a significant and powerful mode of conduct which extends well beyond the lecture hall.

One set of values might be summarised by

> Mathematics is learned by being exposed to definitions, theorems, proofs, techniques and examples, through which one is exposed to formalisation, proof, modelling, techniques etc. The teacher's job is to lay out the material clearly and logically. Students must rehearse many examples in order to develop facility and through facility, gain understanding of the concepts, the techniques and why the techniques work. Hard work is valued, work consisting largely of working through notes and problems to try to understand them.

Another set of values and beliefs might be summarised by

> Mathematics is learned by reconstructing for oneself what others have thought about and tried to expound clearly and logically. Reconstruction is carried out through constructing special and illustrative cases, trying to see generality through the particular, guided by theorems and other exposition. Exposition and practice on exercises is useful, but only as means to reconstruction. Facility and understanding grow together, as each contributes to the other and neither necessarily precedes the other.

There are probably almost as many variants as there are mathematics teachers.

Teaching as a Creative Act

Principles do not emerge without effort, nor do they provide an algorithmic or deductive basis for determining how one teaches. There are no theorems in mathematics education, no hard and fast rules or best practices; rather there are things worth being aware of, that is, sensitivities which enable one to notice or be aware. But every good idea can also be disastrous if employed insensitively or without the requisite awareness of mathematical, psychological and socio-cultural aspects of the situation. For most assertions that are valid in some contexts, the contrary can also be valid in other, often very similar, circumstances. In other words, teaching is a creative response to the structure of attention of human beings, not a mechanical reaction to a pre-specified set of conditions.

4. PRINCIPLES-IN-ACTION

This section briefly summarises responses offered by seven participants to the question: How do you prepare to teach a topic? What emerges are some explicitly articulated principles, as well as a considerable range of principles which appear to be at work underlying then as well.

4.1 Example 1:

Jean Pian took as his topic limits of a function as x tends to infinity. He started with reviewing what students already know: some experience of manipulating and computing limits using rational functions, *exp*, *ln*, etc. Aware that the students have mainly used limits to construct graphic representations of a function, they probably haven't yet met a formal definition nor have they really thought about the meaning of a limit. The objective of the lecture is then to make students feel the need for a precise and formal definition. Questions such as 'what would I like them to understand about limits?' (to which the answer is 'idea of equivalence, bounds and the local nature of limits') drive the preparation. In a lecture there would be description of misconceptions about limits. Students would be asked to consider how a limit might be defined (phrases such as 'very close' would be expected), what a graphic representation might show about a limit and how various formulations of the limit idea are related.

Summarising Comment

Aspects stressed particularly: what do students know, what are their likely perceptions and hence confusions or preconceptions or mis-apprehensions? The teacher addresses the question of where it would be useful for student attention to be directed. The aim of the lecture is to awaken need for a formal definition in order to be certain about finding limits and hence to lead to proof. Tasks are chosen to try to get students to formalise and to see how examples offered by the teacher challenge their perceptions.

4.2 Example 2:

Vivianne Durand-Guerier takes a tutorial group of 30 students on continuity following Pian's 'lecture'. She might start with the question 'if f is continuous on [0, +∞[, strictly positive and with limit zero at +∞, can we affirm that f has a maximum value?' From previous studies and experience we know that most students, perhaps even all, will answer 'yes'. The task is to find a proof. Again from experience and investigation, we know that the arguments which will be offered are likely to be variants of

its obvious;	there are no counter-examples;
we can see on a graph …;	*f* decreases if you look far enough.

While counter arguments one might expect, include

some 'obvious' results may be false;	*f* might not be decreasing (look for an example);
here, examples are not sufficient to be satisfied it is always true;	we can imagine a graph which is not a 'classic' function;

not finding a counter-example does not mean there are none.

Prior analysis of what is involved in locating a proof suggests mathematical content, including students' likely conceptions: particularly the notion that for any epsilon you can go out far enough and find that the function is less than epsilon ever-after. The teacher needs to encourage students to organise the use of the definition and the property they want to use. If the teacher merely tells the students, the whole process may simply seem magical. We were reminded that not all sessions are constructed or conducted this way, but that it is an option.

Summarising Comment

Durand-Guerier's approach is based on a sophisticated perception of the *situation didactique* (Laborde 1989, Artigue 1996, Brousseau 1997). This includes prior analysis of students' past experiences, of what they bring to the session and of what sorts of difficulties, pre-conceptions and mis-apprehensions they are likely to have or to form. These have to be balanced against the aim of the sessions, in terms both of subject matter and also of mathematical thinking processes that students might be exposed to. In order to provoke students to reconstruct their ideas, examples to use as a task for students in order to challenge their conceptions are chosen using what has been learned from research.

Whereas Durand-Guerier works at exposing and challenging conceptions through interaction with small groups of students, Marc Legrand achieves and pursues *scientific debate* in large lectures as well.

Example 3:

Legrand described aspects of his use of *le débat scientifique*, details of which are expanded upon later in this report. (See also Legrand, 1993.)

Example 4:

Chris Rasmussen outlined a class based on first-order differential equations such as $\frac{dy}{dx} = y(3-y)$. He follows Simon (1995) in stressing the importance of contemplating and describing learning goals, learning activities and the kinds of thinking in which students might engage, in preparation for a session.

He also attends specifically to recording student reasoning in a form that is both consistent with their reasoning and is profitable for further mathematical development. In this case, he uses a phase plot to capture and support student thinking about the direction of flow around y = 3.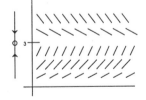

Summarising Comment

In addition to thinking about students' present thinking, aims and goals and the kind of thinking which might arise through choice of task, there are choices to be made about how to deal with what students offer. To prepare oneself to take advantage of useful student suggestions one needs to be sensitised to the issue of student generated notation and to listening to what students say, as well as to be thoroughly familiar with the topic. The idea is to steer students toward useful notation, not simply impose conventional notation upon them. You also need access to a variety of ways of interacting with students in order to choose appropriate responses.

Example 5:

Preparation for work on the Intermediate Value Theorem was presented to the group by Harvey Keynes. An immediate question is Why is it a theorem? Start with rich examples from everyday experience (e.g. heating a circular loop of wire at one point so that symmetrically opposite points have the same temperature). Use the examples to consider why there are hypotheses to the theorem and offer examples of how it fails without those hypotheses. Mathematicians refer to showing that the conditions of a theorem are 'best possible', but there is a more sophisticated pedagogic purpose at work. This is to provoke students into appreciating the role of the conditions in the theorem so that they will think to check them when they want to apply it. Situations are chosen to show how the theorem operates in diverse situations. (Analysis of zeros of functions: graphically for functions from [0, 1] to [0, 1] as the need to cross $y = x$ somewhere, practically where any path from A to B which heads up from A and comes in to B from below, must cross the line AB). Then the foundation of the proof: the idea that there are no holes in the real line.

Summarising Comment

In order to catch students' attention, you need to pose a snappy, catchy problem and to exhibit contexts in which the problem or problem type could arise. To find such problems you need to know what you are looking for, which means developing sensitivities to student perceptions of topics and of mathematics. Most people seek comprehendible examples that students can relate to their experience or imagine readily, but here it is also important that the examples show why hypotheses for the theorem are needed. Furthermore, explicit hypothesis checking may encourage students to do the same when invoking a theorem. Reminding oneself about the fundamental awareness which the theorem captures and indeed the surprise which it encodes (Moshovits-Hadar 1988), as well as the sorts of places where the idea has arisen in the past provide a richness to ones' own awareness.

Example 6:

Siu Man Keung showed us sample slides from his presentation of the Hamilton-Cayley theorem. He includes pictures of the two men and of Frobenius who actually proved the theorem and a brief extract of a letter written by Cayley about it not being necessary to prove the theorem in complete generality (!). Displaying the letter of Cayley to Sylvester produces a warm and stimulating effect on the class as they witness mathematicians talking to each other as friends. He also offers a 'joke-proof' based on mis-interpreting the notation of the theorem in a natural manner. What is perhaps surprising is that the first proof in print by Frobenius published in

1896 is very close in spirit to the 'joke-proof', showing an interplay between idea and technical content. The special case for two-by-two matrices is analysed, and a general proof provided.

Summarising Comment

Here we see use of historical evidence to situate or contextualise the initial problem; use of a 'joke proof' to catch attention (what is wrong with the proposal?) and to lay the conditions for later exposition. Cayley's letter is a boon, for it raises the question of why proofs are needed. Historical difficulties with the theorem provide both interest and a sense of struggle. They also provide a means in this case for highlighting inappropriate interpretations of the notation.

Example 7:

David Henderson offered a brief experience: think of a piece of paper and a bug travelling on the surface. What would the bug consider to be straight, even when the paper was bent, or formed into a cone or cylinder? His approach and style of working is illustrated in his books (Henderson, 1996b and 1998).

Summarising Comments

This typifies Henderson's approach of starting with a challenge based on immediate experience and which can, at least at first, be explored using physical objects. He is prepared to listen carefully to student responses, to follow directions students propose and to offer further examples and situations to bring out and clarify students' ideas. He finds that proceeding in this manner his students are encouraged and empowered to develop their own thinking. He also finds that he learns new mathematics from his students (see Henderson 1996a).

5. PRINCIPLES-INFERRED

One way to locate structure in these brief descriptions is to notice that there are several domains. Explicitly mentioned were

ways of working on mathematics (discussion, scientific debate, conjecturing atmosphere);

designing for activity so that students act upon ideas and mathematical objects and are not just exposed to them;

reviewing ones' own concept images, one's own sense of the topic (typical applications or contexts, surprises, etc.), as well as what students need to work on in order to achieve adequate facility and understanding.

There were many domains that were only tangentially mentioned:

principled perspectives on what mathematics is really about (a collection of results and techniques to be mastered? a disciplined form of enquiry, a way of thinking? Problem posing and resolving? ...);

principled perspectives on how mathematics is most effectively learned;

how one works with different students with different propensities and different mathematical sophistication;

beliefs about who is able to learn mathematics;

and so on.

There is not room here to address all of these domains of concern, so we shall remain with some that emerged most clearly.

Ways of Working

The hardest part of teaching by challenging is to keep your mouth shut, to hold back. Don't *say*, ask! Don't replace the wrong *A* by the right *B*, but ask, "where did *A* come from?" Keep asking "Is that right? Are you sure?" Don't say 'no'; ask 'why?'. (Halmos 1985, p272)

Explicit or implicit in the presented case studies was a notion of ways of working with the students. Aspects of these can be extracted as follows.

Scientific Debate

Marc Legrand has written about the importance of enculturating students into scientific debate and how to do this (see Legrand 1993). Students are seen as participants in a scientific community whose methods of development include conjectures and modifications and proofs and refutations (Lakatos 1976). Scientific debates can arise spontaneously when students query a statement (does $f(A \cap B) = f(A) \cap f(B)$?) and it can be provoked intentionally by asking students to make a conjecture regarding some problematic issue.

Guiding principles for scientific debate include

Disturbance: students must encounter and deal with conflict with ordinary rationality (the need for disturbance to produce conditions for learning can be found in most educational thinkers since Socrates);

Inclusiveness: every human being should have opportunities to understand the deep meanings of what we strive to teach;

Collectivity: collective resolution of issues shows how to work with contradictions and to respect the views of others.

The essence of a scientific debate and of a conjecturing atmosphere is that people are eager to try ideas out and neither embarrassed nor ashamed to make a mistake: everything said is offered as a conjecture, with the intention of modifying it if necessary (in contrast to an ethos in which things are only said when the sayer is confident they are correct). Those who are uncertain about some detail often choose to speak, while those who are confident often choose to listen and then to suggest modifications through counter-examples, images, questions and suggestions. The intention is to create the conditions under which fruitful mathematics is done at any level of sophistication.

Contrat Didactique

The *contrat didactique* which is part of Brousseau's analysis of the *situation didactique (op cit.)* reminds us that the more precisely the teacher indicates the behaviour sought from the students (e.g. on examinations), the easier it is for the students to display that behaviour without generating it. In other words, training in behaviour is fine, but tends to be inflexible. As long as assessment questions are identical to training exercises, students will survive. But if you want students to appreciate mathematics, to think mathematically and to cope with unusual or varied problems, then their awareness needs to be educated as well. They need to be drawn into working-on mathematics, not just working-through routine exercises.

Instead of just expounding and explaining, you can establish a contract with students that you will 'cover' the material and show them how to do standard questions, as long as they agree to work with you in your way for some of the time. Then gradually you will find that students begin to enjoy and appreciate working on mathematical ideas rather than simply working through them. In Brousseau's analysis it is essential that no matter how explicit one tries to be with students about the way of working and the ideas being worked upon, the *contrat didactique* must retain an implicit level, for otherwise it leads to failure. Tahta (1980) put forward the case that every task or activity has an explicit or outer aspect (what people are asked to do, what they do in response to what they construe) and an implicit or inner aspect (what people are likely to encounter in the way of details of the topic, mathematical processes, heuristics and personal propensities which are revealed). If the inner aspect becomes outer (is made explicit) then the whole nature and purpose of the task is lost, just as when you are working on a problem and someone drops a clue or tells you an answer and you feel deflated and uninterested in continuing. Students may then be able to circumvent the work they need to do. They may think they have done what is required by going through the motions (if I get answers to the questions the teacher sets, then somehow I will learn!), but in fact they may fail to make important connections, may fail to form important images, may fail to come across the wrinkles and perhaps even fail to re-construct core ideas to displace previous conceptions. Brousseau points out that what is most important about the implicit contract is that it can be ruptured, for it is when there is a disturbance to established habits that interesting things start to happen (*op cit.* Chapter I, section 3.2).

Being Sensitive to Students

Sensitivity to the experience of students starts by becoming aware of one's own awareness. If you want to direct student attention in useful places (often students attention is focused elsewhere, even when reading an expression, contemplating a problem, task or theorem), then it is valuable to be aware of where one places one's own attention. Only then can you look for devices to attract or direct student attention explicitly or implicitly, using voice tones and emphasis, pausing and rhythm.

Intuitive and Formal

All of the presenters seem to have reached the conclusion that simply presenting formalised mathematics to students is ineffective. Students need help in participating in the formalisation process and this means being explicit about when one is working with intuition, with indefinite or un-rigourous 'sense of' and when one is arguing from first principles, etc.If students are explicitly inducted into the process of formalisation, perhaps even to the extent of the lecturer standing in different parts of the room when working with informal ideas and when being formal, they are more likely to appreciate the formalisation enterprise.

It is fair to say that as well as great similarities, there were some notable differences between the overt descriptions given by our seven contributors. Pian, Durand-Guerier and Legrand are steeped in the French school of didactic engineering, in which to make principled decisions one must have undertaken careful, even formal research. Tasks are designed for a specific *situation didactique,* are tried and evaluated in terms of intended goals and of the inner tasks. Presenters from the USA and China (Hong Kong) were equally demanding and rigorous in their search for suitable tasks to achieve declared and undeclared goals, but do not have the same tradition of conducting explicit research on the effectiveness of tasks. Rather, tasks are tried and modified in a process of ongoing personal-professional development. Sometimes the honed tasks are then presented in texts, but rarely is there any mention of theoretical or empirical foundations or of principles.

6. PRINCIPLES-APPLIED

It is important to remember that any act of teaching has purposes. Choosing to carry it out reflects underlying principles and beliefs. Yet curiously, the same gambit can be used by people holding widely differing beliefs and using it for different purposes.

Talking in Pairs

At various points in a lecture or tutorial, ask students to say to each other something you have just said that involves a few technical terms or is complicated in some way. Ask students to describe in words what they are seeing after you have invited them to imagine something or to look at a diagram. It is amazing how talking in pairs can release some inhibitions people have about contributing to the whole group.

Invoking Mental Imagery

Rather than just drawing diagrams or using multi-media, build up a mental image of some mathematical object through a sequence of verbal instructions: imagine a ...; add in a ... etc. Then get students to draw what they are seeing, or to describe it to someone else.

Being Explicit About Major Mathematical Themes

When a major mathematical theme or thinking process 'comes by', draw students' attention to it explicitly. Mathematics involves a number of innately mathematical ways of thinking and developing an idea. For example, many mathematical theorems seek to capture what is invariant and what is permitted to change in some situation; other theorems are devoted to characterising objects satisfying some property, in which case there is opportunity to explore 'boundary' examples (see Michener, 1978, Watson and Mason, 1998).

Being Explicit about How To Study Mathematics

Siu Man Keung hands out an 8 page set of *HouseKeeping Rules* or *Study Guide*, in which he describes what he expects students to do with/in lectures, the text, assignments and examples. He includes sample problems the course will discuss in order to indicate the scope and range of applications and topics. In the U.K. a handbook is offered to students starting a mathematics course at university which describes a range of study techniques and what to expect from lectures. There has been no research on how effective this is. However, anecdotal evidence from Open University practices suggests that some students pick up advice and are still using it several courses and years later.

Re-entering the Surprise or Significance Afforded by a Theorem

Most theorems are surprising or significant in some way, even if they are confirming intuition or establishing a technique (Moshovits-Hadar, 1988). Locating and re-entering the surprise or significance in what has become routine and obvious and using that to decide what to stress, can make a huge difference to students' appreciation of a topic and can help lecturers and tutors display their interest and enthusiasm. Lecturers who are not enthusiastic about their topic turn students off even more than tutors who are over-enthusiastic!

Generating Discussion

Take two technical terms from different parts of the course and invite students to construct a sequence of sentences which forms a sensible chain of ideas connecting the two words. Give one student a technical term written on a card (perhaps the students generate them themselves) and then get that student to give examples while the other students use these to guess the term on the card. Alternatively, the student can give descriptions or analogies to prompt the others to guess the term. After each 'turn', leave a little time for students to ask questions about ideas that came up. To work well, this activity requires students who are sufficiently confident to try ideas out even if they don't work out. It also requires a mathematical atmosphere in which the tutor does not simply pounce on, or pronounce on, what is incorrect.

Writing to learn; Learning Files

Suggest to students that they keep a file of observations about their struggles and successes. Some tutors take this in every so often and use it to engage in dialogue with individual students. Others find that the action of writing brief notes clarifies issues and helps students articulate their problems to tutors. At the Open University we encourage students in their first course to keep a learning file (they can use it as part of a submission for General National Vocational Qualifications if they wish). See Meier and Rishel (1998).

Easy – Peculiar – General

Ask students to construct an easy question similar to a given exercise. Then ask them to construct a peculiar one, that is, one that no one else in the class is likely to think of. Then ask them to formulate the most general question they can of that type.

The 'peculiar' provokes them into awareness of a range of possibilities open to them and so supports constructions of a generalised version (Bills, 1996).

Student Constructed Tests

Ask students to construct (and do) a collection of questions that they think will probe someone's understanding of a topic. Alternatively, ask them to construct and do questions that show that they understand a topic. What they choose gives an indication of what they are aware of and so what is missing may indicate areas that need further work (see Watson and Mason, 1998). It must be borne in mind that absence of evidence is not evidence of absence.

Historical Features

Make use of the Internet to obtain pictures and information about mathematicians involved in the topic you are teaching. Try to enter the thinking of previous generations of mathematicians and to provoke students to think similarly (some of them may indeed be thinking that way).

Displaying Essentials

Students take time to come to grips with new notation and new concepts. Arrange to keep important definitions and results on display while you are developing a topic, by using them to prove a theorem or to carry through a technique. You can even repeat the 'expanded' meaning of a new symbol or term each time it appears, as this is what students need to do. Then gradually reduce the amount of detail until you are just using the symbol or term.

Half Video, Half Discussion

There are excellent videos and computer animations to be shown as introductions to various topics and to be used as motivation for discussion. Getting students to close their eyes and reconstruct the video, then describe to a neighbour some details which they found striking, can stimulate discussion and reveal the need for a second, more focused viewing, before developing what they have seen or exploring it themselves.

7. ISSUES FOR FURTHER DISCUSSION AND INVESTIGATION

Cultural and social dependency-situatedness of any practice.

- How can one most effectively support colleagues, some of whom are dissatisfied with their teaching (or with the students' learning) and others who are satisfied?
- How can one have an effective conversation with colleagues about teaching rather than about mathematics? How can one teach mathematics humanely, as a human enterprise and disciplined way of thinking rather than as a cold pre-formed collection of results?
- How do computers, calculators and Internet communication influence teaching, learning and professional development?
- How does one cope with 150 students and a single overhead projector?
- How does one support colleagues in questioning and developing their practices?
- What can be done when one is frustrated by students not responding when asked, or by students responding poorly on assessment?
- How can mathematics be taught humanely, that is, as a human endeavour, calling upon human powers and corresponding to individual desires and hopes?
- How can one continue to teach creatively?

Issues mentioned by participants as worthy of attention

- What are student expectations of lectures, tutorials, homework and exercise sessions? What are lecturer-tutor assumptions about student activity (e.g. 1 hour or 2 hours study for each hour in class? Re-reading of notes of session before next session?)
- What are the real differences between lectures and tutorials as far as students are concerned? What could the differences be?
- What is the purpose of set exercises as homework, from the student point of view?
- Thinking of a really good lecture or tutorial we have attended, what made it memorable?
- What implications are there in preparation time for lecturers and tutorial-takers if they are going to develop their teaching?
- Might students be put off if a response is demanded, particularly in a culture in which experts are respected but not challenged?

Contract with students: If you will work my way some of the time (engaging in dialogue with me, trying little tasks and reflecting upon them, exploring mathematics with me), then I will work your way the rest of the time (revising topics, doing worked examples, ...). Seek a balance between different ways of working.

The most one can hope for is that most people are engaged most of the time.

Start with oneself … what is my experience?

Students want formulae and simplicity, a 'royal road'. Our task is to inspire them to want to work and support them in developing their powers to work effectively on mathematics.

What mathematics are we teaching? At a conference most people are not using their own native language to express ideas. And everybody has an experience when you have to talk in a language which is not natural for you and it is hard to find correct words to express what are you thinking. Mathematics also has its own language (may be languages?) and we have to listen to another person carefully to understand what he/she wants to say. Are we so patient with our students? How do we help students to express their ideas in a 'foreign' language? We understand a person talking even with mistakes because we say we understand if we understand the ideas the person is talking about. And there is no damage to the ideas if there are some grammar mistakes. Why do we want mathematics to be an exception? Why is formal language so sacred? See Taimina (1998).

Students have to expect to put in effort too. Indeed, it is a disservice to students to let them think that mathematical understanding comes without effort. Unfortunately consumer-oriented education emphasises making everything easy and enjoyable and laying the full blame on the teacher: if students do not learn, then the fault must lie completely with the teacher in not making the study enjoyable and easy. As Baylis and Haggarty (1988) put it "The professional mathematician will be familiar with the idea that entertainment and serious intent are not incompatible."

8. CONCLUSION

This chapter represents an initial stumbling in the direction of collecting and sharing practices and principles, tasks and perspectives, which is the aim of the MEMEs project mentioned earlier. In order to make sense of the many stimulating contributions in the group it was helpful to me as summariser to organise what I heard in a three-fold structure of frameworks:

> Preparing oneself to teach a topic, based on elaborating the notion of a concept image by interweaving cognition-awareness, affect-emotion and enaction-behaviour.

> Preparing to teach a session or class, based on designing for activity by balancing resources–tasks with student and teacher current state and goals as well as mastering subject knowledge. It is useful to collect possible teaching acts or gambits, to imagine oneself using them in a particular situation and afterwards, to re-enter salient moments and re-affirm the use of specific gambits in the future.

> Preparing to make effective choices in the flow of the event by developing awareness of your own mathematical thinking, so you can be sensitive to what your students are thinking. Becoming aware of the conditions and

ethos which support mathematical thinking for you is likely to lead to working at developing a conjecturing mathematical atmosphere in your sessions, in which everything that is said is taken as a conjecture to be modified if necessary and those who are certain choose to listen to and support others, while those who are uncertain choose to try to express what they understand. Under these conditions, scientific debate and effective mathematical discussion become possible.

Despite different ways of articulating principles and practices, there was considerable agreement. The main difference lay in the degree of formality and explicitness of the underlying theories and principles that guide different teachers' practices.

ACKNOWLEDGMENTS

I am grateful to the participants for their contributions both in Singapore and afterwards, to Jane Williams who scribed all the sessions and to Anne Watson and Celia Hoyles for helpful discussions about structure.

REFERENCES

Artigue, M. (1996). The role of epistemology in the analysis of teaching learning relationships in mathematics education. In Y. Pothier (Ed.), *Proceedings of the Annual Conferences of the Canadian Mathematics Education Study Group*, pp. 7-21. London, Ontario: Mount St. Vincent University.

Bills, E. (1996). *Shifting Sands: Students' understanding of the roles of variables in 'A' level mathematics*. Unpublished PhD Dissertation, Open University.

Brousseau, G. (1997). *Theory of Didactical Situations in Mathematics: Didactiques des mathématiques*, 1970-1990, N. Balacheff, M. Cooper, R. Sutherland, V. Warfield (Trans.). Dordrecht: Kluwer Academic Press.

Dewey, J. (1922, reprint 1957, Modern Library, New York: Random House). *Human Nature and Conduct*. New York: Henry Holt.

Griffin, P. (1989). Teaching Takes Place in Time, Learning Takes Place Over Time. *Mathematics Teaching*, 126, 12-13.

Halmos, P. (1985). *I Want To Be A Mathematician: An Automathography*. New York: Springer-Verlag.

Henderson, D. W. (1996a). I learn mathematics from my students - multiculturalism in action. *For the Learning of Mathematics*, 16, 34-40.

Henderson, D. W. (1996b). *Experiencing Geometry on Plane and Sphere*. Upper Saddle River, NJ: Prentice-Hall.

Henderson, D. W. (1998). *Differential Geometry: A Geometric Introduction*. Upper Saddle River, NJ: Prentice-Hall.

James, W. (1890, reprinted 1950). *Principles of Psychology*, Vol 1. New York: Dover.

Laborde, C. (1989). Audacity and Reason: French research in mathematics education. *For the Learning of Mathematics*, 9, 3, 31-36.

Lakatos, I. (1976). *Proofs and Refutations*. Cambridge: Cambridge University Press.

Legrand, M. (1993). Débate scientifique en cour de mathématiques. *Repères IREM*, no 10, Topiques Edition.

Mason, J. (1998). *Learning and Doing Mathematics* (Second revised edition). York: QED Books.

Mason, J. (1998a). Enabling teachers to be real teachers: Necessary levels of awareness and structure of attention. *Journal of Maths Teacher Education*, 1, 3, 243-267.

Mason, J. (2000). Asking Mathematical Questions Mathematically. *International Journal of Mathematical Education in Science and Technology*, 31, (1), 97-111.

Meier, J. and Rishel, T. (1998). *Writing in the Teaching and Learning of Mathematics*. MAA Notes 48. Washington DC: Mathematical Association of America.

Michener, E. (1978). Understanding Understanding Mathematics. *Cognitive Science*, 2, 361-383.

Moshovits-Hadar, N. (1988). Surprise. *For The Learning Of Mathematics*, 8, 3, 34-40.

Simon, M. (1995). Reconstructing mathematics pedagogy from a constructivist perspective. *Journal for Research in Mathematics Education*, 26, 114-145.

Simpson, A. (1998). The inpact of tensions between views of mathematics and students' worldviews. Paper delivered at the ICMI Study Conference, Singapore, on the teaching and learning of mathematics at university level.

Tahta, D. (1980). About Geometry. *For the Learning of Mathematics*, 1, 1, 2-9.

Taimina, D. (1998). What mathematics are we teaching? In T. Breiteig and G. Brekke (Eds.), *Theory Into Practice in Mathematics Education*, pp.237-243 Research Series No. 13. Kristainsand, Norway: Agder College.

Watson, A. and Mason, J. (1998). *Questions and Prompts for Mathematical Thinking*. Derby: ATM.

John Mason
The Open University, England
j.h.mason@open.ac.uk

LEIGH WOOD

THE SECONDARY-TERTIARY INTERFACE

1. INTRODUCTION

The secondary-tertiary interface is critical to university mathematics. How well students gain skills, concepts and learning strategies from secondary school and then transfer these to their new environment is instrumental to the success of mathematics. The good news is that in nine countries quoted in the TIMSS report (IEA,1995-6), 85% or more of final year secondary students reported that they were currently taking mathematics. By contrast, many fewer students study science.

So, compared to other discipline areas, mathematics is in a privileged position in the school curriculum. In most countries, the study of mathematics is compulsory at least until the minimum school leaving age. There has been a strong message that mathematics is important for university study in a wide range of disciplines and that not studying mathematics can severely limit one's choice of degree programme. Added to this is the success of female students in secondary mathematics study. In the past 20 years, the numbers of female students completing high levels of secondary mathematics has increased dramatically. Even so, gender differences in advanced mathematics achievement favours males (IEA, 1995-6) in a majority of countries.

But university mathematicians are complaining. Students are more numerous and more diverse, at least at first year level. Pedagogical and curriculum changes at pre-university level have had an impact on students' skills and ideas of rigour. Technology is changing the way mathematics is done and changing mathematics itself. There are ongoing questions about the quality of graduates and the integrity of the discipline. It is harder to attract good quality students into higher mathematics study at university. Indeed, it is getting more difficult to attract students to higher-level mathematics study at secondary school. This is in contrast to the pervasiveness of mathematics in other disciplines and in the workplace (Australian Academy of Sciences, 1996).

Much can be done to improve the response of university mathematicians to these changes. There are positive outcomes from the changes to secondary education. The diversity of students in mathematics classes can lead to new ways of working with mathematics. It also raises questions such as: "How necessary are algebraic manipulation skills when computing power is so great?" and "How important is the need for rigour for all students when there is a huge diversity of mathematical careers?"

Derek Holton (Ed.), The Teaching and Learning of Mathematics at University Level: An ICMI Study, 87—98.
© 2001 *Kluwer Academic Publishers. Printed in the Netherlands.*

can be a long time as they take a break from study in general, and mathematics and particular. Others have not even succeeded in finishing secondary school. This is frequently an area of political priority where governments fund special programmes to increase the numbers and the success rate of disadvantaged students in tertiary education.

Therefore, this overview will include bridging programmes for mature-aged students returning to study as well as school leavers who have not studied the appropriate level of mathematics for the degree programme that they wish to undertake.

2. TRANSITIONS

The transition from secondary to tertiary mathematics study involves many adjustments. Firstly there is the mathematical content, secondly the teaching and learning style, and thirdly there are personal and interpersonal adjustments. All contribute to success or failure at tertiary level. Even students who can successfully pass mathematics subjects may not be able to reach their academic potential if they are unable to adjust. This adjustment depends on many factors in the students' background: socioeconomic status, gender, ethnic background, parental expectations, schooling, peer group and so on. It also depends on the motivation of the student to study mathematics.

Changes to the teaching and learning of mathematics at secondary level have been successful in opening mathematics to more students, especially those from non-academic backgrounds. However the profile of mathematics students has changed making it necessary for tertiary mathematicians also to adapt. Two main transition adaptations are bridging courses between school and university and changes in the first year university mathematics curriculum. Systemic changes, such as the creation of senior high schools and junior colleges are briefly discussed in a later section.

3. BETWEEN SCHOOL AND TERTIARY STUDY

3.1 Bridging courses

For students, the time between school and university can be a period of from a few months to many years. The nomenclature here is difficult. Courses between school and university are called foundation courses, bridging courses, access courses, pre-calculus courses, university survival courses and so on. The majority are run by universities or by a closely affiliated college. Some courses are commercial, such as courses for overseas students who wish to study at universities in the host country, some are free, some are compulsory, some last for one year, some for one week. Pre-calculus courses in the USA are now such an industry that the economics of these courses justify the mathematics faculty's size.

A range of programmes can be called bridging courses because they bridge between the academic needs of school and tertiary study. For ease of comparison we will discuss their aims and target groups, rather than their length, cost or who teaches them. Bridging courses prepare students for the academic concepts, skills and attitudes of their degree programmes as opposed to orientation programmes that prepare students for university life. Orientation programmes will be considered in the next section.

Some of the earliest bridging mathematics programmes were introduced after the Second World War for soldiers returning to study. Since then bridging programmes have reflected societal changes and government priorities. In the 1970s, programmes were established to assist women to enter and succeed in tertiary study. These programmes were often for women only and contained a component of confidence building and career information. There were several interesting names, such as WOW (work opportunities for women) and Mathematics Assertiveness. For women in their thirties, forties and fifties, bridging courses opened doors to universities; equal pay and equal opportunity policies opened the workforce to them. This idea of bridging courses as a way of compensating for discrimination is still the dominant philosophy in many universities. Following the success of programmes for women, other programmes were devised for disadvantaged groups, such as Native Americans, Hispanics and so on. With an increased access to universities and an emphasis on equal opportunity for disadvantaged groups, the number and variety of bridging courses has exploded. In countries, such as the Republic of South Africa, political and social goals require greater access to higher education for disadvantaged groups so there is great interest in devising bridging programmes.

Another type of bridging course was very successful with particular groups in the 1980s. There were waves of refugees from the war areas of South East Asia migrating to North America, Europe, Australia and New Zealand. These students had fractured schooling, poor language skills in their adopted country and had suffered the trauma of war. Lecturers devised innovative teaching methods to teach language, mathematics and study skills in the same programme. Conventional syllabuses and materials were not appropriate and lecturers moved to condense years of mathematics into short courses. These were very successful with motivated students.

Bridging courses allow lecturers to be innovative because they are not part of the mainstream mathematics degree programmes and are therefore not contingent on the same constraints. Frequently young, inexperienced and non-tenured lecturers were used to teach the programmes. The teaching and learning ideas developed in teaching these groups are now being applied to mainstream classes.

So bridging programmes are not new. They have been used for political and social purposes to prepare particular groups for tertiary studies. The philosophy and style of the bridging programme depends on the target group and the political climate. There are also bridging courses for students who have just finished school. These courses cater for the many students who have not studied the correct level of mathematics for their tertiary studies either through bad advice, disadvantaged schooling or through changing their minds as to career choice.

3.2 Aims of bridging courses

One of the aims of bridging courses is to fill gaps in knowledge, to have students review (or study) essential facts and skills that have been forgotten, concentrating on those topics that are essential for their first year mathematics courses. These basic skills are necessary for all students to succeed in university mathematics and may be all that is required for students studying non-theoretical mathematics, such as Business or Nursing students.

A second aim of bridging courses is to prepare students for first year mathematics courses in which results are proved. In many secondary schools, proofs have almost completely disappeared. The result is that for many students the confrontation with proofs at tertiary level is difficult. The target groups for these types of programmes are students who have succeeded in secondary studies and wish to move into tertiary mathematics.

A third aim is to develop the attitudes and language to become a successful mathematician or student of mathematics. The target groups for these programmes are students who come from non-academic backgrounds, students who are new to the education system such as international students and students who are studying in a different language to their mother tongue. These students may have excellent mathematical skills but may not be able to read textbooks quickly enough or understand what is required for assessment. Many university mathematicians argue that *all* students should be explicitly taught the discourse of advanced mathematics.

Other aims, not so directly related to mathematics, include the development of computing skills in cases where students will be required to use technology in their mathematics study at university and the development of study skills. Students may be required to partake in more than one bridging programme.

Examples of the first aim are one year or one semester bridging programmes for adults and young students who did not study enough mathematics at secondary school, such as those offered in Linköping University, Sweden and Harvard University, USA. Other universities have short summer programmes to bring students up to the level expected, for example, Vrije Universiteit Brussel, Belgium, University of Cape Town, Republic of South Africa and University of Sydney, Australia.

Some universities have a comprehensive suite of bridging courses aimed at different groups of students, some have diagnostic tests, such as Warwick University, UK, and some have different levels and content depending on which degree programme you will be entering. An example is the Vrije Universiteit, Brussels, which has been offering bridging courses in mathematics for more than 20 years. The bridging courses are offered at different levels of difficulty and with different content, aimed at students entering different majors. They are offered during the month of September, before the start of the academic year, for a period varying from 2 to 4 weeks and they are taken on a voluntary basis (Grandsard, 1996). Part of the bridging programme includes a computer package on methods of proof and introductory problem solving (Grandsard, 1988).

London University (UK) has a bridging course that fulfils the second aim. Students studying mathematics there have high entry grades but it was felt that students were finding difficulty with the transition to formal ideas. For two weeks in September prior to entry to the university, students are offered a voluntary 'top-up' course with the aim of introducing formal ideas informally. Lecturers use a narrative style with informal explanations rather than rigorous proofs. There are lectures in the mornings and problem sessions in the afternoons.

Another trend in bridging courses is that bridging courses are offered for credit towards a degree. Bridging courses have become mainstreamed as more students require the skills and as lecturers realise that careful development of skills and attitudes in first year can lead to more effective learning in later years: indeed it can lead to more students taking mathematics in later years. With the recognition that not all students enter tertiary study with the same background, new courses have emerged to cater for this clientele. Tall (1992) and others have documented the difficulties of the transition to tertiary mathematics. Students who have studied low levels of mathematics at secondary level may be able to study high school mathematics for credit at university but it may slow their progress through a degree or add more years to the length of a normal degree. There is some opposition to this trend, as students receive credit for what is essentially school level mathematics.

This is similar to trends in other subject areas such as physics, which few students now study to a high level at secondary school. What may have been considered school level physics in the past is now taught at the tertiary level. Indeed many disciplines such as economics, psychology and sociology are rarely offered at secondary school and must start from scratch at university. Mathematics needs to adapt to these changing circumstances and make all students who want to study mathematics at tertiary level welcome. Lecturers in statistics have made great strides in making statistics accessible to all students: mathematicians should do the same.

There are examples of subjects that aim to introduce students to advanced mathematical thinking. These subjects can be considered as bridging subjects even though they are for credit as part of a degree. Often there is opposition from mathematics faculty for the introduction of a subject of this type as these 'ideas' subjects take time away from process or content subjects. Mathematicians often find it easier to teach mathematics itself rather than take a step back and help students examine the ideas of mathematics. Faculty at Concordia University in Montreal, Canada have developed a course to assist students to refine their mathematical thinking. This course is a credit course that develops ideas about proof, proof techniques, and the use of precise language and notation (Hillel, 1999). Whilst this could not be considered a secondary level course it takes the role of bridging students to advanced mathematical thinking.

Another example of a subject that aims to bridge students to mathematical thinking and mathematical reading and writing is on offer at the University of Technology, Sydney, Australia. This subject was established to teach students the skills of reading, writing, listening and speaking mathematics. Again this is not a secondary level subject but it does bridge students' academic language for university study. This subject is for all mathematics majors regardless of their background and has been used to orient students to the study of mathematics at university (Wood and Perrett, 1997). There is evidence that students' reading skills in mathematics have improved significantly.

3.3 Success of bridging courses

There are several questions we should be asking. Have we been too successful with bridging courses in the sense that students think that they can avoid mathematics at secondary school and do it all in a short bridging course? If students believe that they can obtain mathematical knowledge quickly this undermines both mathematics at secondary school and the role of secondary teachers. This unintended consequence may have has caused a reduction in the numbers of students taking higher-level mathematics at secondary level. In general, university mathematicians would prefer students to take high levels of mathematics at secondary level because of the better development of skills over longer periods of time in small groups with secondary teachers.

Other questions concern bridging courses themselves. Are they successful in meeting their objectives? Do their graduates succeed in tertiary mathematics studies? In many cases formal evaluations are not done, though there is ample anecdotal evidence of success. All lecturers would have examples of students from disadvantaged backgrounds who have succeeded at tertiary study after completing a bridging course. As in much educational research, evaluation is difficult. Students are often choosing to do the bridging courses. Would these students have been successful anyway? Are other factors more important, such as peer groups and changes in assessment in first year studies? The situation is rarely static enough to allow easy comparisons between treatments.

Reactions by students on course experience questionnaires for bridging programmes have always been very positive. The numbers taking the courses are increasing which can be taken as a positive outcome. However, there are doubts about the long-term effects of remedial programmes. As mastery is required in the long term, it is important that students keep reviewing prerequisites regularly. And finally, the programme does not work for very weak students: no one can make up in one month for deficiencies that may have accumulated over six years.

There are groups of academics with a common interest in bridging programmes. They hold regular sessions at ICME congresses and these often reflect world-wide priorities, such as women returning to study or adults returning to mathematics. Another group that meets regularly is the Bridging Mathematics Network in Australasia. This is a group of university mathematicians involved in Mathematics Centres and bridging programmes. They have been holding biyearly conferences since 1992. The conferences cover many aspects: administration of bridging and support programmes, content, teaching methods and resource development. Teaching methods and resource development for disadvantaged groups are particularly emphasised.

4. FIRST YEAR OF TERTIARY STUDY

Research into students' learning at the beginning of university has influenced curriculum design for mathematics. Crawford, Gordon, Nicholas, and Prosser (1998) showed that many students arrive at university with a surface learning approach to learning mathematics and that these students were not as successful in their studies as those with a deep learning approach. Tall (1992) examined specific areas where students had cognitive difficulties in the transition to advanced mathematical thinking. De Guzmán, Hodgson, Robert, and Villani (1998) listed difficulties with the transition from the secondary to tertiary level, particularly with routine skills. This research has led to changes in curriculum. For example, it is rare for epsilon-delta proofs to be studied in first year mathematics programmes and many topics, such as integration techniques, have been considerably reduced.

There are major adaptations to be made by new tertiary students of mathematics. Classes are often large and impersonal. Students have trouble coping with large amounts of new material in a short time. Academic staff seem unapproachable and there may be little support for students with difficulties. Students are expected to do much of the work by themselves. There are many new computer programs to learn. Tertiary educators are taking note of student difficulties and are making changes but the question remains, are we expecting too much of first year tertiary students?

Pedagogical changes at first year university level.

There have been many imaginative changes to teaching first year university mathematics in response to changes in the student intake. The University of Melbourne, Australia has experimented with the idea of *cognitive conflict* to reduce the incidence of algebraic misconceptions. The Chichester Institute of Higher Education, UK, has used peer teaching to improve the learning for different groups of students. The University of Växjö, in Sweden and the University of the Witwatersrand in South Africa have been using a language approach with oral presentations. The Harvard Consortium and the Lawrence Hall of Science in Berkley, USA have been using modelling and applications to motivate the study of mathematics. Many universities, with the University of Illinois, USA being the most evangelistic, have incorporated extensive use of computers in teaching and learning. The University of Cape Town has been experimenting with a lyrical approach using analogy to make connections with a more diverse student population. Many universities are experimenting with group assignments and peer teaching.

Routine algebraic skills.

There have been consistent negative comments from university mathematicians on the algebraic manipulation skills of entering tertiary mathematics students. The TIMSS study (1995-6) showed that more time spent solving equations was an indicator of high achievement on their advanced mathematics test. It is clear that, in many countries, time spent on algebra has been reduced in secondary schooling. There have been various responses to this. Warwick University in the UK has responded to different entry skills by using diagnostic tests to identify weaknesses and then providing remedial exercises. The Australian Defence Force Academy in Canberra, Australia has an Essential Mathematics Skills programme in the first year and students must achieve mastery before proceeding. Others have written extensive lists of prerequisite knowledge and skills that they expect students to have before embarking on that course or subject. The question of what are essential algebraic skills is hotly debated. What is essential? What algebraic skills do we want students to be able to perform and why? Are algebraic skills necessary for conceptual development?

4.1 Academic support services

In response to large failure rates in university mathematics, many institutions have set up academic support programmes for mathematics and statistics. There was a time when university mathematicians did not have to justify large failure rates in mathematics but now, in many systems, funding is dependent on the numbers of students who pass. In order not to reduce standards, academics are looking towards support for students who have difficulties. Academic support, in the form of drop-in centres, extra tutorials and counselling has expanded. Below we list some of the types of support that are available.

Individual assistance. Some students may require individual assistance. Students with a record of failure or students with low confidence may improve significantly with individual assistance. Students with a disability may also require assistance. The assistance can take the form of counselling by psychologists in student learning or may be individual tuition by a member of the mathematics faculty or a combination of both. Though this is expensive in terms of time and personnel, many students will only require a few sessions to gain enough confidence and skills to succeed.

Enabling tutorials. Where a group with similar problems can be identified, a regular weekly tutorial is provided. This tutorial may be taught by a member of the mathematics faculty or by a professor who may be appointed to work particularly with students who are having difficulties with mathematics and statistics.

Workshops. Short workshops (3-6 hours) are provided for groups of students with particular problems, including examination review. These are generally held on Saturdays or in the evenings. Many universities have found these popular especially with students undertaking mathematics as a service subject in their engineering, business or science degrees.

Review mathematics. Students who clearly will not succeed with their mathematics subjects are encouraged to defer their mathematics study until after completion of some basic mathematics review.

Advice for staff. Because mathematics faculty is frequently dealing with students who have difficulty with mathematics and statistics, academic staff in all faculties may enquire about areas in their curriculum that may cause difficulty for their students.

Self-study materials. Review mathematics materials may be available in print, video or on the Web for students who do not wish to or are unable to attend class.

Academic support has been a very successful way to change attitudes to mathematics and to assist students to succeed. Peer tutoring (the use of later year students as tutors or mentors) in Academic Support Centres has also been successful.

5.   OTHER ASPECTS OF TRANSITION

5.1 Systemic solutions: senior or junior colleges

Many countries have senior or junior colleges (depending on whether they are considered as senior high school or junior tertiary) to orient students to tertiary study. These can act as the final years of secondary school or be considered as the first years of tertiary level. Generally students are given more freedom and less direction than at secondary school in order to prepare them for the freedom and responsibilities of university study. These institutions, often called colleges, are common in North America where 'going to college' is a rite of passage that includes leaving the parental home. Pedagogically the mathematical preparation of these students is similar to senior secondary school.

5.1.1 Extended degree programmes

A common solution to differing entry levels is the addition of extra time to a degree programme. Students may be given 4 years to complete a three-year degree. In the first two years students will do fewer academic subjects than the usual programme and will be given extra support and perhaps some mentoring for their academic subjects. Life skills, study skills and language skills often from part of the programme. The University of Port Elizabeth in the Republic of South Africa is experimenting with such a programme. However, government policies of funding universities may make these types of integrated bridging courses difficult to fund.

5.2 Orientation

Orientation programmes differ from country to country, and from region to region. In Sweden, many universities have elaborate initiation ceremonies organised by later year students. Other educational systems have more formal orientation with speeches from the vice-chancellor; some have team-building camps.

The aim of orientation programmes has moved from information, such as finding the library and classrooms, to making the transition to university a more personal 'whole person' approach that includes how to find friends, join clubs and enjoy studying. There is significant evidence that affective domain variables have an important effect on the transition to university study and the decision to stay as shown in Montague and Hepperlin's (1997) study comparing the experiences of first year undergraduates who leave with those who stay.

In response to this study, the University of Technology, Sydney (Australia) has introduced a *Semester Zero*, a coordinated orientation for new first year students. Students are offered a range of activities in the month between enrolment and commencing classes. Counsellors talk to students about practical issues, such as housing and money, and provide a general introduction to study skills. Campus clubs and the Student's Union provide social and sporting activities and each faculty provides a more academic programme. The University-wide activities start first and narrow down to the student's own faculty just before classes begin. At enrolment, students are given a detailed flow chart and a self-evaluation questionnaire to check whether they have the prerequisite knowledge for their courses. Students are referred to appropriate bridging courses (mathematics, chemistry, English) where necessary.

5.2.1   Interaction with teachers in secondary schools

The transition to tertiary education would not be complete without some discussion about the interaction between teachers at secondary and tertiary levels. The tertiary education sector and secondary education interact very differently in different countries, or even within countries. Some have argued that universities have had too much influence on the curriculum at secondary schools, thus making it too academic for the majority of secondary students. In other educational systems (such as Belgium), there appears to be little influence of university mathematicians on secondary curriculum. How much or how little should university mathematicians interact with the secondary sector?

The dialogue between secondary and university faculty has for too long been in one direction. As secondary school curricula change, reflecting increased use of technology as well as alternate pedagogical approaches, such as group activities, we (at the university level) need to listen to what the high school teachers have to say. This communication needs to be two-way. This section reports on educational systems where successful interaction has occurred.

In France there have been recent initiatives to increase the cooperation between secondary and tertiary mathematics teachers. The meetings included representatives of the academic mathematicians at universities and Grandes Ecoles and the mathematics societies (GRIAM). France also has for many years an *Institut du Recherche sur L'Enseignement des Mathematiques* in each Academy. These institutes are situated in universities. Secondary and tertiary teachers work together to produce innovations, make proposals for syllabuses, provide teacher in-service and are closely connected with didactic research. The institutes work both ways; university teaching is informed by the presence of secondary teachers.

In Colombia many tertiary mathematicians are now engaged in teacher pre- and in-service training for secondary schools. This was a government decision which came with financial incentives for universities to organise and supervise the initial teacher training and the continuing education courses required of teachers to increase their ranking and salary in their profession. Secondary school teachers influence and challenge the university professors with questions derived from their experience and the inventiveness of their pupils, with problems of philosophical, epistemological, historical or didactical import. This cooperation was achieved in a few years by the combined push of economic pressures on the public universities to self-finance part of their budget, and of economic incentives to engage in designing and orchestrating pre- and in-service teacher education. One added advantage to this is that many university and secondary school mathematicians have built an atmosphere of solidarity for the cause of mathematics.

Another cooperative model is at the University Eduardo Mondlane in Mozambique where staff is working in partnership with secondary schools, technical institutes and teacher training colleges to improve education and learning.

5.2.2 School examinations

Changes in the end of schooling examination have profound influences on teaching and learning at secondary level. For example, the Chinese have a centralised system and have just introduced one open-ended question into the final examination. This change is assessment will change secondary teaching and learning and therefore change the preparation of students for tertiary study. Is this change desirable? In many countries, university mathematicians sit on the examination boards that write and oversee the final examinations for secondary schools. Influencing the examinations is an excellent way to ensure that the skills required for university are assessed and therefore taught. This is one way to influence the direction of mathematics teaching and learning at secondary level.

In other school systems, Belgium (Flanders) for example, university mathematicians have little influence on secondary school outcomes, however, recently an entrance examination for medical school has been installed, and this will have some influence on secondary school teaching. In Australia (South Australia), other disciplines on the secondary board governing upper secondary curriculum have worked together to reduce the number of hours for mathematics. This political aspect of mathematics education and control of upper secondary curriculum and examinations has an important influence on tertiary mathematics.

6. CONCLUSION

It is clear that the pre-eminent position of mathematics has been challenged, particularly in Western countries. Its position has been taken over by computing and computer-based technology. Fewer students are studying mathematics at higher levels in secondary schools, due to competition for time in the secondary curriculum. Fewer students are studying mathematics as a major at tertiary level. However the numbers studying some mathematics as part of a degree continues to increase, especially statistics and mathematical computing.

Universities are reacting to this in several ways. Some continue to define mathematics as a discipline associated with proof and analysis, some have moved to more applied areas of statistics, operations research and financial mathematics, some teach mathematics as a service subject to other disciplines. In all of these areas, university mathematicians find their students have changed.

Reactions to changes include: bridging courses, more interaction with secondary teachers, changes in curriculum at first year tertiary level, and support for students with difficulties. Perhaps it is time to stop reacting and being more proactive in the transition to mathematics curriculum at tertiary level.

REFERENCES

Australian Academy of Sciences (1996). *Mathematical Sciences: Adding to Australia*. Report of the National Mathematical Committee for Mathematics. Australian Government Printing Service: Australian Academy of Sciences.

Crawford, K., Gordon, S., Nicholas, J. and Prosser, M. (1998). Qualitatively different experiences of learning mathematics at university. *Learning and Instruction*, 8, 5, 455-468.

de Guzmán, M., Hodgson, B.R. Robert, A. and Villani, V. (1998). Difficulties in the passage from secondary to tertiary education. *Documenta Mathematica*. Extra Volume ICM, 747-762.

Grandsard, F. (1996). Bridging Courses to smoothen the transition from high school to university. Paper presented at ICMI 8, Sevilla.

Grandsard, F. (1988). Computer-assisted courses on Problem Solving and Methods of Proof. In H. Burkhardt, S. Groves, A.H. Schoenfeld and K.C. Stacey (Eds.), *Problem Solving - A World View*, pp. 274-278. Nottingham: Shell Centre.

Hillel, J. (1999). A Bridging Course on Mathematical Thinking. In W. Spunde, R. Hubbard and P. Cretchley (Eds.), *The Challenge of Diversity*, pp. 109-113. Toowoomba: Delta '99.

IEA (1995-96). Third International Mathematics and Science Study (TIMSS).

McInnis, C., James, R. and McNaught, C. (1995). *First Year on Campus*. Canberra: Australian Government Printing Service.

Tall, D.O. (1992). The transition to advanced mathematical thinking: functions, limits, infinity and proof. In D.A. Grouws (Ed.) *Handbook of Research on Mathematics Teaching and Learning*. pp. 495-511. New York: Macmillan.

Wood, L.N and Perrett, G. (1997). *Advanced Mathematical Discourse*. Sydney: UTS.

Leigh Wood
University of Technology, Sydney, Australia
leigh@maths.uts.edu.au

LARA ALCOCK AND ADRIAN SIMPSON

THE WARWICK ANALYSIS PROJECT: PRACTICE AND THEORY

1. INTRODUCTION

First-year analysis is hard.

At the beginning of a university education, students are required to learn more abstract material (Dreyfus, 1991), place previously encountered knowledge into a formal framework (Hanna and Jahnke, 1993), deal with the Weierstrassian construction of infinite processes (Davis and Vinner, 1986) and understand proof as a formal confirmation within an axiomatic system (Moore, 1994). These new mathematical and meta-mathematical ideas are encountered at the same time as students are required to construct new didactic contracts (sets of expectations about the teaching and learning context described by Brousseau and Otte, 1991), to work in new environments and to cope with the many social, personal and emotional changes that accompany the move to university (Perry, 1970).

In this paper we will explore the transition to university mathematics in the context of two different first year analysis courses. The first part of the paper examines the practical issues related to the courses and the second discusses the theoretical basis for the differences we note.

2. CONTEXT

In the UK[1], formal analysis is traditionally one of the main topics in the first year, pure mathematics curriculum, alongside some work on groups and rings, and linear algebra. The syllabus usually contains the formal definitions and rigorous proofs of theorems related to limits of sequences and series (including power series), continuity, differentiation and Riemann integration. Those coming to university to study mathematics will have met some of these concepts, treated less formally, in their 'A' level courses (pre-university courses for 16-18 year olds).

In the 1990s there was a broadening of the both the general 'A' level system and the mathematical content of the national curriculum for students up to the age of 16. This means that the techniques of differentiation and integration are retained within the 'A' level core but have reduced importance. Ideas such as complex numbers and

---

[1] The system discussed in this paper is associated most strongly with the school and university systems of England and Wales. The Scottish system is somewhat different, though many of the ideas translate to it.

Derek Holton (Ed.), The Teaching and Learning of Mathematics at University Level: An ICMI Study, 99—111.
© 2001 *Kluwer Academic Publishers. Printed in the Netherlands.*

vectors are no longer in the core and modules on problem solving and data handling have been introduced or expanded (SMP, 1990, SEAC, 1993). During the same period the UK university system changed so that more than double the proportion of 18 year-olds now move on to degree level education (Dearing, 1997). In this context, concerns grew about the standards of mathematical ability of first-year students (LMS, 1995).

Warwick University is renowned for its high quality mathematics department. The department has the highest available rating for mathematics research and it demands some of the highest grades of students wishing to attend, controversially requiring a special examination as part of its offer. It is also one of the most closely studied in the country: Warwick University is home to the Mathematics Education Research Centre, which specialises in research in advanced mathematical thinking. Many research projects have been centred on Warwick's undergraduate mathematics degree (Alcock and Simpson, 1999; Gray, Pinto, Pitta and Tall, 1999; Yusof and Tall, in press).

While some UK universities have responded to a perceived change in student preparation by abandoning formal analysis or, more commonly, moving it to later in the degree, Warwick retains analysis as a first-term, first-year module. However, the traditional two-term course has been spread out to cover three one-term courses over one-and-a-half years. The first of these courses consists of limits of sequences, completeness and limits of series. The traditional course involves 3 one-hour lectures per week (with classes of over 100 students) in which a lecturer introduces the main definitions, states and proves central theorems and may illustrate these in both their meaning and use. These are complemented by weekly assignments for a small amount of credit, which require students to use theorems from the course in calculations and, to a lesser extent, to prove other minor results. These assignments, and those from the other core courses, are marked by graduate students and covered in small-group help classes (3-5 students) called 'supervisions'.

Most mathematicians are aware that analysis is a difficult subject for students to grasp, and research confirms this. Students often have some concept images which are incompatible with the Weierstrassian concept definitions for infinite processes, and which may therefore act as epistemological obstacles (Sierpinska, 1987; Tall and Vinner, 1981). Formal definitions involving complex, quantified, manipulable logical statements are at odds with students' previous experience of mathematics as procedurally biased and structured around objects described by the teacher (Dubinsky, Elterman and Gong, 1988; Harel and Tall, 1991). Previous experience of a procedural treatment of the topic at schools means that many students balk at being ask to derive results they 'know' to be true.

3. THE WARWICK ANALYSIS PROJECT

The extent to which faculty can respond to perceived student difficulties is constrained by many factors. Openness to possible change is vital to the development of new courses. Recognising that students continued to perform poorly despite his efforts at improved presentation, one member of the Warwick staff with over 30 years of traditional teaching experience chose to adopt a new lecturing style. In an initial modification of a second-year metric spaces course, he encouraged students to develop much of the mathematics for themselves by posing central, illuminating questions. While this course met with limited success and even more limited appreciation from the students, he persevered to initiate the Warwick Analysis project, approaching two members of the mathematics education group to help teach it in the first year.

The project is based on students' completion of a carefully structured sequence of questions, through which the syllabus of the Analysis I course is developed. An excellent textbook bringing together mathematical, pedagogical and mathematics education research knowledge had been developed over a number of years at another university and this text (Burn, 1992) is central to the course.

The text consists mainly of questions, with compressed solutions and a very short summary of the main ideas at the end of each chapter. The questions develop rationales for the main definitions, construct the central arguments that lie behind the main theorems, and allow the students to use those theorems in subsequent arguments. The following question is typical of the style:

---

33 (a) Use a calculator or a computer to evaluate terms of the sequence $(n^2/2^n)$.
Calculate the values for $n = 1, 2, 5, 10, 20, 50$.
Would you conjecture that the sequence is null?

(b) Verify that the ratio $a_{n+1}/a_n = \frac{1}{2}(1 + 1/n)^2$.

(c) Find an integer N such that, if $n \geq N$, then the ratio $a_{n+1}/a_n < \frac{3}{4}$.

(d) By considering that $a_{N+1} < \frac{3}{4} a_N$ and $a_{N+2} < \frac{3}{4} a_{N+1}$, etc. prove that $a_{N+1} < \left(\frac{3}{4}\right)^n a_N$.

(e) Why is $\left(\left(\frac{3}{4}\right)^n a_N\right)$ a null sequence?

(f) Use the sandwich theorem, question 28, to show that $(n^2/2^n)$ is a null sequence.

(g) What numbers might have been used in place of $\frac{3}{4}$ in part (c) (perhaps with a different N) which would still have led to a proof that the sequence was null?

After a pilot project with 35 volunteer students in 1996, all 235 of the single-honours mathematics students in 1997 took the new Analysis I course. The joint-honours students attended a parallel traditional lecture course that provided the opportunity to compare outcomes.

Those on the new course worked in groups of between 3 and 6 students in classes of around 30. Each class had a 'teacher' (a member of staff or graduate student[2]) and two 'peer tutors' (second- or third-year students who had scored well in their own first-year analysis examination). The classes were two hours long and students attended twice a week. They also attended a one-hour lecture each week. This played a similar role to the 'end-notes' in the text: summarising the main ideas of the questions covered that week, placing them in context with other material and indicating the direction in which the material would develop in the following week.

Students also received printed notes that were similar to traditional lecture notes, containing a discussion of the material and concise proofs of the major theorems. To encourage students to work on the questions, they were asked to keep a portfolio of their answers, which was to be submitted for assessment for some credit. The main assessment, however, was a mix of a small amount for weekly assignments with the majority (85%) of the marks coming from formal examinations. This combination of assessments was identical to those taken by the parallel group from the traditional course.

4. IN THE CLASSES

While the teachers occasionally worked with the whole class (typically where they had noted a widespread difficulty), the emphasis of the classes was on allowing the students to work on their portfolio. Some of the groups were strongly collaborative, with every question a matter for discussion. In many groups, however, the students used each other as a resource when they became stuck or to check the correctness of their answers. In a small number of groups, there was virtually no interaction. No pressure was brought to bear to insist on collaborative work. Unlike the C4L project, which appears to put equal weight on 'construction' and 'co-operative learning' (Reynolds et. al., 1995), the individual's development of the mathematics through completion of the questions was seen as more important than the collaborative nature of the work.

The extent to which the groups called for help from the staff also varied. Some felt the need to check every question; some only asked for help when they felt they were stuck; and some only interacted with staff on the staff member's initiation. Prior to the course, the teachers and peer tutors had some instruction in monitoring and intervention techniques. They saw their role principally as helping students to grasp the concepts (which we might see as modifying their concept images to fit more closely the formal definitions, in the sense of Vinner, 1991) and to encourage a much more formal approach to proof [3].

The students generally found the workload of the course extremely challenging. They were asked to complete between 15 and 25 questions from the text, though clearly not all were as hard as the example given above. Short questions may only have taken a few minutes, while longer questions could take a group a substantial

---

[2] The graduate students are normally used for small group supervisions. In this case, the graduate students used were those who were particularly interested in and talented at undergraduate teaching.

[3] The UK school syllabus and, in particular, the 'A' level syllabus has no significant focus on formal proof – for example, there is no emphasis on formal derivations of Euclidean geometry, though through some aspects of the national curriculum, justification and informal reason has a role (Simpson, 1995).

part of a two-hour session to complete, even with help from the peer tutors and teacher.

Certain questions were pivotal. A sequence of questions beginning with the question 33 (given above) and ending with question 36, was the cause of much anxiety.

---

36 Suppose (a_n) and (b_n) are both null sequences, and $\varepsilon > 0$ is given.

(a) Must there be an n_1 such that $|a_n| < \varepsilon$ when $n > n_1$?

(b) Must there be an n_2 such that $|b_n| < 1$ when $n > n_2$?

(c) Is there an n_0 such that when $n > n_0$, then both $n > n_1$ and $n > n_2$?

(d) If $n > n_0$, must $|a_n b_n| < \varepsilon$?
You have now proved that the termwise *product* of two null sequences is null.

(e) If the sequence (c_n) is also null, what about $(a_n b_n c_n)$? And the termwise product of k null sequences?

The students reached this stage at the third or fourth week of the course and complaints about workload reached their peak at this point. It should be noted that these were the first questions to rely heavily on the dual nature of the definition of convergence (Alcock and Simpson, 1998), with the requirement to *use* the fact that the definition holds for two known sequences to *show* that it holds for their product. We claim that such exercises are crucial to the development of an understanding of the way definitions are used in formal mathematical proof. After students completed this section the complaints tailed off, perhaps indicating that a significant proportion of the students were beginning to 'get' university mathematics in the sense discussed in section 8 below.

5. THE TEACHERS' EXPERIENCE

For most of the teachers involved this was their first experience of 'team teaching'. They met for an hour each week to discuss ideas for, and problems with, their classes, and considered this to be a positive use of time. This course involved a completely different environment from those of either lectures or supervisions. The difficulties students encounter and questions they ask when engaged in this type of work are inherently unpredictable. Thus the teachers had to handle moments when they were uncertain of how to respond. Similarly, teachers on this new course were in the 'front-line' when complaints about the workload arose, while in a traditional course the feedback mechanisms are much slower. Also there was time pressure in the classes. Even in two hours and with the assistance of the peer tutors, it was often impossible to adequately monitor the progress of all in the class, or to provide as much assistance as would be desirable.

However, despite these pressures, those involved felt that teaching on the course was a very positive change from their traditional teaching. In lecture classes of over 100 it is nearly impossible to get to know students on an individual level, and new ways of thinking are more likely to be learned through individual study. In the new course, this learning could be seen happening in the class and many of the teachers spoke of experiences of 'seeing the lights go on'.

Some of the mathematics staff not involved in the new course were initially sceptical of it. Its heavy cost in terms of teacher time, student time and financial support was a cause for concern. However, as the course progressed and for the rest of the year, a number began to talk of indirect benefits – students recalled material more readily, seemed more able to construct arguments and were less likely to become disillusioned with learning university mathematics (a prevalent phenomenon called 'cooling-out' by Cooper, 1990).

6. THE STUDENTS' EXPERIENCE[4]

At the beginning of the course some students expressed concerns about the time commitment of the classes:

"They're fun, but they're far too long."

"…most of your friends finish at four."

Given that the social aspects involved in coming to university are of considerable importance to new students, this concern is a significant one.

Other comments demonstrated their need to negotiate a new 'didactic contract', highlighting a continued expectation of the caricature 'A' level pedagogy of the presentation of techniques, followed by exercises:

"What I would really like is if we could have a lecture, and then be given a set of questions based on that lecture, and do it in class."

In the middle of the term some students seemed to be coping better than others, and once again there was concern about the volume of work to be done:

"Actually I'm quite enjoying doing analysis. If there weren't so many questions it would be good."

"I find increasingly there are a few questions each week that I don't get done, and I need to do those outside the classes."

---

[4] As part of a doctoral study, interviews took place with six of the students taking the course every fortnight for the duration of the autumn term. In addition another two groups of three who were attending a standard lecture course covering the same material were interviewed. The interviews consisted of a task-based section in which material from the course was required, a discussion about the mathematics they were learning in the course, and a more general discussion about their attitudes towards mathematics at university.

By contrast the students who were on the lecture course typically complained about the volume of material covered in their courses in general, but not in analysis in particular. They talked about the problem of 'being able to understand the notes but not to do the worksheets', and felt that, unlike in school, there was no one easily accessible to help them.

The students were asked to complete an end-of-course evaluation (a requirement of quality assurance throughout the university). There continued to be some complaints about the length of the classes and the volume of work, although there was a small number who said that having so many questions had been good. Less frequently, students raised concerns about the lack of a coherent set of notes; they felt they had only their own work to refer to and that they would not have 'something to revise from'; that they lacked 'answers' or model solutions. Similarly, some said that more examples of how to write proofs would have been useful. Again, the importance of lecture notes speaks volumes about the types of expectations students have of the pedagogy of university education (Evans and Abbott, 1998). A small percentage also raised similar concerns to those raised by staff before the project, that time had been taken from other courses and/or that they felt they worked better by themselves.

Remarks about the plenary lectures were principally that the lectures helped to consolidate the material covered and that more lectures would be useful. In keeping with earlier remarks, some also commented that having lectures about the material to be covered in future would be preferable to those 'rounding up' the material of the previous week

Despite the workload, a number of positive comments were made. The classes were praised for the help that was available and for helping to bridge the gap between school and university. Some remarked that the course had helped them to understand the material and that working things out for themselves had given them a sense of achievement. Practically, there were suggestions that the ratio of classes to lectures should be changed. Students felt that while the classes were a good idea, they were too much of a time commitment, and shorter classes and an extra lecture would be preferable.

At the end of the academic year (before their examinations) the students had another opportunity to comment on all their first year courses. This time the response to the new course was more positive. More students said that they thought the classes were a good idea, with substantially fewer remarking that the classes were too long/late. It seems that, looking back on the course after six months, the amount of work required was seen as an aid to understanding and learning the material. The help available in the classes may have been compared with its absence in succeeding courses (which were traditional in style). Looking back on the course some students involved said:

> "I think it was quite helpful in lots of ways, because I found I did actually remember all the proofs I didn't think I would. Like you remember a result a bit more if you've worked to get it."

> "If you work on each step yourself then you understand the whole thing as a whole."

"I think being able to figure things out for yourself is a good thing to have when you're doing maths."

In contrast two joint-honours students interviewed from the lecture class talked about their amazement that their friends in the new course could sit down to just 'do' assignments without reference to their notes. They themselves felt that they always had to have their notes nearby, feeling that getting started on a question was a skill they didn't have.

7. ANALYSING THE NEW COURSE

The raw data from the course provide some indication of its success. The mean results from examinations taken by all students on both the new and traditional lecture courses are given in Table 1. Foundations is a traditional course covering some basic set theory, logic and group theory; it ran simultaneously with Analysis I and both were examined at the beginning of the second term. Analysis II is a traditional lecture course which builds on Analysis I and covers power series, continuity and differentiation. It was examined in the third term.

Table 1. Examination Results

	Foundations	Analysis I	Analysis II
New course students	60.0	76.5	63.5
Traditional course students	57.7	64.6	51.1

While these results seem to strongly highlight the benefits of this new way of teaching, there are other factors that might influence these figures. Those not on the single honours mathematics course have slightly lower 'A' level grades, may be considered to be less committed to their mathematics subjects, and have less time devoted to mathematics within their timetables. Thus it is usually expected that there will be some disparity between the sets of results. The marks, however, concur with what we see as a more important difference in the cognitive preparation for advanced mathematics that the courses provide. The new course allows a much more rapid adoption of a new way of thinking required for university mathematics, which we will suggest is at odds with the human brain's natural forms of cognition.

8. COMPARING COGNITIVE PREPARATIONS

The human brain, like any other organ, is the product of an evolutionary process. Just as eyes and hands developed in *homo erectus* to allow it to function better in its environment, so did the brain develop in *homo sapiens*. The development of mathematics and, in particular, formal axiomatic mathematics (such as that required in first year analysis), has taken place over a time scale which is relatively insignificant from the evolutionary perspective. Thus, we should not be surprised to find that the brain is poorly adapted to function in this way, just as it is poorly adapted to function in many other aspects of modern social and cultural life (Barkow, Cosmides and Tooby, 1992).

From an evolutionary psychology viewpoint, the discovery that the mind works by reasoning from specific prototypical examples or making deductions from mental models (Johnson-Laird and Byrne, 1991; Lakoff, 1987) is entirely sensible. Rosch and her colleagues have demonstrated that the categories used by a given culture, quite unlike mathematical sets, are non-classical (Rosch, 1978). This means that it is not necessarily clear what is and what is not in a category. Some elements are considered more important in reasoning (more prototypical) than others are. Many categories have structures of basic, super-ordinate and sub-ordinate levels which naïve mathematical sets do not have (Markman, 1989). In choosing to reason from a personal prototype, the adapted mind sacrifices some accuracy for a substantial gain in efficiency. For example, in saying 'all birds can fly', we are strictly incorrect, but we are accurate over a very substantial part of the category and the conclusion requires much less mental effort than would a formal deduction from definitions of 'bird' and 'flight' (if such definitions existed, which Fodor et. al. 1989, argue they do not).

In the field of mathematics, we link this natural, evolved form of thinking, to reasoning from concept images (Tall and Vinner, 1981). The adaptation of this natural way of thinking (called 'the general cognitive strategy', Alcock and Simpson, 1999) to mathematics works well for much of pre-university mathematics if combined with an ability to follow procedures. Indeed, at a higher level, the use of cleverly chosen prototypes is of immense importance to mathematical creativity: the mental manipulation of appropriate prototypical images guides the mathematician to important constructions and conjectures (Thurston, 1990).

The step from school to university mathematics however, involves the adoption of an additional way of thinking. Mathematical sets (at least in the sense presented in traditional undergraduate mathematics texts like Halmos, 1960) are classical. Definitions precisely determine the extension of a concept. Unlike dictionary definitions, they are not mere descriptions: they are able to be manipulated. In addition to developing powerful personal prototypes which are consistent with the formal definitions, formal mathematics requires the use of deduction from definitions and previously formally deduced theorems which, we claim, is not a natural function of the adapted mind. We term the acquisition of this new form of thinking, adopting 'the rigour prefix'[5].

The newness of this way of thinking is apparent in the fact that it is obvious to a mathematician that 'prove that (x_n) is convergent' is merely a shorthand for 'show that (x_n) satisfies the definition of convergence', but that this is often not so to a student. In the study of the students on the parallel courses described here, students were often seen to argue from particular examples of convergent sequences as if

---

[5] After Lakoff (1987): a prefix changes some of the aspects of the following word, for example 'white' in 'white lie' nullifies the general assumption that lies are bad. In this case we speak of 'rigorous' mathematics, in which the explicit expression of properties and logical structure is required over and above an (albeit crucial) intuitive understanding of the concepts.

they were prototypical of the set, rather than argue from the definition of convergence (Alcock and Simpson, 1999)

There is, therefore, a communication breakdown between lecturer and student at university level. When lecturers give a definition, they intend that this should precisely generate the concept, and when they present formal deductive arguments based on this definition they intend that this argument ensures that the conclusion holds for all objects under consideration. Of course, they may also give examples intending that these should serve to clarify the general arguments, but thinking of these only as illustrations.

Students, however, may see the definition as a description to introduce the concept, but not as something to manipulate in reasoning. It may even seem redundant in cases where the concept is familiar from their previous experience and so a basic concept image is already established. The general cognitive strategy involves establishing a working personal prototype (or prototypes) for a category, hence students may attend mainly to examples, assuming these too are particularly good instances of the concept in question i.e. good ones on which to base a prototype for use in metonymic reasoning.

The difference between metonymic reasoning based on the general cognitive strategy and that using the rigour prefix provides an explanation for some of the difficulties in analysis and, we will show, explains why the new analysis course is a better cognitive preparation for undergraduate mathematics than the traditional course.

9. WHY IS ANALYSIS HARD?

The definitions of analysis are logically complicated and written in unfamiliar notation, whereas the visual images of sequences, functions and other objects are ubiquitous and beguilingly simple (Pinto and Tall, 1999). They are thus ideal for constructing personal prototypes. Students' complaints that 'it's obvious, why do I have to prove it?' indicate the general cognitive strategy at work: it is easy to draw or imagine a picture to convince oneself that, say, an increasing and bounded sequence must converge, without any need for recourse to definitions or the axiom of completeness. Thus there is little incentive to move to a rigorous way of working.

By comparison in an elementary course in group theory, for example, definitions tend to be logically simple (if long) and pictures less readily available. We may argue, then, that students are more likely to produce something that looks formal, even if they may not understand why it is appropriate to work in this way.

This availability of images in analysis means that even students who haven't previously been inclined to use visualisation in mathematics are likely to adopt some personal prototypes as a way of gaining access to the material. More are therefore likely to encounter the problems associated with the general cognitive strategy as applied to visual prototypes. The first of these is that students sometimes fail to 'see the general in the particular' – where a teacher sees a diagram as representative of a whole class of cases, the student may fail to be aware of possible variations. This may tie thought to irrelevant details or even introduce false data, as well as preventing a student from seeing a proof using a picture as general (Chazan, 1983 and Presmeg, 1986). Similarly the link between visualisation and rigorous analytic thought is often absent in students. This is often a problem of translation – the student may have an understanding of a situation in the form of an image (with or without the type of error described above) but be unable to express this (Dreyfus, 1991, Presmeg, 1986, and Zazkis, Dubinsky and Dautermann, 1996).

Freyd's work suggests that communication between individuals encourages the meanings of terms to be constructed and clarified in a way consistent with the notion of a definition (Freyd, 1983). A requirement to work in co-operation with other students is likely to highlight the need to make explicit what properties are used and how they logically combine to give the required results. In conjunction with guidance in understanding the definitions agreed upon by the mathematical community, this should therefore lead to a quicker acquisition of the thought habits of the rigour prefix in the majority of students. Preliminary results indicate that this is indeed the case for the two courses studied: those on the new course develop the rigour prefix relatively quickly (perhaps as they begin to negotiate questions like 33 and 36 given earlier). In the traditional course, where there is less student-student and student-teacher interaction, the students who were interviewed appeared largely to remain unaware of the status of definitions. They often provided informal, example-based arguments and were rarely able to turn these into proofs. Of course, students without the rigour prefix can sometimes produce arguments based on prototypical images with the patina of formality added without understanding, or those which are simply memorised, which can easily be mistaken in examination marking as a good attempt at a formal proof.

10. CONCLUSIONS

The number of new concepts in analysis, coupled with the new standards of rigour in university mathematics, makes the learning of analysis difficult. No course aiming to cover the standard amount of work in the standard amount of time could hope to change this. However, the intuitive ideas of the mathematicians who developed and taught on this course have led to a new pedagogy: small classes, collaborative learning, questions that encourage students to develop the mathematical content and arguments for themselves. The new course provides for negotiation of a new didactic contract, fast feedback from fellow students and from experienced and sensitive staff and the answering of questions which emphasise the manipulation of definitions. Through this it encourages students to amend their evolutionarily developed general cognitive strategy, which is such a powerful way of thinking outside formal mathematics, with a new awareness vital to understanding university mathematics: the rigour prefix.

REFERENCES

Alcock, L. and Simpson, A. (1998). Definitions in university students' perceptions of structure. *International Conference on the Teaching of Mathematics*, Samos: University of the Aegean, 19-21.

Alcock, L.J. and Simpson, A.P. (1999). The rigour prefix. In O. Zaslavsky (Ed.), *Proceedings of the 23rd International conference for the Psychology of Mathematics Education*, Vol 2, 17-24.

Barkow, J., Cosmides, L. and Tooby, J. (1992). *The Adapted Mind: Evolutionary Psychology and the Generation of Culture*. Oxford: Oxford University Press.

Brousseau, G. and Otte, M. (1991). The fragility of knowledge. In A.J. Bishop, S. Mellin-Olsen and J. van Dormolen (Eds.), *Mathematical Knowledge: Its Growth Through Teaching*, pp. 13-36. Dordrecht: Kluwer Academic Publishers.

Burn, R.P. (1992). *Numbers and Functions: Steps into Analysis*. Cambridge: Cambridge University Press.

Chazan, D. (1993). High school geometry students' justification for their views of empirical evidence and mathematical proof. *Educational Studies in Mathematics*, 24, 359-387.

Cooper, B. (1990). PGCE students and investigational approaches to secondary mathematics. *Research Papers in Education*, 5(2), 127-151.

Davis, R.B. and Vinner, S. (1986). The notion of limit: Some seemingly unavoidable misconception stages. *Journal of Mathematical Behaviour*, 5(3), 281-303.

Dearing, R. (1997). *Higher Education in the Learning Society: Report of the National Committee of Inquiry in Higher Education*. London: HMSO.

Dreyfus, T. (1991). Advanced mathematical thinking processes. In D. Tall (Ed.), *Advanced Mathematical Thinking*, pp. 25-41. Dordrecht: Kluwer Academic Publishers.

Dubinsky, E., Elterman, F. and Gong, C., (1988). The student's construction of quantification. *For the Learning of Mathematics*, 8(2), 44-51.

Evans, L. and Abbott, I. (1998). *Teaching and Learning in Higher Education*. London: Cassell.

Fodor, J.A., Garret, M.F., Walker, E.C. and Parley, C.H. (1980). Against Definition. *Cognition*, 8, 263-267.

Freyd, J.J. (1983). Shareability: the social psychology of epistemology. *Cognitive Science*, 7, 191-210.

Gray, E., Pinto, M., Pitta, D. and Tall, D. (1999). Knowledge Construction and diverging thinking in elementary and advanced Mathematics. *Educational Studies in Mathematics*, 38 (1-3), 111–133.

Halmos, P. (1960). *Naïve Set Theory*. London: Van Nostrand.

Hanna, G. and Jahnke, H.N. (1993). Proof and application. *Educational Studies in Mathematics*, 24(4), 421-438.

Harel, G. and Tall, D.O. (1991). The general, the abstract and the generic in advanced mathematical thinking. *For the Learning of Mathematics*, 11(1), 38-42.

Johnson-Laird, P.N. and Byrne, R.M.J. (1991). *Deduction*, Hillsdale. Lawrence Erlbaum Associates.

Lakoff, G. (1987). *Women, fire and dangerous things – what categories reveal about the mind*. Chicago: University of Chicago Press.

LMS (1995). *Tackling the Mathematics Problem*. London: London Mathematical Society.

Markman, E. (1989). *Categorization and Naming in Children: Problems of Induction*. Cambridge, Mass: MIT Press.

Moore, R.C. (1994). Making the transition to formal proof. *Educational Studies in Mathematics*, 27, 249-266.

Perry, W.G. (1970). *Forms of Intellectual and Ethical Development in the College Years: A Scheme*. New York: Holt Rinehart and Winston.

Pinto, M and Tall, D. (1999). Student constructions of formal theory: giving and extracting meaning. In O. Zaslavsky (Ed.), *Proceedings of the 23rd International Conference for the Psychology of Mathematics Education*, Vol. 4, pp. 65–73.

Presmeg, N.C. (1986). Visualisation in high school mathematics, *For the Learning of Mathematics*, 6(3), 42-46.

Reynolds, B., Hagelgans, N., Schwingendorf, K., Vidakovic, D., Dubinsky, E., Shahin, M and Wimbish, G.J. (1995). *A Practical Guide to Co-operative Learning in Collegiate Mathematics*. Washington, D.C.: Mathematical Association of America.

Rosch, E. (1978). Principles of Categorisation. In E. Rosch, and B. Lloyd (Eds.), *Cognition and categorisation*, pp. 27-48. Hillsdale: Lawrence Erlbaum Associates.

SEAC (1993). *GCE Advanced and Advanced Supplementary Examinations: Subject core for Mathematics*. London: Schools Examinations and Assessment Council.

Sierpinska, A. (1987). Humanities students and epistemological obstacles related to limits. *Educational Studies in Mathematics*, 18(4), 371-387.

Simpson, A.P. (1995). *Developing a proving attitude. Justifying and Proving in School Mathematics.* London: Institute of Education.

SMP (1990). *Schools Mathematics Project: 16-19 Mathematics.* Cambridge: Cambridge University Press.

Tall, D. and Vinner, S. (1981). Concept image and concept definition in mathematics with particular reference to limits and continuity. *Educational Studies in Mathematics*, 12(2), 151-169.

Thurston, W. (1990). *Three Dimensional Geometry and Topology.* Princeton: Princeton University Press.

Vinner, S. (1991). The role of definitions in teaching and learning. In D. Tall (Ed.), *Advanced Mathematical Thinking*, pp. 65-81. Dordrecht: Kluwer Academic Publishers.

Yusof, Y. and Tall, D. (in press). Changing Attitudes to University Mathematics through Problem-solving. *Educational Studies in Mathematics.*

Zazkis, R., Dubinsky, E. and Dautermann, J. (1996). Co-ordinating visual and analytic strategies: a study of students' understanding of the group D4. *Journal for Research in Mathematics Education*, 27(4), 435-457.

ACKNOWLEDGEMENTS

The Warwick Analysis Project has involved a large number of staff, without whose dedication and drive it would not have survived. Prof. David Epstein initially developed the course and taught it with the help of Adrian Simpson and Prof. David Tall. The main course was evolved by Prof. Epstein, Alyson Stibbard and Roger Tribe, with the help of Heather MacCluskey, Lara Alcock, Andrew Clow and Bertold Weist.

Lara Alcock and Adrian Simpson
University of Warwick, Coventry, CV4 7AL
lja@maths.warwick.ac.uk
a.p.simpson@warwick.ac.uk

HARVEY KEYNES AND ANDREA OLSON

PROFESSIONAL DEVELOPMENT FOR CHANGING UNDERGRADUATE MATHEMATICS INSTRUCTION

1. INTRODUCTION

This paper discusses a model for professional development used in a new reformed calculus sequence for science, engineering, and mathematics students created for the Institute of Technology, University of Minnesota. This model uses a team approach and different modes of mentoring to provide explicit and implicit professional development for all team members - senior faculty, graduate and undergraduate teaching assistants (TAs), and teaching specialists. We discuss why this model has been effective in having senior faculty use and support changes in pedagogy, including instructional teamwork and student-centred learning, students working cooperatively in small groups, and exploring mathematical ideas using appropriate technologies. We indicate how other members of the instructional team are also encouraged and mentored in using these approaches. A key feature is the development of materials on complex mathematics topics, which can be most effectively taught using these modern approaches. The future implications of this model, including expanded uses of these approaches in more advanced mathematics coursework, are also addressed. Finally, some directions for future research suggested by this model are discussed.

A central issue facing undergraduate mathematics instruction is finding ways to encourage innovation and improvement of both content and instructional methods. Major forces both within and outside of higher education have created the need for very serious consideration of the use of modern approaches to college instruction and an increased focus on active student learning. Some of these forces include the changing and increasing demands for high-level training in mathematics in many disciplines, the growing role of technology, the changes in content and pedagogy at the primary and secondary school levels, and the changing composition of the undergraduate student population. In addition, current knowledge from the cognitive sciences and educational research about how to improve learning for large numbers of students has created the need for faculty to carefully consider changes in some of the traditional pedagogical approaches. However, many mathematics faculty are unwilling to embrace such changes unless adoption of these methods shows improved student learning of the content that *faculty* believe is essential. For example, many mathematics faculty have been rather hesitant to use technology in the classroom due to their skepticism about the validity of evidence that it improves student learning of important concepts.

Derek Holton (Ed.), The Teaching and Learning of Mathematics at University Level: An ICMI Study, 113—126.
© 2001 *Kluwer Academic Publishers. Printed in the Netherlands.*

With these assumptions in mind, one of the objectives of the Calculus Initiative (CI) at the University of Minnesota, a project which successfully revitalized the undergraduate calculus sequence for engineering students, was to introduce changes in pedagogy and practice that made faculty aware of the value of such efforts. The CI emphasized (i) how the active learning approaches enhanced the faculty's own success as teachers; and (ii) how these methods improved student motivation and learning of important classical calculus topics. In this sense, many of the Initiative's efforts were devoted to innovative ways of providing professional development for the diverse members of the CI instructional teams—senior faculty, post-doctoral fellows, visiting faculty, graduate students, teaching specialists (many of whom were outstanding high school teachers on sabbatical), and undergraduate teaching assistants. A major objective was to provide a mentoring environment that helped each of these groups to be accepting of and successful in both short- and long-term implementation of these changes, which incorporated modern instructional approaches. The results of a four-year study of the CI are given in Keynes, Olson, O'Loughlin and Shaw (2000).

Several of the effects of the CI are documented in this paper. First, some examples of the materials created for its curriculum and their intended use in the classroom are given, followed by a discussion of the goals and influences of the instructional team approach. Then a description of the essential restructuring process that took place at Minnesota is presented. A special emphasis is placed on how the materials were designed and introduced within the coursework in order to motivate the instructional team. Statistical outcomes and the challenges of implementing this approach are given in Keynes and Olson (2000).

In particular, the CI curriculum was developed to include coverage of deep mathematics at a fairly high computational and conceptual level, in order to entice senior faculty to put efforts into trying similar modern pedagogical approaches for other complex topics. The major premise was that if faculty were successful at teaching relatively sophisticated topics in calculus using group work, projects and technology which resulted in improved student learning, this success would motivate them to use similar materials and approaches throughout the undergraduate curriculum. Furthermore, since a team approach to instruction was used to discuss and review the coursework and pedagogy throughout the Initiative, the enthusiasm of the senior faculty would naturally lead to the professional development and mentoring of all of the instructors. In fact, this increased instructional interaction between the faculty and the post doctoral students, teaching specialists, and graduate and undergraduate teaching assistants, increased all of the members' strength and interest in the teaching of undergraduate mathematics, in addition to enhancing the leadership skills of the faculty.

2. MODERN APPROACHES FOR CLASSICAL TOPICS

Even at a large research university, many senior faculty have a genuine interest in improving their ability to teach, most especially at the content and conceptual levels which they believe are relevant. Thus, a key aspect of having senior faculty try to utilize modern pedagogical approaches is to enable them to personally experience improvements in their professional goals for satisfactory teaching, as well as to see improvement in student learning as a result of these changes.

Based on one of the author's experiences that a wide range of materials involving conceptually, computationally, and geometrically challenging classical topics can be effectively taught using group work and mini-projects, a large collection of materials was gathered for use in the CI classroom. These materials were primarily developed for both new and traditional topics covered in Stewart (1997). Most of this group work and projects are now contained in the companion Instructor's Guide for this text (Keynes, Stewart, Shaw and Hesse, 1999). The guide also provides suggestions regarding the use of these approaches, and specifically, how to implement the projects (see Appendix A of the guide). Several examples of advanced approaches involving traditional topics that have been used effectively in the CI workshops are provided below. These selections might serve as exemplars of innovative approaches to active learning, which could be developed for more advanced mathematics courses.

Example 1
The first example shows how to create group work to clarify some of the theory and rigour behind the definition of limits. Choosing three functions, two with a limit and one not so obviously without a limit at the designated point, students attempt to find δs for various values of ε. The function without a limit $h(x) = \dfrac{|x|}{100x}$, might appear to have the limit $\dfrac{1}{100}$ at $x = 0$ until you choose $\varepsilon < \dfrac{1}{50}$, while students can, of course, always find a δ for the other 2 functions. This type of conceptually challenging group work cautions students not to just check $\varepsilon = .1$ and $\varepsilon = .05$ to see if a limit exists, and has helped to make calculus students more aware of the issues in the formal δ–ε process.

Example 2
Another intriguing problem is to investigate Euler's constant $\gamma = \lim_{n \to \infty} \gamma_n$, where $\gamma_n = \left(\sum\limits_{j=1}^{n} \dfrac{1}{j} \right) - \ell n\,(n + 1)$. The solution to this problem involves a broad range of approaches - unusual geometric arguments to estimate lower and upper bounds for γ, analytic arguments to show γ_n is increasing, and some numerical estimates for the value of γ (whose precise value is still unknown). This material is certainly at the advanced level for calculus, and yet has been successfully presented to the large CI audience using carefully developed group work.

Example 3
In studying limits of functions, it is nice to be able to visualize functions that are 'naturally' discontinuous despite having a single equation. A group work was

developed around analyzing the behaviour of $f(x) = \dfrac{1}{1 - 2^x}$ near $x = 0$. This

function satisfies $\lim_{x \to 0^-} f(x) = 1$ and $\lim_{x \to 0^+} f(x) = 0$. While one can easily visualize these one-sided limits on a graphing calculator, the effort required to analytically show these results requires some careful work with exponential functions.

Example 4
It has always been intriguing for students to see that Newton's method applied to

finding $\sqrt{a}$ gives the 'divide and average' algorithm $x_{n+1} = \dfrac{1}{2}\left(x_n + \dfrac{a}{x_n}\right)$, a very fast

and efficient method easily implemented on even a simple calculator. A nice group work extends this to $\sqrt[3]{a}$ or $\sqrt[k]{a}$, for any $k \geq 2$. Students learn that Newton's Method now requires $(k - 1)$ parts of the prior guess and one part of the division checking

accuracy: $x_{n+1} = \dfrac{1}{k}\left((k-1)x_n + \dfrac{a}{x_n^{k-1}}\right)$, rather than just simple averaging, and also

realize that the case $k = 3$ is more precise in showing the correct method of generalization than $k = 2$.

Example 5
A very important idea in mathematics is the use of different representations to solve problems stated in one context using techniques from another context. An interesting group work recast substitution problems for definite integrals in terms of showing whether areas under graphs with very different shapes are equal. For example, students are asked to determine which of the areas under the following graphs are equal and why: $y = 4xe^{(-x2)}$, $0 \leq x \leq 2$; $y = 2e^{-x}$, $0 \leq x \leq 4$, $y = -2e^{-x}$, $-4 \leq x \leq -1$ and $y = 2e^x$, $1 \leq x \leq 2$.

One of the main objectives for developing group work materials for classical mathematics topics is to demonstrate the value of using cooperative learning and group work even with complex and challenging concepts. This means that all levels of instructors—senior faculty, post-doctoral fellows, and visiting faculty, graduate students, teaching specialists, and undergraduate teaching assistants—need to participate in discussing and analyzing appropriate materials and practices, and in formulating and testing conjectures about the usefulness of these materials and approaches. When the CI instructors at all levels became more involved in these processes, their focus on using quality modern pedagogical approaches in high-level mathematics increased, and their focus on using it just to acquire routine computational skills and 'mechanical' instructional techniques decreased. However, while this shift in their focus may itself be a very important reason supporting adoption of modern approaches, ongoing affirmation and mentoring by senior faculty and continuing professional growth are essential to reinforce its success.

3. THE INSTRUCTIONAL TEAM APPROACH

A critical feature of the CI model was a consistent environment that supported a unified mathematical approach among the instructional team members. A built-in mentoring aspect was created by having the lecturers visit the workshops and recitations, and the recitation leaders and teaching assistants actively participate in the lectures. The weekly meetings encouraged discussions about mathematics in general in addition to specific course materials and teaching practices.

The structure of the CI was deliberately designed to provide opportunities for the use of various modern pedagogical techniques in both the large and small group sessions. Drill, practice, and conceptual discovery were integrated into both lecture and workshop, as opposed to reserving lectures for concepts and workshops for drills. Moreover, materials frequently foreshadowed a topic in one venue, which was then formally taught in the other. This necessitated the senior faculty to assume a role of academic leadership, and become involved in the careful orchestration of the workshops and lectures, with cooperative involvement of all of the instructional team members.

Another important instructional goal was to increase the personal interaction between students, senior faculty, and other instructional team members. One outcome was having senior faculty mentoring the instructors on curricular and content objectives, and explaining why they presented classical mathematics concepts for their intrinsic value as well as their applications. Adjunct faculty hired specifically for teaching, such as outstanding high school instructors, provided other team members with their unique insights on how to interact with students who were experiencing college-level courses for the first time. The ability of the graduate teaching assistants to assist high school instructors on content questions and the enthusiasm of the undergraduate teaching assistants led to many informal mentoring relationships among the instructional team members.

One major issue in using the team approach is whether the value added is outweighed by the increased time and effort for the instructional staff. The CI addressed this problem by effectively minimizing the need for additional efforts and resources. By providing mentoring for the teaching specialists and undergraduate teaching assistants within the team structure, they were able to lead the small group recitations/workshops with modest additional instructional development costs. Some initial analyses indicated that a positive relationship existed between using the team approach and improved instructional and learning outcomes. Three factors appeared to be critical to sustaining the quality of the CI team approach: (i) ongoing faculty interest and motivation; (ii) ongoing academic leadership by the senior faculty; and (iii) ongoing professional development including investigating modern curriculum approaches, conducting curriculum reviews, and maintaining discussions about faculty development.

4. THE UNIVERSITY OF MINNESOTA MODEL

In the mid-1990s, the IT Center for Educational Programs and the School of Mathematics, University of Minnesota, developed and successfully implemented a restructuring of the calculus sequence for undergraduate science and engineering students primarily based on the principles of (i) modern pedagogical approaches that emphasized active student learning; (ii) enhanced curricular materials that supported modern approaches without compromising the quality of the content; and (iii) the instructional team approach to support effective use of components one and two. The planning and development process began in 1993 with the support of an NSF grant, and the Calculus Initiative was piloted in the fall of 1995. The model was designed to incorporate the common key aspects of successful calculus reform efforts, and had as its objectives to become the standard format for the engineering and science calculus instruction at Minnesota, as well as to serve as a model for other large, public research institutions. While it has undergone several modifications as it evolved and became fully institutionalized at Minnesota in the fall of 1999, the initiative continues to reflect many of the important initial principles. This section of the paper discusses the main aspects of the restructuring process, which proved to be essential for the successful implementation of the CI.

4.1 Restructuring

Starting with a calculus sequence that was primarily computational and taught in a traditional lecture/recitation format, the initial goal of the development process was to design a core undergraduate calculus sequence that emphasized conceptual and geometric/visual reasoning as well as the use of mathematics in other disciplines. The necessary restructuring for this core sequence was the reconfiguration of the large lecture/recitation format to permit additional time for approaches that increased personal interaction between the faculty and students, encouraged more student-centred pedagogy, and allowed for the incorporation of technology as the curriculum progressed. The table below lists the major components of the core sequence and its structure.

Table 1. Six main components of the CI

1. Two 50-minute weekly class sessions of 100 students are held primarily using a lecture format but also include up to 15 minutes of group work per session.
2. Each week, two workshop sessions (one 100-minute session and one 50-minute session) of 25 students are conducted consisting primarily of group work (in year one (Y1)) or computer labs (in year two (Y2)), to encourage active learning and peer collaboration. Some homework is discussed and short quizzes are given in the workshops.
3. Periodic visits to the workshops by the lecturer, and visits by the workshop leaders to the lectures, help facilitate the group work.
4. Three large-scale team projects (some including computer lab applications) are

	assigned in Y1. in Y2, weekly computer labs using appropriate applications take place. Half of the labs require a written report, and one lab is assigned over several weeks as a group project.
5.	In Y1, 'gateway exams' on differentiation and integration consist of standard computations to be completed at a higher level without the use of a calculator.
6.	Texts that contain rich applications. (The actual texts used in both Y1 and Y2 were: Ostebee and Zorn,1995, 1997; Stewart, 1997; Blanchard, Devaney and Hall, 1998; and Barr, 1997.)

While the initial structure was very successful, it required significant faculty and administrative staff time commitments. Thus, it became necessary to reduce these efforts, while maintaining the key responsibilities. Table 2, which is listed on the following page, reflects the changes in the structure/staff as the enrolment increased. First, the amount of administrative effort to manage the CI was decreased by systematizing the materials and procedures into a format that allowed undergraduate student workers and part-time coordinators to handle most administrative duties. Second, both cost factors and the lack of interested and pedagogically qualified mathematics graduate students led to the increased use of teaching specialists (primarily outstanding secondary teachers), and eventually in 1998 to the use of several outstanding undergraduate upper division students as workshop leaders under careful mentoring by the lecturers and workshop administrators. However, maintaining the level of administrative effort necessary to effectively manage the sequence, and having a pool of qualified workshop instructors/TAs willing to participate as instructional team members appear to be two critical components in maintaining the quality of this model.

Table 2. Composition of the CI Staff

1995-96	1996-97	1997-93	1993-99
YI-1	YI-2	YI-3	YI-4
100-students	200-students	300-students	400-students
2 co-lecturers	2 lecturers	3 lecturers	4 lecturers
1 wkshp adm (post-doc)	2 co-wrkshp adms (graduate students)	1 wkshp adm (post-doc)	1 wkshp adm (post-doc)
4 wkshp leaders (2 post-docs, 2 grad TAs)	8 wkshp leaders (grad TAs)	12 wkshp leaders (3 post-docs, 3 grad TAs, 6 TS)	16 wkshp leaders (3 post docs, 4 grad TAs, 1 ugrad TA, 8 TS)
	Y2-1	Y2-2	Y2-3
	75 students	120 students	220 students
	1 wkshp adm (grad TA)	1 wkshp adm (post doc)	1 wkshp adm TS)
	3 wkshop leaders (grad TAs)	5 wkshop leaders (grad TAs)	9 wkshp leaders (1 post doc, 5 grad TAs, 1 ugrad TA, 2 TS)
1 sr. adm (33%)	1 sr. adm (25%)	1 sr. adm (15%)	1 sr. adm (10%)

Legend: wkshp admin is workshop administrator; wkshp leader is workshop leader; grad TA is graduate student teaching assistant, TS is teaching specialist; and ugrad TA is undergraduate teaching assistant.

4.2 Restructuring Beliefs About Teaching/Learning

In addition to procedural and class format restructuring, a key element was the development of improved instructional and learning methods that used group work for teaching classical mathematical concepts and their integration into the curriculum. The critical time for implementing group work is during the workshop activities that focus on instructional experimentation. The impact of active student learning was heavily dependent on the senior faculty understanding and valuing its importance to the learning of complex topics. This understanding was initiated and reinforced during all of the phases of the CI by providing:

- opportunities for materials development by the senior mathematics faculty in the core sequence and other coursework;
- co-development of interdisciplinary project-based modules with faculty from disciplines outside of mathematics;
- expansion of group work and laboratory approaches in the workshops to topics and teaching approaches used in the lectures;
- clear and consistent support and rewards from the School valuing such efforts by faculty, including faculty who successfully participated in the sequence restructuring.

Essential components for maintaining effective pedagogical changes and helping new faculty and other instructional team members to effectively utilize these changes are:

- establishment and maintenance of some type of faculty and TA development programme that consistently supports ongoing change. Such programmes should lead to increased willingness to consider the exploration of complex mathematics through experimentation and technologies;
- programmes that increase faculty awareness and appreciation of the value of modern pedagogical approaches. Much of this can take place within the instructional team meetings.

The best practices of the CI continue to be implemented after institutionalization, and have successfully survived several changes in faculty leadership and a more diverse student population. The outcomes of the Initiative have been carefully studied over four years and are given in Keynes and Olson (2000) and Keynes, Olson, O'Loughlin and Shaw (2000). The next section of this paper summarizes the main impact of these findings on the instructors and students.

5. THE INFLUENCE OF MODERN PEDAGOGICAL APPROACHES ON INSTRUCTORS AND STUDENTS

5.1 Senior Faculty

A major goal of introducing advanced group work was to encourage senior faculty to consider active student learning even when addressing complex and challenging mathematics concepts. While it is too early to comment on any long-term effects, several senior faculty who have taught in the CI have indicated in various evaluations that they are now investigating the use of these approaches in more advanced courses.

The exploration of complex topics using group work and other modern approaches did provide opportunities for senior faculty to adopt some of these techniques for limited use within a traditional lecture, without sacrificing the overall quality of the lecturing method. This resulted not only in improved interactions between senior faculty and students within the lecture, but also led to increased satisfaction about the teaching process by the lecturers—teaching was simply more engaging when using these approaches. In addition, the student ratings about the importance and quality of the lectures rose significantly, as well as student ratings of the lecturer's interest in student learning. With this more friendly and rewarding environment for teaching and learning, many faculty showed more interest in working with students on group work and projects during the extended workshops and laboratories. Together with introducing other similar changes, senior faculty felt comfortable that the teaching and learning environment for both themselves and the students had improved. In the end, the faculty were more receptive to investigate if this approach could pass the 'acid test'—the improvement of student learning, resulting in increased understanding and long-term retention of the central concepts of calculus and its major applications to other disciplines. While such initiatives require more faculty engagement and leadership, the CI model has exemplified the value-added to student learning, faculty development, and the culture of mathematics by these extra efforts in the appropriate settings.

5.2 Adjunct Faculty, Teaching Specialists, and Graduate Student Teaching Assistants

The issues of how to provide effective mentoring and professional development for adjunct faculty, teaching specialists, and graduate student teaching assistants to encourage their use of modern approaches to teaching the undergraduate mathematics curriculum are quite complex. Among the issues are the mathematical background and student/work experiences for instructors at this level. These aspects necessitated some procedural changes in the CI to accommodate their diverse backgrounds. For example, when outstanding high school instructors were hired as teaching specialists to lead the CI recitations/workshops, the teachers insisted on more time to look through the workshop materials, and requested solutions for all of the group work. Several requested more frequent meetings with the course chair and senior faculty to discuss both the subject matter and pedagogical approaches. Graduate teaching assistants who were part of the CI felt time pressures from their degree programmes, and nearly all had very limited teaching experience using modern practices. In addition, their own undergraduate experiences (especially for international TAs) frequently resulted in questioning the need to modify teaching methods from the traditional calculus course and changing the nature of their instruction. These concerns served to amplify the need for advocacy, leadership and mentoring by senior faculty in order to implement modern approaches in teaching undergraduate mathematics. On the other hand, concerns about the quality of some of their own current instruction in mathematics and the realization that improved pedagogy would be an important feature in their own hiring and promotion opportunities typically resulted in a generally open attitude to modifying their instructional practices. For example, the high school teachers involved with the CI reported that in addition to learning deeper mathematics, they were able to improve their own high-school courses by incorporating many of the techniques and materials of the Initiative.

Two features seemed to have had the most impact on supporting changes in pedagogy for this group. The first was the dynamics and interactions of the weekly instructional team meetings. Graduate TAs generally experienced improvements in their teaching, and more comfort with using modern pedagogy as a result of working closely with senior faculty and experienced high school teachers on teaching and content issues. For example, one international TA, who initially was not considered to be a particularly effective teacher, improved sufficiently after teaching two quarters of the CI to the point where she personally was more confident and much more successful as a teacher, as indicated by student evaluations and requests to continue in her section. She attributed much of her success to the team approach. The second important feature was the unexpected pedagogical 'mentoring' role assumed by several of the high school teachers who were workshop leaders. These excellent high school teachers brought a wealth of teaching experiences and student knowledge to the CI. On the other hand, their knowledge of some of the newer and more conceptual calculus content was marginal, and they were not very familiar with some of the approaches in the group work developed for this course. Thus, the graduate TAs could frequently help them with the content while they could help the TAs with their pedagogical skills. As an example, one graduate TA, who was having difficulty with using group work, had a high school teacher visit his class to provide advice. His techniques and student outcomes showed marked improvement based on the suggestions from the high school teacher. In fact, several of the workshop activities were modified for the entire course based on input from the high school teachers.

5.3   Undergraduate Teaching Assistants

After three years of mentoring graduate TAs and teaching specialists using the team approach, the CI extended this model to include the involvement of undergraduate TAs as workshop/recitation leaders. Professional development and carefully constructed mentoring of these young teaching assistants, especially in the areas of small group work, development and implementation of workshop projects, and effective use of technologies, were the most important components that enabled the undergraduates to be successful. The undergraduates felt many challenges similar to the graduate TAs: time and scheduling pressures and the lack of teaching experience, and these constraints needed to be addressed in their mentoring. While these aspects required additional engagement and leadership from all of the other team members—graduate TAs, teaching specialists, and faculty—the success of these undergraduate TAs exemplified the value added by the team approach and increased attention to, and interest in, quality teaching by the entire team as a result of these efforts.

An unexpected but very important outcome of using undergraduate TAs as part of the instructional team was cultivating the interests of these academically strong mathematics, science, and engineering students to seriously consider mathematics education as part of their career goals. The CI structure and TA experiences motivated these students to become interested in the teaching of mathematics. The weekly team meetings and lecture and workshop/recitation visitations proved to be powerful tools for keeping academically strong undergraduates focused on rich mathematics content and high quality, innovative teaching methods. Faculty participation at the workshop/recitation sessions sent a powerful message to the young teaching assistants about the importance of teaching and learning and the relevance of establishing a dynamic classroom environment. Several of these undergraduate TAs are now in graduate programmes leading to teaching careers in secondary mathematics classrooms and two-year colleges. This model for encouraging academically strong mathematics, engineering, and science students to pursue teaching careers in primary and secondary schools deserves further attention.

5.4 Student Perceptions

CI student evaluations clearly showed that a key factor in having students more effectively learn central concepts and more conceptual reasoning in calculus was in the quality of the small group work and collaboration with peers. Overall, these aspects were rated very highly by students, and the quality of informal student collaborations increased with the quality of the formal collaborations. Moreover, the willingness of students to put greater efforts into the coursework appeared to be correlated to the quality of the student-student and student-faculty interactions during small group work and collaborative projects. Finally, the quality of the materials themselves seemed to have a significant influence on the student interactions during small group work and projects. Materials which challenged students both individually and collectively and showed the value of quality classroom collaboration to both the students and the teacher convinced students as well as faculty that these approaches were genuine alternatives leading to improved learning of important content. These findings again point to the potential instructional improvements by developing challenging group work for undergraduate mathematics coursework.

During the course of the CI, a 'wish list' of characteristics that calculus students should have by the time they completed the undergraduate sequence was developed. These skills included the ability to:

- carefully carry out computations;
- think geometrically and conceptually;
- explore and understand complex concepts;
- work independently and with others;
- communicate important mathematical ideas clearly.

It was gratifying that even by the end of the first-year calculus course, many students had started to acquire these attributes. It appears that the design of the particular methods in the CI for teaching complex topics, along with the structure of the team approach, provided a different interconnection of instructional approaches to calculus and improved student opportunities to obtain these skills. One student noted on an evaluation, "The instructors did such a great job in getting students to explore every detail of the material, to be interactive, and to learn so well that other classes which used calculus were made greatly less difficult."

6. CONCLUSIONS AND FUTURE IMPLICATIONS

The experiences with the CI illustrate the need to address several key issues in order to create a successful professional development experience which changes current teaching practices and maintains its influence on instruction. These issues are:

1. An understanding of the deeply held beliefs and values of most senior faculty about what constitutes important mathematical content and essential computational techniques, and what students need to know mathematically.
2. An approach to using modern pedagogy which allows senior faculty to adapt different teaching approaches to more effectively teach the content that they *as individuals* value, as well as the general content of the course.
3. An appreciation of the general lack of understanding on the part of senior faculty about the influence of their teaching and mathematical approaches on undergraduate students' learning and views about mathematics. Research on the pre-service preparation of secondary mathematics teachers has documented the major influence of instructional practices of mathematicians in content courses for pre-service teachers on how students came to learn mathematics, their subsequent beliefs about what is important in mathematics, and how they believe their students should learn mathematics. Just as with the graduate TAs in the CI, these instructional practices of faculty strongly influence most undergraduate students who will be using mathematics in some type of professional capacity.
4. An appreciation about the general lack of understanding by senior faculty on the need and value of cooperative and team instructional approaches and

discussions, and the need for mentoring workshop instructors on content issues and directions.

5. An understanding of the general lack of knowledge and frequently deeply-held skepticism by senior faculty on the roles and value of technology to help students learn the mathematics that the faculty member values. For example, while many faculty are familiar with symbolic algebra programs and software in a research context, they are generally not aware of the issues involved in using them in a learning context. Thus, some faculty will use technology to visualize only complex or exotic mathematical graphs and structures, rather than using it to illustrate mainstream ideas and central examples. In addition, very few senior faculty are truly familiar with graphing calculators and their potential use in instruction.

To help improve professional development, more research and materials are needed to create additional ways to address these issues. Careful studies on the precise ways in which content and pedagogical beliefs of senior faculty influence and possibly hamper efforts to change undergraduate instruction and pedagogical approaches would be very useful. Even more critical would be extensive studies on the real roles, advantages and disadvantages of using various types of technology to improve the teaching and learning of all aspects of calculus—conceptual, visual and computational. These studies should attempt to develop a better understanding of the relationships between the complexity, depth and quality of the design of the software, both at the user interface and mathematical levels, and the resulting improvements in student learning and improved understanding of mathematics. With positive results from carefully designed studies on both of these aspects, the opportunities for designing effective professional development programmes for senior faulty and instructors at all levels would be greatly improved.

REFERENCES

Barr, T.H. (1997). *Vector Calculus*. UpperSaddle River, NJ: Prentice-Hall Inc.

Blanchard, P., Devaney, R.L. and Hall, G.R. (1998). *Differential Equations*. Pacific Grove, CA: Brooks/Cole Publishing.

Chickering, A. W. and Gamson, Z. F. (Eds.), (1991). *Applying the Seven Principles of Good Practice in Undergraduate Education* in New Directions for Teaching and Learning. No. 47. Jossey-Bass Publishers. San Francisco.

Keynes, H. and Olson, A. (2000). Redesigning the Calculus Sequence at a Research University: Issues, Implementation, and Objectives. *International Journal of Mathematical Education in Science, and Technology*, 31 (1), 71-82.

Keynes, H., Olson, A., O'Loughlin, D. and Shaw, D. (2000). Redesigning the Calculus Sequence at a Research University: Faculty, Professional Development, and Institutional Issues. In S. Ganter (Ed.) *Calculus Renewal: Issues for Undergraduate Mathematics Education in the Next Decade*, pp. 103-120. New York: Plenum Publishers.

Keynes, H., Stewart, J., Shaw, D. and Hesse, R. (1999). *Instructor's Guide for Stewart's Calculus: Concepts and Contexts*. Pacific Grove, CA: Brooks/Cole Publishing.

Ostebee, A. and Zorn, P. (1995, 1997). *Calculus From Graphical, Numerical, and Symbolic Points of View*. Orlando, FL: Saunders College Publishing, Harcourt Brace & Company.

Smith, D. (1998). *Renewal in Collegiate Mathematics Education*. Documenta Mathematica. Extra Volume ICM. p. 777-786.

Stewart, J. (1997). *Calculus Concepts and Contexts*. Pacific Grove, CA: Brooks/Cole Publishing.

Tucker, A. and Leitzel, J. (Eds.), (1995). *Assessing Calculus Reform Efforts*. Washington, D.C.: MAA.

Wilson, R. (1997). A Decade of Teaching Reform Calculus Has Been a Disaster, Critics Charge. *The Chronicle of Higher Education*, 43, Feb. 7, p. A12-A13.

Wu, H. (1998). *The Joy of Lecturing—With a Critique of the Romantic Tradition in Education Writing*. Department of Mathematics #3840. University of California. Berkeley, CA. p. 1-12.

Ostebee, A. and Zorn, P. (1995,1997). *Calculus From Graphical, Numerical, and Symbolic Points of View*. Orlando, FL: Saunders College Publishing, Harcourt Brace & Company.

Harvey B. Keynes
University of Minnesota, Minnesota, USA
keynes@math.umn.edu

Andrea M. Olson
University of Minnesota, Minnesota, USA
olson@math.umn.edu

MARC LEGRAND

SCIENTIFIC DEBATE IN MATHEMATICS COURSES

1. INTRODUCTION

Most of the time during lectures and tutorials students are not scientifically engaged. They impassively listen to impersonal assertions, in the form of definitions, theorems, and proofs, which are considered to be true. Scientific debate, on the other hand, is a teaching approach which aims to change this passive attitude by making students become the authors of mathematical statements, such as conjectures, propositions or proofs.

In this paper, I will first describe the nature of scientific debate and give the epistemological and pedagogical bases for this approach. This will be followed by a discussion of implementation as well as examples of the types of mathematical situations which have been used to provoke debates in the classroom. I will then signal some potential problems with the approach and the benefits, as expressed by students who have taken part in the approach. I will conclude by underlying the essential features of the scientific debate.

2. PRINCIPLES OF CLASSROOM-BASED SCIENTIFIC DEBATE

Underlying the scientific debate approach is a basic assumption that students can neither enunciate conjectures nor propose proofs unless they have their own opinion about what is scientifically reasonable and what is not. Instead of feeling obliged to address the teacher in an orthodox way: "I learned that ...", "I read in a book that ...", "I have been taught that ... and hence I affirm that ...", without assuming the epistemological responsibility of what is being said, the student who wants to participate in the scientific debate organised by the lecturer is invited to speak directly to the other students in the following somewhat iconoclastic way : "I do think that this idea is valid ... that this argument proves or contradicts an idea defended by me or another ... and here are my reasons."

Part of the scientific debate contract is that no one assumes either that conjectures written by the lecturer on the blackboard are theorems or that the proposed proofs are valid. In the first stage of the debate, the lecturer is not asked to give his or her opinion about the relevancy of what is said; the role of the latter is to facilitate the expression of ideas and to allow opposing views to be clearly stated. During the debate, students must defend their own ideas as long as they feel that they are more reasonable that the other explanations. Contrary to what happens in a polemic debate, in this case, students that have been convinced by opposing

127

Derek Holton (Ed.), The Teaching and Learning of Mathematics at University Level: An ICMI Study, 127—135.
© 2001 *Kluwer Academic Publishers. Printed in the Netherlands.*

arguments, have to abandon their own ideas and explain the reasons for changing their mind.

In this 'debate of ideas and explanations', everyone knows that he or she profits not when the debate shows that they are right but rather through the clarifying of others' convictions which subsequently lead the whole group to a deeper understanding of the situation.

During such debate, the role of the teacher is to answer for the scientific aspect of the debate but not to vouch for the truth nor for the validity of the results and arguments that are proposed during the debate. Only at the end of the debate does the instructor institutionalise the results by giving the appropriate definitions and theorems, and by identifying appealing but wrong results, the kind of recurrent mistakes which need to be addressed over and over again. The instructor also highlights procedures which have produced ideas and which have contributed to the differentiation of truth and falsehood.

3. EPISTEMOLOGICAL AND DIDACTICAL BASIS

Our basic principle is the following:

- when we learn mathematics, we do not necessarily become professional mathematicians.

Conversely

- to learn and retain mathematics effectively, we must temporarily become a mathematician, and the classroom must act as a scientific community!

We claim that a student who does not act mathematically, cannot master mathematics. We consider that mathematics essentially reflects a particular view of reality, a specific way of thinking about the world. When mathematicians try to understand a part of reality, or want to solve a specific problem, not only do they not stop dreaming, but rather they nurture their dreams and simultaneously try to structure their imagination, leading to some conjectures. There is a certain element of risk-taking when a mathematician tells himself or herself and then a community of colleagues: "I think that is true."

But, in order not to fall under the spell of dreams and hopes in place of reality, paradoxically very often a mathematician first attempts to discover whether the conjecture is false and it is only when these efforts fail, the business of proving that it is impossible for the idea to be false begins.

There is a specific way of thinking at work here, which is unlike what happens in everyday life. When conjectures are proved, mathematicians are naturally happy. But even if conjectures are shown to be false, there is still a sense that the work that went into proving this falsehood was useful rather than a waste of time; it helps to clarify what was misunderstood, and where one's thinking went awry.

Thus, scientific debate in mathematical courses is a type of a 'didactical engineering' (Artigue, 1994) which attempts to put this epistemological principle into practice. It also rests on two other principles:

- *a socio-constructivist principle*: if an essential part of scientific knowledge is to give us the capacity to grasp situational complexity and act rationally, students must encounter and deal with the conflict between everyday rationality (common sense) and scientific rationality.
- *a socio-ethicist principle*: every human being should have the opportunity to understand the deep meaning of what we strive to teach.

Ultimately, we realize that school is not just a place to obtain knowledge and diplomas, but also an important place to develop one's potential and acquire the habit of community life: the ability to understand other people's arguments, to create and develop one's own arguments, defend one's own thesis, even if the interlocutor is more knowledgeable, more powerful, older and wiser.

We believe that school education may be a good vehicle to discover the richness of communal work, and that the mathematical approach, in particular, may be a powerful tool to acquire the autonomy essential for cooperation without domination. But, in order to achieve this goal, we must establish a new didactical contract (Brousseau, 1997) whereby students and teachers must consider examinations as a means, a necessary tool for assessing competence, and not as the central goal.

4. IMPLEMENTATION OF THE SCIENTIFIC DEBATE

During a course which emphasizes the principles of the scientific debate, students are considered as a community of mathematicians. The teacher proposes topics and the students are invited to formulate conjectures and to engage in a scientific debate pertaining to the relevance and truth of their conjectures.

What is radically different here compared to conventional education is the status of truth; for the student who honours this new contract, truth does not reside in a professor's words or is discovered in books. Rather, truth is to be constructed jointly with other classmates. This requires that:

- first, the student must believe in what he or she conjectures;
- next, the student must develop rational arguments so as to be convinced that the conjectures are not false;
- finally, the students must find words, formulas, theorems and metaphors capable of persuading not only the professor and the classmates, but also everyone else who understands mathematics.

To build truth from one's own assertion requires the ability to listen and understand the objections of classmates until, at last, one is able to disprove counterarguments if they are incorrect. To construct truth from a conjecture requires that students accept their own mistakes, and understand the thinking that got them there. Consequently, the professor has to try to avoid revealing any opinion about the truth of the subject under discussion – a very, very difficult challenge for a mathematics professor. Instead, the task of the professor is to foster diversity of opinions with the objective of revealing conflicts in rationality.

For all these reasons, a teacher should:

- give sufficient time (neither too short nor too long – a delicate choice) for students to develop their own opinions based on scientific reasons, and record on the blackboard the global results of this individual work (for example: True 15, False 17, Other 40);
- write on the blackboard the arguments presented by students, without alteration, (also not an easy task), repeating aloud what may have been said too softly;
- strive to maximize the number of students involved in discovering a rational solution. The debate should be stopped only when a majority of students are well engaged in the problem.

Another challenge to the professor is at the end of the debate, when synthesizing results and focussing the students' attention on what is important to bear in mind. The professor should avoid congratulating those who gave correct arguments, and avoid putting down those who gave incorrect solutions. Instead, the role should be one of tracing the community's erratic path and showing how mistakes, once analyzed, can yield good ideas and useful solutions.

This is essentially the same schema that Lakatos proposes in Proofs and Refutations (Lakatos, 1976). The difference is that in a classroom, the actors are not professional mathematicians and the teacher has an important role to be the memory of the debate, if he or she wants to maximize the participation of the students in the debate over a long time.

5. TYPES OF DEBATE SITUATIONS

Several types of situations which are particularly suitable for the scientific debate approach arise.

5.1 The 'unplanned' debate

Sometimes the beginning of a scientific debate can be triggered by a question asked during class. For example, a student asks: "May we write $f(A \cap B) = f(A) \cap f(B)$?" Rather than answering the question, the professor simply writes on the blackboard:

Conjecture: If f is a function, $f: E \to F$, and if A and B are two subsets of E, then $f(A \cap B) = f(A) \cap f(B)$.

If after five minutes of individual reflection all students reach a similarly accurate conclusion, the professor will decide to go on with the lecture. But if (as is usually the case) opinions differ, the scientific debate becomes necessary to eliminate mistakes and construct a significant answer.

5.2 Planned situation intended to introduce a new concept or to overcome an epistemological obstacle

We give below two examples of situations which are pre-planned by the instructor in order to provoke a debate and to help resolve some well-documented epistemological obstacles (Bachelard, 1938).

5.2.1 The Riemann-Lebesgue transition

M. Artigue (this volume, pp. 207-220) has described the 'bar situation' that was designed in order to introduce first year students to the concept of the Riemann integral (see also Legrand, 1993). If we now want to introduce the Lebesgue integral to third year students familiar with the Riemann integral, we are faced with an obstacle of a much higher complexity; and the following two situations can help to introduce the necessary change of point of view:

1) Find the area of the subset $A = \{(x,y)\in \Re x\Re: x = 1/n, n \in N^*, 0 \leq y \leq 1\}$ of the unit square.

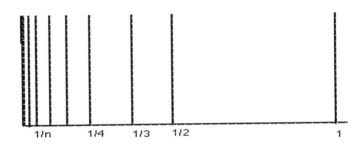

2) Find the area of the subset $B = \{(x,y)\in \Re x\Re ; x = n/m, (n,m)\in NxN^*$ and $n \leq m,$ $0 \leq y \leq 1\}$ of the unit square.

For the subset A, it is enough to find the area by using approximations by rectangles as is done for the Riemann integral. A can be placed within a finite number of rectangles with the sum of their areas arbitrarily small. For the subset B, if one measures B by the interior of a finite number of rectangles, one finds 0 and if one measures B by the exterior one finds an area always greater than 1, and similarly for its complement in the square $[0,1] \times [0,1]$.

Out of the students' debate emerges, bit by bit, the following paradox: there are relatively simple sets that are not 'Riemann measurable'. This pathology gives meaning to the introduction of the Lebesgue integral, which otherwise can appear to be a very unusefully complicated notion.

5.2.2 Relation between linear algebra and analysis

At university, if we want to establish relations between linear algebra and analysis, we have to face the obstacle of the triangle inequality: the norm of the sum is rarely equal to the sum of the norms and, by necessity, is often 'abnormally' small compared to the latter.

Initial situation: If $e_1, e_2, \ldots, e_n$ is the basis of a normed vector space E and if X is a vector of E, then X can be uniquely written as $x_1e_1 + x_2e_2 + \ldots + x_ne_n$. Which inequalities can be written between the absolute values $|x_i|$ of the components of X and the norm $\|X\|$ of X?

The answer given by almost all of the students is that $|x_i| \leq \|X\|$. Their proposed proofs include: $\|X\|^2 = \sum |x_i|^2$, and $\|X\| = \|x_1e_1\| + \|x_2e_2\|\| + \ldots + \|x_ne_n\|$. To overcome this obstacle, one can then look at the following situation, one that we have used in the past at secondary level to introduce the concept of vector.

The Blue Jeans situation: The blue jeans hanging on this rope weighs 3 kg.

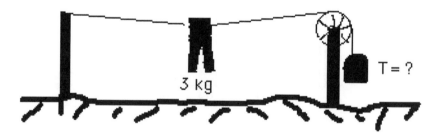

$$T = ?$$

$$3 \text{ kg}$$

Question : What is the least tension in kg needed to keep the jeans in this position (i.e. how heavy is the required counterweight)?

If given a choice of weights for the answer ranging from 1.5kg to 100kg, the great majority of the students choose values at the low end of the range[1]. This is due

---

[1] This was tried with both third year students majoring in mathematics and Phd students in various science disciplines. The vast majority of both groups choose values in the 1.5 to 6 kg range.

to the fact that most students think, without reflecting, that the norms of vectors making up a sum are smaller than the norm of the sum. In this situation, thinking of 'tensions' only as numbers draws the students' attention away from the crucial parameter: Students eventually begin to see that the variable 'angle' plays a crucial role since as this angle tends to 180°, the tension on each arm must tend to infinity.

Through a scientific debate the students are eventually in a position to appropriate a more complex notion of magnitude, one in which the sums are angle dependent. Following this way of looking at things, one can show that 'the parallelogram rule' for vector addition is the appropriate model for the Blue Jeans situation (more details about this situation can be found in Legrand, 1995.)

5.2.3 The deepening of a concept or theory

After the construction of the integral is covered in class, one can work on maps defined by an integral, on continuity, or on differentiability. One can ask students the following: If $F(x) = \int_a^x f(t)\, dt$, make some conjectures that link the properties of f to those of F. Students generally end up with conjectures such as: "f continuous $\Rightarrow$ F continuous", or "f increasing $\Rightarrow$ F increasing". However, eventually one ends up modifying these conjectures and coming to correct theorems such as "F is always continuous", or "$f \geq 0 \Rightarrow F$ increasing." (For further elaboration, see Grenier, Legrand and Richard 1985, and Legrand, 1986, 1991.)

6. DIFFICULTIES AND STUDENTS' REACTIONS

Setting up a scientific debate in the classroom is a slow process which requires a lot of investment of time and effort from the professor, as well as an agreement by the students and the professor to take risks. Questions such as: "Will something interesting happen today?", "Are we going to cover the syllabus?", "Are we going to learn what we need to succeed later?" are always looming in the backgroumd..

If this didactic practice is not an immediate success for everyone, it has the capacity of transforming most students' understanding of science. It also enables them to imagine the concept of truth and proof. Included below are certain observations by students after three months of scientific debate which demonstrate the nature of the changes they underwent.

Students in the first year of University:

- This year I found it possible to think in mathematics!
- In October, maths was a simple tool, holding no interest for me. I have discovered that it is possible to take some pleasure from studying maths.
- I never thought of mathematics as playing a role in reality; I viewed mathematics as a science entirely separated from reality. Scientific debate has changed my point of view.

- I tell myself: "I want to speak," because if you want to speak, you have to be clear in your mind.
- I surprised myself in taking pleasure!
- I used to consider proofs as magician's tricks; scientific debate shows that one can produce that sort of reasoning if one thinks about it with method.
- I used to think that scientific debate was wasted time during which we learned too little, but I realize that we have learned much.
- What most annoyed me at first has now become most interesting: don't think of mathematics as a completed science, don't expect to learn miraculous formulae or to perform impressive calculus.

Students in the last year before they become teachers:

- What matters and is absent from the traditional mathematics lesson is creativity, the possibility for everyone to be active, and not merely a passive spectator.
- We do not cover so much material but ... in this way we can go deeper, we can question ourselves because it is legitimate. In a conventional class, if the best questions seem to arise out of the mathematics course, it is because we don't know if our own questions are good or absurd and the answers obvious or not. The act of formulating questions is a good way to reveal that you don't understand!
- In the scientific debate there is no reason to hide what you have not understood because you are not ashamed when you make mistakes.
- Scientific debate is both simple and very complicated. Simple, because what could be more natural than to engage in speculation which moves us to uncover new objects better adapted to our questions ...; it is the way of research, however, a complex daily practice! What seems to me fundamental is this: at the base of democracy lie three principles, liberty, equality, fraternity. The most forgotten is the third. Here, we succeed in remembering it.

7. WHAT IS ESSENTIAL TO THE SCIENTIFIC DEBATE?

Our goal is not to introduce the mystic that "you are able to discover alone all theorems in mathematics". Our goal is to concretely show students the need to act mathematically in order to understand mathematics: "your questions, your solutions, your arguments, bad or true, are necessary to build sense of abstract theories, are useful to transcend natural naïvety, to build a bridge between theory and practice, between mathematics and other sciences, between concepts and algorithms ..."

The act of formulating a conjecture compels its author to engage scientifically: "if I want to assert this or that, is it reality? Do I deeply think this or that is true?" The act of collectively resolving conjectures shows us how useful it is to be able to work with contradiction, to respect those who don't think as we do. If the professor is really neutral, every student can be led to believe that it is useful to have, to dare, and to defend ideas because different points of view can help everyone to better understand what is not obvious.

What matters for students is not to show what they know, and hide what they don't, but to do what helps everyone to understand better. What matters for professors is not to show that they know everything, but to show how they know, how they discover their own mistakes, how they succeed in working with their imagination and with the imagination of others. The unavoidable teacher's authority is now based on another system of values, and some parts of the didactic contract must be made explicit: scientific debate is possible if everybody respects the professor because he or she is competent, but if also his or her magisterial knowledge doesn't make students shy. The didactic contract is accepted when differences between students' knowledge do not create situations of domination or defensiveness.

Finally, scientific debate requires a collaborative spirit in order to enable one to better understand and explain. It needs and induces another way of looking at knowledge. Instead of considering knowledge as being exterior to a person (knowledge is usually what is taught, what is doubtless true, and what has to be known!), one can, on the contrary, see knowledge as internal to the knower. It then becomes a tool to transform oneself by thinking.

REFERENCES

Artigue M. (1994). Didactical engineering as a framework for the conception of teaching products. In R. Bielher, R.W.Scholz, R. Straesser, B.Winkelmann (Eds.), *Didactics of Mathematics as a Scientific Discipline*, pp. 27-39. Dordrecht: Kluwer Academic Publishers.

Artigue, M. (2001). What can we learn from educational research at the university level? , this volume, pp.207-220.

Bachelard G. (1938). *La Formation de l'Esprit Scientifique*. Paris : Librairie J. Vrin.

Brousseau G. (1997). *The Theory of Didactic Situations*. Dordrecht: Kluwer Academic Publishers.

Grenier, D., Legrand, M and Richard, F. (1985). *Une séquence d'enseignement sur l'intégrale en DEUG A première année*. Cahiers de Didactique des Mathématiques, n° 22. Paris: IREM Paris VII.

Lakatos, I (1976). *Proofs and Refutations, The Logic Of Mathematical Discovery*. Cambridge: Cambridge University Press.

Legrand, M. (1988). Genèse et étude sommaire d'une situation codidactique: le débat scientifique en situation d'enseignement. In C. Laborde (Ed.), *Actes du Premier Colloque Franco-Allemand de Didactique des Mathématiques et de l'Informatique*, pp. 53-66. Grenoble: La Pensée Sauvage.

Legrand, M. (1991). Les compétences scientifiques des étudiants sont-elles indépendantes de la façon dont nous leur présentons la science? *Gazette des Mathématiciens*, Supplément n° 48, 44-58.

Legrand, M. (1993). Débat scientifique en cours de mathématiques. *Repères IREM n°10*, 123-158.

Marc Legrand,
IREM de Grenoble, France
marc.legrand@ujf-grenoble.fr

KENNETH C. MILLETT

MAKING LARGE LECTURES EFFECTIVE:
AN EFFORT TO INCREASE STUDENT SUCCESS

1. INTRODUCTION

As is the case with many college and university professors, I regularly 'teach' groups of 100 or more students gathered together several times each week in a lecture hall. More precisely, I give lectures on the relevant material, supervise the work of graduate teaching assistants, and assign course grades intended to indicate the level of achievement of those enrolled in the class. These are to be accomplished in a manner satisfactory to the students, at least as measured by their evaluations of the course or reflected by an absence of complaints to the chair or dean. Without any concrete course goals or departmental standard of achievement, I am the sole arbiter of the grading standard applied in the course. The identification of legitimate and concrete measures of effectiveness with which to evaluate the course is left to me. Since the course 'product' is, arguably, the mathematical achievement of the students, I believe that any measure of success must be principally constructed of measures of student performance. This, then, is where I start in trying to figure out how to teach any course. It is this process that I wish to discuss in this report on my continuing effort to increase the effectiveness of my teaching in large classes, (see Millett, 1996, 1997 and 1999).

While the following maxim may be debated by many college and university faculty members, the foundation of my teaching efforts is the statement: *"If students haven't learned, you've not taught."* Beginning with an assessment of the context, I try to establish statements of the course goals that are understandable and with which my students and I can determine whether or not they have been achieved. Depending upon these goals and the context, instructional strategies are selected. Of course, these are often modified during the instruction to take experience into account. For this purpose, lots of information is collected throughout the course. The success or failure of the implementation is determined by this data as are the final grades. An analysis of this data and the experience with the students provides the basis of modifications of the course goals and strategies in the future.

In this note, I describe the experience with a 'pre-calculus' class during the Winter Quarter period, January through March, 2000, following the approach described in the previous paragraph. Although I had not taught this course for quite a while, I am scheduled to teach it again next year. This course provides the occasion for a review of some strategies applied to large class situations, on one hand, and will allow me to describe how one can try to systematically improve the effectiveness of such courses. First I will describe the context and the goals set for

137

Derek Holton (Ed.), The Teaching and Learning of Mathematics at University Level: An ICMI Study, 137—152.
© 2001 *Kluwer Academic Publishers. Printed in the Netherlands.*

employed in the effort to achieve the goals. Finally, I will discuss the outcome of this effort and what this suggests to me about changes to make when I next teach the course in January 2001.

2. CONTEXT

During the Winter Quarter, 2000, at the University of California, Santa Barbara (UCSB), I was the instructor for Mathematics 15, Precalculus. The official objective of the course is to prepare students for either a social and life science calculus course or a physical science and engineering calculus course. Nevertheless successful completion of the course, with a grade of C or better, is not sufficient to enter the physical science course. Nor is it required for entrance to the social and life science calculus course. For the former, a gateway examination must be passed with a minimum score. Taking the exam is required for the social and life science course but no minimum score is required as there is substantial emphasis on pre-calculus material during the first quarter of that course. In a new policy decision, future enrolment in Precalculus will require a minimum score on the gateway exam. To further complicate a student's enrolment decision, the departmental policy recommends pre-calculus enrolment for many students who have exceeded the standard required for enrolment in the physical science and engineering calculus course. The students enrolling in Precalculus represent a wide range of UCSB degree programs, the largest category being 'undeclared.' Since many UCSB students explore but do not select majors prior to their third year of university, they are commonly thought to enrol in the course in order to provide a calculus option if required for their major or to prepare for other quantitatively oriented courses such as statistics that are required in their majors. This view is consistent with discussions with students enrolled in the Winter Quarter 2000 course.

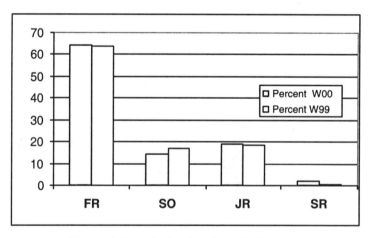

Figure 1. Precalculus Enrolment by Year

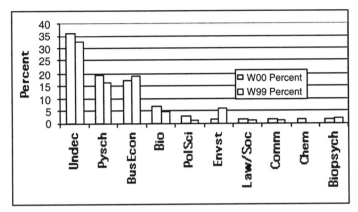

Figure 2. Precalculus Enrolment by Major

As there are substantial differences between the student profile depending on the quarter, I will focus this report on Winter Quarter data. For example, the Winter Quarter 2000 course consisted in 64% first year students, in contrast to a typical Fall Quarter enrolment of more than 80%, c.f. Figure 1. Approximately one-third of the students has not declared a major. The three largest majors at UCSB, Biological Science, Business Economics and, Psychology, are the dominant majors in the class, c.f. Figure 2. While a significant number, about one-third, of these students report having a positive attitudes about mathematics, this is not the case for most of them. On the average, Math 15 students are ambivalent about mathematics as reflected in

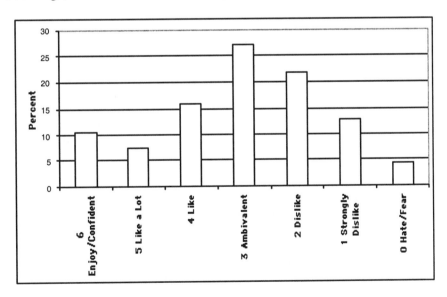

Figure 3. Students Attitudes about Mathematics

an average of 3.01 on a six point scale, c.f. Figure 3. Despite the title 'Precalculus,' a study of the Fall 1996 cohort showed that about 31% of the students took no further mathematics classes, 12% repeated the course, 27% enrolled in the social and life science calculus course and, 28% enrolled in the physical science and engineering calculus course. Students may repeat a course if they have obtained a grade of C- or lower. The final grade distribution suggests that, despite the elementary level of the course, few students attain a mastery of the material in the view of their instructors and many fail to attain a grade of at least a C, c.f. Figure 4.

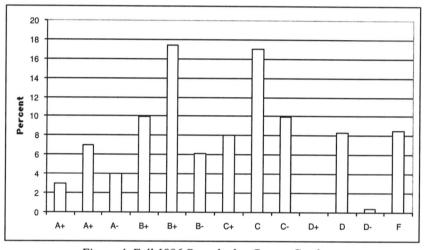

Figure 4. Fall 1996 Precalculus Course Grades

3. COURSE GOALS FOR WINTER 2000

The overarching goal for Winter 2000 was to increase the overall level of student performance as measured by the course grades and performance on the calculus gateway examination. Since this is the 'terminal mathematics course' for about a third of the enrolled students, the secondary course goals were broader than simply preparing students for the physical sciences calculus course. While the official course description is "A functional approach integrating algebra and trigonometry. Topics include: one-to-one and onto functions; inverse functions; properties and graphs of polynomial, rational, exponential, and logarithmic functions; properties and graphs of inverse trigonometric identities; and trigonometric equations", the secondary goals were selected with the broader needs of students in mind. The first of these goals was to expand the range of student mathematical work or activities to include reading the text (many students appear to rarely or unproductively read mathematics texts) and to devote more productive time and effort to working on learning and doing mathematics, either alone or with other students. The second goal was to expand the students' understanding of mathematics from a collection of facts and procedures to be memorized and applied in familiar contexts, to an understanding of mathematics in which thinking played the primary role. One specific manifestation of this goal would be a productive reaction to opportunities to apply concepts and methods in new settings. The third goal was to improve the students' attitudes about mathematics and their ability to 'do mathematics.'

4. WINTER 2000 STRATEGIES

An excellent basic resource on teaching at the college and university level is Steven Krantz's book, How to Teach Mathematics (Krantz, 1999). In this book, Krantz provides solid practical advice about the art of talking to a large group of students in a manner that will not alienate too many of them and, perhaps, even make it a lively experience for at least some of them. But, he says "There are virtually no data to support the contention that small classes are 'better' than large classes for mathematics teaching. Statistics do not indicate that students in small classes perform better or retain more. The statistics *do* indicate that students feel better about themselves and the class, and enjoy the situation much more if the class is small. Another way to say this is that large lectures are not a good tool for *engaging students in the learning process*" One recommendation, of a practical nature, is to have every detail prepared in advance so that nothing disrupts the smooth flow of the class meeting or examination. In short, minimize the unexpected, at least from the perspective of the student. Krantz asks, "What is it that makes teaching a large class seem to be difficult? Is it the size of the room, the populations of students, the use of a microphone, the feeling that there is less room for error, or what? What can we as math teachers - not preachers or performers - do to make a large class still seem like a family? What can we do to make a large class still feel like a place where learning is taking place?"

Among those contributing appendices to the book are proponents or defenders of large classes, those who see themselves as 'sages-on-the-stage.' Frequently advocates for the model of 'direct instruction' argue that the true cause of a students' failure lies with the student and not with the teaching. For example, George Andrews (1999), states that "students are not studying enough" and outlines several factors the he believes contributes to this situation: a lack of concern for this in reform efforts, a lack of concern for the amount of time devoted to homework, large lectures, computerized student evaluations, limited importance of teaching in promotion, and a lack of support for 'high standards.' "Too plentiful student success" represents, he says, a failure to "maintain standards." In another appendix, Richard Askey (1999), discusses problems associated with the use of graduate student assistants, their role in the instruction. He also affirms the necessity of 'maintaining standards' but offers no advice on ways to insure that effective teaching occurs. H. Wu (1999), another avowed 'sage on the stage,' states in his appendix that "There are situations where lectures are very effective, and in fact, there are even circumstances that make this method of instruction mandatory." How Wu proposes to measure effectiveness and against what standard it is measured is left unstated. His focus is on the role of the teacher in the implementation of a 'contract' with the student. "Lecturing is one way to implement this contract. It is an efficient way for the professor to dictate the pace and to convey his vision to the students, on the condition that students would do their share of groping and staggering toward the goal on their own." While acknowledging that "Lecturing can take many forms," Wu provides no alternative to the 'traditional lecture.' One benefit of the lecture approach, he tells us, is that instructors can share ideas and insights not available in the text. "I see no reason why students cannot profit from such insight by making an effort to understand the lectures instead of adamantly resisting them by coming to class unprepared, or for that matter, leaving it without making an effort to understand it later."

An assumption underlying the praise of the traditional lecture is that the cause of any 'failure' lies principally with the student. The sole responsibility of the teacher is the 'transmission of information.' I do not agree. Too often the nostalgic paeans for the 'good old days' ignore the widespread failure of large numbers of students to become mathematically proficient. They wish for a time when professors were treated like divinity, were not expected to be sensitive to insulting, offending or, otherwise alienating students, especially women or members of minority groups; when 'good lectures' meant only inspiring or entertaining repetitions of the material found in the text. Some, typically the textbook authors themselves, would go so far as to recite the 'jokes' that were included in the texts. Another common feature is the attitude that only the very few, persons such as themselves, were capable or worthy of success common among college and university faculty members. And, indeed, this opinion of students is still found across the entire K-16 educational landscape. Such thinking is not, I believe, compatible with making student achievement the principal measure of success of the course. Therefore, a more careful analysis of the large class setting is required, whether or not, a traditional lecture is the instructional method that is ultimately selected.

With few exceptions, the students come to Math 15 with a history of limited mathematical success, with little appreciation of their own ability to learn or do mathematics, without successful strategies for leaning mathematics and, with 'passing the course' as their principal goal. Few recognize the need to change their approach to learning, even with a record of previous failure. Nor do they seek an understanding of mathematics sufficient to allow them to apply its concepts and methods in new settings. In order to place increased focus on the student, I decided to assign homework and grade on a weekly basis. They are passive listeners. While, perhaps, physically present during class meetings, they come unprepared to learn or unprepared to participate actively in the class. As a consequence, the structure of the class and its meetings needed to address these facts in order to make learning feasible. In order to place greater focus on student performance, I decided to assign and grade homework problems on a weekly basis. Graded by an undergraduate assistant, the homework score was combined with scores from 'surprise quizzes' to give 25% of the comprehensive course grade. There were two course exams (25%) and a comprehensive final exam (50%) that also formed portions of the comprehensive course grade. Students who achieved a 'physical calculus qualifying grade' on the gateway exam before the end of the quarter received a bonus award of 15 points (out of a scaled score of 100) added to the comprehensive score. Finally, students were told prior to the comprehensive final examination that their course grade would be no lower than the grade received on that final examination. The later option was instituted in order to encourage sustained effort for those who may have faltered on some earlier aspect of the course.

The text for the course was Functions Modeling Change: A Preparation for Calculus by Connally, Hughes-Hallett, Gleason, et. al., John Wiley and Sons, Inc., 2000. It was selected in order to promote a larger understanding of mathematics, its concepts, its methods and, its application both to calculus and more widely. During the ten-week quarter , reading assignments and problems were taken from the first nine chapters, with topics on rational functions only briefly discussed during the last week of the term. Some supplementary material was added, mostly of a historical nature. Students were encouraged, but not required, to use graphing calculators in all aspects of the course. While some of the assigned homework problems required the use of a calculator, the course examinations were constructed so as to be calculator neutral. Furthermore, some questions explicitly forbid the use of calculators.

Students attended 150 minutes of 'lecture' and 50 minutes of mandatory 'discussion section' with a graduate teaching assistant designated by the department. The discussion meetings focused on returning graded homework and answering a few questions on the course material. Some students elected to join the Achievement Program's academic workshops sponsored by the California Alliance for Minority Participation. These workshops met 150 minutes each week. They included specific attention to the needs of individual students such as an informal assessment of their preparation for the course, their study skills or approaches to reading the text as part of the learning process. The workshops were staffed by a graduate student and a highly successful undergraduate. They both attended weekly staff meetings at least partially devoted to increasing the effectiveness of the workshops through diagnosis of individual student progress, promotion of productive mathematics based interactions among students, and in-depth review of returned homework, quizzes, and supplementary work associated with the workshop groups.

In response to prior student experience and attitude, the seventy-five minute 'lecture' did not follow the traditional formula but employed a format that was primarily 'interactive' in style:

- the first segment consisted of a brief administrative reminder of assignment and exam content and dates followed by an invitation for questions from students;
- the second segment was devoted to a discussion of any systematic mathematical problems that had been identified by course assistants or the instructor;
- the third segment was frequently, but not always, launched with a group 'surprise quiz' consisting of a question taken from the reading assigned for preparation for the day's lecture. These 'quizzes' produced written responses from groups of students, typically two or three. These were given to the instructor following a plenary discussion and student presentations of various alternative approaches to the question, including any underlying concepts and strategies or algorithms employed in the response. This discussion was at the 20

minute mark (of the 75 minute lecture period) and represented a bit of a 'change of pace' marking the move from old to new ideas or methods;

- the fourth and longest segment concerned new concepts and/or methods. They grew out of the problem or the surprise quiz and were developed in a manner based on the subsequent discussion. No more than one or two new key elements formed the focus of the work;
- time permitting, any secondary new material was incorporated in the presentation and discussion of a closing question/problem situation;
- in closing, students were reminded of homework, quizzes, material to be reviewed, etc.

Beginning with the first meeting and throughout the ten weeks of the quarter, I recognized an increasing number of students by name during the lecture. In order to effectively shrink the apparent size of the class, an effort was made to identify students sitting in all areas of the lecture hall, specifically those sitting in the very last rows. As students typically are very consistent in when and with whom they sit, a mental map can be developed and used quite reliably. Those students who ask questions are requested to provide their names, which are then used in the course of responding to their question. Each student is thanked for the question and is provided with explicit appreciation of the usefulness of the question in addressing course goals. For example, in an attempt to begin encouraging students to interact mathematically with their classmates, I asked students to find a partner to whom they were to describe a 'real life' example of a function that they had encountered within the last day. After about four or five minutes of conversation, I selected a student from the middle of the class and asked them their name and major. Using their name, I asked them to introduce me to their partner. Most often they do not know or recall the name and so I ask everyone to take another few minutes to introduce themselves to the persons seated on each side of them. Once that is done, using their name, I ask the first person selected to describe the example of their partner's function. Rarely can they do this. If this is the case, I ask them to repeat the conversation with their partner and to be prepared to not only describe the example but explain why it is an example. Finally I ask about five students to describe their partner's example and to compare it with their own. This exercise requires a fair bit of time but is used to lead into an explanation of what I am seeking from them by way of interaction with me during the class meeting. Specifically, I will be often telling them about some mathematics or answering a question posed by another student. I expect that they should be able to explain what I have said before we agree to move on to another matter. If not, they are responsible for asking questions or for ensuring that their questions are answered at another time. Failure to follow through on this is a significant contributing obstacle to student learning and requires reminders regularly through out the course.

Another example of my effort to shrink the apparent class size occurs during the 'surprise quizzes.' I regularly walked about the lecture hall and engaged groups of students, by name, in discussions about the question. I am seeking information as to their preparation of the material and to identify aspects of the material about which there may be questions shared by large numbers of students. This information was used to evaluate and to revise the lecture material for the day. As is often the case, students have not prepared for the meeting by reading the assigned material. I remind them that not doing so impacts their course grade through the quiz grade and the reduced usefulness of our time together. Attending class means coming prepared to do the work of the day!

Three 'office hours' were scheduled each week by my graduate assistant and myself. Few students actually made use of this opportunity to discuss the course work. Most such questions arose during the lecture or discussion session or immediately following them. A substantial amount of interaction occurred via email, most of which could be characterized as non-mathematical: classes missed due to personal problems: requests to submit 'late' homework, requests for 'make-up-exams', or complaints about the effectiveness of the graduate assistants and the accuracy of the grading on homework and exams. As a matter of policy, no 'late' homework was graded and 'make-up-exams' were not given. I did agree to make a notation in my grade record that homework was done by students even though it was not graded. This was an effort to keep them productively engaged in the class and doing the homework, even if late. These notations did not, ultimately, have any influence on the course grades as the involved students had grades that were quite clearly within each final grade cohort.

All individual pieces of course work were evaluated on a six point scale with six being assigned to technically correct and fully explained mathematical work and zero assigned to work that provided no contribution to progress on the associated problem or question. Five corresponded to minor technical errors or slight weaknesses in explanations for which no further mathematical discussion would be necessary in order that a student revision be fully correct. A four was associated to work with somewhat more important errors, mathematical gaps or missing elements of an explanation for which only a modest discussion or reminder of some aspect of the work should be adequate to support a successful revision. A three corresponded to a submission for which some important aspect was absent or inadequate while others were present. In the case of a three, some additional mathematical learning might be required before a complete solution could be provided. Two indicated that some productive work and justification was provided, but that much was missing and further instruction was clearly needed. A score of one indicated that there was some potentially productive work provided but only a very modest amount with principal elements missing. This grading scheme caused some confusion for the graduate assistants, the undergraduate grader, and the students enrolled in the course. As a consequence, the evaluation of student work required a much more careful reading and, as a consequence, the reading required more training and supervision by the instructor to ensure an acceptable quality of evaluation. While students seemed to welcome the 'partial credit', they were quite unaccustomed with the requirement that each answer be supported by an explanation in order to receive full credit. Technically correct calculations and numerical answers with absent explanations would receive a maximum score of three.

Finally, much more and different information about the student experience and performance was collected than is the custom. I did this in order to begin a systematic effort to improve the effectiveness of the course. This report represents one dimension of that effort.

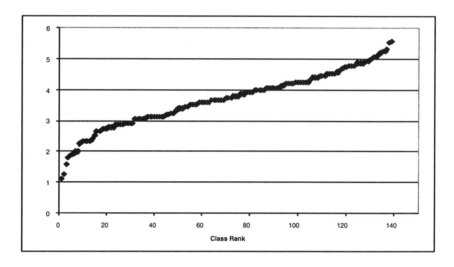

Figure 5. Examination I Grades by Class Rank

The distributions of scores on homework, course exams and, comprehensive exams are shown in Figures 5 - 9. One dimension of overall student performance is represented by average scores of 3.690 on Exam I, Figure 5, 2.205 on Exam II, 3.646, Figure 6, on Homework/Quiz, Figure 7, 2.759 on the Final Exam, Figure 8,

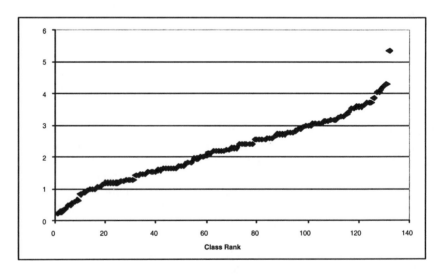

Figure 6. Examination II Grades by Class Rank

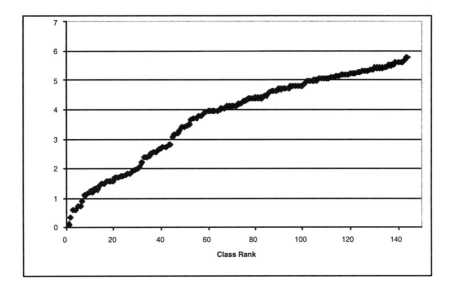

Figure 7. Homework/Quiz Grade by Class Rank

and 2.744 on the Course Composite grade, Figure 9. These data suggest that the learning of major mathematical components of the course was not demonstrated by the performance of students on the final exam. Indeed, the second examination

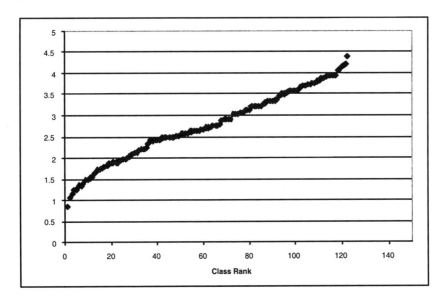

Figure 8. Final Exam Grades by Class Rank

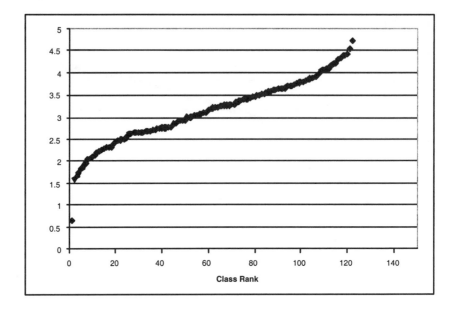

Figure 9. Composite Grade by Class Rank

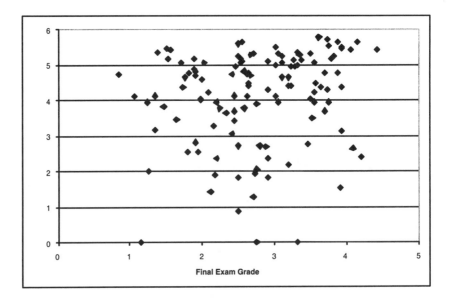

Figure 10. Homework Grade versus Final Examination Grade

performance, covering more complex material than that found on the first exam, foreshadows this fact. The average final course grade was 2.593, on a four point scale, compared to the 2.45 recorded in Winter Quarter 1999. In the absence of concrete achievement standards for the course, it is not possible to determine the extent to which the increased average grade is evidence of actual increased student achievement or reflects different expectations among the instructors.

5. ANALYSIS

One of the important questions that can be asked about the meaning of this data concerns the value of collecting and grading homework. For me, the purpose of graded homework is to give value to work designed to promote student learning of the concepts and methods. Visually and numerically, as represented by a covariance of 0.175 and an R-Square of 0.0263, Figure 10, there is strong evidence that the homework did not make a significant contribution to student performance on the comprehensive final examination. One could conclude, therefore, that the homework component did not make a measurable contribution to student learning and, thus, to the achievement of the course goals.

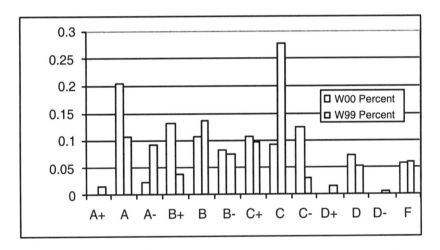

Figure 11. Final Grade Comparison between Winter 1999 and Winter 2000

Another innovation was the offer of 'bonus' credit for passing the calculus gateway examination. Historically, 28% of the students taking Precalculus enroll in the physical science calculus course. In Winter 2000, 32.7% of the Precalculus students achieved the admission standard on the gateway examination necessary for enrolment in this course. Of these, 22 received A's, 11 B+'s, 4 B-'s and 4 C+'s, Figure 11. Of these 41 students, roughly one-third of the final course enrolment, seven students, or 5.7% of the class, received an improved grade as a consequence of the bonus. One interpretation is that the effort dedicated to passing the gateway examination also contributed to student success on the comprehensive examination despite the multiple choice structure of the later and the requirement for explanations in the former.

Several consistent themes arose during the course of interactions with students enrolled in the course. They are (1) a desire that the course be graded 'on a curve'; (2) the opinion that asking for solutions to 'unfamiliar' questions was not appropriate; (3) a confusion between the memorization of formulae and methods and 'understanding' them; and (4) the belief that course grades should be based on effort rather than performance. In interviews with students the goal of working problems and studying for the course in order to make it possible to solve problems without 'having to think' arose frequently. Mathematical accomplishment, it appeared, was not a question of being able to use the concepts and methods to figure out how to solve a problem but, rather, a question of being able to recognize the problem type and apply the solution procedure appropriate to the question. In looking at student work on examinations, it seems likely that a careful reading of the problem statement and the question is not occurring. Frequently, students would state that they understood how to solve the problem but were unable to explain either the problem setting, the question, or their proposed solution method. Indeed, when asked about this, students frequently apologized for their mathematical ineptitude by claiming they were not 'mathematical' persons or by blaming their prior mathematical courses. In the latter case these courses ranged from those of a hierarchically structured curriculum with a focus on algorithmic practice to those that integrated areas of mathematics and its applications. In common was the statement that "I never understood what I was doing."

6. CONCLUSIONS CONCERNING NEXT STEPS

Despite an intensive effort to actively engage all students in productive work on the curriculum, many of the students remained on the sidelines unconcerned gaining the expected understanding of the mathematics. Preparation for class meeting was inadequate and, even though physically present, class participation in the work was uneven. One possible strategy to address this problem would be to increase the contribution of the 'surprise quizzes' to the course grade. Greater emphasis could be placed on the individual responsibility to read the material in advance. Some instructors, not in mathematics, have made this preparation a requirement of class attendance. While this seems a bit extreme, it does make clear the student responsibility to participate actively in the course. I will certainly devote more time to addressing the lack of understanding of what learning mathematics will mean in the course and to suggesting strategies that will help students learn this mathematics successfully.

In particular, I will try to more clearly describe the purpose of collecting and grading homework in relationship to the overall course goal. In short, the purpose of the homework is not to "do the homework, but to understand the math involved." I will also arrange meetings with the graduate students and the various campus agencies that provide tutoring to help them understand ways in which they can better help students in my course be more successful. Another dimension of this is the need to have a better understanding of the character of the standards used in the assessment of student work. As I had done for calculus courses earlier, I will develop a small collection of problems and a variety of 'solutions' together with evaluations and explanations of these evaluations. These will help students, tutors, graders, and graduate students have a more concrete understanding of the mathematical performance sought in the course. Finally, as I have in other classes, I will replace office hours by weekly hours in a 'mathematics laboratory' during which time each student will be required to personally give me their homework. I will scan a selected problem to assess the quality of the work and give the student a quick assessment of the acceptability of their submission. In the event that it is below the standard, I will point out difficulties and invite them to immediately revise their submission. This strategy has worked well in calculus classes in the past by leading to greater individual responsibility for the quality of the work and, consequently, higher levels of overall class success.

Working on knowing individual students did seem to promote more open interaction and questions. It is uncertain if this had the desired 'halo effect' of shrinking the apparent class size. Extending the interaction to regularly include each student, may help address this aspect. I will keep closer records of interaction and of student performance next year in order to gain a clearer understanding of the impact of this feature of the course.

The successful teaching of Precalculus is, I believe, a very challenging task if one is willing to accept the notion that success is measured by the mathematics learned by the students. I have tried to illustrate, in the description of several aspects of my effort, how it is possible to teach large classes at a college or university level in a manner that is student centred. I have not gone into my analysis of how students fared on more specific mathematical tasks, such as the use of inverse functions in the context of trigonometric and exponential/logarithmic functions, which involves an item analysis of problems on examinations. Rather, I have tried to give a larger grain overview of key strands of how I approach this task so as to be able to systematically improve the effectiveness of the class. In doing so, I encourage you to join in the effort to move the spotlight from the 'sage-on-the-stage' to the 'students-in-the-class.' This is where our successes and our failures to teach are really determined.

REFERENCES

Andrews, George E. (1999). The Irrelevance of Calculus Reform; Ruminations of a Sage-on-the-Stage. An appendix to Steven G. Krantz, *How to Teach Mathematics, Second Edition* , pp. 157-159. Providence, RI: American Mathematical Society.

Askey, Richard (1999). Mathematical Content. An appendix to Steven G. Krantz, *How to Teach Mathematics, Second Edition* , pp. 161-171. Providence, RI: American Mathematical Society.

Krantz, Steven G. (1999). *How to Teach Mathematics, Second Edition.* Providence, RI: American Mathematical Society.

Millett, Kenneth C. (1996). Teaching and Making it Count. In Yvonne M. Pothier (Ed.), *Proceedings of 1995 Annual Meeting, Canadian Mathematics Education Study Group*, pp. 23-42. Halifax, Nova Scotia: Mount Saint Vincent University Press.

Millett, Kenneth C. (1997). Talking is Not Teaching, Listening is Not Learning, *The California Alliance for Minority Participation in Science, Engineering and Mathematics Quarterly*, Vol. 6, No. 1, 17-19.

Millett, Kenneth C. (1999). Teaching or Appearing to Teach: What's the Difference? An appendix to Steven G. Krantz, *How to Teach Mathematics, Second Edition* , pp. 241-252. Providence, RI: American Mathematical Society.

Wu, H. (1999). The Joy of Lecturing-with a Critique of the Romantic Tradition of Educational Writing. An appendix to Steven G. Krantz, *How to Teach Mathematics, Second Edition* , pp. 261-271. Providence, RI: American Mathematical Society.

Kenneth C. Millett
University of California, Santa Barbara, USA
millett@math.ucsb.edu

MOGENS NISS[1]

UNIVERSITY MATHEMATICS BASED ON PROBLEM-ORIENTED STUDENT PROJECTS: 25 YEARS OF EXPERIENCE WITH THE ROSKILDE MODEL

1. INTRODUCTION

Since its inception, and inauguration in 1972, as a small university designed to innovate higher education, Roskilde University (RUC), Denmark, has based all its programmes on problem-oriented project work performed by students. This is especially true for mathematics and the sciences. In these programmes, half of the students' time is devoted to project work, whereas the other half is devoted to systematic courses taught by university teachers, in more or (sometimes) less traditional ways. By a ***problem-oriented project*** in the RUC sense, henceforth simply referred to as a ***project***, we understand a study activity in which a small group of students, typically 2-8, gather in accordance with their interests in order to investigate a problem, or a complex of problems, of their own choice, i.e. identified, specified, and studied by themselves, under the supervision of a faculty member. So what we are dealing with is a combination of two components which in principle are independent: a certain form of study, i.e. project work (which could be put to use towards any sort of content, including systematically organised subject matter), and a perspective on content, i.e. problem-orientation (which could be pursued by any kind of study activity, including one-way lectures). The normal duration of an RUC project is half the time of a semester (roughly four working months), but sometimes a project takes two semesters to complete. The project students themselves do all the work. They discuss and formulate the problems they want to deal with, find and read relevant literature, conduct empirical or theoretical analyses, do experiments or make calculations, interview resource persons, visit institutions or companies, and so forth, but of course under guidance of their project supervisor, with whom they meet on average once per 1-2 weeks. In an idealised description, a problem-oriented project typically contains the following phases:

- identification and specification of the problem field to be investigated, leading to a formulation of the final problem to which an answer is going to be sought;

[1] The author was a member of the founding staff of Roskilde University and has remained their ever since, today as a professor of mathematics and mathematics education, thus being (co-)responsible for a non-negligible part of the programmes and their underlying ideas.

Derek Holton (Ed.), The Teaching and Learning of Mathematics at University Level: An ICMI Study, 153—165.
© 2001 *Kluwer Academic Publishers. Printed in the Netherlands.*

- the finding, selecting and reading of potentially relevant literature;
- the establishment of the theoretical core of the project and the acquisition of the knowledge necessary to deal with it – the latter usually involves the assistance of the supervisor;
- the working with external contacts, if relevant;
- the carrying out of analyses of the specified problem;
- the synthesis and analysis of the findings of the project problem;
- the production the project product, which will always include a written report but may also contain the design of an exhibition, a text book, a video programme, etc.

As stated, one visible outcome of a project is a written report, of 50-100 pages, to which all students in the group have contributed. As well as the report, the project work itself is assessed at the end of term (in various ways, according to the circumstances), by one or more independent assessors from inside or outside the university. The supervisor is always involved in this assessment.

In principle the students themselves decide when a project is finished, and they are responsible for all the consequences that follow from this decision. In practice, however, they consult their supervisor to obtain his or her opinion on the degree of completion of the project. If, for some reason, a project is not finished (which happens very seldom), or the students do not pass, they will have to improve the project, or to undertake another one, and have the new outcome tried before assessors. So, it is not possible for a student to omit or miss out a project that is required in the programme.

2. POSITION AND STRUCTURE OF THE MATHEMATICS PROGRAMME

We shall begin by outlining the structure and organisation of the mathematics programme and its embedding in the general study structure of the university.

Firstly, the normal road for a mathematics student at RUC (the deviations from 'normality' are of a technical nature and are omitted for the sake of simplicity and clarity) is to enrol in the university's two-year *basic programme in the sciences*. During this programme each student will choose and take 8 'ordinary' courses in the sciences and mathematics, the choices being subject to certain constraints. The mathematics courses offered in the programme are calculus, linear algebra, complex numbers and functions, differential equations, statistics, and mathematical modelling in the sciences (as to the latter course, see Ottesen, (this volume pp.335-346). Students with (evolving) mathematical inclinations will typically choose to take two or three of these courses. Moreover, in the basic programme, each student will carry out four projects, each of a duration of one semester. At the end of the two-year programme each project group will defend its (normally) final project at an oral examination conducted by their supervisor, with the involvement of an external examiner from another university, institution, or organisation. This examination will be based on the project report that both examiners will have read in advance.

The two-year basic programme in the sciences does not lead to a degree but is a common platform from which students embark on so-called further studies to obtain a Bachelor's or, much more frequently, a Master's degree. These programmes were instigated in continuation of the basic programme in 1974. Only after having completed the basic programme do students have to choose which subjects to pursue further. It is a remarkable and recurrent empirical fact that the majority of those students who at this stage decide to study mathematics did not have that option in mind when they enrolled at university. A student who wants to go on to study mathematics will have to choose one more subject to be studied on a par with mathematics: physics, chemistry, computer science, biology, geography, history, philosophy, psychology, English, German, or French, etc. As, for historical reasons, almost all students in Denmark want to obtain a Master's degree, we shall confine ourselves to considering these programmes. Here, each of the two subjects has to be studied for nominally 1.5 years, which yields a nominal total of 5 years for a Master's degree. Thus, in Roskilde we don't distinguish between majors and minors, the two subjects have a parallel position. The Master's programme is completed with a thesis to be written in either of the two subjects, or across them if this makes sense and the student so prefers. The Master's thesis takes the form of yet another project, typically carried out by students individually or in a group of two, but usually much more depth is expected than is the case with the other projects. Finally, it is a peculiar feature of the Danish – including the Roskildean – academic tradition, which again has historical roots, that almost no students in classical academic disciplines complete their studies within the nominal five years. Usually they spend 1-2 years more before graduation.

Also in the further mathematics programme, students spend half of their time taking courses in systematically organised subject matter, and half of their time on problem-oriented projects. Without going into judicial details, students are typically required to become familiar with the following topic areas (but not necessarily by taking courses), in continuation of the ones they encountered in the basic programme: linear structures from algebra and analysis, advanced mathematical analysis 1 & 2, discrete mathematics, geometry, probability theory and statistics, basic structures in mathematics, and at least one 'advanced topic' (such as optimisation theory, measure and integration, partial differential equations, functional analysis, fluid dynamics, wavelets, stochastic processes).

Each student has to carry out three projects in mathematics (and three in his/her other subject, giving a total of 10 projects throughout an entire Master's programme). It is important to note that the sorts of problems that students' projects deal with certainly need not be specifically mathematical problems in the classical sense. They can be, of course, but normally the problems dealt with are intellectual problems, of a complex and general nature. More specifically, each student has to take part in one project which focuses on one (or more) *mathematical model(s)* in an area outside of mathematics, either by investigating and examining existing models from mathematical perspectives, e.g. with respect to their properties and behaviour, or – less frequently – by constructing a new/modified model. Naturally, if possible and appropriate, students will tend to study or construct models which are relevant to their other subject, whether, say, physics, biology, chemistry or computer science. Secondly, each student has to perform one project on the nature of *mathematics as a discipline*. Such a project tends to be fairly theoretical. Typical projects in this category deal with aspects of the foundations and architecture (structure) of mathematics, with the history of mathematics, or with philosophical issues related to the position and nature of mathematics. Finally, each student in his/her third project has to choose between three major options: a second *modelling* project, in which an original mathematical model is to be constructed in cooperation with professionals in other disciplines or fields of practice; a project on and within *pure mathematics*; or a project on aspects of *mathematics education*. These options will typically reflect students' career expectations or intentions, thus serving to prepare them for their future professions as users of mathematics, researchers or teachers.

These project types have not been designed at random. On the contrary, they are meant to provide students with *representative experiences*, based on selected cases identified and shaped by the project groups themselves, of the essential facets and properties of mathematics as a pure science, an applied science, an instrument for societal practice, an educational subject, and a field of aesthetics.

Although students have to undertake projects within certain prescribed categories, their freedom in choosing and specifying the problem field they want to investigate is infinite, and the projects actually carried out within each category are in fact very different, as will be illustrated below. Once again, each project group will have to present itself and its work at an oral examination, based on the project report, and conducted by their project supervisor and an external examiner.

3. EXAMPLES OF MATHEMATICS PROJECTS

In order to illustrate the spectrum of possible projects within each category, we give a few examples of themes of projects actually undertaken over the years.

Analysis of mathematical models: Stochastic models in population genetics; Models of gonorrhoea; Mathematical models of the Belousov-Zhabotinskij reaction in chemistry; The models involved in RSA cryptography; The application of the Radon Transform to CT scanning; Models of voting and election methods; An analysis of the World Bank's global population forecast model; An analysis of terrain models; Structure models.

The nature of mathematics as a discipline: The role of the number domains in the edifice of mathematics; The trinity of Bourbaki - the general, the mathematician, the spirit; The place and role of non-standard analysis in mathematics; The mathematical foundation and position of dynamic programming; "Is not the title of this project' is not the title of this project': Self reference in mathematical systems; Mathematical symbols and their roles in mathematics; The significance of pathological examples in the development of mathematics; Euler and Bolzano: Mathematical analysis from the perspective of science philosophy; Internal and external factors behind the genesis and development of the standard methods of statistics; The impact of hyperbolic geometry on the philosophical interpretation of mathematics; Galois' contribution to the development of abstract algebra; The genesis and development of almost periodic functions; Is mathematics a natural science? Cayley's problem of Newton iteration: An historical analysis 1870-1918; Proofs in mathematics.

Building of mathematical models: Statistical methods to determine safe doses of carcinogenic substances; Analysis of multi-spectral satellite pictures; A model for periodic selection in E-coli bacteria; Modelling calcium transportation through cell membranes; Monte Carlo simulation of non-equilibrium Ising models in molecular dynamics; A linear programming approach to maintenance schemes for public bridges; Modelling the cardiovascular system with respect to neural pulse control; Modelling 'viscous fingers'.

Pure mathematics: The ellipsoid method in linear programming; Numerical methods for the solution of rigid differential equations; Contrafactual conditionals in the language HOL; Knots, links, and algebraic invariants; Wavelets, Graph theory and multidimensional contingency tables; Proof theory, exemplified by Gentzen's proof of the consistency of the theory of natural numbers; Self-avoiding random walks.

Mathematics education: The perception of mathematics amongst students in 11th grade; Around the world on flat maps – a textbook unit for the teaching of spherical geometry and map projections at the upper secondary level; Competencies of modelling – developing and testing of a conceptual framework; The justification problem in upper secondary school mathematics; Problem solving and modelling in secondary mathematics education; Rhetoric or reality? Mathematical applications in upper secondary school mathematics in Denmark, 1903-88; 'Model talk' – An upper secondary textbook on differential equation models of dynamical systems; The position of proof – Proof and proving in upper secondary mathematics teaching.

4. AN ILLUSTRATIVE CASE, CLOSE-UP

For an illustration, let me describe in some detail one of the projects in the above list. The project, 'Proofs in Mathematics', was conducted by a group of five students (three female and two male) during the autumn semester of 1999. I have chosen to present this project for a variety of reasons. It is a recent project and one that I supervised myself, so I can report on it on the basis of first-hand experiences. It is a project that, at the final examination, was given a grade slightly above average, so it is not a special project in terms of quality. In fact it is quite typical in many respects. Three of the five students did this project as their very first in the mathematics part of the programme, i.e. after having completed the basic programme, while one student did it as her second project in the mathematics programme. For these four students, the project belonged to the category in which the nature of mathematics as a discipline is investigated. The last student was in a slightly different situation as this was his last project in mathematics, categorised as a so-called 'profession-oriented project' of the variant focusing on mathematics education. This particular student therefore had to append a small additional report of his own (on epistemological obstacles to proof and proving in the teaching and learning of mathematics) to the main report, of which he was a co-author on a par with the other students. However, for simplicity I shall confine myself to presenting only the main project. The fact that this project group consisted of students from different intake cohorts is not unusual.

This group of students responded to a general observation made by me at the introduction seminar for the autumn 1999 semester, that a multitude of project possibilities were unexploited at RUC within the domain of proof and proving. A few projects had been done in the past within this theme, but a host of aspects had been left untouched. Several students then became engaged in discussing possible projects on proof and proving. When the project group was eventually established it comprised five students.

The students spent a fair amount of time discussing which *problématique* they wanted to focus on, while reading general literature on proof and proving at a not too technical level (e.g. Brandt and Nissen: 'QED: An introduction to mathematical proof' (in Danish), Davis and Hersh: 'The Mathematical Experience', Garnier and Taylor: '100% Mathematical Proof', Hanna and Jahnke: 'Proof and proving', Howson: 'Logic with trees'), some of which were pointed out by me, and some of which were discovered by the students themselves. It was difficult for the group to identify the problem they wanted to investigate. First, they thought of focusing on different types of proof but gave this up because, on closer inspection, they found the issue too shallow. Then they thought of concentrating on investigating the difference between proofs that prove and proofs that explain, but that idea was abandoned because it seemed to lead to an educational rather than an epistemological project, which was not in accordance with the students' wishes. During their discussions, the students time and again returned to the issues: How can we know that, and when, a mathematical statement has really been proved? How much can be removed from the presentation of a proof without removing its 'proofness'? To what extent is a proof identical to a written formulation of it? Against that background, students settled on the following simple problem formulation 'What does it take for a mathematical statement to be proved?' Thus, ultimately the project concentrated on proof as a product, not on proving as a process.

From the beginning, students had decided that they didn't want to produce a philosophical treatise of a general nature but to perform a specific analysis based on concrete cases. Moreover, they wanted these cases to be drawn from mathematics that was either part of what they had already learnt or were about to learn, or was not so demanding that they would need to spend large parts of the project time on studying new mathematical topics. In other words, students saw it as essential to be able to concentrate on the analysis of proof without being distracted by too difficult mathematics. (It should be noted that this was a decision made by this particular group of students. It equally frequently happens with project groups – though usually with more experienced ones – that one criterion for the choice of a project theme is its capacity to make students familiar with mathematical topics that are new to them.)

As three of the students had just started in the mathematics programme it was agreed to concentrate on subject matter that pertained to elementary number theory, linear algebra, Weierstrassian real analysis (they were taking courses in both concurrently with this project) and (Euclidean) geometry. It was not an easy task to select the cases to be investigated. In addition to the ones actually included in the project, several other cases were taken into consideration as possible candidates but were discharged for some reason or the other. The resulting cases were the following:

The irrationality of $\sqrt{2}$. Here, students came to realise that there are actually two statements involved in this. One concerning the rationals only, in which context we have an impossibility theorem: There is no rational number whose square is 2. And one concerning the reals: There is a real number whose square is 2, and that number is not rational, which is an existence theorem combined with an impossibility theorem. Many textbooks, including the one they were studying, do not attempt to make a distinction between these two statements.

The Hippokrates moon on a quarter circle chord can be squared. This theorem in classical plane geometry was investigated in the context of squaring problems at large, in particular 'where exactly does a generalisation of Hippokrates' method to the 60 degree chord fail to prove (!) that a circle can be squared?'.

Every real, bounded sequence has a convergent sub-sequence. Two proofs of this theorem were studied, one based on the notion of 'lead elements' of tail sequences, and the classical one based on bisection. In both cases the 'point of appearance' of the completeness axiom of the reals was given special attention.

The dimension theorem on linear transformations (for a linear transformation of a finite-dimensional vector space V, the dimension of V equals the sum of the dimensions of the null space and the image space of the transformation).

The complex expontial function exp satisfies exp(z+w) = exp(z)exp(w). The exponential function is often introduced, as it was to these students, by means of the real exponential, sine and cosine functions. The proof of the complex functional equation is then based on well known properties of these functions, in particular the addition formulae. This example was discussed by studying (consistent) ways to introduce these real functions so as to yield the desired properties, primarily by means of power series.

The binomial formula. Here the point was to compare the nature, foundation, and scope of two proofs, a combinatorial proof and an induction proof.

The bulk of the project work was spent on analysing these cases. This was done by means of a general template that was applied to all the cases. It took a long time for the group to develop this template and specify its components. These are: What type of theorem (existence, uniqueness, impossibility, ascription of properties (including identities))? What type of proof (see below)? In what mathematical universe(s) is the theorem embedded and what conceptual foundation does it invoke? What are the basic ideas in the proof? This template was then used to guide the specific analysis of each case. It turned out that it was not always easy to clearly classify the aspects of the proofs examined into the categories anticipated by the template. The most difficult part was to identify all relevant mathematical universes actually implicated in the proof. However, by and large the template did work rather well. Moreover, as students had not taken any courses in mathematical logic, they wanted to survey the fundamentals of logic so as to make sure that all aspects of logical inference were appropriately taken into account in their analyses. Besides, they wanted this survey to serve as a basis for an initial identification of different sorts of proof (contra-positive, indirect, induction, combinatorial).

The final project report, of 86 pages, was written jointly by the students and in several steps. On average each section had been revised a handful of times before it reached its final stage. One way of achieving both collective ownership to the project report and a fair degree of homogeneity of the text is to rotate the sections between the members of the project group. This procedure, adopted here, took place through a collective editorial process. Firstly, initial drafts were produced for the individual pieces by one or two students. These drafts were then discussed in the whole group and another member was asked to take care of the revision on the basis of the comments given. Such cycles were repeated several times until no more suggestions for improvements were made.

Now what role did the supervisor, i.e. myself, have in the entire project process? A role similar to the one I usually have with projects. In this case, the initial inspiration for the project was a so-called 'project germ' that I put forward to students at the beginning of the semester. A project germ is nothing else than an indication of a potential source and generator of project issues and themes, which a group of students can develop into a project problem in infinitely many different ways. For a project germ to result in a project it has to be cultivated, developed, tailored, specified, etc. At the beginning of each semester not only the supervisors but also the students, on a par with the faculty, propose possible project germs. The project germs put forward, constitute a 'project germ market' which is used as the base for generating new projects for that particular semester.

Once the project group was formed from a common interest in the germ, my main task was to assist students in identifying the problem they wanted to investigate. This task is of a Socratic nature, because it's a matter of helping students to examine their own minds. I usually ask them 'What insight do you want to have gained when this project is finished; what questions that challenge you today do you wish to be able to answer at the end?' The importance and the difficulty of arriving at a problem that is both tractable and corresponds to all project members' real wishes shouldn't be underestimated. If this process runs sloppily it is rather likely that serious problems will emerge at a later stage. In the case of this project the process was rather lengthy, in fact to the point of frustration with some of the group members. Nevertheless, the problem to be investigated was eventually found and formulated.

In addition to insisting that students reflect on what they want to do and why, it was my task to come up with initial suggestions of the literature that might serve as inspiration, give rise to ideas, and provide information that could prove useful. This literature was supplemented with literature found by the students themselves in libraries, on the Internet, etc.

When it came to selecting the cases, my role was to anticipate the consequences of choosing the candidate cases and to give advice as to their eventual feasibility *vis-à-vis* the outcomes students wanted their project to have. As always, the final selection of cases was completely the responsibility of the students.

During the remainder of the project, I met with students to listen to what they had done, to help them understand difficult points in the literature they were using, to help them overcome obstacles in the analysis of cases, to point out traps and perspectives that they hadn't seen themselves, etc. When drafts of project report sections began to appear I commented on them in general terms, indicating weak or missing points, more or less like a referee for a research journal.

When the project report was completed copies were submitted to the supervisor and to the external examiner (in this case a colleague at the University of Copenhagen). After a couple of weeks, during which the supervisor and the external examiner read the report, students presented themselves, as a group, at an oral exam. The main point is to identify the depth and extent of each student's ownership to the project. The aim is *not* to examine students' knowledge of traditional subject matter. This is done only to the extent that such subject matter happens to form a significant part of the project. The subject of the exam is the project in its entirety, not the project report only which is just one ingredient in the project, albeit an important one. The oral exam for a group of this size typically takes 2-3 hours. At the beginning each student makes a short presentation of one aspect of the project (usually not one which has been exhaustively dealt with in the report). Then the supervisor conducts a general discussion by raising issues and posing questions concerning the project, and the external examiner joins in. Sometimes questions are addressed to the whole group sometimes to individual members, as the circumstances warrant. The idealised model of the oral exam is one of a group defence of a thesis, although there is no requirement of scientific originality in the sense that results should be produced that are publishable. Emphasis is placed on the relevance of the problem posed, correspondence between questions and conclusions, depth of analysis, choice of methodology, difficulty of the task undertaken, independence and autonomy in the project work, etc.

When the oral exam is over (i.e. when no important issues are left to be discussed), the students leave the room and the supervisor and external examiner discuss the project, the project report and the examination. Typically, they have already arrived at a common impression of the project report prior to the oral exam, and the subsequent discussion has two objectives. Firstly to determine whether the events during the examination changed that impression in a positive or a negative direction, and secondly to see if significant differences amongst students could be detected in their exam achievements. The latter is a fairly delicate matter. For if, on the one hand, too much emphasis is placed on performance, for instance activation of factual knowledge, alertness, eloquence, initiative, etc., then the project work proper gains lesser emphasis, which will have a negative backwash effect on the whole enterprise. On the other hand, it is also important to ensure that no student can hide behind the work of others and get credit that he or she does not deserve. In principle it is possible to encounter a project group in which one student has done the main or the most important part of the work while the others are in on a 'free ticket', but this happens only very seldom. Or the other way round, one can encounter a group (and this happens more often, but still not frequently) in which one or two students are in on a free ticket whereas the others have shared all the work on a fairly equal basis. It is one of the tasks of the oral examination to detect if either is the case and to assign credits accordingly. Usually the outcome is either that all students get the same mark or that they deviate by one or two marks from one another. Only pretty rarely do we see a larger spread of marks within one project group, but this does happen from time to time.

In the present case, the project was judged to be rather homogenous as was the written report, so all five students got the same mark, a little above average, as mentioned before.

No systematic research has been done on the outcomes of project studies in mathematics at RUC. However, at a recent Danish national (but non-scientific) evaluation of university studies in mathematics, physics and chemistry, conducted by a Scandinavian evaluation panel, the Roskilde model was assessed in very positive terms. Apart from that, judgements of the outcomes have to rely on impressions gained by concrete experience by those involved, including the external examiners who submit brief reports on every exam to a central agency of external examiners. The students in this project have stated in several places in the project report, at the oral exam, and on other occasions, that they felt they had learnt a lot from it. In fact, they insisted that this project not only changed their conception of the nature, role and significance of mathematical proof but also of the nature of mathematics as a discipline. After the completion of the project and the oral exam, this group decided to include their project report in the department text series. We usually encourage a project group whose project achieved a minimum level of success to do so, in order to make it generally available to present and future students (and faculty) at the department. This has proved very useful in creating a project culture in which students can get inspiration concerning what to do and what not to do, by consulting previous project reports in their area of interest.

5. ISSUES AND COMMENTS

One of the questions which is often raised concerning the approach described above is what mathematical knowledge and skills students (need to) have when embarking on a project. How can they perform a project without possessing all the necessary prerequisites in advance?

In some cases, such as the one described in the previous section, students more or less know the mathematical subject matter which is going to be involved in the project – but they certainly do not know it from the perspectives adopted in the project. In most cases, however, students do not know all the mathematics they are going to encounter in the project, and then it is part of the project work to study it. In other words, in such cases they will have to acquire the prerequisites underway, and not the mathematical prerequisites only, also the ones that are to do with, say, subject matter knowledge in a discipline to which mathematics is being applied, the history or philosophy of mathematics, and so forth. Needless to say, this makes it very important for students to strike a balance between what they already know and what they have to become familiar with. It is a key task of the supervisor to give advice as to what is tractable and what is beyond reach for the particular project group at issue. This being said, project groups are sometimes capable of digesting a remarkable amount of difficult mathematics as part of a project. But of course, the spectrum of variation is wide and strongly depends on the capacities of group members.

It is fair to say that *on average* projects will not involve very advanced or very specialised branches of mathematics. Mostly, the topics invoked are not too distant from the topic areas studied in the courses offered. However, it should be borne in mind, once again, that the aim of project work is not to be a new means for students to learn systematically organised subject matter. Instead, the aim is that students should encounter, and gain first hand experiences with, mathematics from a variety of different perspectives with the ultimate end of deepening their mathematical understanding, insight, and competence. Thus, students in the RUC programme will typically have been exposed to a smaller set of mathematical branches, a smaller syllabus, than forms part of Master's programmes in more traditional universities. We are ready to pay this price if the outcome is graduates who have a deeper reflective knowledge and a more multi-faceted competence concerning the mathematics they have actually learnt. Our experiences during the past 25 years suggest that this is the case with the majority of the graduates.

Another question that often arises is what demands this approach puts on the supervisors. As may appear from what has been said above, the supervisor will normally be able to supervise a project on the basis of his or her general knowledge and experience in mathematics at large. This should not be taken to imply that supervisors will never encounter issues, topics, branches of mathematics, material, and so forth, which is almost as new to them as it is to the students. In fact this happens in any project. For instance, I once supervised a project on dynamic programming of which I had only little prior knowledge. This made me engage in a fair bit of reading of the literature in this field. The outcome of a project is always original and new in the sense that its particular *problématique*, its perspectives, the particular objects it investigates, or its combination of fields of knowledge, and suchlike, did not exist in that form before the project was carried through. Of course, this is not meant to say that project outcomes are new in a research sense (although occasionally they are), only that you will not be able to find the resulting 'package' on libraries' bookshelves (at least not as far as the project group and its supervisor know). So, supervisors will never be able to exert their supervision by just pouring from their taps of knowledge. Indeed, if they could the result would not be a genuine project but an extended and possibly demanding exercise. In other words, the supervisor also learns something new from any project he or she supervises.

What expertise, then, does it require to be a good project supervisor? Apart from a broad and solid knowledge of mathematics, not only from inside but from outside as well (including the history, philosophy, and sociology of mathematics and its applications), the two most important things are a broad outlook and a fair amount of 'mathematical culture'. This means the ability to perceive and reflect on mathematics from a variety of perspectives, on the one hand, and the ability to detect students' capacities, interests, potentials and limitations, on the other hand, so as to assist them in shaping their projects in ways that provide suitable challenges which match their backgrounds and ambitions. It follows that supervisor expertise grows considerably with experience. This is also reflected in the fact that novice supervisors need to spend a non-negligible amount of time to 'learn the trade', while experienced supervisors can be quite efficient without an undue investment of time. Therefore, new faculty hired to the department are expected to be willing to engage in this type of teaching, even if they may have to learn to do it – not in formal courses, though, but by actually doing it. Usually, applicants are fully aware of this when they apply for a vacant post.

Although we are not able to offer research evidence for the success of the Roskilde model – as I said before its outcomes have never been investigated from a research perspective – 'eating of the pudding' evidence based on experience abounds. Denmark has a national admission system to higher education. Students enrolling at RUC know in advance, from official information distributed to high school students, and from public debates on education, that this university differs from the traditional universities. Most of them have deliberately chosen to study at RUC as their first priority, the rest have it as their second priority. In general, the graduates from the mathematics programme have no difficulty in finding jobs within a multitude of sectors. They are well received, and not infrequently explicitly sought after, by employers, and some of them have interesting and impressive careers, in research, education, industry, popularisation institutions such as science centres and museums, and administration. But, naturally, career perspectives also vary considerably with individual characteristics. Our department maintains contact with our graduates after they have left the university. Once a year we have a meeting at which graduates, present students, and staff meet to exchange information and experiences. Graduates report on their work and careers and discuss to what extent their university education prepared them for that sort of employment. It is a consistent feature of those reports that the general abilities fostered by doing project work under conditions that are meant to be fuzzy, confusing, frustrating, demanding, but also inspiring, challenging, rewarding, and gratifying, are essential to them in their jobs. The specific material they studied and dealt with is less important than the competencies they acquired by doing it.

If the RUC model is reasonably successful, as I claim it is, why has it not spread to other universities to a considerable extent? To fully answer this question would take another paper. Let me concentrate on the main points.

Firstly, there is no unique, superior way of structuring and organising studies of mathematics. Every mathematics programme is a high-dimensional vector, and there is no single programme that is 'the best' with respect to every component. Every programme has its assets and drawbacks in comparison to other programmes. So, even if for obvious reasons the Roskilde programme is very dear to my heart, I believe in diversity and would not want all university programmes in mathematics to be organised according to the Roskilde model. At other universities other priorities are given emphasis, and often the price that has to be paid for making project work a substantial component of the programme is perceived to be too high in relation to those priorities. Moreover, students are different and have different learning styles. Some thrive and prosper in a learning environment like that at RUC, in which they are expected to be active, autonomous, and take initiatives in contexts and settings which by definition are not pre-structured. Others need plans, schedules, well-defined requirements, and prefer to be told what to do. So, we need different kinds of universities. On top of that, however, there is no reason to hide the fact that many Danish universities and their faculty have held considerable scepticism, of a somewhat xenophobic nature, towards the Roskilde approach simply because of its novelty and its being at odds with traditional modes of operation. It also cannot be excluded that some of the scepticism has relied on a less than complete knowledge base.

Secondly, as is probably evident from this presentation, the approach taken by RUC is not restricted to mathematics but is a full-scale scheme permeating the entire university. It is not easy to copy one component, the mathematics programme, and insert it into a completely different setting. That being said, after an initial one-and-a-half decade of scepticism towards the Roskilde model, several Danish universities have adopted in their mathematics programmes, in a modified form, content and structural elements similar to the ones at RUC.

6. CONCLUDING REMARKS

Problem-oriented project work has proved to be a an excellent *pedagogical* tool to foster and develop: students' initiative and independence; responsible activity in mathematics; analytical and reflective approaches and attitudes; ability to pose mathematical questions and to think creatively in, with and about mathematics; ability to communicate on matters mathematical in a non-rudimentary language; collaborative abilities. It follows that such work serves to generate and further, students' *enthusiasm*.

Furthermore, problem-oriented project work has proved to be an excellent *strategic* tool to help students create a multi-faceted and balanced perception of mathematics and its roles and functions in nature, society, and culture. The different types of projects students carry out provide them with first hand experiences of different representative dimensions of and perspectives on mathematics which can hardly be accessed through traditional modes of teaching and study. In other words, project work is not primarily a new motivational gadget to teach students standard material from the text book shelves, but a means to accomplish new ends. Such project work cannot stand alone, however. Only seldom does it seem to be an appropriate means for the acquisition of usual, systematically organised subject matter. For that purpose, classical courses organised and taught by academic teachers remain relevant – which certainly does not imply that the way such courses are taught is of minor importance; on the contrary, but that's not on our agenda in this context.

These remarks show that the mathematics programme at RUC is designed to grasp the essence of mathematics by means of a pair of pinchers, with one jaw consisting of problem-oriented studies, cast in the form of project work, while the other consists of the study of systematically organised subject matter, typically through participation in courses. As was said before, this way of structuring and organising the programme has costs. Measured in terms of the amount of systematically organised subject matter students are exposed to, this type of programme will contain a smaller component of such subject matter than is the case with traditional programmes. However, any mathematics programme can contain only a small selection of the branches and topics of contemporary mathematics. So, a selection has to be made anyway, and what matters at the end of the day is not so much how much food is being served but rather the selection and quality of the food, and how much is digested and how well.

Mogens Niss
Roskilde University, Denmark
mn@mmf.ruc.dk

DAVID A. SMITH

THE ACTIVE/INTERACTIVE CLASSROOM

1. CREATING A CLASSROOM ENVIRONMENT: HARDWARE AND SOFTWARE

For the past year, I have had the pleasure of teaching in a new Interactive Computer Classroom (ICC), a 'studio' classroom designed for hands-on student work with computers. The students refer to this room as a 'lab', but it has several characteristics that distinguish it from a more traditional computer laboratory:

- each table accommodates two students but only one computer;
- the computer tables are arranged in a double-U facing the back and side walls, so the instructor can see all of the screens. As a result, students are not distracted by the computers when they turn around to see the front of the room;
- there are empty tables to facilitate group work without computers, as well as data-gathering experiments that require some physical setup (e.g., a spring mass system and a motion detector);
- in addition to traditional chalkboard and overhead projector, there is a sophisticated lecture podium from which the instructor can project materials from the built-in computer or a laptop, videotape, laser disk, or DVD;
- all of the computers are served by a local network and by high-speed access to the Internet.

For more details about the classroom, including pictures of students at work, see ICC Task Force (2000).

In addition to the physical facility, our environment is enhanced by the Blackboard CourseInfo software for Web delivery of course materials. This product enables an instructor who knows little about Web pages to construct and maintain a substantial course Web site. You may visit any of the CourseInfo course sites (on any campus) as a guest by logging in as 'guest' with password also 'guest'. You will be able to see everything on the site except pages that identify students or that enable the instructor to control the content. In particular, please visit my most recent course sites at Smith (1999), (2000a), and (2000b).

The principal categories on every CourseInfo home page are

- Announcements
- Staff Information
- Assignments
- External Links

- Course Information
- Course Documents
- Communication
- Student Tools

Derek Holton (Ed.), The Teaching and Learning of Mathematics at University Level: An ICMI Study, 167—178.
© 2001 *Kluwer Academic Publishers. Printed in the Netherlands.*

This structure enables an instructor to easily manage a multifaceted course with multiple classroom strategies, multiple assessment tools, multiple due-date structures, and a mix of individual and group work. It also enables students to stay in touch with each other and with the instructor, which is very valuable in building a sense of community. Students quickly learn that they need to visit the course page every day, where the current Announcements are always displayed on the home page. The instructor can use this page for announcements made or forgotten in class, reminders of due dates or test dates, corrections of mistakes in the text or lectures, pointers to new content on other pages, and many other things. The examples that follow are taken from the home page for my differential equations course. The other course pages are not identical, but they use CourseInfo in similar ways.

My Course Information page contains what I used to provide on many pieces of paper at the beginning of each course: textbook and other materials, course goals, syllabus, requirements, and policies. Students generally lost my paper handouts and could never remember what I had told them about the course. Now I rarely get questions about course matters, and when I do, I, or another student, can reply "Look on the Web page." Most of this information is static, but it resides on live Web pages, so it is easy, for example, to make a change in the syllabus if that becomes necessary. It is also easy to include external links in these documents. For example, where the syllabus specifies an on-line computer module for a lab exercise, there is a link that will bring up the module. My goals for engineering courses are usually quoted from the ABET 2000 criteria, and there is a link to the standards page of the Accreditation Board for Engineering and Technology (1999).

Requirements and policies are organized in folders to enable easy retrieval during the semester without scrolling through lengthy documents. The Requirements folder contains documents titled Reading, Writing, Homework, Computer Assignments, Test Information, Electronic Journal, and others. I will quote some of these to indicate the structure of a typical course.

> **Writing.** You will be expected to ... write clear, direct English prose about your thought processes and your work on various sorts of problems. You will find it useful to read *A Guide to Writing in Mathematics Classes* [live link] by Annalisa Crannell (Franklin and Marshall College) ...
>
> **Homework.** ... In general, **no solution will be given full credit unless you have written an explanation of why you** *know* **it is correct**. For example, an *acceptable* explanation for a solution of an equation is that you have substituted the proposed solution into the original equations and found that it satisfies the equation identically – and you have to show your work. Three examples of *unacceptable* explanations:
> - It matches the answer in the back of the book.
> - I did the work again and it came out the same.
> - This is what the computer gave me.
>
> **Tests.** All tests will be open-book and take-home...

There are, of course, some policies imbedded in the Requirements documents. Others are in the Policies folder – again, I quote:

> **Teamwork.** Computer work will be done by teams of two students, and some class work will be done by teams of four, usually by pairing two computer teams. The instructor will make team assignments and will change them about once a month... You will find useful information about working in groups in *Ten Guidelines for Students*

Doing Group Work in Mathematics [live link] by Anne E. Brown, Indiana University South Bend. ...

Cooperation. ... Learning is a cooperative activity – and the 'real world' demands workers who have learned to cooperate in order to solve more substantial problems than individuals can solve working alone. Therefore, some tasks in this course will be assigned to designated groups, and such tasks will receive a group grade.

Open- vs. Closed-Books. The 'real world' is ... a very large collection of 'open books'. Success does not require memorizing the contents of those 'books' but rather understanding how to use available resources in an intelligent way. Closed-book tests do little to measure such understanding, so all activities in this course – including tests – will be open-book.

Grading. There is no intellectually respectable justification for pretending that understanding of mathematics ... can be measured by numerical point scales. ...[E]ach graded activity will also be assigned a letter grade relative to the instructor's expectations ... Such grading is sometimes called 'subjective.' The alternative to subjective grading is not 'objective' – it's arbitrary.

Your letter grades from the different components of the course will be weighted approximately in the following way.

Team work		Individual work	
Computer reports (9 to 10)	25-30%	Homework	15%
Class participation and projects	10-15%	Electronic journal	5%
		Take-home tests (2)	20%
		Final Exam	20%
Total	40%	**Total**	60%

The next important segment of CourseInfo is Course Documents. The instructor can post any type of document that Microsoft environments recognize – text, Word, HTML, PowerPoint, jpeg, and so on. With a little care in opening and saving, this facility can also be used to provide unrecognized file types, such as Maple. I find it useful to construct my lectures (yes, there are some) in Maple so that I can vary examples on the fly, with instant access to precise and dynamic graphics, and little possibility of an arithmetic or algebraic error. The 'lecture notes' are immediately available to students, and they can manipulate the Maple code to enhance their understanding, further modify the examples, or adapt code to address their homework exercises. Because these notes are available in the classroom, I can build follow-up exercises and worksheets that will be used in the same class period. The most important folder in Course Documents is the Lecture Support file.

The Course Documents page also contains assignments of students to teams (often more than once per semester), a summary of the class demographics and individual goals, explanations of troublesome concepts, pictures taken in the classroom, slides from lectures given elsewhere that relate to the course, and any other documents that may be useful.

There is much more in CourseInfo, but I will add just a few words about my Assignments pages. The major folders here are Weekly Plans, Worksheets (for planned group classroom activities), and Review Problems (plus solutions that are posted after students have worked on the problems in the classroom with their teammates). The Weekly Plans are the organizing heart of the course. Each plan announces the major concepts for the week, makes connections to work already completed and to the overarching themes of the course, sets the daily homework assignments, maps out what will happen in the classroom meetings, and sets and reviews due dates for homework, in-class work, computer modules, and any other assignments.

2. CREATING A CLASSROOM ENVIRONMENT: PEOPLE

Only so much can be done with hard and soft course structures. The most important characteristic of a classroom in which everyone can succeed is *trust*. And trust is possible only if the environment is cooperative and collegial, not competitive. If the highest grades – the 'coin of the realm' for students – are available only to a small percentage of the class, then trust and cooperation are not likely. If the instructor is seen as an adversary – always 'taking points off' rather than supporting student growth – trust and cooperation are simply impossible.

Students often go to great lengths to 'psych out' a course and its instructor, optimizing strategies to achieve high grades with (what they think is) minimal effort. If they can postpone study until the week of an exam, cram for one or two nights, and get a good grade, that is what they will do, no matter what the instructor says to do. Our elite colleges, such as my own, tend to be populated with students who have succeeded with such strategies – and who are reluctant to give them up. However, they *will* give them up if they can be convinced that a) that won't work in this class, b) there are quite different strategies that will lead to good grades, and c) real learning can be a lot more fun than periodic cramming and regurgitation – just as real dining is a lot more fun than bingeing and regurgitation.

The policies announced on my CourseInfo pages make point a) – there are no available 'points' for memorization. I reinforce that frequently in class by reminding students that the only rule for my tests is that they will see problems they have never seen before. Furthermore, the tests account for only 40% of the course grade. I have to teach to points b) and c). I address b) with a variety of assessed activities, most of them with grades that 'count', but some not, most of them likely to lead to success and to a sense of growth and preparation for that still-distant exam. Point c) is addressed mainly by careful selection of quality materials and interesting problems, plus frequent interaction with working groups to keep them moving forward and to help them learn group skills.

The most important fact about trust is that someone has to go first. Since I'm the one who really wants that atmosphere of trust, that would be me. I give out my home phone number on day 1, and I don't mind if they use it. I generally respond to email in a reasonable time, and I pay attention to the electronic journal entries. I make it very clear to my students that I trust them totally, that I simply don't believe it is possible that they would cheat, because there is essentially nothing that can be gained that way, and it would be destructive to oneself and one's teammates.

Once it is settled that I have refused the role of adversary, I see the bonds of trust gradually strengthen. Students often sign up for my section (so they tell me) because they have heard that I give good grades. A few back off when they realize how much work I expect, but most stick around because they believe the good grades are achievable, even if not the way they expected. Almost all of my students are in class almost all of the time, and almost all earn a B or better – even if they have never done that before in a maths class. And the small number of students who do not achieve at that level generally concede that it was their own fault, not mine.

3. SUPPORT IN THE RESEARCH LITERATURE

There is a vast and growing literature supporting the use in college classrooms of the practices outlined above: active learning, group work, continuous and mixed assessment, reading and writing to learn, intelligent use of technology, and above all, community. A useful resource for pointers to this literature is *The National Teaching And Learning Forum*; some of the articles I have found helpful are those by McLeod (1996), Rhem (1995) and Zull (1998). A concise summary of research findings prior to the reform efforts of the 1990s, applied specifically to mathematics education, is *Everybody Counts* (NRC, 1989). A more recent and more general survey, which includes results of both cognitive psychology and neurobiology in the last decade, is in Bransford, Brown and Cocking (1999) and its companion volume Donovan, Bransford and Pellegrino (1999). For the reader unfamiliar with the learning implications of recent neurobiological research, Hannaford (1995) and Sylwester (1995) are also helpful.

I have written at some length elsewhere (see Smith, 1998, 2000c) about some aspects of this literature and their relevance to college mathematics classrooms and curricula. I will summarize here some key points that support my current classroom practices.

In the late 1980s, Chickering and Gamson (1991) surveyed "50 years of research on the way teachers teach and students learn" and distilled their findings into Seven Principles of Good Practice in Undergraduate Education:

Good practice:
1. encourages student-faculty contact;
2. encourages cooperation among students;
3. encourages active learning;
4. gives prompt feedback;
5. emphasizes time on task;
6. communicates high expectations;
7. respects diverse talents and ways of learning.

They summarize, "While each practice can stand on its own, when they are all present, their effects multiply. Together, they employ six powerful forces in education: Activity, Cooperation, Diversity, Expectations, Interaction, and Responsibility."

Kolb's research in experiential learning (see Wolfe and Kolb, 1984, pp. 128-133) led to his description of a learning cycle as a two-dimensional way to think about students' preferred learning styles and about ways to build better learning environments (see Figure 1). The four stages of this cycle, interpreted in terms of brain research, are

- Concrete Experience (CE): input to the sensory cortex of the brain: hearing, seeing, touching, body movement.
- Reflection/Observation (RO): internal, mainly right-brain, producing context and relationship needed for understanding.
- Abstract Conceptualization (AC): left-brain activity, developing interpretations of our experiences and reflections.
- Active Experimentation (AE): external action, requires use of the motor brain.

The ideal learner cycles through these stages in each significant learning experience. The AE stage represents testing in new situations the implications of concepts formed at the AC stage. Depending on the results of that testing, the cycle starts over with a new learning experience or with a revision of the current one. The ideal learning environment is designed to lead the learner through these stages and not allow 'settling' in a preferred stage.

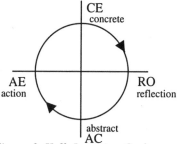

Figure 1. Kolb Learning Cycle

In fact, there are very few ideal learners. Most students have preferred learning activities and styles, and they are not all alike. This is one reason why learning experiences work better for everyone in a diverse, cooperative, interactive group.

Learning takes place by construction of neural networks. External challenges (sensory inputs) select certain neural connections to become active, and this is a random selection among many possible connections that could occur, not something that happens by deterministic design. The sensory input can trigger either memory, if it is not new, or learning, if it is new. The cognitive psychology term for this process is *constructivism*: The learner builds his or her own knowledge on what is already known, but only in response to a challenge or 'disequilibration.' In particular, knowledge is not a commodity that can be transferred from knower to learner.

Deep learning, learning based on understanding, is *whole brain* activity – effective teaching must involve stimulation of all aspects of the learning cycle (Rhem, 1995). Research by Entwistle, Marton, Ramsden, Schmeck, and others (see especially Chapters 1, 2, 3, 7, and 12 in Schmeck, 1988), has shown that a student's learning *approach* or *strategy* may be more important than learning style – a student may exhibit different approaches in different courses at the same time, depending on perceptions of what is expected for success. Furthermore, some course characteristics have been found to foster surface approaches and some to foster deep approaches:

Course characteristics that encourage *surface* approaches:
- excessive amount of material;
- lack of opportunity to pursue subjects in depth;
- lack of choice over subjects and/or method of study;
- threatening assessment system.

Course characteristics that encourage *deep* approaches:
- interaction – peers working in groups;
- well-structured knowledge base – connecting new concepts to prior experience and knowledge;
- motivational context – choice of control, sense of ownership;
- learner activity plus faculty connecting activity to abstract concept.

The first of these lists describes the way I used to teach, more than a decade ago. The second describes what I am trying to do now. When we free ourselves from the tyranny of the packed syllabus, we can give our students many freedoms, with or without computers. There is much we don't know yet about the World Wide Web as a learning tool, but having an on-line studio classroom allows us to conduct and observe some of the necessary experiments.

The research results mentioned here all reinforce each other, and together they constitute a clear and consistent message about good design and execution of curricula and pedagogy.

4. A SAMPLE LEARNING ACTIVITY

I will illustrate implementation of the research findings in the preceding section by way of a learning activity from a differential equations course. In the sixth week of the course we were studying undriven harmonic oscillators and their representation by second-order linear, homogeneous, constant-coefficient differential equations. (See Section 1 for instructions for viewing the syllabus and the Week 6 Plan on the course Web page.)

The activity begins with students gathering position data on a spring-mass system with a Calculator-Based Laboratory (CBL) and motion detector. At (Interactive Computer Classroom Pictures 2000), the last six pictures show this experiment in progress. As students observed the spring motion under various damped and undamped conditions, the data were displayed in real time by way of a Texas Instruments TI-89 calculator with a view screen on an overhead projector. The calculator is quite capable of completing the analysis of the data, but we transferred the data sets to a computer in order to make them available in CourseInfo documents for easy copying into a computer worksheet. In this example, the physical experiment is the principal part of the CE phase of the Kolb cycle, but well-designed computer simulations can play this role as well.

Next the students worked in teams of two on a Maple-based Connected Curriculum Project module (Moore and Smith, 2000). We devoted a 50-minute class period to this part of the activity, and then students had several days to finish the module on their own time.

The module begins with student entry of the undamped data and an attempt to fit the data with a cosine function, which they already know from assigned reading and classroom discussion should be an appropriate model. The following figures and text illustrate work in progress. Student inputs to the Maple worksheet and their comments are in boldface, and Maple outputs are centered.

>**y0:= (0.835273-.6787664300);**
>**Y:=t->y0*cos(sqrt(K/m)*t + Pi/10);**

$$y0 := .1565065700$$

$$Y := t \rightarrow y0 + \cos\left(\sqrt{\frac{K}{m}} t + \frac{1}{10}\pi\right)$$

We add the factor of amplitude of y0 to our model, taking the highest peak from our data and subtracting the mean from it. We do this, instead of merely taking the value at t=0 because at t=0, we see that the spring is not at its peak. Hence, we adjust for the actual amplitude by taking the highest peak of the data. We then had to adjust for the period (knowing that the data graph does not start at its peak at t=0). Thus, we adjusted for this by adding a delay factor of Pi/10, (which seems to be sufficient).

Next we graph the data and the model function together.
>**graph2:=plot(Y(t), t=0..7,color=red):**
>**display({graph1,graph2});**

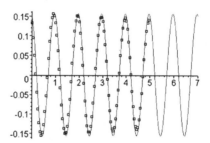

Our model seems to fit the data pretty well. With our adjustments for amplitude and our delay (because the data did not exactly start at the peak for t=0), we have a model that seems to show an undamped spring motion. The only difference comes where as t grows larger, the amplitude of the data graph decreases. This is because the data was not ideally undamped (it had a CD attached to the weight), and so air resistance was damping the spring as time went on.

This student sample needs a few words of explanation.

1. Some details that we don't want students to waste time on are already provided in the worksheet. For example, they didn't have to do anything with the graphing commands except execute them. But the plot command here does not do anything until the model function Y(t) has been defined, and that's the key thing we want students to think about at this stage. We (the authors of the module)

chose to display the data as discrete boxes and the model function as a continuous curve so the distinction would be instantly clear.

2. Our intention was that students would start their data gathering at the instant the extended spring was let go, so y_0 would be one of the minimum points on the plot, eliminating the need to think at this stage about 'delay factor.' Since the experiment did not go according to plan, students had to think about *something* that would account for not starting at a peak or valley. Some students discarded the first eight data points, and some (like the team quoted here) more or less invented phase shift. The first quoted paragraph is their justification of this procedure. Note the revelation of 'prior knowledge': They felt compelled to express that horizontal shift as a multiple of pi.

3. In the second quoted paragraph, the students are tentatively taking credit for a good fit. They are also noticing the obvious fact that some damping is present and attributing that to the cause they could see: a CD attached to the weight. (Don't throw away all your 'Free AOL' CD-ROMs. They make excellent lightweight targets for a motion detector.) In fact, there is just as much frictional damping without the CD, but 'seeing is believing.' Their conclusion here may also be influenced by a later stage of the experiment, in which we used a one-foot square of cardboard (in place of the CD) to induce significant air damping.

At this early stage of the module, students have already completed the Kolb cycle at least once. They have reflected on the concrete experience and visualized the motion as something familiar, namely, a sinusoid. They have connected this with their prior experience (in particular, reading) to abstract the motion into a formula, and they have tested and adjusted their formula until they are confident it is right. We don't actually see this last phase in their report – that's where the instructor comes in. I *know* that formula was not copied from a book or from another student team because I watched the students working in class and interacted with them as they did so. Collectively, the class had as many *different, correct* formulas as there were teams, and the learned concept was the same for all of them.

The second Kolb cycle in this module uses the damped data and proceeds in a similar manner, but success is more difficult to achieve, because there are now four parameters in the model, only one of which can be estimated directly from the data. We provide some direction about the order in which students should attempt to find the other three. Here are the results from the same student team – note their judgment about quality of the data:

$$R := .2541494750$$
$$L := .1$$
$$\theta := 1.82\pi$$
$$\delta := .3571428571\,\pi$$
$$y := t \rightarrow R\,e^{(-Lt)}\cos(\theta t - \delta)$$

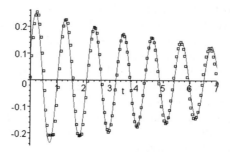

Our approximation is pretty good. It fits the period and the same damping and the same amplitude. One difference is in the first valley, where we see our model goes down farther than the data. However, looking at the graph of the data, we see that this seems to be a data error, since the graph of the data is not very smooth here. The device did not seem to collect enough data here to make the graph smooth.

To reinforce abstract conceptualization about the harmonic oscillator, the module has a third part on critical damping – which cannot be tested by the physical apparatus at hand, but for which computer testing of the model can still be carried out. The outcome of this exercise is that most students have a clear understanding of the significance of the constant, exponential, and trigonometric factors in the solution formula and of the parameters as well.

Each CCP module ends with a Summary section in which students respond with paragraphs to integrative questions that force them to think through internal and external connections among the various experiences and their prior understandings. In addition to the explicit connections with the Kolb cycle, it should be clear that this learning activity resonates with the characteristics that foster deep learning and with Chickering and Gamson's Seven Principles of Good Practice in Undergraduate Education.

The key findings of the National Research Council study *How People Learn* (Bransford, Brown and Cocking, 1999) are summarized in Donovan et al., (1999, pp. 10-15):

1. Students come to the classroom with preconceptions about how the world works. If their initial understanding is not engaged, they may fail to grasp the new concepts and information that are taught, or they may learn them for purposes of a test but revert to their preconceptions outside the classroom.

2. To develop competence in an area of inquiry, students must: (a) have a deep foundation of factual knowledge, (b) understand facts and ideas in the context of a conceptual framework, and (c) organize knowledge in ways that facilitate retrieval and application.

3. A 'metacognitive' approach to instruction can help students learn to take control of their own learning by defining learning goals and monitoring their progress in achieving them.

A lot of implications are packed into those three statements, and one purpose of the NRC study is to unpack them. Suffice it to say that the CCP learning environments and my other classroom activities are consciously designed to align with these findings. The statement about metacognition brings us to some final observations.

5. PROGRESS TOWARD GOALS

On homework and tests, I refuse to give full credit even for correct answers unless accompanied by a correctness argument. I also use frequent and varied assessments, formal and informal, to assist students in monitoring their progress. The CCP materials referred to in the preceding section contain many self-monitoring steps.

Of course, one cannot monitor progress toward goals if there are no goals. Early in each of my courses, I engage the students in a goal-setting exercise. Not all of the expected learning gains are necessarily identified among students' own goals, but their goals usually have a substantial overlap with my own – which I post on the course Web page before the semester starts. Here are the goals for the differential equations course – as noted in section 1, these goals are stated in the language of the ABET 2000 criteria.

Students who complete this course should be able to:

- apply knowledge of mathematics, specifically, ordinary differential equations and systems of differential equations, especially linear equations and systems;
- conduct mathematical experiments and analyze and interpret data from those experiments;
- identify, formulate, and solve problems;
- function effectively as a member of a team;
- communicate mathematical understandings effectively, both orally and in writing;
- engage in life-long learning;
- use the techniques, skills, and modern tools necessary for engineering practice, especially a computer algebra system.

Of course, we don't have the opportunity to observe whether students actually do engage in life-long learning, but progress toward the other six goals could be observed and measured during the course. I set high expectations for my students, and almost all of them respond with a high level of performance.

REFERENCES

Accreditation Board for Engineering and Technology (1999). *ABET Engineering Criteria 2000*. URL: http://www.abet.org/eac/EAC_99-00_Criteria.htm#EC2000. [2000, June 21].
Bransford, J. D., Brown, A. L. and Cocking, R. R. (Eds.). (1999). *How People Learn: Brain, Mind, Experience, and School*. National Research Council. Washington, DC: National Academy Press.
Chickering, A. W. and Gamson, Z. F. (Eds.). (1991). *Applying the Seven Principles of Good Practice in Undergraduate Education*. New Directions for Teaching and Learning No. 47. San Francisco: Jossey-Bass Publishers.
Donovan, M. S., Bransford, J. D. and Pellegrino, J. W. (Eds.). (1999). *How People Learn: Bridging Research and Practice*. National Research Council. Washington, DC: National Academy Press.
Hannaford, C. (1995). *Smart Moves: Why Learning is Not All in Your Head*. Arlington, VA: Great Ocean Publishers.
Interactive Cluster Classroom Task Force(2000). URL: http://www.aas.duke.edu/admin/icc/. [2000 June 21].

Interactive Computer Classroom Pictures (2000). URL: http://www.math.duke. edu/education/icc/ iccpix.html. [2000, June 21].

Materials for Engineering Mathematics (2000) [Online]. Connected Curriculum Project. URL: http://www.math.duke.edu/education/ccp/materials/engin/. [2000, June 21].

McLeod, A. (1996). Discovering and Facilitating Deep Learning States. *National Teaching And Learning Forum*, 5 (6), 1-7.

Moore, L. C. and Smith, D. A. (2000). *Spring Motion* [Online]. Connected Curriculum Project. URL: http://www.math.duke.edu/education/ccp/materials/diffeq/spring/. [2000, June 21].

National Research Council (1989). *Everybody Counts: A Report to the Nation on the Future of Mathematics Education*. Washington, DC: National Academy Press.

Rhem, J. (1995). Deep/Surface Approaches to Learning. *National Teaching And Learning Forum*, 5 (1), 1-5.

Schmeck, R. R. (Ed.). (1988). *Learning Strategies and Learning Styles*. New York: Plenum.

Smith, D. A. (1998). Renewal in Collegiate Mathematics Education, *Documenta Mathematica* Extra Volume ICM 1998 III, 777-786. URL: http://www.math.duke.edu/~das/essays/renewal/. [2000, June 21].

Smith, D. A. (1999). *Applied Mathematical Analysis II*. URL: http://cinfo.aas. duke.edu/courses/ MTH114.01/. [2000, June 21].

Smith, D. A. (2000a). *Elementary Differential Equations*. URL: http://cinfo.aas. duke.edu/courses/ MTH131.03/. [2000, June 21].

Smith, D. A. (2000b). *Introduction to Linear Programming and Game Theory*. URL: http://cinfo.aas.duke.edu/courses/MTH126.01/. [2000, June 21].

Smith, D. A. (2000c). Renewal in Collegiate Mathematics Education: Learning from Research. Chapter 3 in S. L. Ganter (Ed.), *Calculus Renewal: Issues for Undergraduate Mathematics Education in the Next Decade*, pp. 23-40. New York: Kluwer/Plenum.

Sylwester, R. (1995). *A Celebration of Neurons: An Educator's Guide to the Human Brain*. Alexandria, VA: Association for Supervision and Curriculum Development.

Wolfe, D.M. and Kolb, D.A. (1984). Career Development, Personal Growth, and Experiential Learning. In D. A. Kolb, I. M. Rubin and J. M. McIntyre (Eds.), *Organizational Psychology: Readings on Human Behavior in Organizations* (4th ed.), pp. 128-133. Englewood Cliffs, NJ: Prentice-Hall.

Zull, J. E. (1998). The Brain, The Body, Learning, and Teaching. *National Teaching And Learning Forum*, 7 (3), 1-5.

David A. Smith
Duke University, North Carolina, USA
das@math.duke.edu

JOEL HILLEL

CONCORDIA UNIVERSITY, MONTREAL, CANADA

1. INTRODUCTION

The Department of Mathematics and Statistics currently has 28 full-time faculty members, most of whom hold doctoral degrees and are actively engaged in research. It offers undergraduate, masters and doctoral programmes in mathematics, statistics, and actuarial mathematics, as well as an in-service masters for secondary and college teachers.

2. ENTRANCE REQUIREMENT

Students in Québec complete 13 years of pre-university studies consisting of 11 years of elementary and secondary education, followed by a 2-year college. It then takes three years for them to complete most undergraduate programmes.[1] Students applying to a university need to specify a faculty (e.g. Arts and Science, Commerce, Engineering and Computer Science, Fine Arts) and programmes (e.g. psychology, electrical engineering, or statistics) which they wish to enter, ranked in order of preference. They do not necessarily get into their first-choice programme. The general admission standards are sufficiently moderate that nearly every student who has completed their collegial studies and wants to enter some university programme can do so. Fully qualified students entering mathematics or the sciences would have already completed two introductory calculus courses and a basic linear algebra course at college. They need a 70% average in their college mathematics courses in order to enter an undergraduate mathematics programme.

3. UNDERGRADUATE PROGRAMME AND PROGRAMME STUDENTS

Most undergraduate degrees (bachelors) consist of 90-credits where each one-semester course is valued at 3 credits (corresponding to about 40 hours of classroom instruction, spread over 13 weeks). A student studying full-time would normally take 15 credits (5 courses) each semester and so would complete the bachelors degree in three years.

---

[1] On most Canadian provinces, the normal route is 12 years of pre-university studies followed by a 4-year undergraduate programme.

Derek Holton (Ed.), The Teaching and Learning of Mathematics at University Level: An ICMI Study, 179—183.
© 2001 *Kluwer Academic Publishers. Printed in the Netherlands.*

The Department offers two undergraduate programmes, a major programme in mathematics and statistics, and an honours programme (in pure and applied, statistics, or actuarial mathematics). These programmes have different goals and cater to different student clientele. Each programme has its own unique set of 'core courses'. Several non-core courses can be taken by students in either programme.

3.1 The major in mathematics and statistics

This consists of 45 credits (15 one-semester courses) and students are free to choose the remaining 45 credits more or less at will. Students enrol in the major for a whole variety of reasons. For many, mathematics and statistics is, at best, a second-choice subject: they were not successful getting into engineering, computer science, or commerce, where competition is a lot tougher. Other students have a vague notion that a degree in mathematics and statistics will enhance their job opportunities. In general, the students enrolling in the major are job-oriented and have no intention of pursuing graduate work in mathematics. Their mathematical skills barely meet the minimum requirement for entering the majors. The major programme is therefore focused on the applicable nature of the mathematical sciences as tools for solving, and ways of thinking about, a wide range of problems. It consists of a core of 10 courses (mathematical thinking and proofs, multivariate calculus I and II, linear algebra and its application I and II, applied probability, mathematical modelling, introduction to stochastic methods of operations research, introduction to mathematical computing, and a programming course given by computer science) and five other courses from a list which includes, among others, applied statistics, numerical analysis, and investment mathematics[2].

3.2 The honours programme

This can be taken in pure and applied, statistics, or actuarial mathematics and consists of 60 credits (20 one-semester courses), and a 6-credit honours project. It has more stringent admission requirements and it is intended for students who have more ambitious career goals in mathematics and statistics, or who want to become actuaries. There is a core of 10 courses for all options which include four analysis courses (the first two at the level of Spivak: Advanced Calculus, and the last two being multivariate calculus), linear algebra I and II, numerical analysis, probability, statistics, and operation research. These courses are more theoretical and with more emphasis on proofs than their counterparts in the majors. Beyond the core courses, the compulsory and optional courses vary with the option chosen. The honours is overwhelming populated by students choosing the actuarial option, a subject which has become very popular with career-minded students in the past 10 years. In contrast, there is a dearth of students who choose the pure and applied, or the statistics options.

---

[2] A description of all courses can be found by accessing http://www-cicma.concordia.ca/math/

Students who wish to switch from the honours to the major get credited for the courses they have taken as part of the honours programme and are given the obvious exemptions for courses in the major. Switching in the opposite direction is rare and is more problematic. If undertaken sufficiently early on, it is done on a more-or-less *ad hoc* basis.

3.3 Enrolment of programme students

In 1999-2000, 185 new students enrolled as programme students in the department: 113 in the major and 72 in the honours. Only 14 students in the major had mathematics and statistics as their first choice. Among the honours students, 50 entered into the actuarial option, 18 in pure and applied, and 7 in statistics. In all, there are approximately 500 undergraduate programme students in mathematics and statistics, counting all three years.

3.4 Non-programme students

By far, the majority of students taught by the department are not mathematics programme students. For instance, the department teaches all of the first year engineering mathematics courses. The courses, multivariable calculus and differential equations, and matrices and advanced calculus, are courses specifically designed for the needs of engineering students. Other engineering mathematics courses are taught either by engineers or by the department. Computer science students in the software systems option take several courses in common with the honours students (analysis I and II, linear algebra I, statistics, and probability). The physics department, on the other hand, offers its own mathematics courses, and students from the other natural sciences do not generally need more mathematics for their programmes and would only occasionally take mathematics as an option.

A large non-programme clientele consists of students who enter the university without the proper mathematical prerequisites. They are either local students who have not taken the full mathematics programme in their collegial studies or they have come from different educational systems. They are generally trying to get into a science, engineering or computer science programme but are missing the first two calculus courses and linear algebra. The department also offers two special mathematics courses: fundamental mathematics I and II, designed for commerce bound students. On the other hand, statistics for social science and psychology students is taught by the home departments.

Finally, there is a special clientele consisting of 'mature students' who are those over 21 years of age who have only completed high school. They are admitted into a special 4-year programme so the department offers some courses that are a review of high school mathematics, including: analytic geometry, basic algebra, and functions.

3.5 Teacher training

In the past, the route to becoming a high-school mathematics teacher was to complete an undergraduate degree and then take a special year to get teaching certification. However, the new system in Québec requires potential teachers to enter into a 4-year Bachelor of Education degree (which our university does not offer). Consequently, there are rarely any undergraduate students in the department who are intending to have teaching as a profession.

4. TEACHING: PRACTICE AND INNOVATION

In a typical year, the department offers about 190 one-semester courses (about 70 undergraduate programme courses, 15 graduate courses, 25 for engineers, 25 for commerce students, and the rest at the 'pre-university' level). None of the courses are of the large-lecture format. Rather, class sizes tend to be no larger than 60 students, and many courses have multisections (up to 10 sections of the same course). Offering small classroom size rather than the large lecture format has been historically the strong 'selling point' of the university, but the pressure has been mounting yearly to increase the ceiling (ten years ago this was 45 students).

The undergraduate and graduate courses are all taught by full-time faculty, visiting professors and post-doctoral visitors. The normal teaching load of a full time faculty member is 2 courses per semester, which amounts to about 5 hours of classroom contact time per week. Faculty are expected to hold 'office hours', usually, about 2 hours per week.

Courses at the 'pre-university' level are mostly taught by graduate students and part-time lecturers (mathematics instructors from the colleges, retired faculty, and others).

4.1 Transition Issues

To deal with insufficient mathematical skills, the department runs a Maths Help Centre for about 20 hours a week. It is staffed by graduate students and deals mostly with elementary courses, up to and including one-variable calculus. Furthermore, a course entitled introduction to mathematical thinking and proofs has been designed with the specific goal of enculturating students into the practice and expectations of university mathematics. The course is part of the core in the major, taken by students in their first year. An 'advisor system' whereby students are attached to a faculty member who acts as an advisor-mentor is in place but does not work very well. (Students are supposed to contact their advisor twice a semester and talk about their progress, but they rarely do.)

4.2 Technology

The Department's philosophy is that, whenever appropriate, computer algebra systems and other tools should be integrated into courses (the university holds a general Maple license, available on all platforms). In practice, the use of technology has depended very much on the interest of the instructor. Some sections have been taught with a lot of emphasis on students using software tools such as Maple or C^+. In other instances, instructors have used the technology for demonstration purposes and have given students the option of doing their assignments with computer tools. Some instructors present all their lectures using a computer in class in an interactive format (this requires having access to some of the classrooms which are specially equipped), and one linear algebra course has a complete web-based version.

One recent initiative taken by the department is to institute a special course on the use of Maple. The course covers topics from calculus, linear algebra, differential equations, and so on. It aims to make students sufficiently fluent in using the software that they are able to use it in their other courses. This course is now part of the core of the major programme. Coupled with the course on mathematical thinking and proofs, it is taken by students in their first or second semester of studies.

4.3 Assessment practices

There is a long-standing tradition of giving regular (often weekly) assignments to students. These assignments are graded (either by the instructor or by a marker) and can (optionally) count for the final grade (10-20% of the final grade in most cases, more in others). Coupled with the assignments, there are mid-term tests (one or two) that can also be part of the final grade. Most courses still rely on a 3-hour final examination as the main component of assessing students' understanding. In some upper level courses, take-home exams are also acceptable.

The department has always struggled with policy regarding assignments. On the one hand, it recognizes that actively and continually engaging students in solving problems is essential for learning so, in assessing students, one would like to give a lot of weight to the students' work on assignments. On the other hand, such practice invariably leads some students to try to get the credit without doing the work themselves. There is also the difficulty that the marking of assignments tends to be relegated to graduate students who don't tend to give students real feedback except to mark the questions as right/wrong.

Joel Hillel
Concordia University, Montreal, Canada
jhillel@alcor.concordia.ca

URS KIRCHGRABER

EIDGENÖSSISCHE TECHNISCHE HOCHSCHULE ZURICH, SWITZERLAND

1. INTRODUCTION

The Department of Mathematics at Eidgenössische Technische Hochschule (ETH) Zurich has 26 permanent professorships, which constitute nearly 30% of all professorships in mathematics in Switzerland[1]. Affiliated with some of the chairs are 10 permanent senior scientists, their duty being to support teaching, research and consulting. There are also 6 non-tenured assistant professorships[2], about 60 teaching assistant positions[3] and about 25 staff members (secretaries, the IT support group, a librarian). The main areas of research are Algebra and Topology, Arithmetic and Geometry, Geometric Analysis, Mathematical Physics, Probability and Mathematics of Finance and Insurance, Statistics, Computational and Applied Mathematics, Operations Research.

The teaching load for a professor is 4-6 hours per week where this number may include a student seminar.

The department's most important instrument to enhance research is the mathematics research institute (Forschungsinstitut für Mathematik FIM). It organizes visitors' programmes as well as scientific activities over a broad spectrum of fields in mathematics. Each year about 130 guests from all over the world visit the FIM and stay for periods ranging from one week to a year. In 1998 the Department of Mathematics at ETH Zurich hosted 65 Ph.D. students, which amounts to a third of all Ph.D.s in Mathematics in Switzerland.

The Department is evaluated regularly by an internationally recruited group of peers appointed by ETH's President. Based on a report prepared by the Department Chair, professors are reelected periodically by the Board of the Eidgenössischen Technischen Hochschulen. On the recommendation of the Department, ETH's

---

[1] There are two federal institutes of technology (ETH Zurich and ETH Lausanne) in Switzerland and eight classical universities administrated by the Kantons, i.e. the counties. Four of them are situated in the German speaking area and four in the French speaking part. Besides ETH, Zurich hosts the University of Zurich. The number of students of both institutions together is about 30000, which amounts to the order of 30% of all university students in Switzerland.

[2] Assistant professorships have been established at ETH to promote the careers of young scientists. The initial appointment is for three years, with the possibility of renewal for three additional years.

[3] Teaching assistants hold a diploma degree in mathematics from ETH or elsewhere. Their dual key duty is to work towards a Ph.D. and to conduct weekly tutorials, grade student work, prepare examinations etc.

185

Derek Holton (Ed.), The Teaching and Learning of Mathematics at University Level: An ICMI Study, 185—190.
© 2001 *Kluwer Academic Publishers. Printed in the Netherlands.*

2. ENTRANCE REQUIREMENT

Typically freshmen at ETH Zurich are 19 to 20 years old. They start their studies after 12 to 13 years of (mostly public) school. In general they have graduated from high school (Gymnasium). The high school diploma - the so-called Matura - certifies that the holder has a basic expertise in his native language (which may be German, French, Italian or Raeto-Romanisch), in at least two foreign languages, in international and domestic history, in the natural sciences and in mathematics. About 17.5% of nineteen-year olds nation wide hold a Matura diploma. This varies from 10% in the eastern and central part to about 30% in Geneva.

In the final two years of high school the syllabus in mathematics covers the following subjects: vectors and spatial geometry, an informal introduction to calculus, an introduction to probability and statistics. There are several different programmes which differ by the specialty they emphasize, e.g. modern languages, classical languages, etc. In the programmes emphasizing mathematics the weekly number of lessons in mathematics is six to eight, while there are about four weekly lessons in all other programmes. In all programmes graphing calculators[4], like the TI-83, TI-85 or TI-89/92, are widely used and most schools allow them to be used in examinations including the Matura examination.

Because the Swiss secondary school system is rather decentralized[5], because of differences due to the rural or urban background of students and not least due to the different Matura programmes described above,[6] the degree of preparedness of freshmen differs markedly.

In spite of this fact, everyone who holds a matura certificate is permitted to enroll at ETH Zurich to whatever study programme he or she is interested in.

Every student upon entering ETH Zurich specifies the programme he or she plans to pursue (e.g. Architecture, Mechanical Engineering, Chemistry, Pharmacy, Mathematics, etc.)

3. DIPLOMA STUDIES IN MATHEMATICS AT ETH ZURICH

Roughly speaking the skeleton of all study programmes at ETH Zurich is as follows. The (theoretical) duration is 8 semesters, i.e. 4 years. The first 4 semesters make up the basic studies (the so-called Grundstudium); the second two years cover the main studies (the so-called Fachstudium).

While the Grundstudium is well structured with little freedom to make choices, the Fachstudium opens a large variety of different tracks, so students can specialize in many directions. While some students favour Pure Mathematics, others emphasize Applications; the large variety of specialized courses offered by the

---

[4] It is supposed that Gymnasium students can afford to buy such a calculator.

[5] In many schools each teacher has the freedom to design his own matura examination problems; in other schools the teachers agree on a particular years' examination problems.

[6] TIMSS has revealed that the difference in average scores of the two types of students is about 100 points, which is of the order of magnitude of the difference between The Netherlands and the US! But in spite of this fact, Swiss Gymnasium students rank in the TIMSS-3 top group even in the average.

Department can support either. In 1998/99 the Department offered 50 advanced courses, i.e. courses directed to students in the Fachstudium. All courses have a strong emphasis on precise notions and proofs.

The following courses are *compulsory* (the figures given are weekly hours; about two thirds are covered by lectures, one third by exercises and tutorials; a semester lasts for 14 weeks).

First Semester: Analysis (9), Linear Algebra (6), Geometry[7] (6), and Computer Science (4).

Second Semester: Analysis (9), Linear Algebra (6), Numerical Analysis (6), and Experimental Physics (6).

Third Semester: Complex Analysis (5), Mathematical Methods in Physics (6), Algebra (4), Numerical Analysis (3), and Experimental Physics (6).

Fourth Semester: Analysis (6), Complex Analysis (3), Algebra (4), Probability and Stochastics (6), and Computer Science (3).

Fifth Semester[8]: Differential Geometry (5), Functional Analysis (4), and Theoretical Mechanics (6).

Sixth through Eighth Semester: Differential Geometry (3), Functional Analysis (3).

The following are additional requirements for *Semesters 5 to 8:*

- One out of the following applied fields (total 8 weekly hours during one semester) called Kernwahlfächer: Theoretical Physics, Stochastics, Computer Science, and Numerical Analysis.
- Three subjects from a list of 17 fields (8 in pure, 9 in applied mathematics or physics), the so-called Wahlfächer. To fulfill the requirements for a Wahlfach students have to take 2 courses (2 weekly hours at least each) and to deliver a seminar talk or to prepare a written paper.

Even if ETH Zurich as a technical university emphasizes engineering and the natural sciences, from the very beginning there was general agreement that the other 'culture' must not be neglected. For this reason, the Humanities, Social Science and Political Department, as it is called today, was installed. It offers a rich programme of courses from such fields as philosophy, psychology, German, French and English literature, art and music, history, sociology, politics and law and students have to fulfil a minimum of requirements prior to graduation.

Actuarial mathematics: The Department offers a number of courses in actuarial mathematics in the context of the Fachstudium. By suitably combining these courses a student qualifies for a special certificate which is highly respected by industry and

---

[7] To give a flavour of the content of the courses, the main topics of the geometry course of the first semester are: symmetry groups, conics, quadrics in R^3, hyperbolic geometry, geometry of $SO(3)$.

[8] The Fachstudium is presently under revision.

recognized in partial fulfilment of the diploma as Aktuar of the SAV (Schweizerische Aktuar Vereinigung, i.e. the Swiss Actuarial Society).

Interdisciplinary programmes: In 1997 the Mathematics Department in collaboration with several other departments (Physics, Computer Science, Electrical and Mechanical Engineering, Chemistry) instituted a new interdisciplinary programme called Rechnergestützte Wissenschaften (in the Anglo-Saxon area referred to as Scientific Computing). Students enter the programme completing the 4-semester Grundstudium, in any of the collaborating departments. Their Fachstudium consists of an individual selection of courses carefully chosen by the student and his advisor. There is a common core of advanced applied mathematics and computer science courses. A first group of 12 students has recently finished their studies.

Enrolment of programme students: The number of freshmen in mathematics has increased from about 40 in 1993 to about 70 in 1998. The total number of diploma students in mathematics in 1998 was about 320. About a third of all mathematics students in Switzerland study at ETH Zurich.

Non-programme students: All programmes at ETH Zurich include some mathematics. Typically it is required that students take three or more courses in undergraduate mathematics. The courses cover single and multivariable calculus, linear algebra, numerical methods and probability/statistics. The courses are specifically designed for the types of students to which they are addressed and the level of maturity required depends on the addressees, with students from such programmes as computer science or electrical engineering engaging in more advanced material than, say, environmental scientists.

It is a specific feature of ETH Zurich that physics students undergo almost the same mathematics training as the mathematics students in the first two years. (And vice-versa, mathematics students are requested to take a certain number of physics courses.)

Teacher Training: The Department offers a one-year training program for future Gymnasium mathematics teacher. It is supplementary to the Fachstudium and consists of four parts: a) two courses to introduce the students to the educational sciences (Allgemeine Didaktik und Pädagogik), b) two courses on mathematics education (didactics of mathematics), c) a small amount of guided teaching practice, d) a Wahlfach (see above), which relates Gymnasium to University mathematics. The key feature of the programme is tying together as closely as possible mathematics as a scientific discipline, general didactics, didactics of mathematics and teaching practice. A diploma in mathematics is a prerequisite for obtaining ETH's Gymnasium mathematics teacher certificate.

4. TEACHING: PRACTICE AND INNOVATION

Members of the Mathematics Department (i.e. professors, permanent senior scientists, a number of graduate students and a few part-time lecturers) teach virtually all mathematics courses at ETH Zurich. In 1998/99 the Department was responsible for more than 50 undergraduate courses in which some 8000 students participated. The number of students per course varies from about 30 to close to 300. The number of client students (i.e. neither mathematics nor physics students) was about 6200. The number of students taught by the Mathematics Department has increased by about 10% since 1995/1996.

On average, a professor conducts 1-2 diploma theses (see below) per semester.

Teaching is traditional in the sense that two thirds of undergraduate mathematics and even a larger percentage of advanced mathematics is taught in the form of lectures. The emphasis is on judiciously selected material that is carefully presented. Students' activity is relegated to the exercise sessions, tutorials and homework. It is expected that students spend a considerable amount of time preparing for the examination sessions, which are quite consciously placed towards the end of the summer and winter vacations.

All courses are evaluated regularly by the students and the results are reported in part to the Department Chair and ETH's Vice President for Teaching Affairs.

Technology: The very day a student is matriculated at ETH Zurich he or she has access to ETH's extensive computer facilities, in particular to e-mail, the Intra- and Internet, etc. Students are encouraged to buy their own PC or Laptop. As can be seen from the details of the study programme presented above, every Mathematics Programme student has to attend at least two computer science courses. Students are strongly encouraged to use program packages such as MATLAB and Computer Algebra Systems like Maple and Mathematica. Depending on the taste and personal experience of the instructors, such tools are integrated to a smaller or larger degree in undergraduate courses. Increasingly student assignments and their solutions, supplementary exercises, handouts, and a wealth of information are made available on the Web. Most recently both the Board of the Eidgenössischen Technischen Hochschulen and the Presidency of ETH, have made funds available to support projects that explore the use of the Web for teaching and to enhance distant learning.

Assessment practices: As mentioned earlier, the first two years form the Grundstudium, which is followed by the Fachstudium. During the Grundstudium weekly exercise sessions and possibly a tutorial accompany virtually each course. A teaching assistant or one of the 80 students that are regularly hired by the Department on a part time basis during the semester runs these sessions with 20-25 students. The assistant's job is to introduce the weekly problems, to grade the students' works and to provide additional explanations. It is required that students show reasonable performance on about 80% of the assignments. This is a prerequisite for the upcoming exams.

Before the second year starts a first set of oral and written exams takes place (the so-called 1. Vordiplom[9],[10]), and there is a second such set prior to the third year, the 2. Vordiplom. The organization of the written exams and the heavy burden of carefully marking students' work are part of the duty of the teaching assistants.

The final examination leading to ETH's Diploma in Mathematics is an oral exam (which may be split into two parts). In addition students have to prepare a diploma thesis under the supervision of one of the professors. It is to be completed within four months. Most students choose to fulfill this requirement in their ninth semester. The oral examination and the diploma thesis make up what we call the Schlussdiplom.

In written exams of about 3 hours duration, students are asked to solve problems similar to assignments encountered in the weekly exercise sessions. In oral exams of about 20 minutes duration, students usually present a bit of theory from the course the exam is about, for example, an important theorem and the key elements of its proof. Diploma theses range from working out published research results to original and sometimes even publishable contributions.

To provide a rule of thumb, the passing rate in the 1. Vordiplom is about 70%; it is considerably higher in the 2. Vordiplom and it is close to 100% in the Schlussdiplom. The Schlussdiplom is considered to be equivalent to a Master's Degree. Due to the fact that students who do not complete the programme do not earn a degree of any kind, ETH is presently discussing the introduction of Bachelor's and Master's diplomas, a topic which is hotly discussed throughout Europe these days.

Urs Kirchgraber
ETH-Zentrum, Zürich
kirchgra@math.ethz.ch

---

[9] In a sense the first year of the Grundstudium is designed in part to facilitate the transition from Gymnasium to university and to some extend the 1. Vordiplom serves as an entrance exam to ETH Zurich.

[10] Passing a Vordiplom is (but) a prerequisite for admission to the next exam. Not passing does not prevent a student from continuing his/her studies. A student can attempt to pass a Vordiplom (but) twice.

NESTOR AGUILERA AND ROBERTO MACIAS

UNIVERSIDAD NACIONAL DEL LITORAL, SANTA FE, ARGENTINA

1. THE INSTITUTION

Most universities in Argentina are organized by 'Facultades', having their own mathematics departments (if necessary). In the Universidad Nacional del Litoral there are several facultades, such as agronomy, architecture, biochemistry, chemical engineering, economics, science teacher training, hydraulic sciences, law and veterinary science. There are no facultades of medicine or science.

Here we describe the programme for mathematics majors in the Universidad Nacional del Litoral, which is taught in the Facultad de Ingeniería Química (Chemical Engineering). Other programmes, for Chemistry, Industrial Engineering, and Food Engineering, are carried out in the same Facultad. On the other hand, the training of future high school teachers in mathematics is done in the 'Facultad de Formación Docente en Ciencias' (Faculty of Science Teacher Training).

The Department of Mathematics has currently 23 faculty members. 12 of them have full-time positions, 15 hold doctorate degrees, 3 hold master's degrees, and most are actively engaged in research. It offers undergraduate, masters and doctoral programmes in mathematics.

The normal teaching load for a full-time faculty member is 2 courses per semester, which amounts to about 8 hours (480 minutes) of classroom contact time per week. Faculty are expected to hold 'office hours', usually about 2 hours per week.

In a typical year, the department offers about 50 semester courses (about 30 undergraduate courses, 10 graduate courses, and 10 for engineers and chemists).

2. DEGREE STUDIES AT UNIVERSIDAD NACIONAL DE LITORAL

2.1 Entrance requirements.

Any student, who has completed 12 years of elementary and secondary school, may continue with university studies in mathematics without any further requirements. There are currently 'mandatory' entrance courses that are delivered by distance learning over a period of about four weeks. One of these is common to all students. Some disciplines, such as mathematics, have their own courses. The mathematics course is very much a pre-calculus course. However, students do not have to pass

Derek Holton (Ed.), The Teaching and Learning of Mathematics at University Level: An ICMI Study, 191—194.
© 2001 *Kluwer Academic Publishers. Printed in the Netherlands.*

these, or even sit the exam, to be allowed entry into university. At the moment discussions are under way to change 'mandatory' to 'compulsory'.

2.2 *Licenciatura en Matemática Aplicada (Undergraduate Programme in Applied Mathematics)*.

The course work for the students in this programme, based on semesters, consists of 24 courses, 21 of which are mandatory and 3 of which are chosen from a wide list. There are also 4 mandatory workshops ('talleres') and 1 workshop to be chosen from a number of options. The student must also fulfil some English language requirements (for non-native speakers). In all, the student must take about 2700 hours of classes to obtain the degree.

The content of the mandatory workshops is transversal to the courses, in the sense that they either integrate their contents, they work on applications or underlying techniques, or they consider aspects common to different courses. Usually the teachers who are concurrently giving the mandatory courses during the corresponding term, work together in the design and teaching of these workshops.

After the first fundamental course, called basic mathematics, the mandatory courses, in three broad areas, are:

Analysis: calculus I, II and III, introduction to analysis, complex variables, measure and integration, ordinary differential equations.

Algebra and Geometry: discrete mathematics, linear algebra, euclidean plane geometry, geometry of curves and surfaces.

Applied Mathematics: mathematical methods in physics, (computer) programming, numerical calculus I and II, probability, statistics, linear programming, and mathematical modelling.

The required workshops are: reading and production of texts (this is a non-mathematical workshop and is a required workshop for all the students in the University), introduction to computers (working with a computer, Word, Excel, etc.), mathematical reasoning, and algebra and calculus.

Among the optional courses we have: introduction to functional analysis, linear dynamical systems, series expansions, partial differential equations, algorithms and data structures, discrete mathematics II.

The optional workshop may be on subjects such as problem solving, geometry or fractals.

It must be noticed that this Licenciatura has one of the most applications oriented profiles in Argentina. Most other programmes are much more abstract. For instance, they may not have courses such as numerical calculus or mathematical programming but will have courses on topology or algebraic structures.

2.3 Programme students.

In the 2000 academic year (we are in the Southern Hemisphere), 40 new students were enrolled. Overall there are about 100 students pursuing the mathematics major, and about 10 received their degree this year. Most of the graduating students (6) in 2000 are pursuing the master's or doctorate degree, i.e. following a research path.

2.4 Non-programme Students.

As is the case in other places, most of the teaching done by the Department of Mathematics is as a service to other careers.

The Department has about 1000 students per semester who are not in the mathematics majors track. Such students study chemical engineering, chemistry, industrial engineering, and food engineering. All the mathematics courses for these students are taught by the Mathematics Department.

Very few of the courses for mathematics majors are shared with other majors. One course that is shared by mathematics and industrial engineering majors is the linear programming course. In the second half of 2000, a pilot trial will be carried out with the discrete mathematics course. It will be available for both mathematics and industrial engineering majors.

3. TEACHING PRACTICES AND ASSESSMENT

Due to the increasing number of students entering the university, the courses that are not strictly for mathematics majors have increased their size considerably. The larger sections will have up to 80 students.

Usually, assessment is by two short (one hour) midterm exams, and a final (three hour) exam. If the class size allows it, quizzes and homework assignments are given. Especially in the initial courses, sometimes a minimum attendance of 80% is required.

The assessment given varies depending on the teacher and the course, from one extreme with quizzes, homework, midterms and final exams all being counted in some proportion, to the other extreme where just the final exam determines the grade.

4. USES OF TECHNOLOGY

The philosophy of the Department is to introduce the students as early as possible to the use of computer tools which will be valuable in applying what they are learning. Besides simple tools such as Excel (introduced in the first semester), mathematics students use Matlab (also introduced during the first semester), Mathematica and SPSS.

A computer programming course is taught during the first year, and usually includes the learning of the language Pascal, recursion, arrays and linked lists structures.

Some of the courses that use the computer are fully given in labs (such as the computer programming course); others have a 2 hour laboratory session per week (such as calculus or linear algebra). Up to 3 students work together simultaneously on the same computer and if this limit is exceeded generally computer work will not be required for the course.

5. TEACHER TRAINING

Usually in Argentina, the training of teachers for the elementary and secondary level are done in tertiary (as opposed to university) institutions. In Santa Fe, the university offers a degree for teaching at the secondary level, which is done in the Faculty of Science Teacher Training.

There is an on-going project of merging the courses for mathematics majors (licenciatura) and teachers (profesorado), as much as possible, so that students in either programme may switch more easily from one to the other. The programme for teachers has, of course, several courses in the social sciences (including didactics, psychology, etc.), and fewer mathematical courses. Hopefully, more than 50% of the courses will be in common to both programmes in the near future.

Nestor Aguilera and Roberto Macias,
Universidad Nacional del Litoral, Santa Fe, Argentina
aguilera@fermat.arcride.edu.ar
rmacias@fiqus.unl.edu.ar

UNIVERSITI TEKNOLOGI MALAYSIA, MALAYSIA

1. INTRODUCTION

The Department of Mathematics at the Universiti Teknologi Malaysia (UTM), has currently 73 full-time faculty members, all of whom are actively engaged in research and hold at least a Masters degree. The department is divided into 4 panels; algebra and analysis, operations research, applied mathematics and statistics. It offers undergraduate, masters and doctoral programmes in Mathematics and provides Mathematics and Statistics service teaching for other faculties in the university.

2. ENTRANCE REQUIREMENT

Students who hold the 'Sijil Pelajaran Malaysia'[1] (SPM, Malaysian Certificate of Education) or 'Sijil Pelajaran Malaysia Vokesional' (SPVM, Malaysia Vocational Certificate of Education) are usually eligible to enrol in UTM. They are also subject to certain specific requirements such as a:

- credit pass in Bahasa Malaysia;
- credit pass in Modern Mathematics; and
- credit pass in three other subjects required by the chosen major.

The subjects at SPM/SPVM are awarded at a range of levels from distinctions (A1 & A2) to credits (C3 to C6). However, several new programmes have been introduced to increase the number of students enrolling in the university. Among these programmes are:
a) relevant diploma holders who are placed directly into second year;
b) MARA Express: selected students pursuing their studies prior to knowng their SPM results;

---

[1] SPM is equivalent to GCE 'O' Levels (UK)

Derek Holton (Ed.), The Teaching and Learning of Mathematics at University Level: An ICMI Study, 195—198.
© 2001 Kluwer Academic Publishers. Printed in the Netherlands.

c) UTM Prepatory Course: students who did not secure places in any other local institution are registered as off-campus students. They may enrol in the first year of a course upon completion of their studies;

d) Ministry of Defense: cadets from various training colleges are selected by the Ministry of Defense to enrol in a university.

3. UNDERGRADUATE PROGRAMMES AND ENROLMENT

The department is currently offering the degree of Bachelor of Science in Industrial Mathematics for its students. It also collaborates with the Faculty of Education in offering two other degrees: Science and Computing with Education (Mathematics) and Science with Education (Mathematics). The graduates from these degrees will teach in secondary and higher secondary schools. (Primary teachers get their training from teachers' colleges.) In the 2000/2001 academic year, a total of 180 students registered for the above courses. In a future session, the department hopes to offer two newly designed programmes: B. Sc (Computational Mathematics) and B. Sc (Statistics and Operations Research).

The distribution of students by degree programmes is shown in Table 1. Here 1 credit is equivalent to 1 hour of lectures per week.

Table 1. Distribution of Credits

Programmes	No. of Students	No. of Credits (Mathematical & Sciences Subjects Foundation)	No. of Credits (Core Mathematics Subjects)	No. of Credits	Total Credits
Industrial Mathematics	61		75	62	137
Education with Mathematics	55	29	36	72	137
Computing with Education and Mathematics	64	29	36	64	129

Students who are pursuing a degree in Industrial Mathematics must take the subjects below.

Basic Mathematics   *Basic Calculus*
Advanced Calculus   *Vector Calculus*
Numerical Analysis   *Real Analysis*
Complex Analysis   *Inferential and Applied Statistics*
Operations Research   *Ordinary Differential Equation*
Partial Differential Equations

Furthermore, they must take some elective subjects worth of 15 credits from the list below.

Modern Algebra	*Decision Theory*
Experimental Design	*Optimization*
Discrete Mathematics	*Calculus of Variations*
Further Numerical Analysis	*Further Operations Research*
Optimal Control	*Time Series*
Mathematical Software	

On the other hand, students who are pursuing the two Education programmes will select some mathematics core subjects as part of their studies.

4. NON-PROGRAMME STUDENTS

By far the majority of students taught by the department are in service mathematics programmes. They come mostly from Civil, Mechanical, Electrical and Chemical Engineering faculties but some come from the Faculty of Education, and some from the Faculty of Computer Science and Information Technology. The department offers subjects such as basic mathematics, basic calculus, multi-variable and vector calculus, statistics, numerical analysis and differential equations for the engineering students. On the other hand, the department also teaches subjects such as linear algebra and discrete mathematics along with basic mathematics, basic calculus, multi-variable and vector calculus to the students from the Faculty of Computer Science. The department is currently teaching a total number of 9597 students, counting all students in the different years.

5. TEACHING: PRACTICE AND INNOVATION

The average teaching load for the majority of staff is 10-12 hours. This is made up of 6-9 hours of lectures and 3-4 hours of tutorials.

Several teaching issues have been identified by the department. These issues have been discussed and strategies have been taken to minimise their impact on the students.

Methods of teaching. The main teaching methods used by the departments are:
a) Lectures: Lectures are conducted to a large number of students (70 – 240 students).
b) Tutorials: Tutorials are intended to support lectures. They are designed to accommodate a smaller number (60) of students.

Transition. An added difficulty for the students in the first year is the change of learning environment from the term structure common in school to the semester system. They have to assimilate their learning in a shorter period before examinations. They also need to cope effectively with the change of teaching methods and classroom settings. The number of students in a lecture is usually very large and this is vastly different from their school experience. There they were given more individual attention; at university they are on their own in the anonymity of the lecture hall. Their lessons were more guided in school; at UTM they have to cope with independent learning.

Students' mathematical background. The first year subjects offered are designed to reinforce students' basic knowledge in mathematics. The subjects are carefully designed in order to prepare them for the mathematics taught in subsequent years. This is done due to the fact that UTM's entry requirements only demand credit passes in Modern Mathematics. On the other hand, students who have taken Additional Mathematics, an extra mathematics subject at SPM, tend to cope better with our programmes. Consequently, students in their first year who have only passed Modern Mathematics or who have very poor results in Additional Mathematics, are given extra classes to help them cope with first year mathematics.

Learning materials. The availability of learning materials solely depends on the lecturers' initiative. The lecturer will normally use one or all of the following:

i.   Modules: They are specially written for certain subjects to help the students in their learning.
ii. Web-based notes: Students have Internet access to learning materials prepared by their lecturers. Nevertheless, it is still compulsory for students to attend lectures and tutorials.
iii. Use of technology: The department encourages technology integrated teaching but it is left to the discretion of staff to implement it.

6. ASSESSMENT

Students are assessed on the basis of their coursework and a final examination. The coursework is in the form of quizzes, tests or assignments, and this carries at least 50% of the final mark. On the other hand, the final examination carries not more than 50% of the final mark. The duration of the final examination depends on the assigned credit for each subject. For example, 2 – 2.5 and 2 - 3 hour examinations are set for subjects with 2 and 3 credits, respectively. Students' achievement in each semester and throughout out their studies are evaluated using a Grade Point Average (GPA) and a Cumulative Point Average (CPA), respectively. Their CPA determines their academic status.

The department recognizes the importance of assignments in the learning process. However, due to the large number of students in a given class, some lecturers resort to tests and quizzes as a means of assessments in order to keep marking to a manageable level.

Contact person: *Roselainy, Abdul Rahman*
Universiti Teknologi Malaysia, Malaysia
lainy@mathsun.fs.utm.my

MARTTI E. PESONEN

UNIVERSITY OF JOENSUU, FINLAND

1. OVERVIEW OF THE DEPARTMENT

The department of Mathematics at the University of Joensuu has a teaching staff of 13 consisting of six professors, two lecturers, three senior assistants and two assistants. All but two younger assistants hold doctorate degrees. The academic staff also includes one full-time research fellow and seven full-time Ph.D. students.

The focus of undergraduate education is on mathematics subject teacher education. Students from the teacher education programme make up the largest group of those graduating. Students graduate with the Master of Science degree and, on average, there are about 20 graduates each year.

The main research interests of the faculty are function spaces, complex analysis, Clifford algebras and potential theory, and the numerical solutions of ordinary differential equations. The department has also research activities related to learning and teaching mathematics.

The department participates actively in the Erasmus chapter of the Socrates programme and has mobility agreements of student and teacher exchange with several European universities. In applied mathematics, the department also participates in the Leonardo exchange programme and the European Consortium for Mathematics in Industry.[1]

2. ENTRANCE REQUIREMENTS

In Finland, students entering universities have usually completed 12 years of pre-university studies. These include 9 years of elementary and lower secondary education and a three-year upper secondary school, finished with a national student examination. In upper secondary school, pupils can choose an extensive or a general syllabus in mathematics. In practice, only students who have completed the

---

[1] Socrates, a European Community action programme in the field of education, seeks to promote language learning, mobility and innovation in Europe. Erasmus is the higher education chapter of Socrates.

Leonardo da Vinci is a European Union Educational Programme, whose general scope is initial and continuous management and vocational training for adults, young people and the unemployed.

The European Consortium for Mathematics in Industry was founded in 1986 by mathematicians from ten European universities. Its aims are to promote the use of mathematical models in industry and to educate industrial mathematicians to meet the growing demand for such experts.

Derek Holton (Ed.), The Teaching and Learning of Mathematics at University Level: An ICMI Study, 199—204.
© 2001 *Kluwer Academic Publishers. Printed in the Netherlands.*

Towards the end of the 1990s the national policy has been to increase student enrolment in universities, especially in science and technology. Although nation-wide efforts have been initiated to increase the quality of science teaching and learning in schools, to attract students to mathematical subjects, and to improve teachers' and parents' attitudes towards mathematics and science, we still face serious recruitment problems.

The admission standards for studying mathematics in Finland are nowadays moderate. School grades and student examination results are the major factors in determining admission. Although there is a voluntary admission test in Joensuu for those who want to raise their achievement level, its role today is marginal. Almost everyone who has passed the senior secondary school with reasonable marks in the extensive syllabus in mathematics, is admitted to a mathematics department.

Those who enter to study mathematics make a choice between the three programmes described below. The choice is usually made after the first year but it can be delayed. However, in the mathematics teacher programme there is an additional aptitude test, containing an interview of the student and some observed activities in groups. Emphasis is on students' attitudes, communication skills and suitability for the teaching profession.

3. UNDERGRADUATE PROGRAMMES

Undergraduate degrees (Master of Science degree) consist of a minimum of 160 credits. One credit is worth 1.5 ECTS credits[2] and corresponds roughly to one week's workload. A full-time student normally completes the degree in five to six years. The duration of studies varies since many students work part-time during the last years of their studies.

The department offers three programmes: the mathematics teacher programme, the applied mathematics programme and the pure mathematics programme. The 1999 enrolments in these programmes are shown in Table 1:

Table 1: Mathematics major students 1999

Mathematics teacher	117
Pure mathematics	64
Applied mathematics	36
Total	217

There are several basic mathematics courses common to these three programmes. All mathematics students study an introductory mathematics course, 3 courses in analysis and a linear algebra course (these courses add up to 24 credits). There are also two or three elective courses that are common for all mathematics students.

Mathematics teacher programme. Students who graduate from the mathematics teacher programme are qualified for teaching positions at secondary level schools. Students aiming for the teaching profession have to pass an aptitude test, which is

---

[2] ECTS is the European credit transfer system to be applied within the Erasmus mobility scheme.

organised jointly by the Faculty of Education and the Faculty of Science. The structure of the degree in the mathematics teacher programme is described in Table 2.

Table 2: The structure of the mathematics teacher programme

Mathematics	60 credits
Minor subject	35 credits
Pedagogical studies	35 credits
Masters thesis	12 credits
Languages and communication	3 credits
Free choice	15 credits

There are 40 credits of fixed core courses of mathematics that are compulsory for all students in the programme. The remaining 20 credits can be chosen at will. The minor subject is usually physics, chemistry or computer science. Pedagogical studies take more or less than one year and they include some practical teacher training.

The mathematics and physics teacher programme has been selected twice during the 1990s as a national Centre of Excellence in university education. Most of the efforts in developing teaching and enhancing student engagement described below have started within the teacher programme.

Applied mathematics programme. The structure of the degree is described in Table 3.

Table 3: The structure of applied mathematics programme

Mathematics	68 credits
Minor subject	35 credits
Masters thesis	19 credits
Language and communication	3 credits
Free choice	35 credits

The courses in the applied mathematics programme are mainly concentrated on differential equations and their numerical analysis, and signal processing. Students can choose any minor subject they wish but most students choose computer science, physics or economic science. There is also a specialised signal processing curriculum developed in collaboration with the department of computer science and the department of physics. Applied mathematics students taking this option have to take the corresponding signal processing oriented topics from physics and computer science as their minor subjects. The signal processing study line has only been established for a few years and it is hoped that it will attract more students in the future.

Pure mathematics programme. The structure of the degree is the same as in Table 3. Again, students can choose any minor subject. In most cases, however, it is one of the following: computer science, physics, statistics, economic science or chemistry.

The pure mathematics programme is mainly intended for potential postgraduate students. In practice, however, the students taking pure mathematics form a heterogeneous group, including some very good students and some unmotivated students. The latter group are students who don't want to become school teachers and who are not willing to study computer science, so they take pure mathematics as a default option. There is a serious drop-out rate with this group.

Non-programme students and teaching of students from other departments. Almost one half of all teaching involves non-programme students. Most of the students in the physics teacher programme and some chemistry/computer science students take the 35-credit package of courses in mathematics. These students mainly participate in the same courses as the major students. The department also offers three elementary courses for non-programme students. These are partially aimed at those students of different subjects who just want to improve the quality of their mathematical prerequisites.

The department organises special courses as a part of the primary school teacher education. Annually, approximately 50 students participate in this programme. The number of primary school teacher students specialising in mathematics is large by national standards.

4. TEACHING PRACTICE AND INNOVATION

In a typical year, the department offers 26-28 one-semester programme courses, three elementary mathematics courses for students of different subjects and several short courses for primary school teacher students. Among the ordinary courses there are both seminars where students introduce their work and frequent special courses given by visitors from different countries. The number of students participating in particular courses varies from almost one hundred in basic courses to about 10 in advanced courses.

In the University of Joensuu a flexible work allocation system has been in use since 1992. The faculty members have a 1600 hour annual working load, fixed yearly together in discussion between by the staff member and the department. Consequently, the amount of teaching varies considerably among the faculty and may even vary from year to year for a particular member of staff. The weekly teaching load can vary from a professor's 4 lecture hours to a lecturer's 14 teaching hours (of which 10 are exercise hours).

Fifteen years ago there was only one programme for all major students and teaching used the very traditional "chalk-and-talk" approach. Students were assigned homework problems and presented their solutions on the blackboard during exercise sessions. Since then there has been an increasing tendency to vary the teaching practices and to organise special courses for the students in different programmes.

An overview of the changes in teaching methods. Traditional teaching by lecturing and the basic mathematical content have not been abandoned but they have been partially replaced by some alternative approaches suggested by pedagogical research and by the priorities of the particular teachers in charge. The most important changes are given below.

4.1 Tutorial study groups and homework tutoring.

The basic analysis courses form the core of the mathematics studies. Most first and second year students consider the compulsory analysis courses very demanding and difficult. In these courses so-called tutorial study groups have been especially arranged for the students in the teacher programmes. The basic idea behind the tutorial study groups is to encourage students to discuss mathematics. The students are not only listeners but also teachers themselves. The learners, working in pairs, are teaching each other, trying to solve problems or to construct proofs. An important task is also to learn to work at the blackboard, explaining while writing and drawing. Beyond the content of analysis, the interrelations between school mathematics and the structure of university mathematics are dealt with. One of the purposes is also to increase the appreciation of the teacher's profession.

Among the tutorial study groups, another kind of group work was trialled during the autumn 1998 analysis course. Groups were given some rather tough problems, which they partially solved at home. The problems were then studied under the guidance of the teacher, and the final written solutions were graded.

The newly established daily homework tutoring hours offer freshmen a guided opportunity for doing their homework, while the tutors, older students, can practise their teaching.

4.2 Profession-oriented mathematics courses for pre-service teachers.

Examples of courses, which had not belonged to the traditional syllabus, are: *School Mathematics* - a course for solving school mathematics exercises, designing exercises and examinations, and practising problem solving skills; *Pro-seminar* where students report the work that they have done in their project. The topics for these projects are often taken from publications like 'College Mathematics Journal' and 'Mathematics Magazine'; *Geometry,* a compulsory advanced course with an analytic approach to plane geometry; *Structural Questions in Mathematics,* an overview of mathematical structures, their development and mutual relations, and their meaning for school mathematics;

Advanced courses in mathematics specially tailored for the teacher programme students, but generally available for all students are, *Number Theory, Mathematical Computing in Schools, History of Mathematics, Geometry of Fractals.*

4.3 Mathematics subject studies with pedagogical aspects.

While general didactics and special content for pre-service teachers are taught both in pedagogical studies and in the courses above, the subject studies can provide an authentic environment for adopting attractive discipline-specific teaching and learning methods. Several mathematics courses have a significant emphasis on 'Didactical Mathematics' which, roughly speaking, means teaching rather traditional mathematics content using certain pedagogical approaches, in order to offer various models of learning platforms to the students. In axiomatic geometry 'playing' with toy sets like 'balks and bolts' is used to cement the meaning of axioms and properties. In linear algebra, identification and production tasks concerning the connections between multiple representations – verbal, symbolical, graphical – of objects like functions and binary operations, both standard and even weird ones, are used extensively in learning the axiomatics of general vector spaces. Dynamic geometry computer programs like Geometer's Sketchpad, Cabri Géomètre or their Java applet based counterparts, can make this approach more concrete and illustrative, and initiate students into an investigative mode of learning. The process nature of objects becomes visible through interactive animations. A new approach to the elements of complex analysis is a Web-based course, which contains text and traditional exercises. These are enriched by multiple choice questions and identification problems checked automatically by the technical system, animations produced using Maple V graphics, and interactive or automated animations using the dynamic geometry applet JavaSketchpad. This approach requires special kinds of pedagogy and new ways of interaction.

Since 1989, we have used Matlab-programming in teaching discrete mathematics algorithms, especially those arising in graph theory. This positive experience, together with the idea of co-operative learning, led us to try teaching abstract algebra by combining the two coinciding methods in 1997[3].

5. COMPARISON WITH OTHER UNIVERSITIES IN FINLAND

The main institutes offering undergraduate and graduate mathematics education in Finland are the universities in Helsinki, Joensuu, Jyväskylä, Oulu, Tampere and Turku, and the Helsinki University of Technology in Espoo. In Joensuu, pre-service teachers comprise about 10% of the national total. There are lower percentages of students in the pure and applied mathematics.

The Universities of Joensuu, Oulu and Helsinki have been the most active in developing and updating undergraduate mathematics teaching, especially in teacher education. The mathematics teacher education programme at the University of Joensuu was chosen to be one of the first national Centres of Excellence in university teaching in 1995, and again in 1996. An important factor was undoubtedly the department's serious contribution to primary school teacher education that was mentioned above. At present, the Department of Mathematics in Oulu is a Centre of Excellence, and also the Mathematics Department in the University of Helsinki has had considerable success, especially with its subject teacher education programme.

Martti E. Pesonen
University Of Joensuu, Finland
Martti.Pesonen@Joensuu.Fi
(The departmental Web site is http://www.joensuu.fi/sciences/mathematics/index.html)

[3] This is described in more detail in the ICMI study special issue of *Int. J. Math. Educ. Sci. Technol.*, 2000, vol. 31, no. 1, pp. 113-124.

SECTION 3

RESEARCH

Edited by Michèle Artigue and Alan Schoenfeld

MICHÈLE ARTIGUE

WHAT CAN WE LEARN FROM EDUCATIONAL RESEARCH AT THE UNIVERSITY LEVEL?[1]

1. INTRODUCTION

For more than 20 years, educational research has dealt with mathematics learning and teaching processes at the university level. It has tried to improve our understanding of the difficulties encountered by students and the dysfunction of the educational system; it has also tried to find ways to overcome these problems. What can such research offer to an international study? This is the issue I will address in this article, but first I would like to stress that it is not an easy question to answer, for several reasons including at least the following:

1. Educational research is far from being a unified field. This characteristic was clearly shown in the recent ICMI study entitled "What is research in mathematics education and what are its results?" (See Sierpinska and Kilpatrick, 1996.) The diversity of existing paradigms certainly contributes to the richness of the field but, at the same time, it makes the use and synthesis of research findings more difficult.
2. Learning and teaching processes depend partly on the cultural and social environments in which they develop. Up to a certain point, results obtained are thus time- and space- dependent, their field of validity is necessarily limited. However, these limits are not generally easy to identify.
3. Finally, research-based knowledge is not easily transformed into effective educational policies.

I will come back to this last point later on. Nevertheless, I am convinced that existing research can greatly help us today, if we make its results accessible to a large audience and make the necessary efforts to better link research and practice. I hope that this article will contribute to making this conviction not just a personal one. Before continuing, I would like to point out that the diversity mentioned above does not mean that general tendencies cannot be observed. At the theoretical level, these are indicated, for instance, by the dominating influence of constructivist approaches inspired by Piaget's genetic epistemology, or by the recent move

---

[1] A shorter version of this paper, Artigue (1999), was published in the *Notices of the American Mathematical Society*.

Derek Holton (Ed.), The Teaching and Learning of Mathematics at University Level: An ICMI Study, 207—220.
© 2001 *Kluwer Academic Publishers. Printed in the Netherlands.*

attempt to take more account of the social and cultural dimensions of learning and teaching processes (see Sierpinska and Lerman, 1996). But within these general perspectives, researchers have developed a multiplicity of local theoretical frames and methodologies, which differently shape the way research questions are selected and expressed, and the ways they are worked on – thus affecting the kind of results which can be obtained, and the ways they are described. At the cultural level, such general tendencies are also observed. Strong regularities in students' behaviour and difficulties as well as in the teaching problems met by educational institutions, have been observed. These, up to a point, apparently transcend the diversity of cultural environments.

In the following, after characterizing the beginnings of the research enterprise, I will try to overcome some of the above-mentioned difficulties presenting research findings along two main dimensions of learning processes: qualitative changes, reconstructions and breaches on the one hand, cognitive flexibility on the other hand. These dimensions can to some degree, be considered 'transversal' with respect to theoretical and cultural diversities as well as to mathematical domains. No doubt this is a personal choice, induced by my own experience as a university teacher, as a mathematician, and as a education researcher; it shapes the vision I give of research findings, a vision which does not pretend to be objective or exhaustive.

2. FIRST RESEARCH RESULTS: SOME NEGATIVE REPORTS

The first research results obtained at university level can be considered negative ones. Research began by investigating students' knowledge in specific mathematical areas, with particular emphasis on elementary analysis (or calculus in the Anglo-Saxon culture), an area perceived as the main source of failure at the undergraduate level. The results obtained gave statistical evidence of the limitations both of traditional teaching practices and of teaching practices which, reflecting the Bourbaki style, favoured formal and theoretical approaches. The structure and content of the book, *Advanced Mathematical Thinking* (Tall, 1991), gives clear evidence of these facts, noting that:

- by the early eighties, Orton (1980), in his doctoral thesis, showed the reasonable mastery English students had of what can be labelled as 'mere algebraic calculus': calculation of derivatives and primitives (anti-derivatives), but the significant difficulty they had in conceptualizing the limit processes underlying the notions of derivative and integral;
- at about the same time, Tall and Vinner (1981), highlighted the discrepancy between the formal definitions students were able to quote and the criteria they used in order to check properties such as functionality, continuity, derivability. This discrepancy led to the introduction of the notions of concept definition and concept image in order to analyze students' conceptions;

- very early, different authors documented students' difficulties with logical reasoning and proofs, with graphical representations, and especially with connecting analytic and graphical work in flexible ways.

Schoenfeld (1985), also documented the fact that, faced with non-routine tasks, students – even apparently bright students – were unable to efficiently use their mathematical resources.

Research also showed, quite early, that the spontaneous reactions of educational systems to the above-mentioned difficulties were likely to induce vicious circles such as the following. In order to guarantee an acceptable rate of success, an increasingly important issue for political reasons, teachers tended to increase the gap between what was taught and what was assessed. As the content of assessments is considered by students to be what should be learnt, this situation had dramatic effects on their beliefs about mathematics and mathematical activity. This, in turn, did not help them to cope with the complexity of advanced mathematical thinking.

Fortunately, research results are far from being limited to such negative reports. Thanks to an increasing use of qualitative methodologies allowing better explorations of students' thinking and the functioning of didactic institutions (Schoenfeld, 1994), research developed and tested global and local cognitive models. It also organized in coherent structures the many difficulties students encounter with specific mathematical areas, or in the secondary/tertiary transition. It led to research-based teaching designs (or engineering products) which, implemented in experimental environments and progressively refined, were proved to be effective. Without pretending to be exhaustive, let us give some examples, classified according to the two main dimensions given above. (For more details, the reader can refer to the different syntheses in Artigue, 1996, Dorier, 2000, Schoenfeld, 1994, Tall, 1991 and 1996; to the special issues dedicated to advanced mathematical thinking by the journal *Educational Studies in Mathematics* in 1995 edited by Dreyfus; by the journal *Recherches en Didactique des Mathématiques* in 1998 edited by Rogalski; to some of the diverse monographs published by the Mathematical Association of America about calculus reform, innovative teaching practices; and to research about specific undergraduate topics to be found in the MAA Notes on Collegiate Mathematics Education.)

3. QUALITATIVE CHANGES, RECONSTRUCTIONS AND BREACHES IN THE MATHEMATICAL DEVELOPMENT OF KNOWLEDGE AT UNIVERSITY LEVEL

One general and crosscutting finding in mathematics education research is the fact that mathematical learning is a cognitive process that necessarily includes 'discontinuities.' But, depending on the researcher this attention to discontinuities is expressed in different ways. In order to reflect this diversity and the different insights it allows, I will describe three different approaches: the first one, in terms of process/object duality, the second one in terms of epistemological obstacles, the third one in terms of reconstructions of relationships to objects of knowledge.

3.1 Qualitative changes in the transition from processes to objects: APOS theory

As mentioned above, research at the university level is the source of theoretical models. The case of APOS theory, initiated by Dubinsky (see Tall 1991) and progressively refined (see Dubinsky and McDonald, this volume, pp. 275-282), is typical. This theory, which is an adaptation of the Piagetian theory of reflective abstraction, aims at modelling the mental constructions used in advanced mathematical learning. It considers that "understanding a mathematical concept begins with manipulating previously constructed mental or physical objects to form actions; actions are then interiorized to form processes which are then encapsulated to form objects. Objects can be de-encapsulated back to the processes from which they were formed. Finally, actions, processes and objects can be organized in schemas" Asiala et al, 1996. Of course, this does not occur all at once and objects, once constructed, can be engaged in new processes and so on. Researchers following this theory use it in order to construct genetic decomposition of concepts taught at university level (in calculus, abstract algebra, etc.) and design teaching processes reflecting the genetic structures they have constructed and tested.

As with any model, the APOS model only gives a partial vision of cognitive development in mathematics, but one cannot deny today that it put to the fore a crucial qualitative discontinuity in the relationships students develop with respect to mathematical concepts. This discontinuity is the transition from a process conception to an object one, the complexity of its acquisition and the dramatic effects of its underestimation by standard teaching practices.[2] Research related to APOS theory also gives experimental evidence of the positive role which can be played by programming activities in adequate languages (such as the language ISETL, cf. Tall, 1991) in order to help students encapsulate processes as objects.

Breaches in the development of mathematical knowledge: Epistemological obstacles. The theory of epistemological obstacles, firstly introduced by Bachelard (1938) and imported into educational research by Brousseau (1997), proposes an approach complementary to cognitive evolution, focussing on its necessary breaches. The fundamental principle of this theory is that scientific knowledge is not built in a continuous process but results from the rejection of previous forms of knowledge: the so-called epistemological obstacles. Researchers following this theory hypothesize that some learning difficulties, often the more resistant ones, result from forms of knowledge which are coherent and have been for a time effective in social and/or educational contexts. They also hypothesize that epistemological obstacles have some kind of universality and thus can be traced in the historical development of the corresponding concepts. At the university level,

---

[2] Note that a very similar approach was developed independently by Sfard, with more emphasis on the dialectic between the operational and structural dimensions of mathematical concepts in mathematical activity (Sfard, 1991).

such an approach has been fruitfully used in research concerning the concept of limit (cf. Artigue 1998 and Tall 1991 for synthetic views). Researchers such as Sierpinska, (1985), Cornu, (1991) and Schneider, (1991) provide us with historical and experimental evidence of the existence of epistemological obstacles, mainly the following:

- the everyday meaning of the word 'limit', which induces resistant conceptions of the limit as a barrier or as the last term of a process, or tends to restrict convergence to monotonic convergence;
- the overgeneralization of properties of finite processes to infinite processes, following the continuity principle stated by Leibniz;
- the strength of a geometry of forms which prevents students from clearly identifying the objects involved in the limit process and their underlying topology. This makes it difficult for students to appreciate the subtle interaction between the numerical and geometrical settings in the limit process.

Let us give one example (taken from Artigue, 1998) of this last resistance, which occurs even in advanced and bright students. In a research project about differential and integral processes, advanced students were asked the following non-standard question: "How can you explain the following: using the classical decomposition of a sphere into small cylinders in order to find its volume and area, one obtains the expected answer for the volume $\frac{4}{3}\pi R^3$, but $\pi^2 R^2$ for the area instead $4\pi R^2$?" It was observed that, faced with this question, the great majority of advanced students tested got stuck. And, even if they were able to make a correct calculation for the area (which they were not always able to do) they remained unable to resolve the conflict.

As the students eventually said, because the pile of cylinders, geometrically, tends towards the sphere, the magnitudes associated with the cylinders behave in the same way and thus have as a limit the corresponding magnitude for the sphere. Such a resistance may look strange but it appears more normal if we consider the effect produced on mathematicians by the famous Schwarz counterexample showing that, for a surface as simple as a cylinder, limits of areas of triangulations when the size of the triangles tends towards 0, can take any value greater than or equal to the area up to infinity, depending on the choices made in the triangulation process, an effect nicely described by in Lebesgue, (1956). The historical and universal commitments of the theory which leads to such results can be discussed and are presently discussed (see, for instance, Radford, 1997). However, what cannot be negated is the fact that the above-mentioned forms of knowledge constitute resistant difficulties for today's students; moreover, that mathematical learning necessarily implies partial rejection of previous forms of knowledge, which is not easy for students.

3.2   Reconstructions in the secondary/tertiary transition: The case of calculus

Qualitative changes in the relationships students develop with respect to mathematical concepts can be approached in a less radical way: in terms of necessary reconstructions. In this section, we illustrate corresponding research findings, by focusing on reconstructions which have been proved to play a crucial role in calculus at the secondary/tertiary transition, at least in the educational situation which tends to predominate now where an intuitive and pragmatic approach to calculus in the secondary curriculum, precedes the formal approach introduced at university. Some of these reconstructions deal with mathematical objects already familiar to students before the official teaching of calculus. Real numbers are a typical example. They enter the secondary curriculum early as algebraic objects with a dense order, with a geometrical representation as the real line, and with decimal approximations that can be easily obtained with pocket calculators. Nevertheless, many pieces of research show that, even upon entering university, students' conceptions remain fuzzy, incoherent, and poorly adapted to the needs of the calculus world. For instance, the ordering of the real numbers is recognized as a dense order. However, depending on the context, students can reconcile this property with the existence of numbers just before or after a given number (0.999... is thus often seen as the predecessor of 1). More than 40% of students entering French universities consider that, if two numbers A and B are closer than $\dfrac{1}{N}$ for every positive N they are not necessarily equal, just infinitely close. Relationships between irrational numbers and their decimal approximations remain fuzzy. There is no doubt that reconstructions are necessary for understanding 'calculus thinking modes'. Research shows that these are not easily induced by the kind of intuitive and algebraic analysis which is the main focus of calculus instruction at the high school level, and that the constructions of the real number field introduced at the university level have little effect if students are not faced with the incoherence of their conceptions and the resulting cognitive conflicts.

A second category of reconstructions results from the fact that only some facets of a mathematical concept can be introduced at a first contact with it. The concept of integral illustrates this case fairly well. In many countries the first contact with integrals occurs at the upper secondary level via the notion of anti-derivative and a pragmatic approach to the Fundamental Theorem of Calculus which allows the anti-derivative to be connected with an intuitive notion of area. Only at university is a theory of integration developed, first as the theory of Riemann integrals, then, at a more advanced level, as the Lebesgue theory. All of this requires successive reconstructions of the relationships that students have with the integral concept. Much research has been devoted to this theme with great consistency in the results obtained all over the world, documenting the limitations of standard teaching strategies. These results clearly show that reconstruction cannot result from a mere presentation of the theory of Riemann integrals. Through standard teaching practices, students become reasonably successful on standard tasks, but no more. For example, if students are asked in modelling tasks to decide by themselves whether a problem requires an integral process for being solved, they get completely stuck or base their answers on the linguistic hints, if any, that they have learnt to notice in the standard versions of such tasks. Most students think that the safest way to deal successfully with this domain is not to try to understand, but to just function mechanically. I would like to add that we don't have to see this as a sort of cognitive fate. We merely observe our students' economic ways of adaptation to inadequate educational practices.

Research, as was stressed above, is not limited to such negative reports. I would like now to present a situation created by Legrand (1997), in the context of a research project involving mathematicians and physicists with the goal of making first-year university students really feel by themselves the necessity of the integral concept. The situation is based on the following apparently very simple problem (the most effective situations found by researchers are very often apparently simple ones). A linear bar of M_1 and a point mass M_2 are located as shown. Students are asked to calculate the intensity of the attraction between the two masses.

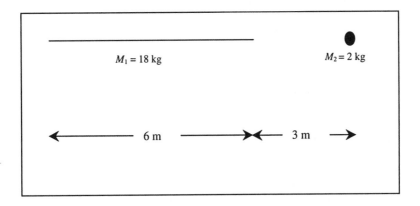

Figure 1. Attraction between a bar and point mass

This situation has been shown effective in various experiments in different contexts. Why is it effective? To answer this question, we need a brief didactic analysis. When asked this question without any linguistic hint, first-year students don't recognize it as an integral problem. But the first important point is that they are not stuck because they can rely on a strategy often used in physics: concentrating the mass of the bar at its centre of gravity and applying the familiar attraction law between two point masses. In experiments, this strategy has always predominated. But, in a group of reasonable size, as is easily the case at university level, there are always students who have some doubts. "Is the gravity principle valid in that particular case?" A second strength of the situation results from the fact that one can test the validity of the gravity principle, simply by applying it in another way. Students generally suggest that the bar be cut into two halves and the gravity principle be applied to each half. Of course, this does not give the same result and the gravity principle is shown to be invalid in that particular case. But this negative answer is also a positive one because it makes salient an essential fact: the contribution of a piece of the bar to the attraction force depends on its distance to the mass x. This allows students to propose upper and lower bounds for the required intensity. Moreover, the technique which was the basis for the invalidation process can be then used in a progressive refinement process, which leads students to the conviction that the force, whose existence is physically attested, can be approximated as accurately as desired. What underlies this is simply the fundamental integral process. In the didactic design elaborated by Legrand, this is just the starting point. Students have then to work on situations that, in different contexts, require the same solution process. Then they have to look for and discuss the analogies between the solutions in order to make the integral process an explicit tool (in the sense of the distinction between the tool and object dimensions of mathematical concepts introduced by Douady, 1987). Only at that point does the university teacher connect this with the theory of Riemann integrals and develop the notion of integral as a mathematical object that will be then reused in more complex situations.

Before leaving this point, let me stress the following: efficiency here is not only linked to the characteristics of the problem which I have just described, it strongly depends on the kind of scenario developed in order to organize students' encounter with this new facet of the integral concept. In a crucial way, this scenario plays on the social character of learning processes. It is through group discussion that the initial strategy is proved to be erroneous. It is the collective game which allows a solution to be found in a reasonable amount of time and which fosters regularities in the dynamic of the situation which could not be ensured if students were faced with the same problem individually or in very small groups. (A similar point is made by Stigler and Hiebert, 1999, p. 164.) No doubt also that the effect would be different if the teacher simply presented this particular example during a lecture.

This example might appear idyllic. But I must confess that educational research does not so easily provide us with effective means to deal with all necessary reconstructions. For instance, differences are evident if one considers the concept of limit, central to calculus. With this particular example, we come to a third category of reconstructions, reconstructions necessary because, as was already acknowledged at the beginning of the last century by the famous mathematician Poincaré (1904), concepts cannot be necessarily taught from the start in their definitive form. At high school level, in most countries today, the impossibility of entering the field of analysis, formally, has been acknowledged. Teaching relies both on a dynamic conception of the limit, based on graphical and numerical explorations, and on techniques of an algebraic nature (Artigue, 1996). These allow students to solve simple but interesting problems of variation and optimization. The transition towards more formal approaches, which takes place at university, represents a tremendous gap both conceptually and technically.

From a conceptual point of view, one crucial point is the following: what is at play through the formalization of the limit concept is, above all, an answer to foundational, unification and generalization needs (see Dorier, 1995, Robert, 1998, or Robert and Speer, this volume, pp. 283-301). It is not easy to make young students sensitive to such concerns because those concerns are not really part of their mathematical culture. From a technical point of view, the following is essential: in the algebraic analysis of the first contact, technical work does not really break with ordinary algebraic work. This is no longer the case when one enters the field of formal analysis. For example, students must reconstruct the meaning of equality and understand that it doesn't necessarily result, as in algebra, from successive equivalencies, but from ε-proximity for every positive ε. Another point is that inequalities become more frequent than equalities, generating a strong increase in technical complexity, especially so as associated modes of reasoning most often rely on sufficient conditions. These new modes require a carefully controlled loss of information based on a good awareness of the respective orders of magnitude of the different parts of the expressions students have to deal with. In brief, students have a completely new technical world to identify and learn to master. This is far from being easy and is necessarily a long-term process.

3.3 Some concluding remarks: From calculus to linear algebra

Up to now, I have focussed on qualitative changes and more or less radical reconstructions. As stressed above, research shows that teaching practices underestimate both the conceptual and technical costs of these changes. Teaching tends to leave the responsibility for most of the corresponding reorganization to students, with dramatic effects for the majority of these, especially at the secondary/tertiary transition. Research also shows that alternative strategies can be developed fruitfully. Examples have been given for calculus, a domain extensively explored by research. But the growing body of research in linear algebra attests to the existence of similar phenomena (see Dorier and Sierpinska, this volume, pp. 253-272). For instance, the concept of abstract vector space in its axiomatic form, from an epistemological point of view, has been proved to share some common characteristics with the formal concept of limit. When it entered the mathematical scene, its value as a generalizing, unifying, and formalizing concept was stronger than its potential for solving new problems and it was not easily accepted by mathematicians. The same situation occurs with our students who do not need this abstract construction to solve most problems in a first linear algebra course. In France, some researchers have developed specific didactic strategies which aim at making it possible for students to do the necessary reflective and cultural mathematical work (see Dorier et al. 2000). In other countries, these difficulties tend to be removed by reducing topics in first linear algebra courses to those in spaces isomorphic to R^n and by emphasizing matrix calculus and applications Carlson et al (1993). Recent Canadian research (Hillel and Sierpinska, 1994) suggests that this choice is not so benign as it might appear at first sight. Living in a linear algebra world built on the structure of R^n makes it difficult to differentiate vectors and transformations from their canonical representations and can induce further obstacles.

3.4   Cognitive Flexibility in Learning and Teaching Processes

The result just mentioned above is linked with a more general issue, that of relationships between mathematical concepts and their semiotic representations, an issue to which educational research pays increasing attention. This fact does not seem independent of the global evolution of theoretical frames mentioned at the beginning of this article, because socio-cultural and anthropological approaches are especially sensitive to the role played by the material and symbolic tools of mathematical activity in learning processes. Depending on the theoretical perspective, this attention is expressed in different ways, but the fundamental point is that it breaks with a common vision of instrumental and semiotic competencies as a by-product of conceptualization and hypothesizes strong dialectic relationships in their mutual development. This is of particular importance, especially if one has in mind the current technological evolution of the instruments of mathematical activity. More generally, mathematical learning can no longer be seen, as is often the case, only as a regular ascension towards higher levels of abstraction and formalization. Connections between mathematical fields of experience, points of view, settings, and semiotic registers are a crucial part. With such considerations in mind, we enter a wider domain that could be labelled the domain of cognitive flexibility, which is increasingly investigated by research (see, for instance, Dreyfus and Eisenberg, 1996).

I will use some examples taken from recent research in linear algebra in order to illustrate this point. As stressed by Dorier (2000), historically linear algebra helped to unify different pre-existing mathematical settings: geometry, linear systems in finite and infinite dimensions and determinants, differential equations, and functional analysis. This unifying role and power is an essential epistemological value of linear algebra that has to be understood and used by students. But this cannot be achieved without the development of complex connections among reasoning modes, points of view, languages, and systems of symbolic representations. Once more, research helps our understanding of the complexity of the necessary cognitive constructions and, at the same time, shows the insensitivity of the educational system to this complexity. In Dorier (2000), for instance, on the one hand, Hillel points out the necessary interaction in linear algebra between three different levels of language and representations: those of the general theory, of geometry, and of R^n. On the other hand, Sierpinska et al. show the necessary interaction between three different reasoning modes, respectively labelled as synthetic and geometric, analytic and arithmetic, analytic and structural.[3] Both show the inadequacy of the different teaching practices documented, from lectures to tutorials. Alves Dias (1998), in her recent doctoral thesis, analyses the relationships between two fundamental points of view in linear algebra: the parametric and

---

[3] In the synthetic mode, mathematical objects are, in some way, directly given to the mind, which tries to grasp and describe them. In the analytic mode, they are given indirectly: built through definitions and properties of their elements. This analytic mode is divided by researchers into two different sub-modes: the analytic-arithmetic where objects are given by a formula which makes it possible to calculate them, and the analytic-structural where objects are defined by a set of properties.

Cartesian points of view.[4] She clearly shows that, even if the conversion between parametric and Cartesian representations of vector subspaces is, a priori, easily achieved thanks to ordinary techniques for solving systems of linear equations, when dealing with vector spaces of finite dimensions, a flexible connection between these two points of view is far from being mastered by advanced French and Brazilian students. Mathematical symbols such as matrices can foster errors in the use of those formal representations because students operate on the formal symbols without checking to see if the operations they perform are meaningful in terms of the objects the symbols represent. This often leads to absurd results which are not recognised by students because they do not interpret or check their findings through geometrical or dimensional arguments. The detailed analysis of textbooks Alves Dias carried out, shows that they don't pay attention to these questions or develop theoretical arguments, for instance in terms of duality, which remain too far away from the technical level to make students able to control the connection.

These are examples in linear algebra. As documented by research, mutatis mutandis, there are similar examples in calculus. In that more extensively explored area, research also provides experimental evidence that computer technologies, if properly used (which is not so easy) can play a crucial role in fostering flexible connections among semiotic representations. For instance, among graphical, numerical, and symbolic representations of functions, and help graphical representations to become effective tools of the mathematical work (see Tall, 1991 and Dubinsky and Harel, 1992). Research also shows that the effective use of computer technologies requires the development of specific mathematical knowledge, a requirement which is not easily accepted by an educational institution whose values have been traditionally defined with respect to paper and pencil environments.

4. POTENTIAL AND LIMITS OF RESEARCH FOR ACTION ON THE EDUCATIONAL SYSTEM

As we have tried to show in this article, research carried out at the university level helps us better understand the learning difficulties our students have to face, the surprising resistance of some, and the limitations and dysfunction of some of our teaching practices. Moreover, in various cases, research has led to the production of teaching designs that have been proved to be effective, at least in experimental environments. But we must also recognize that research does not give us a general way to easily improve the learning and teaching processes. Some reasons can be found in the current state of research: up to now, efforts have been concentrated on a few domains taught at university level. Also, the training of future mathematicians at the expense of the great diversity of students taking university mathematics courses, has more or less implicitly been assumed. Research remains thus very partial due

---

[4] A parametric point of view is adopted with a vector subspace for instance if the subspace is characterized by some set of generators. A Cartesian point of view is to characterize a subspace as the solutions of a linear system or as the null space of a linear operator.

both to the content it explores and to its vision of the expected form and content of knowledge. In my opinion, the way the issue of computer technologies has been generally addressed evidences this fact. It mainly focuses on the ways computer technologies can support conceptualization and the cognitive flexibility recognized as an essential component of this conceptualization. It does not give the same attention to what is really a professional mathematical activity assisted by computer technologies, and the specific and non-specific mathematical needs, depending on professional specialty, required to become an efficient and critical user and how the corresponding knowledge can be constructed in ordinary or service mathematics courses. Nevertheless, this is also a real challenge we must face today, taking into account the fact that, at university, our main concern is no longer the development of some kind of general mathematical culture.

Other reasons such as the following seem more fundamental: it is rare that research allows us to think that through minimal and cheap adaptations we could obtain substantial gains. On the contrary, most research-based designs require more engagement and expertise from teachers, and significant changes in practices (see for instance Dubinsky, Mathews and Reynolds, 1997 as regards collaborative learning). One essential reason is this. What has to be reorganized is not only the content of teaching (it is not enough to write or adopt new textbooks), but more global issues such as the forms of students' work, the modes of interaction between teachers and students, and the forms and contents of assessment. This is not easy to achieve and is not just a matter of personal good will.

Another crucial point is the complexity of the systems where learning and teaching take place. Because of this complexity, the knowledge that we can infer from educational research is necessarily very partial. The models we can elaborate are necessarily simplistic ones. We can learn a lot even from simplistic models but we cannot expect that they will give us the means to really control didactic systems. So we must be realistic in our expectations and careful about generalizations. This does not mean, in my opinion, that the world of research and the world of practice must live and develop separately. Far from it. But it does mean that finding ways of making research-based knowledge useful outside the communities and experimental settings where it develops cannot be left as the sole responsibility of researchers. It is our common task.

REFERENCES

Alves Dias, M. (1998). *Les problèmes d'articulation entre points de vue cartésien et paramètrique dans l'enseignement de l'algèbre linéaire*. Ph.D. Thesis. Université Paris 7.

Artigue, M. (1996). Learning and Teaching Elementary Analysis. In C. Alsina, J.M. Alvarez, M.Niss, Aérez, L.Rico, A.Sfard (Eds.), 8th International Congress on Mathematics Education – Selected Lectures, pp. 15-30. Sevilla: S.A.E.M. Thalès.

Artigue, M. (1998). L'évolution des problématiques en didactique de l'analyse. *Recherches en Didactique des Mathématiques*, vol. 18.2, 231-261.

Artigue, M. (1999). The teaching and learning of mathematics at the university level: Questions for contemporary research in education. *Notices of the American Mathematical Society*, 46(11), 1377–1385.

Asiala, M., Brown A., DeVries D., Dubinsky E., Mathews D. and Thomas K. (1996). A framework for Research and Curriculum Development in Undergraduate Mathematics Education. *CBMS Issues in Mathematics Education*. vol. 6. 1-32.

Bachelard , G. (1938 *La formation de l'esprit scientifique*. Paris: J. Vrin.

Brousseau, G. (1997). *The Theory of Didactic Situations*. Dordrecht: Kluwer Academic Publishers.

Carlson, D., Johnson C., Lay D. and Porter, A. (1993). The Linear Algebra Curriculum Study Group recommendations for the first course in linear algebra. *College Mathematics Journal*, 24.1, 41-46.

Cornu, B. (1991). Limits. In D. Tall (Ed.), *Advanced Mathematical Thinking*, pp. 153-166. Dordrecht : Kluwer Academic Publishers.

Dorier, J.L. (1995). Meta level in the teaching of unifying and generalizing concepts in mathematics. *Educational Studies in Mathematics*, no. 29.2, 175-197.

Dorier, J.L. (1998). The role of formalism in the teaching of the theory of vector spaces. *Linear Algebra and its Applications*, 275-276, 141-160.

Dorier, J.L. (Ed.) (2000). *On the teaching of linear algebra*. Dordrecht: Kluwer Academic Publishers.

Dorier, J.L., Robert, A., Robinet, J., Rogalski M. (2000). The meta lever. In J.L. Dorier (Ed), *On the teaching of linear algebra*, pp. 151-176. Dordrecht: Kluwer Academic Publishers.

Dorier, J.L. and Sierprinska, A. (2001), Research into the Teaching and Learning of Linear Algebra, this volume pp.255-274.

Douady, R. (1987). Dialectique outil/objet et jeux de cadres. *Recherches en Didactique des Mathématiques*, vol. 7.2, 5-32.

Dreyfus, T. (Ed.) (1995). Special issue on Advanced Mathematical Thinking. *Educational Studies in Mathematics*, vol. 29.2.

Dreyfus, T. and Eisenberg, T. (1996). On different facets of mathematical thinking. In R.J. Sternberg and T. Ben-zeev (Eds.), *The Nature of Mathematical Thinking*, pp. 253-284. Mahwah, NJ: Lawrence Erlbaum Associates, Inc..

Dubinsky, E. and Harel, G. (Eds) (1992). *The Concept of Function: Some Aspects of Epistemology and Pedagogy*. MAA Notes no. 25. Washington D.C.: Mathematical Association of America.

Dubinsky, E. and MacDonald, M.A. (2001). APOS: A Constructivist Theory of Learningin Undergraduate Mathematics Education Research, this volume pp.275-282.

Dubinsky, E., Mathews, D. and Reynolds, B.E. (1997*). Readings in Cooperative Learning for Undergraduate Mathematics*. MAA Notes no. 44. Washington D.C.: Mathematical Association of America.

Hillel, J. and Sierpinska, A. (1994). On One Persistent Mistake in Linear Algebra. In J. Pedro da Ponte and J.F. Matos (Eds.), *Proceedings of the 18th International Conference of the International Group for the Psychology of Mathematics Education*, vol. III, pp. 65-72. Lisbon : Universidade de Lisboa.

Lebesgue, H. (1956). *La mesure des grandeurs*. Paris: Gauthier Villars.

Legrand, M. (1997). La problématique des situations fondamentales et l'approche anthropologique. *Repères IREM*, no. 27, 81-125.

Orton, A. (1980). *A cross-sectional study of the understanding of elementary calculus in adolescents and young adults*. Ph.D. thesis, University of Leeds, England.

Poincaré, H. (1904). Les définitions en mathématiques *L'Enseignement des Mathématiques*, no. 6, 255-283.

Radford, L. (1997). On psychology, historical epistemology and the teaching of mathematics : towards a socio-cultural history of mathematics. *For the Learning of Mathematics*, vol. 17.1, 26-30.

Robert, A. (1998). Outils d'analyse des contenus mathématiques à enseigner dans l'enseignement supérieur. à l'université. *Recherches en Didactique des Mathématiques*, vol. 18.2, 139-190.

Robert, A. and Speer, N. (2001), Research on the Teaching and learning of Calculus/Elementary Analysis, this volume pp.283-300.

Rogalski, M. (Ed.) (1998). Analyse épistémologique et didactique des connaissances à enseigner au lycée et à l'université. *Recherches en Didactique des Mathématiques*, Special issue, vol. 18.2.

Schneider, M. (1991). Un obstacle épistémologique soulevé par des découpages infinis de surfaces et de solides. *Recherches en Didactique des Mathématiques*, vol. 11/2.3, 241-294.

Schoenfeld, A.H. (1985). *Mathematical Problem Solving*. Orlando: Academic Press.

Schoenfeld, A.H. (1994). Some Notes on the Enterprise (Research in Collegiate Mathematics Education, That Is*). CBMS Issues in Mathematics Education*, vol. 4, 1-19.

Sfard, A. (1991). On the dual nature of mathematical conceptions. *Educational Studies in Mathematics*, no. 22, 1-36.

Sierpinska, A. (1985). Obstacles épistémologiques relatifs à la notion de limite. *Recherches en Didactique des Mathématiques*, vol. 6.1, 5-68.

Sierpinska, A. and Kilpatrick, J. (Eds.) (1998). *Mathematics education as a research domain: A search for identity*. Dordrecht: Kluwer Academic Publishers.

Sierpinska, A. and Lerman, S. (1996). Epistemologies of mathematics and of mathematics education. In A.J. Bishop, K. Clements, C. Keitel, J. Kilpatrick and C. Laborde (Eds.), *International Handbook of Mathematics Education*, pp. 827-876. Dordrecht: Kluwer Academic Publishers.

Stigler, J. and Hiebert, J. (1999). *The Teaching Gap*. New York: Free Press.

Tall, D. (Ed.) (1991). *Advanced Mathematical Thinking*. Dordrecht: Kluwer Academic Publishers.

Tall, D. (1996). Functions and Calculus. In A. J. Bishop, K. Clements, C. Keitel, J. Kilpatrick and C. Laborde (Eds.), *International Handbook of Mathematics Education*, pp. 289-325. Dordrecht: Kluwer Academic Publishers.

Tall, D. and Vinner, S. (1981). Concept image and concept definition in mathematics with particular reference to limits and continuity. *Educational Studies in Mathematics* 12-2, 151-169.

Michèle Artigue
Université Paris 7, France
artigue@math.jussieu.fr

ALAN H. SCHOENFELD

PURPOSES AND METHODS OF RESEARCH IN MATHEMATICS EDUCATION[1]

Bertrand Russell has defined mathematics as the science in which we never know what we are talking about or whether what we are saying is true. Mathematics has been shown to apply widely in many other scientific fields. Hence, most other scientists do not know what they are talking about or whether what they are saying is true.

Joel Cohen, *On the nature of mathematical proofs*

There are no proofs in mathematics education.

Henry Pollak

1. INTRODUCTION

The first quotation above is humorous, the second serious. Both, however, serve to highlight some of the major differences between mathematics and mathematics education – differences that must be understood if one is to understand the nature of methods and results in mathematics education.

The Cohen quotation does point to some serious aspects of mathematics. In describing various geometries, for example, we start with undefined terms. Then, following the rules of logic, we prove that if certain things are true, other results must follow. On the one hand, the terms are undefined – i.e., "we never know what we are talking about." On the other hand, the results are definitive. As Gertrude Stein might have said, a proof is a proof is a proof.

Other disciplines work in other ways. Pollak's statement was not meant as a dismissal of mathematics education, but as a pointer to the fact that the nature of evidence and argument in mathematics education is quite unlike the nature of evidence and argument in mathematics. Indeed, the kinds of questions one can ask (and expect to be able to answer) in educational research are not the kinds of questions that mathematicians might expect. Beyond that, mathematicians and education researchers tend to have different views of the purposes and goals of research in mathematics education.

This paper begins with an attempt to lay out some of the relevant perspectives, and to provide background regarding the nature of inquiry within mathematics

[1] A closely related paper (Schoenfeld, 2000a) was published in the *Notices* of the American Mathematical Society.

Derek Holton (Ed.), The Teaching and Learning of Mathematics at University Level: An ICMI Study, 221—236.

© 2001 *Kluwer Academic Publishers. Printed in the Netherlands.*

enterprise? That is, what are the purposes of research in mathematics education? What do theories and models look like in education, as opposed to those in mathematics and the physical sciences? What kinds of questions can educational research answer? Given such questions, what constitute reasonable answers? What kinds of evidence are appropriate to back up educational claims? What kinds of methods can generate such evidence? What standards might one have for judging claims, models, and theories? As will be seen, there are significant differences between mathematics and education with regard to all of these questions.

2. PURPOSES

Research in mathematics education has two main purposes, one pure and one applied:

- Pure (Basic Science): To understand the nature of mathematical thinking, teaching, and learning;
- Applied (Engineering): To use such understandings to improve mathematics instruction.

These are deeply intertwined, with the first at least as important as the second. The reason is simple: without a deep understanding of thinking, teaching, and learning, no sustained progress on the 'applied front' is possible. A useful analogy is to the relationship between medical research and practice. There is a wide range of medical research. Some is done urgently, with potential applications in the immediate future. Some is done with the goal of understanding basic physiological mechanisms. Over the long run, the two kinds of work live in synergy. This is because basic knowledge is of intrinsic interest and because it establishes and strengthens the foundations upon which applied work is based.

These dual purposes must be understood. They contrast rather strongly with what, from the perspective of many mathematicians, is, or should be, the single purpose of research in mathematics education:

- "Tell me what works in the classroom."

Saying this does not imply that mathematicians are not interested, at some abstract level, in basic research in mathematics education – but that their primary expectation is usefulness, in rather direct and practical terms. Of course, the educational community must provide useful results – indeed, usefulness motivates the vast majority of educational work – but it is a mistake to think that direct applications (curriculum development, 'proof' that instructional treatments work, etc.) are the primary business of research in mathematics education.

3. ON QUESTIONS

A major issue that needs to be addressed when thinking about what mathematics education can offer is "What kinds of questions can research in mathematics education answer?"

Simply put, the most typical educational questions asked by mathematicians – "What works?" and "Which approach is better?" – tend to be unanswerable in principle. The reason is that what a person will think 'works' will depend on what that person values. Before one tries to decide whether some instructional approach is successful, one has to address questions such as: "Just what do you want to achieve? What understandings, for what students, under what conditions, with what constraints?" Consider the following examples.

One question asked with some frequency by faculty and administrators is "Are large classes as good as small classes?" I hope it is clear that this question cannot be answered in the abstract. How satisfied one is with large classes depends on the consequences one thinks are important. How much does students' sense of engagement matter? Are students' feelings about the course and toward the department important? Is there concern about the percentage of students who go on to enrol in subsequent mathematics courses? The conclusions that one might draw regarding the utility of large classes could vary substantially, depending on how much weight these outcomes are given.

Similar issues arise even if one focuses solely on the mathematics being taught. Suppose one wants to address the question, "Do students learn as much mathematics in large classes as in small classes?" One must immediately ask, "What counts as mathematics?" How much weight will be placed (say) on problem solving, on modelling, or on the ability to communicate mathematically? Judgements concerning the effectiveness of one form of instruction over another will depend on the answers to these questions. To put things bluntly, a researcher has to know what to look for, and what to take as evidence of it, before being able to determine whether it is there.

The fact that one's judgements reflect one's values also applies to questions of the type "Which approach works better (or best)?" This may seem obvious, but often it is not. Consider calculus reform. Soon after the Tulane 'Lean and Lively' conference, whose proceedings appeared in Douglas (1986), the National Science Foundation (NSF) funded a major calculus reform initiative. By the mid-1990s NSF program officers were convinced that calculus reform was a 'good thing', and that it should be a model for reform in other content areas. NSF brought together mathematicians who had been involved in reform with researchers in mathematics education, and posed the following question: "Can we obtain evidence that calculus reform worked (that is, that reform calculus is better than the traditional calculus)?" What they had in mind, basically, was some form of test. They thought it should be easy to construct a test, administer it, and show that reform students did better.

Those who advocated this approach failed to understand that what they proposed would in essence be a comparison of apples and oranges. If one gave a traditional test that leaned heavily on the ability to perform symbolic manipulations, 'reform' students would be at a disadvantage because they had not practised computational skills. If one gave a test that was technology-dependent or that had a heavy modelling component, traditional students would be at a disadvantage because technology and modelling had not been a large part of their curriculum. Either way, giving a test and comparing scores would be unfair. The appropriate way to proceed was to look at the curriculum, identifying important topics and specifying what it means to have a conceptual understanding of them. With this kind of information, individual institutions and departments (and the profession as a whole, if it wished) could then decide which aspects of understanding were most important, which they wanted to assess, and how. As a result of extended discussions, the NSF effort evolved from one that focused on documenting the effects of calculus reform to one that focused on developing a framework for looking at the effects of calculus instruction. The result of these efforts was the 1997 book *Student Assessment in Calculus* (Schoenfeld, 1997).

In sum, many of the questions that would seem natural to ask – questions of the type "What works?" or "Which method works best?" – cannot be answered, for good reason.

Given this, what kinds of questions can research in mathematics education address? I would argue that some of the fundamental contributions from research in mathematics education are the following:

- theoretical perspectives for understanding thinking, learning, and teaching;
- descriptions of aspects of cognition (e.g., thinking mathematically; student understandings and misunderstandings of the concepts of function, limit, etc.);
- existence proofs (evidence of cases in which students can learn problem solving, induction, group theory; evidence of the viability of various kinds of instruction)
- descriptions of (positive and negative) consequences of various forms of instruction.

Michèle Artigue's paper (this volume pp.207-220) describes many of the results of such studies. I will describe some others and comment on the methods for obtaining them in the section after next.

4. ON THEORIES AND MODELS (AND CRITERIA FOR GOOD ONES)

When mathematicians use the terms 'theory' and 'models', they typically have very specific kinds of things in mind – both regarding the nature of those entities, and of the kinds of evidence used to make claims regarding them. The terms 'theory' and 'models' are sometimes used in different ways in the life sciences and social sciences, and their uses may be more akin to those used in education. In this section I shall briefly walk through the examples indicated in Table 1.

Table 1. Theories and models in mathematics/physics, biology, and education/psychology[2]

	Subject		
	Mathematics/Physics	Biology	Education/Psychology
Theory of...	Equations; Gravity	Evolution	Mind
Model of ...	Heat Flow in a Plate	Predator-Prey Relations	Problem Solving

In mathematics theories are laid out explicitly, as in the theory of equations or the theory of complex variables. Results are obtained analytically – we prove that the objects in question have the properties we claim they have. In classical physics there is a comparable degree of specificity; physicists specify an inverse-square law for gravitational attraction, for example. Models are understood to be approximations, but they are expected to be very precise approximations, in deterministic form. Thus, for example, to model heat flow in a laminar plate we specify the initial boundary conditions and the conditions of heat flow, and we then solve the relevant equations. In short, there is no ambiguity in the process. Descriptions are explicit and the standard of correctness is mathematical proof. A theory and models derived from it can be used to make predictions, which, in turn, are taken as empirical substantiation of the correctness of the theory.

Things are far more complex in the biological sciences. Consider the theory of evolution, for example. Biologists are in general agreement with regard to its essential correctness – but the evidence marshalled in favour of evolution is quite unlike the kind of evidence used in mathematics or physics. There is no way to prove that evolution is correct in a mathematical sense; the arguments that support it consist of (to borrow the title of one of Pólya's books) 'patterns of plausible reasoning', along with the careful consideration of alternative hypotheses. In effect, biologists have said the following: "We have mountains of evidence that are consistent with the theory, broadly construed; there is no clear evidence that falsifies the proposed theory; and no rival hypotheses meet the same criteria." While predictions of future events are not feasible given the time scale of evolutionary events, the theory does support an alternative form of prediction. Previously unexamined fossil records must conform to the theory, so that the theory can be used to describe properties that fossils in particular geological strata should or should not have. The cumulative record is taken as substantiation for the theory.

In short, theory and supporting evidence can differ substantially in the life sciences and in mathematics and physics. The same holds for models, or at least the degree of precision expected of them: nobody expects animal populations modelled by predator-prey equations to conform to those models in the same way that heat flow in a laminar plate is expected to conform to models of heat flow.

Finally, it should be noted that theories and models in the sciences are always subject to revision and refinement. As glorious and wonderful as Newtonian gravitational theory was, it was superseded by Einstein. Or, consider nuclear theory.

---

[2] Reprinted with permission from Schoenfeld, 1998a, page 9.

Valence theory, based on models of electrons that orbited around nuclei, allowed for amazing predictions, such as the existence of as-yet-undiscovered elements. But, physicists no longer talk about electrons in orbit around nuclei; once-solid particles in the theory such as electrons have been replaced in the theory by probabilistic electron clouds. Theories evolve.

Research in mathematics education has many of the attributes of the research in the physical and life sciences described above. In a 'theory of mind', for example, certain assumptions are made about the nature of mental organization – e.g., that there are certain kinds of mental structures that function in particular ways. One such assumption is that there are various kinds of memory, among them working or 'short-term' memory. According to the theory, 'thinking' gets done using working memory: that is, the 'objects of thought' that people manipulate mentally are temporarily stored in working memory. What makes things interesting (and scientific) is that the theory also places rather strong limits on working memory: it has been claimed (see, e.g., Miller, 1956) that people can keep no more than about nine 'chunks' of information in working memory at one time.

To see that this claim might actually be true, one could try to multiply 379 by 658, with eyes closed. Most people will find it difficult if not impossible. (In a recent meeting I gave a group of about 75 mathematicians this task. None of them succeeded within a few minutes.) The reason is that the number of things a person has to keep track of – the original numbers, and the various subtotals that arise in doing the multiplication – exceeds nine. Now, a person is better able to do the task mentally after rehearsing some of the subtotals – e.g., a person can calculate 8×379 = 3032 and repeat '3032' mentally until it becomes a 'chunk' and occupies only one space (a 'buffer') in working memory. That leaves enough working space to do other computations. By using this kind of chunking, people can transcend the limits of working memory. Now, consider the truth status of the assertion that people's working memory has no more than about nine slots. There will never be absolute proof of this assertion. First, it is unlikely that the researchers will find the physical location of working memory buffers in the brain, even if they exist; the buffers are components of models, and they are not necessarily physical objects. Second, the evidence in favour of this assertion is compelling but can not be definitive. Many kinds of experiments have been performed in which people are given tasks that call for using more than nine slots in working memory, and people have failed at them (or, after some effort, performed them by doing what could be regarded as some form of chunking).

As with evolution, there are mountains of evidence that are consistent with this assertion; there is no clear evidence to contradict it; and no rival hypothesis meets the same criteria. But is it proven? No. Not in the mathematical sense. The relevant standard is, in essence, what a neutral jury would consider to be evidence beyond a reasonable doubt. The same holds for models of, say, problem solving, or (my current interest) models of teaching (see Schoenfeld, 1998b, 1999, 2000b). I am currently engaged in trying to construct a theoretical description that explains how and why teachers do what they do, on the fly, in the classroom. This work, elaborated at the same level of detail as a theory of memory, is called a 'theory of teaching-in-context'. The claim is that with the theory and with enough time to model a particular teacher, one can build a description of that person's teaching that characterizes his or her classroom behaviour with remarkable precision. When one looks at this work, one cannot expect to find the kind of precision found in modelling heat flow in a laminar plate. But (see, e.g., Schoenfeld, 1998b) it is not unreasonable to expect that such behaviour can be modelled with the same degree of fidelity to 'real-world' behaviour as with predator-prey models.

We pursue the question of standards for judging theories, models, and results in the section after next.

5. METHODS

In this paper I can not provide even a beginning catalogue of methods of research in undergraduate mathematics education. As an indication of the magnitude of that task, consider the fact that the Handbook of Qualitative Research in Education (LeCompte, Millroy, and Preissle, 1992) is nearly 900 pages long! Chapters in that volume include extensive discussions of ethnography (how does one understand the 'culture of the classroom', for example?), discourse analysis (what patterns can be seen in the careful study of conversations?), the role of culture in shaping cognition, and issues of subjectivity and validity. And that is qualitative work alone – there is, of course, a long-standing quantitative tradition of research in the social sciences as well. My goal rather, is to provide an orientation to the kinds of work that are done, and to suggests the kinds of findings (and limitations thereof) that they can produce.

Those who are new to educational research tend to think in terms of standard experimental studies, which involve 'experimental' and 'control' groups, and the use of statistics to determine whether or not the results are 'significant'. As it turns out, the use of statistics in education is a much more complex issue than one might think. For some years from mid-century onward, research in the social sciences (in the United States at least) was dominated by the example of agriculture. The basic notion was that if two fields of a particular crop were treated identically except for one 'variable', then differences in crop yield could be attributed to the difference in that variable. Surely, people believed, one could do the same in education. If one wanted to prove that a new way of teaching X was superior, then one could conduct an experiment in which two groups of students studied X – one taught the standard way, one taught the new way. If students taught the new way did better, one had evidence of the superiority of the instructional method.

Put aside for the moment the issues raised in the previous section about the goals of instruction, and the fact that the old and new instruction might not focus on the same things. Imagine that one could construct a test fair to both old and new instruction. And, suppose that students were randomly assigned to experimental and control groups, so that standard experimental procedures were followed. Nonetheless, there would still be serious potential problems. If different teachers taught the two groups of students, any differences in outcome might be attributable to differences in teaching. But even with the same teacher, there can be myriad differences. There might be a difference in energy or commitment: teaching the 'same old stuff' is not the same as trying out new ideas. Or, students in one group might know they are getting something new and experimental. This alone might result in significant differences. (There is a large literature showing that if people feel that changes are made in their own best interests, they will work harder and do better – no matter what the changes actually are. The effects of these changes fade with time.) Or, the students might resent being experimented upon.

Here is a case in point. Some years ago I developed a set of stand-alone instructional materials for calculus. Colleagues at another university agreed to have their students use them. In all but two sections, the students who were given the materials did better than the students who were not given them. However, in two sections there were essentially no differences in performance. It turns out that most of the faculty had given the materials a favourable introduction, suggesting to the students that they would be helpful. The instructor of the sections that showed no differences had handed them out saying "They asked me to give these to you. I don't know if they're any good."

In short, the classical 'experimental method' can be problematic in educational research. To mention just two difficulties, 'double blind' experiments in the medical sense (in which neither the doctors nor the patients know who is getting the real treatment, and who is getting a placebo treatment) are rarely blind; and many experimental 'variables' are rarely controllable in any rigorous sense. (That was the point of the example in the previous paragraph.) As a result, both positive and negative results can be difficult to interpret. This is not to say that such studies are not useful, or that large-scale statistical work is not valuable – it clearly is – but that it must be done with great care, and that results and claims must be interpreted with equal care. Statistical work of consistent value tends to be that which

1. produces general findings about a population. For example, Artigue (this volume, pp. 207-220) notes that "more than 40% of students entering French universities consider that if two numbers A and B are closer than N for every positive N, then they are not necessarily equal, just infinitely close."
2. provides a clear comparison of two or more populations. For example, the results of the Third International Mathematics and Science Study document the baseline performance of students in various nations on a range of mathematical content.
3. provides substantiation, over time, of findings that were first uncovered in more small-scale observational studies. For example, the general notion of 'concept image' was introduced by Tall and Vinner about 20 years ago (see, e.g., Tall and Vinner, 1981). The cumulative weight of studies since then has provided a 'large n' substantiating the widespread nature of the phenomenon. The same can be said about the presence of 'epistemological obstacles' (see, e.g., Brousseau, 1997), the impact of poor 'control' or of mathematical beliefs on students' problem solving (see, e.g., Schoenfeld, 1985), or the constructs related to APOS theory (see, e.g., Asiala, Brown, de Vries, Dubinsky, Mathews and Thomas, 1996).

What you will find, for the most part, is that research methods in undergraduate mathematics education – in all of education for that matter – are suggestive of results, and that the combined evidence of many studies over time is what lends substantiation to findings.

In a broad sense, for example, all of the work mentioned in 3 immediately above, is grounded in a particular set of assumptions related to the nature of thinking and learning – specifically, that we do not perceive reality directly, but rather that we build mental structures that shape the ways in which we perceive reality. The point is a simple one: if we perceived reality directly, we wouldn't fall prey to optical illusions. But it is also fundamentally important, because it says that what 'counts' in instruction is not simply what the teacher lays out for the students, but rather the ways in which students do (or don't) make sense of it. (See the example of elementary school students' mis-construction of simple subtraction algorithms in the next section.) Once this is understood, one has a powerful tool for thinking about student learning: one must think about how students build their own understandings of mathematical topics, rather than 'simply' worrying about how to present material clearly.

It should be noted that most of the main findings listed in Artigue's paper are of the type discussed above: individual papers are suggestive, and findings are substantiated through the cumulative weight of evidence. Moreover, a variety of methods have been used to generate such results. All of the researchers mentioned above conduct interviews with students, looking closely at the ways in which students 'make sense' of particular mathematical ideas. Often, on the basis of these observations, they hypothesize mental structures – the presence or absence of particular mental constructs. They might then engage in 'teaching experiments', to see if they can help students develop particular mental constructs – and if so, will look to see if there are differences in student performance.

As indicated above, research results in education are not 'proven', in the sense that they are proven in mathematics. Moreover, it is often difficult to employ straightforward 'experimental' or statistical methods of the type used in the physical sciences, because of complexities related to what it means for educational conditions to be 'replicable'. In education one finds a wide range of research methods. A look at one of the first volumes on undergraduate mathematics education, Tall's 1991 *Advanced Mathematical Thinking*, suggests the range. If anything, the number and type of methods have increased, as evidenced in the three volumes *of Research in Collegiate Mathematics Education*. One finds, for example, reports of detailed interviews with students, comparisons of 'reform' and 'traditional' calculus, an examination of calculus 'workshops', and an extended study of one student's developing understanding of a physical device and graphs related to it. Studies employing anthropological observation techniques and other 'qualitative' methods are increasingly common.

How 'valid' are such studies, and how much can we depend on the results in them? That issue is pursued immediately below.

6. STANDARDS FOR JUDGING THEORIES, MODELS, AND RESULTS

There is a wide range of results and methods in mathematics education. A major question, then, is the following: how much faith should one have in any particular result? What constitutes solid reason, what constitutes "proof beyond a reasonable doubt?"

The following list puts forth a set of criteria that can be used for evaluating models and theories (and more generally, any empirical or theoretical work) in mathematics education:

- descriptive power
- explanatory power
- scope
- predictive power
- rigour and specificity
- falsifiability
- replicability
- multiple sources of evidence ('triangulation')

I shall briefly describe each.

6.1 Descriptive power

By descriptive power I mean the capacity of a theory to capture 'what counts' in ways that seem faithful to the phenomena being described. As Gaea Leinhardt (1998) has pointed out, the phrase "consider a spherical cow" might be appropriate when physicists are considering the cow in terms of its gravitational mass – but not if one is exploring some of the cow's physiological properties! Theories of mind, problem solving, or teaching should include relevant and important aspects of thinking, problem solving, and teaching, respectively. At a very broad level, fair questions to ask are "Is anything missing?" Do the elements of the theory correspond to things that seem reasonable? For example, say a problem solving session, an interview, or a classroom lesson was videotaped. Would a person who read the analysis and then saw the videotape, reasonably be surprised by things that were missing from the analysis?

6.2 Explanatory power

By explanatory power I mean providing explanations of how and why things work. It is one thing to say that people will or will not be able to do certain kinds of tasks, or even to describe what they do on a blow-by-blow basis; it is quite another thing to explain why. It is one thing, for example, to say that people will have difficulty multiplying two three-digit numbers in their heads. But that does not provide information about how and why the difficulties occur. The full theoretical description of working memory, which was mentioned above, comes with a description of memory buffers, a detailed explanation of the mechanism of 'chunking', and the careful delineation of how the components of memory interact with each other. The explanation works at a level of mechanism: it says in reasonably precise terms what the objects in the theory are, how they are related, and why some things will be possible and some not.

6.3 Scope

By scope, I mean the range of phenomena 'covered' by the theory. A theory of equations is not very impressive if it deals only with linear equations. Likewise, a theory of teaching is not very impressive if it covers only straight lectures!

6.4 Predictive power

The role of prediction is obvious: one test of any theory is whether it can specify some results in advance of their taking place. Again, it is good to keep things like the theory of evolution in mind as a model. Predictions in education and psychology are not often of the type made in physics.

Sometimes it is possible to make precise predictions. For example, Brown and Burton (1978) studied the kinds of incorrect understandings that students develop when learning the standard U.S. algorithm for base 10 subtraction. They hypothesized very specific mental constructions on the part of students – the idea being that students did not simply fail to master the standard algorithm, but rather that students often developed one of a large class of incorrect variants of the algorithm, and applied it consistently. Brown and Burton developed a simple diagnostic test with the property that a student's pattern of incorrect answers suggested the false algorithm he or she might be using. About half of the time, they were then able to predict the incorrect answer that the students would obtain to a new problem, before the student worked the problem!

Such fine-grained and consistent predictions on the basis of something as simple as a diagnostic test are extremely rare, of course. For example, no theory of teaching can predict precisely what a teacher will do in various circumstances; human behaviour is just not that predictable. However, a theory of teaching can work in ways analogous to the theory of evolution. It can suggest constraints, and even suggest likely events.

(Making predictions is a very powerful tool in theory refinement. When something is claimed to be impossible and it happens, or when a theory makes repeated claims that something is very likely and it does not occur, then the theory has problems! Thus, engaging in such predictions, is an important methodological tool, even when it is understood that precise prediction is impossible.)

6.5   Rigour and specificity

Constructing a theory or a model involves the specification of a set of objects and relationships among them. This set of abstract objects and relationships supposedly corresponds to some set of objects and relationships in the 'real world'. The relevant questions are:

How well-defined are the terms? Would you know one if you saw one? In real-life? In the model? How well-defined are the relationships among them? And, how well do the objects and relations in the model correspond to the things they are supposed to represent? As noted above, one cannot necessarily expect the same kinds of correspondences between parts of the model and real-world objects as in the case of simple physical models. Mental and social constructs such as memory buffers and the 'didactic contract' (the idea that teachers and students enter a classroom with implicit understandings regarding the norms for their interactions, and that these understandings shape the ways they act) are not inspectable or measurable in the ways that heat flow in a laminar plate is. But, we can ask for detail, both in what the objects are and in how they fit together. Are the relationships and changes among them carefully defined, or does 'magic happen' somewhere along the way? Here is a rough analogy. For much of the eighteenth century the phlogiston theory of combustion – which posited that in all flammable materials there is a colourless, odourless, weightless, tasteless substance called 'phlogiston' liberated during combustion – was widely accepted. (Lavoisier's work on combustion ultimately refuted the theory.) With a little hand-waving, the phlogiston theory explained a reasonable range of phenomena. One might have continued using it, just as theorists might have continued building epicycles upon epicycles in a theory of circular orbits.[3] The theory might have continued to produce useful results, good enough "for all practical purposes." That may be fine for practice, but it is problematic with regard to theory. Just as in the physical sciences, researchers in education have an intellectual obligation to push for greater clarity and specificity, and to look for limiting cases or counterexamples, to see where the theoretical ideas break down.

Here are two quick examples. First, in my research group's model of the teaching process we represent aspects of the teacher's knowledge, goals, beliefs, and decision-making. Sceptics (including ourselves) should ask, how clear is the representation? Once terms are defined in the model (i.e., once we specify a teacher's knowledge, goals, and beliefs) is there hand-waving when we say what the teacher might do in specific circumstances, or is the model well enough defined so that others could 'run' it and make the same predictions? Second, the 'APOS theory' as expounded in Asiala et al. (1996) uses terms such as Action, Process, Object, and Schema. Would you know one if you met one? Are they well defined in the model? Are the ways in which they interact or become transformed well specified? In both cases, the bottom line issues are "What are the odds that this too is a phlogiston-like theory? Are the people employing the theory constantly testing it, in order to find

---

[3] This example points to another important criterion, *simplicity*. When a theory requires multiple "fixes" such as epicycles upon epicycles, that is a symptom that something is not right.

out?" Similar questions should be asked about all of the terms used in educational research, e.g., the 'didactical contract', 'metacognition', 'concept image', and 'epistemological obstacles'.

6.6 Falsifiability

The need for falsifiability – for making non-tautological claims or predictions whose accuracy can be tested empirically – should be clear at this point. It is a concomitant of the discussion in the previous two subsections. A field makes progress (and guards against tautologies) by putting its ideas on the line.

6.7 Replicability

The issue of replicability is also intimately tied to that of rigour and specificity. There are two related sets of issues: (1) Will the 'same thing' happen if the circumstances are repeated? (2) Will others, once appropriately trained, 'see' the same things in the data? In both cases, answering these questions depends on having well-defined procedures and constructs.

The phrasing of (1) is deliberately vague, because it is supposed to cover a wide range of cases. In the case of short-term memory, the claim is that people will run into difficulty if memory tasks require the use of more than nine short-term memory buffers. In the case of sociological analyses of the classroom, the claim is that once the didactical contract is understood, the actions of the students and teacher will be seen to conform to that (usually tacit) understanding. In the case of 'beliefs', the claim is that students who hold certain beliefs will act in certain ways while doing mathematics. In the case of epistemological obstacles or APOS theory, the claims are similarly made that students who have (or have not) made particular mental constructions will (or will not) be able to do certain things.

In all of these cases, the usefulness of the findings, the accuracy of the claims, and the ability to falsify or replicate, depends on the specificity with which terms are defined. Consider this case in point from the classical education literature. Ausubel's (1968) theory of 'advance organizers' postulates that if students are given an introduction to materials they are to read that orients them to what is to follow, their reading comprehension will improve significantly. After a decade or two and many, many studies, the literature on the topic was inconclusive: about half of the studies showed that advance organizers made a difference, about half not. A closer look revealed the reason: the very term was ill-defined. Various experimenters made up their own advance organizers based on what they thought they should be – and there was huge variation. No wonder the findings were inconclusive! (One standard technique for dealing with issues of well-definedness, and which addresses issue (2) above, is to have independent researchers go through the same body of data, and then to compare their results. There are standard norms in the field for 'inter-rater reliability'; these norms quantify the degree to which independent analysts are seeing the same things in the data.)

6.8 Multiple sources of evidence ('triangulation')

Here we find one of the major differences between mathematics and the social sciences. In mathematics, one compelling line of argument (a proof) is enough: validity is established. In education and the social sciences, we are generally in the business of looking for *compelling evidence*. The fact is, evidence can be misleading – what we think is general may in fact be an artefact or a function of circumstances rather than a general phenomenon.

Here is one example. Some years ago I made a series of videotapes of college students working on the problem, "How many cells are there in an average-size human adult body?" Their behaviour was striking. A number of students made wild guesses about the order of magnitude of the dimensions of a cell – from "let's say a cell is an angstrom unit on a side" to "say a cell is a cube that's 1/100 of an inch wide." Then, having dispatched with cell size in seconds, they spent a very long time on body size – often breaking the body into a collection of cylinders, cones, and spheres, and computing the volume of each with some care. This was *very* odd.

Some time later I started videotaping students working problems in pairs rather than by themselves. I never again saw the kind of behaviour described above. It turns out that when they were working alone, the students felt under tremendous pressure. They knew that a mathematics professor would be looking over their work. Under the circumstances, they felt they needed to do *something* mathematical – and volume computations at least made it look as if they were doing mathematics! When students worked in pairs, they started off by saying something like "This sure is a weird problem." That was enough to dissipate some of the pressure, with the result being that there was no need for them to engage in volume computations to relieve it. In short, some very consistent behaviour was actually a function of circumstances rather than being inherent in the problem or the students.

One way to check for artefactual behaviour is to vary the circumstances – to ask, do you see the same thing at different times, in different places? Another is to seek as many sources of information as possible about the phenomenon in question, and to see whether they portray a consistent 'message'. In my research group's work on modelling teaching, for example, we draw inferences about the teacher's behaviour from videotapes of the teacher in action – but we also conduct interviews with the teacher, review his or her lesson plans and class notes, and discuss our tentative findings with the teacher. In this way we look for convergence of the data. The more independent sources of confirmation there are, the more robust a finding is likely to be.

7. CONCLUSION

The main point of this article has been that research in (undergraduate) mathematics education is a *very* different enterprise from research in mathematics, and that an understanding of the differences is essential if one is to appreciate (or better yet, contribute to) work in the field. Findings are rarely definitive; they are usually suggestive. Evidence is not on the order of 'proof', but is cumulative, moving toward conclusions that can be considered to be "beyond a reasonable doubt." A scientific approach is possible, but one must take care not to be scientistic. What counts is not the use of the trappings of science, such as the 'experimental method', but the use of careful reasoning and standards of evidence, employing a wide variety of methods appropriate for the tasks at hand. It is worth remembering how young mathematics education is as a field. Mathematicians are used to measuring mathematical lineage in centuries, if not millennia; in contrast, the lineage of research in mathematics education (especially undergraduate mathematics education) is measured in decades. The journal *Educational Studies in Mathematics* dates to the 1960s. The first issue of Volume 1 of the *Journal for Research in Mathematics Education* was published in January 1970. The series of volumes *Research in Collegiate Mathematics Education* – the first set of volumes devoted solely to mathematics education at the college level – began to appear in 1994. It is no accident that the vast majority of articles cited by Artigue in her review of research findings (this volume, 207-220) were written in the 1990s; there was little at the undergraduate level before then! There has been an extraordinary amount of progress in recent years – but the field is still very young, and there is a very long way to go.

Because of the nature of the field, it is appropriate to adjust one's stance toward the work and its utility. Mathematicians approaching this work should be open to a wide variety of ideas, understanding that the methods and perspectives to which they are accustomed do not apply to educational research in straightforward ways. They should not look for definitive answers, but for ideas they can use. At the same time, all consumers and practitioners of research in (undergraduate) mathematics education should be healthy sceptics. In particular, because there are no definitive answers, one should certainly be wary of anyone who offers them. More generally, the main goal for the decades to come is to continue building a corpus of theory and methods that will allow research in mathematics education to become an ever more robust basic and applied field.

REFERENCES

Artigue, M. What can we learn from Educational Research at the University Level , this volume, pp.207-220.

Asiala, M., Brown, A., de Vries, D., Dubinsky, E., Mathews, D., and Thomas, K. (1996). A framework for research and curriculum development in undergraduate mathematics education. In J. Kaput, A. Schoenfeld, and E. Dubinsky (Eds.), *Research in Collegiate Mathematics Education*. II, pp. 1-32. Washington, DC: Conference Board of the Mathematical Sciences.

Ausubel, D. P. (1968). *Educational psychology: A cognitive view*. New York: Holt. Reinhardt, and Winston.

Brown, J. S. and Burton, R. R. (1978). Diagnostic models for procedural bugs in basic mathematical skills. *Cognitive Science*, 2, 155-192.

Brousseau, G. (1997). *The theory of didactic situations*. Dordrecht: Kluwer.

Cohen, J. (1969). On the nature of mathematical proofs. In R. A. Baker (Ed.), *A stress analysis of a topless evening gown* (pp. 93-99). Garden City, NY: Doubleday.

Douglas. R. G. (Ed.). (1986). *Toward a lean and lively calculus*. (MAA Notes Number 6). Washington, DC: Mathematical Association of America.

LeCompte, M., Millroy, W., and Preissle, J. (1992). *Handbook of Qualitative Research in Education*. New York: Academic Press.

Leinhardt, G. (1998). On the messiness of overlapping goals in real settings. *Issues in Education*, Volume 4, Number 1, 125-132.

Miller, G. (1956). The magic number seven, plus or minus two: some limits on our capacity for processing information. *Psychological Review*, 63, 81-97.

Schoenfeld, A. H. (Ed.). (1997). *Student Assessment in Calculus*. (MAA Notes Number 43). Washington, DC: Mathematical Association of America.

Schoenfeld, A. H. (1985). *Mathematical problem solving*. Orlando, FL: Academic Press.

Schoenfeld, A. H. (1998a). On Theory and models: the case of Teaching-in-Context. Plenary address. In Sarah B. Berenson (Ed.), *Proceedings of the XX annual meeting of the International Group for Psychology and Mathematics Education*. Raleigh, NC: PME.

Schoenfeld, A. H. (1998b). Toward a theory of teaching-in-context. *Issues in Education*, Volume 4, Number 1, 1-94.

Schoenfeld, A. H. (2000a). Purposes and Methods of Research in Mathematics Education. *Notices of the American Mathematical Society*, 47(6), 641-649.

Schoenfeld, A. H. (2000b). Models of the teaching process. *Journal of Mathematical Behavior*, 18 (3), 243-261.

Tall, D. and Vinner, S. (1981). Concept definitions and concept images in mathematics with particular references to limits and continuity. *Educational Studies in Mathematics*, 12, 151-169.

Tall, D. (Ed.) (1991). *Advanced Mathematical Thinking*. Dordrecht: Kluwer Academic Publishers.

Alan H. Schoenfeld
University of California, Berkeley, USA
alans@socrates.berkeley.edu

ANNIE SELDEN AND JOHN SELDEN

TERTIARY MATHEMATICS EDUCATION RESEARCH AND ITS FUTURE

1. WHAT IS MATHEMATICS EDUCATION RESEARCH?

Tertiary mathematics education research is disciplined inquiry into the learning and teaching of mathematics at the university level. It can be conducted from an individual cognitive perspective or from a social perspective of the classroom or broader community. It can also coordinate the two, providing insight into how the psychological and social perspectives relate to and affect one another.

In the case of individual cognition, one wants to know how students come to understand aspects of mathematics or how they develop effective mathematical practices, good problem-solving skills, or the ability to generate reasonable conjectures and to produce proofs. What goes on in students' minds as they grapple with mathematics and how might we influence that positively? More specifically, consider the difficulties that students have with the concept of limit. Does the everyday notion of speed limit as a bound present a cognitive obstacle? Does the early introduction of monotone increasing sequences constitute a didactic obstacle? (See Artigue, this volume pp.207-220.) Are there some, as yet neglected, everyday or school-level conceptions that university mathematics teachers might effectively build on? What is the influence of affect, ranging from beliefs through attitude to emotion, on effective mathematical practice? What roles do intrinsic and extrinsic motivational factors play?

From a social perspective, whether of a single classroom or some broader community, one seeks information on how social interactions affect the group as well as the individuals involved. For example, how might one change the classroom culture so students came to view mathematics, not as passively received knowledge, but as actively constructed knowledge? Or, how might one restructure an entire curriculum to achieve this effect? What are the effects of various cooperative learning strategies on student learning? What kinds of interactions are most productive? Are some students advantaged while others are disadvantaged by the introduction of cooperative learning? Which students succeed in mathematics? Which students continue in mathematics and why? What is the effect of gender, race, or social class upon success in mathematics? In coordinating the psychological and social perspectives, any of the above questions might be asked, along with inquiry into the relationship between the two perspectives. For example, how does an individual's contribution affect a whole class discussion and conversely?

237

Derek Holton (Ed.), The Teaching and Learning of Mathematics at University Level: An ICMI Study, 237—254.
© 2001 *Kluwer Academic Publishers. Printed in the Netherlands.*

Because mathematics education research investigates both individuals and groups in the process of learning mathematics, it has adapted methods from a variety of social sciences ranging from cognitive psychology to anthropology. Yet, because the inquiry often concerns questions directly involving the understanding of mathematical concepts, education researchers need to be well versed in that underlying mathematics.

Mathematics education research is a relatively young field with an applied, or an applicable, character. Interesting questions, some suitably modifiable for investigation, arise when one encounters learning difficulties in the classroom. For example, are one's remedial college algebra students handicapped because they do not understand the notation? Are students of real analysis having difficulty because they are unable to generate suitable generic examples on which to build proofs? (See Dahlberg and Housman, 1997). Would visualization help analysis students construct proofs? (See Gibson, 1998.)

Despite often arising from such everyday questions about the practice of teaching, mathematics education research does not often provide immediately applicable prescriptive teaching information, rather it provides general guidance and hints that might help with teaching and curriculum design. In a recent article, Hiebert (1999) pointed out that, while mathematics education research can inform us, it cannot tell us which curricula or pedagogies are 'best' because such decisions necessarily involve value judgments about what students should know. For example, in this age of technology, when engineering students use computer algebra systems such as Mathematica or Maple, research cannot tell us which calculus and matrix algebra computations students should be able to perform flawlessly by hand. It can, however, inform us on the extent to which certain curricula, once implemented, have succeeded in attaining their goals.

A great variety of topics has been, and could be, investigated. These include aspects of mathematics (functions, analysis, proofs), mathematical cognition (problem solving, students' alternative conceptions), psychological factors (motivation, affect, visualization), teaching methods (lecturing, cooperative learning, uses of technology and writing), change (individual teacher and institutional), programs (new and existing), and culture (gender, equity, classroom culture, cross-cultural comparisons). The nature and size of studies varies from case studies of individuals to large-scale studies of hundreds or thousands of students (Schoenfeld, 1994). (For an overview of results at the university level, see Dreyfus, 1990; Tall, 1991, 1992; Selden and Selden, 1993.)

Mathematics education research cannot provide results having the character of certainty found in mathematics (Schoenfeld, this volume, pp. 221-236). Even very carefully conducted observations can only suggest general principles and yield evidence, rather than proofs. Thus, corroboration of results by subsequent studies is important. Furthermore, the discovery of new results can sometimes make what previously appeared firmly established, less so later. Nevertheless, research in mathematics education shares a powerful 'pyramiding' characteristic with other sciences; results rely on careful observations, are separated from investigators' opinions, and are subject to community scrutiny, so that subsequent work can be based on them.

2. MAKING PROGRESS IN TERTIARY MATHS EDUCATION RESEARCH

When a field, such as tertiary mathematics education research, is beginning, it is perhaps appropriate to survey the landscape to 'see what's out there' by gathering many, sometimes rather isolated, bits of information. There is room for studies of misconceptions, as well as case studies of exemplary teachers or programs. Just as in mathematics, to get informative results, one needs to investigate interesting, yet accessible, questions. As Schoenfeld (1999) says in reference to education research, "The hard part of being a mathematician is not solving problems; it's finding one that you can solve, and whose solution the mathematical community will deem sufficiently important to consider an advance. In any *real* research (in particular, education research), the bottleneck issue is that of problem identification – being able to focus on problems that are difficult and meaningful but on which progress can be made."

A criterion often employed by reviewers (i.e., referees) and editors of mathematics education research journals for the selection of manuscripts is that they should 'push the field forward', that is, establish something new and worth knowing. While some purely observational studies, for example, ethnographic reports, are seen as informative, increasingly it is important to place studies within the established body of research, citing how they fit in and extend or modify the work of others. (See Hanna, 1998; Lester and Lambdin, 1998.)

Indeed, it is widely thought that to make significant progress, there is a need for theory building. Thus information is often gathered, analyzed, interpreted, and synthesized within an explanatory structure, usually called a *theoretical framework*, that provides some coherence, and perhaps even some predictive power, to the results. (See Schoenfeld, this volume, pp. 221-236.)

Research reports are expected to mention which of these concepts and theoretical frameworks are being used. That is, the researcher's assumptions, as well as the questions that were investigated, should be made clear. Beyond that, research results should be based on evidence (from data) and an analysis. (See Lester and Lambdin, 1998.) Findings are also often required to be generalizable, that is, reports should provide enough detail about students, teaching, tests, and the like, so readers, who may be teachers of mathematics, can gauge how similar their situation is to the one being described, and hence, judge whether it has relevance for them.

2.1 Some Basic Concepts and Theoretical Frameworks

Philosophical views play a fundamental role in the perspectives researchers decide to take in their work and in the kind of research they do. Constructivism, the idea that individuals actively construct their own knowledge, can be traced back to Piaget and beyond and leads to emphasizing an individual cognitive perspective. In contrast, those taking a sociocultural view, emphasize the idea that culture mediates individual knowledge through tools and language, an idea that has roots in

Vygotsky's work and leads to taking a social perspective.[1] However, such philosophical views apply to all knowledge, not just to mathematics and are often not particularly conspicuous in the research findings themselves.

Mathematics education research, being domain specific, has developed its own ways of conceptualizing mathematics learning. Robert B. Davis (1990) pointed out that one of the major contributions of mathematics education research has been to provide new conceptualizations and new metaphors for thinking about and observing mathematical behavior. It is very difficult to notice patterns of behavior or thought without having names (and the corresponding concepts) for them. As is often said of other empirical disciplines, one needs a lens (i.e., theoretical framework) with which to focus on (i.e., frame) what one is seeing.

2.1.1   Concept Definition versus Concept Image

The mismatch between concepts as stated in definitions and as interpreted by students is well-known to those who teach university level mathematics. The terms *concept definition* and *concept image* distinguish between a formal mathematical definition and a person's ideas about a particular mathematical concept, such as function. An individual's concept image is a mental structure consisting of all of the examples, non-examples, facts, and relationships, etc., that he or she associates with a concept. It need not, but might, include the formal mathematical definition and appears to play a major role in cognition. (See Tall and Vinner, 1981.) Furthermore, while contemplating a particular mathematical problem, it might be that only a portion of one's concept image, called the *evoked concept image*, is activated. These ideas make it easier to understand and notice various aspects of a student's thinking, for example, to understand the thinking of a student who conceives of functions mainly graphically or mainly algebraically without much recourse to the formal definition. A teacher or a researcher can investigate what sorts of activities might encourage students to employ the definition when that is the appropriate response, as in making a formal proof. It might also be helpful to investigate how students develop their concept images or how such images affect problem-solving performance.

2.1.2   Obstacles to Learning

One set of related ideas that has proved powerful is that of *epistemological, cognitive,* and *didactic obstacles*[2] When applied to the learning of mathematics, these refer, respectively, to obstacles that arise from the nature of particular aspects of mathematical knowledge, from an individual's cognition about particular

---

[1] Cobb (1994) describes the constructivist and sociocultural views as complementary with the former using terms like accommodation, and the latter, terms like appropriation.

[2] The idea of epistemological obstacle was introduced by G. Bachelard, and subsequently Brousseau (1983) imported this idea into mathematics education research via his theory of didactical situations. B. Cornu, and somewhat later A. Sierpinska (1985), analyzed epistemological obstacles in the development of the limit concept and linked these to pupils' behaviour.

mathematical topics, or from particular features of the mathematics teaching. An obstacle is a piece of, not a lack of, knowledge, which produces appropriate responses within a frequently experienced, but limited context, and is not generalizable beyond it (Brousseau, 1997). Using the historical development of function as a guide, it has been found that one epistemological obstacle that students need to overcome is the idea of function as expression, just as was the case with Euler (Sierpinska, 1992).

An example of a didactic obstacle – one that can be traced back to an aspect of teaching – is the geometric definition of tangent line, often taught to junior secondary students, as a line that touches a circle at precisely one point and is perpendicular to the radius at that point. In subsequent calculus/analysis courses, students must reconstruct, often with considerable difficulty, this idea more generally so that a tangent line to a curve at a point is the limit of secants and its slope is the value of the derivative at that point (Artigue, 1992). Other reconstructions – of the real numbers, of integrals, of limit – are necessary as students proceed from the intuitive and algebraic view of calculus presented at secondary school to the more formal presentations at university. (See Artigue, this volume, pp. 207-220.)

2.1.3 Views of Concept Development

Another kind of distinction that has proved productive for tertiary mathematics education investigations has been the ***action-process-object-schema*** (APOS) view of an individual's mental construction of a concept like that of function. (For details, see Artigue, and Dubinsky and McDonald, this volume, pp. 275-282.)

Somewhat similarly, Sfard (1991) has described an individual's journey from an ***operational*** (process) to a ***structural*** (object) conception as reification. She considers these to be dual modes of thinking and observes that previously constructed structural conceptions (objects) can be used to build new operational conceptions (processes). Furthermore, a single mathematical notation can reflect both a process and object conception; $2+3x$ stands for the process of adding 2 to the product of 3 and x, as well as for the result of that process. In order to be mathematically flexible, an individual seems to need both the process and object views of many concepts and the ability to move between these views when appropriate. Tall (1991) calls such concepts, that can be seen as both processes and objects, ***procepts***.

2.1.4 Mathematical Definitions as Contrasted with Everyday Definitions

A further idea that appears to be very useful, but which may not yet be widely found in the literature, is the distinction between ***synthetic*** and ***analytic definitions***. Synthetic definitions are the everyday definitions that are commonly found in dictionaries – they are often ill-specified descriptions of things that already exist. It is often unclear when such everyday definitions (e.g., of democracy) are 'complete' or whether attention to all aspects of them is essential for their proper use. Analytic definitions, by contrast, bring concepts into existence – the concept (e.g., of group as used in mathematics) is whatever the definition says it is, nothing more and nothing less. One cannot safely ignore any aspects of such definitions. There is a sense in which synthetic definitions can be 'wrong', whereas analytic definitions cannot. Some of the difficulties that university students have with formal mathematics might well be viewed as stemming partly from an unawareness that mathematical definitions tend to be analytic, rather than synthetic.

2.1.5 How Might These Concepts be Used by University Teachers

Such ideas (e.g., concept image, epistemological obstacle, action-process-object-schema, synthetic vs. analytic definitions) help frame, not only research, but also discussions of teaching and learning toward more insightful, and ultimately, more productive ends. They help one view students' attempts at mathematical sense-making and understanding as somehow hindered by their current, somewhat limited, conceptualizations – instead of merely emphasizing that some university students don't do their homework, aren't motivated, or are just plain lazy (some of which may also be true).

In general, the pedagogical challenge is to figure out how to help students come to genuine mathematical understanding. Which instructional efforts might be more productive of genuine mathematical understanding? Here various techniques have been tried – computer activities that provoke students to reflect on mathematical situations (and to explicitly construct actions, processes, objects, and schemas), group projects that require students to grapple with, discuss, and focus on mathematical ideas, and process writing to help students clarify their mathematical thoughts (by explicitly describing the evolution of their thinking whilst wrestling with problem situations). When 'calculus reform' took hold in the U.S. in the late 1980s and 1990s, some combination of these pedagogical strategies was included in many reform efforts (Tucker, 1990). However, very few were based on research ideas such as those mentioned above. (See Robert and Speer, this volume, pp. 283-300.)

2.2 Some Theories of Instructional Design

It would be useful for mathematics education research to play an increasing role in informing the development of curricula. There have been some efforts at curriculum design using the results obtained so far. Here are four examples, the last of which did not arise from the research literature, but is in considerable agreement with it. All four teach through student-solved problems and avoid providing worked, template examples.

2.2.1 APOS Theory and the ACE Teaching Cycle

The learning of many university level mathematical topics has been investigated using the APOS (Action-Process-Object-Schema) theory and instructional sequences have been designed reflecting it; see the papers by Artigue and Dubinsky and McDonald (this volume, pp. 275-282) for details. Envisioned as an iterative process, the instructional design begins with a theoretical analysis, called a *genetic decomposition*, of what it means to understand a concept and how that understanding might be constructed or arrived at by the learner. An analysis of instructional attempts may then lead to several revisions of both the theoretical analysis and the instructional sequence.

In this approach students are intentionally put into disequilibrating situations (in which they see their lack of understanding) and, individually or in cooperative groups, they try to make sense of these situations, e.g., by solving problems, answering questions, or understanding ideas. A particular strategy used is the *ACE Teaching Cycle*, consisting of three components: Activities, Class discussion, and Exercises. The intent is to provide students with collaborative experiences that promote their construction of a particular concept and build upon their developing understandings of it in subsequent discussions. The implementation of this pedagogical strategy is somewhat unconventional. Its designers have created textbooks to support it in discrete mathematics, pre-calculus, calculus, and abstract algebra. These textbooks do not contain template solutions to problems and no answers are given to the exercises. Students are encouraged to investigate mathematical ideas for themselves and are allowed to read ahead in their own or other textbooks.[3] (See Asiala, et al, 1996.)

2.2.2 *Didactical Engineering and the Method of Scientific Debate*

Another approach to research and curriculum design is *didactical engineering*, a method favored by many French mathematics education researchers. Based on both Chevallard's *theory of didactical transposition* and Brousseau's *theory of didactical situations*[4] didactical engineering employs the results of research in the elaboration of pedagogical strategies and curriculum materials specifically for mathematics. In this method, an essential part of the teacher's job is to present students with problems whose solutions involve the development of substantive mathematics. Learning is the result of students' adaptation to these mathematical situations (Brousseau, 1997).

Artigue (1991) has used didactical engineering in an instructional approach to differential equations (for first-year university students of mathematics and physics) that, from the beginning, coordinated algebraic, numerical, and graphical approaches with the solution of an associated differential equation.

The *method of scientific debate* is a somewhat similar French approach to conducting mathematics courses with the aim of having students act mathematically, rather than merely covering the syllabus. Students are encouraged to become part of

---

[3] See, for example, the Dubinsky and Leron (1994) textbook for abstract algebra; the course and its rationale have been described in an article for mathematicians (Leron and Dubinsky, 1995).

[4] The theory of didactical transposition aims at describing rules which govern the transformation of scholarly knowledge (mathematics) into taught knowledge, as well as the evolution of taught knowledge (Chevallard, 1991). Some of Brousseau's work on the theory of didactical situations, originally published in French, has recently been translated into English (Brousseau, 1997). Brousseau himself worked primarily at the school level but other French researchers have applied and extended his ideas to the university level.

a classroom mathematical community in which they propose conjectures and debate their relevance and truth. For the method to be successful, the professor should refrain from revealing her or his opinion, should allow time for students to develop arguments, and should encourage maximum student participation. A renegotiation of the 'didactic contract' is necessary so that students come to understand and accept their responsibilities as active participants in the knowledge-building process. (See Legrand, 1993, this volume, pp. 127-137.)

2.2.3 Realistic Mathematics Education and Local Instructional Theories

A third approach to the interaction between research and curriculum design, more often practiced at the elementary and secondary levels but recently being tried at the university level, is that of *realistic mathematics education*. From this perspective, students learn mathematics by mathematizing the subject matter through examining 'realistic' situations, i.e., experientially real contexts for students that draw on their current mathematical understandings. In this approach, the problems precede the abstract mathematics, which emerges from the students' collaborative work towards solutions. Curricula, as well as the instructional theory and its justification, are mutually developed and refined in a gradual, iterative process.

In this approach, curriculum design tends to be integrated with research, perhaps because it is difficult to predict how students will tackle problems for which they have no model solutions. Beginning with realistic mathematics education as the global perspective, the aim is to develop *local instructional theories*. This developmental process begins by positing hypothetical learning trajectories, along with a set of instructional activities. After an instructional sequence has been implemented and observed, researchers engage in retrospective analysis that leads to refinement and revision of the conjectured learning trajectory. Three heuristics are used in designing curricula: (1) the *reinvention principle*, whereby students are guided to construct at least some of the mathematics for themselves, (2) *didactic phenomenology*, whereby researchers analyze practical problems as possible starting points for the reinvention process, and (3) the construction of *mediating,* or *emergent models* of students' informal knowledge and strategies in order to assist students in generalizing and formalizing their informal mathematics. (See Gravemeijer, 1998; Rasmussen and King, 2000.)

2.2.4 The Moore Method of Teaching

This distinctive method of teaching has developed into an informal method of curriculum design and has evolved naturally without calling on research or theory in mathematics education. However, although it arose prior to, and independently of, didactical engineering and the work of Brousseau, some of its aspects are derivable from that work.

The method developed out of the teaching experiences of a single accomplished U.S. mathematician, R. L. Moore, and has been continued by his students (several of whom went on to become presidents of the American Mathematical Society or the

Mathematical Association of America) and their mathematical descendants. It has been remarkably successful in producing research mathematicians, but has also been used in undergraduate university classes. In many versions, students are given definitions and statements of theorems or conjectures and asked to prove them or provide counterexamples. The teacher provides the structuring of the material and critiques the students' efforts. Since apparently no one knows how to effectively tell someone else how to prove a theorem or even how to very usefully explain to a novice what constitutes a proof, students just begin. This sounds a little like throwing someone into a lake in order that he or she learn to swim. However, once a student proves the first small theorem (stays afloat mathematically), it is often possible for her or him to make very rapid progress.

Moore method courses could provide interesting opportunities for research in mathematics education and the method would probably benefit from an analysis in terms of didactical engineering. Even the teaching of Moore himself has not been well researched.[5] However, there is currently an effort to document both the life and teaching method of Moore and to encourage 'discovery learning'.[6] The project is based at the University of Texas, where Moore taught for many years. It is called the R. L. Moore Legacy Project (http://www.discovery.utexas.edu/rlm/).

3. WHO DOES TERTIARY MATHEMATICS EDUCATION RESEARCH?

While there is a substantial amount of research in mathematics education at the school level (Grouws, 1992), the amount at the tertiary level is still modest. Some tertiary studies, e.g., those investigating the effects of gender or the kinds of students who succeed in mathematics, have been conducted by mathematics education researchers without a particularly strong background in tertiary mathematics. However, many studies require that researchers, or at least some members of a research team, have an extensive knowledge of tertiary mathematics itself.

3.1 Getting into the Field

Although a few universities have developed graduate programs with a specialty in tertiary mathematics education leading to the Ph.D., many current researchers have come from the ranks of mathematicians. For example in France, someone interested in learning to conduct such research typically already has a teaching position, joins a research team and a research project, and is trained by 'compagnonage'. Others have made the transition themselves. In the U.S., there has been an effort by the Research in Undergraduate Mathematics Education Community (RUMEC) to mentor interested mathematicians into the field; this group is interested in both doing and promoting research in tertiary mathematics education.

---

[5] However, Moore's students have been interviewed regarding his teaching method (Forbes, 1980) and the method itself has been described by Jones (1977), a mathematician who used it extensively.

[6] The term 'discovery learning' has only recently been applied to Moore method teaching. It refers to the fact that students discover the proofs, and possibly, some of the theorems. It does not appear to be derived from the discovery learning of the 1960s, espoused by Harvard psychologist J. S. Bruner and others.

(See http://www.maa.org/data/features/rumec.html.) It is expected that such mentoring work will be continued by the Special Interest Group of the Mathematical Association of America on Research in Undergraduate Mathematics Education (SIGMAA on RUME).

Where will such researchers come from in the future? Retrained mathematicians? Post-graduate departments of mathematics or mathematics education? What should such researchers know? It seems apparent that they should know a substantial amount of both mathematics and mathematics education, but what topics or courses are most appropriate to their preparation? Would the development of researchers and the furtherance of research be aided by the publication of a 'research agenda' series (e.g., Charles and Silver, 1989), much like the National Council of Teachers (NCTM) did for mathematics education research at the school level?

3.2 Preparation of Future Tertiary Mathematics Education Researchers

Current employment opportunities, combined with the need for more tertiary mathematics education research, suggest some desirable features of Ph.D. programs. In the U.S. and some other countries, new mathematics education Ph.D.s can find employment in mathematics departments, provided they have sufficient knowledge of mathematics – usually the same knowledge as a mathematics Ph.D. except for the dissertation specialty. Often they are expected to teach mathematics to preservice teachers, so a knowledge of school level, as well as tertiary level, mathematics education research is important. If they wish to survive in academia, however, they need to be prepared to produce publishable papers – which means having experience in posing and solving problems, and in writing up results for publication (Schoenfeld, 1999).

Beyond graduate programs in tertiary mathematics education, what less formal training might be useful for would-be researchers? Intensive graduate level summer seminars might help; in France, for example, there is an extensive Research Summer School (*Didactique des Mathématiques*). Perhaps editors of mathematics education research journals might target promising new researchers to review (referee) several manuscripts, thereby introducing them to the criteria for acceptance. Normally, editors try not to overburden individuals, but reviewing papers can be very educational, especially where reviewers are ultimately provided with the editor's and other reviewers' reports.

Also, more mentoring programs, like those of the RUMEC group and SIGMAA on RUME might be beneficial. One or two-day short courses, such as those given at meetings of the American Mathematical Society (AMS) and the Mathematical Association of America (MAA), can help but long-term engagement is really needed. Unfortunately, a productive program of postdoctoral fellowships in the U.S. (the National Science Foundation's PSFMETE program), which supported Ph.D.'s in science, mathematics, engineering, and technology as they learned to do educational work, has just been discontinued.

3.3   The Placement of Tertiary Mathematics Education Researchers

Where will tertiary mathematics education researchers find their academic 'home'? Although there are a number of Ph.D.-granting institutions around the world producing researchers in tertiary mathematics education, unless these individuals already have teaching posts in universities, it is not always clear who will hire them. In United Kingdom, with its current emphasis on research within the academic disciplines, it is almost impossible for new Ph.D.s specializing in tertiary mathematics education research to obtain employment in university mathematics departments (Adrian Simpson, personal communication). In France, didactics is now considered a legitimate applied mathematics research specialty and tertiary mathematics education researchers are "nationally evaluated with the same criteria as other applied mathematicians"; yet this acceptance is still somewhat fragile (Artigue, 1998). Most new French Ph.D.s in didactics find places in IUFM, university institutes for teacher training.

In the U.S., because of the large number of four-year liberal arts colleges and comprehensive state universities, most of whose mathematics departments teach pre-service teachers, the problem of academic 'home' seems less acute. However, only a few mathematics departments in major research universities currently hire persons whose primary research area is tertiary mathematics education although this may be changing.

The problem of academic 'home' is sometimes exacerbated by the low status accorded education research by some mathematicians. Those who specialize in mathematics education (didacticians) are sometimes seen as a "kind of sub-mathematician who finds, in didactic research, a diversion from his or her lack of mathematical productivity" (Artigue, 1998). In Malaysia and China, mathematicians also seem to value only mathematics research, while denigrating mathematics education.

4.   THE APPLICATION OF RESEARCH TO TEACHING PRACTICE

Mathematics education research cannot normally be expected to tell anyone precisely how to teach or to predict whether a teaching method or curriculum will be effective with particular students. What it can do is provide information useful in teaching and for curriculum development. It can give insights into students' intuitive views of limits, functions, logic, etc., which might either be obstacles to new knowledge or useful as bridges to new knowledge. It can help develop ways of teaching specific mathematical topics that arise from an understanding of both mathematics and pedagogy. It can bring out non-obvious barriers to changing teachers' beliefs or pedagogical techniques and a great deal else useful in teaching and curriculum design.

4.1 Dissemination: Getting the Word to Those Who Teach University Mathematics

In order to be useful, research results on the teaching/learning of university level mathematics need to come to the attention of those who teach it, primarily mathematicians, but also including the ever-growing cadre of community college mathematics teachers, lecturers, tutors, etc. Currently, there are several ways of bringing research results to the mathematics community.

In the U.S., the Mathematical Association of America (MAA), has supported dissemination through a number of channels – in the now discontinued newsletter *UME Trends,* in the Teaching and Learning Section of *MAA Online* (www.maa.org), and in *The College Mathematics Journal.* Unfortunately, many universities undervalue exposition as compared to research when it comes to tenure and promotion; hence there is not significant incentive at present for producing such articles.

Other avenues for dissemination include the annual meetings of AMS/MAA at which SIGMAA on RUME organizes mathematics education research paper sessions, along with an expository talk. However, at best this effort can only reach the few thousand mathematicians who attend such meetings. A similar role might be played in the U.K., by the Advanced Mathematical Thinking Working Group. Another modest start towards dissemination and recognition of the field, is the fact that *Zentralblatt für Didaktik der Mathematik (ZDM)* and *Mathematical Reviews (MR)* now both include categories for abstracting research articles in undergraduate mathematics education.

It would be especially beneficial to find ways to bring tertiary mathematics education research results to the attention of graduate students in mathematics, many of whom will take up teaching posts in universities. In the U.S., the Exxon-funded Project NEXT (New Experiences in Teaching) has given several hundred new mathematics faculty members the opportunity to meet and network at annual meetings of MAA, while also attending special workshops on technology and teaching that include some mention of tertiary mathematics education research results. NEXT fellows are linked electronically, and their pedagogical discussions often result in the dissemination of research findings.

4.2 Integrating Research Results into Teaching Practice

The most effective way of bringing tertiary mathematics education research into teaching practice, seems to be via new research-based curricula. In Section 2.2 there are three examples of ways that research has been systematically integrated with curriculum design. Generally speaking, mathematics faculty are busy, and do not have the time to pursue research results in education for their own sake. Hence, to be widely accessible, results in mathematics education must be user-friendly – either in the form of ready-to-wear curricula, or easily accessible digests of useful information. CD-ROMS and Web sites may be effective ways of disseminating information.

4.3 Some Suggestions for Reaching University Mathematics Teachers

Perhaps team teaching, departmental seminars on teaching, or other local efforts could facilitate the incorporation of research results and generally improve pedagogy. One might try intensive summer workshops, such as that on Cooperative Learning in Undergraduate Mathematics Education (CLUME), or the Park City Mathematics Institute (PCMI). (See http://vms.www.uwplatt.edu/~clume/; http://www2.admin.ias.edu/ma/park.htm.) Such workshops and institutes tend to be expensive; the two mentioned here were funded by the U.S. National Science Foundation (NSF).

Currently, there are 'research into practice' sessions (Wilson, 1993) at NCTM meetings; perhaps similar sessions at meetings of university mathematics teachers, such as those of the AMS/MAA, would be beneficial. It is also important to have mathematics Ph.D. students, who will become tomorrow's university mathematics teachers, take research in tertiary mathematics education seriously. Perhaps, in those Ph.D. programs where coursework is required, one could insist that students take a course on tertiary mathematics education research, or even conduct a mini-research project on some aspect of students' mathematical thinking. To facilitate the teaching of such courses, it might be helpful to develop a list of expository and other readings in tertiary mathematics education research, post it on the Web, and update it regularly.[7]

5. NEW DIRECTIONS FOR MATHEMATICS EDUCATION RESEARCH

Clearly, the existing tertiary mathematics education research barely 'scratches the surface.' While some topics of interest in both secondary and tertiary teaching, like the function concept, have benefited from being considered by a number of researchers (see Artigue, this volume, pp. 207-220), other topics such as real analysis or the learning of post-graduate students in mathematics are just now beginning to be studied.

Many areas such as students' learning, teaching and teacher change, problem solving and proofs, social structures like departments, views of mathematics, theoretical frameworks, and pedagogical content knowledge are ripe for further investigation.

5.1 Some Ideas that Might be Worth Pursuing

In his plenary address, Hyman Bass (1998) pointed to four areas of mathematics education, with some associated questions, that are critically in need of systematic research: the secondary/tertiary transition, instructional use of technology,

---

[7] There is currently such a list under "Readings" in the article "Research on Undergraduate Mathematics Education: A Way to Get Started" (http://www.maa.org/data/features/rumec.html); however it is not now regularly updated.

university-level teaching, and the context of the university with respect to teaching. Here is a potpourri of additional questions.

5.1.1 Beginning University Students: The Secondary/Tertiary Interface

In the U.S., many students entering junior colleges and comprehensive state universities are unprepared to take calculus, and much teaching occurs at the pre-calculus level. Some of these students are older, non-traditional students whose secondary mathematical preparation needs renewal. Would it be useful to catalogue the many, and possibly interacting, difficulties of these pre-calculus students? Indeed, even for successful secondary school graduates, there are a number of problems concerning the secondary/tertiary interface. Often the mathematics curriculum at secondary school encourages the development of informal, intuitive ideas, whereas university mathematics courses tend to be more formal and rigorous. Typical of the difficulties encountered is the necessity to reconstruct one's understanding of the real numbers to allow for the equality of 0.999... and 1 or to reconstruct one's notion of equality to allow two numbers a and b to be equal if $|a - b| < 1/n$, for all natural numbers n. (See Artigue, this volume, pp. 207-220.) While some of the difficulties are well known, what to do about them is not.

5.1.2 Learning to Understand and Validate Proofs

Once students get beyond calculus, they move into courses that are more formal and often require them to construct and validate proofs. Validating a proof, i.e., reading it to determine its correctness, involves mentally asking and answering questions, inventing supplementary proofs, etc. (Selden and Selden, 1995). This validation process is part of the implicit curriculum and appears to be a principal way that mathematicians learn new mathematics. But in normal circumstances, it is largely mental, and hence, unobservable. Both conceptual reflection on it and mimicking the validation processes of others may be difficult for students. How does one learn this validation process? How are various kinds of prerequisite knowledge, e.g., logic, related to learning to construct proofs?

5.1.3 Teaching Service Courses for Non-specialists

Much research at the tertiary level has, often implicitly, taken the view that universities train future mathematicians, whereas a large amount of university mathematics teaching occurs in 'service courses' for 'client disciplines', a trend that may well increase (Steen, this volume, pp. 303 312). There have been a few studies of how practicing professionals – architects, biologists, bankers, nurses – use mathematics, with the ultimate aim of improving the teaching of such courses. The classic view of mathematical modelling, which involves identifying and simplifying a problem, solving a decontextualized mathematical version, and mapping the solution back, does not agree with workplace experience. (See Pozzi, Noss and Hoyles (1998), Smith, Haarer and Confrey (1997), Noss and Hoyles (1996).) More workplace studies of mathematics use, especially as those relate to curriculum development, would be helpful.

In addition, the teaching of mathematics to pre-service elementary and secondary teachers comprises another large share of the courses taught in some mathematics departments. Analyzing the kinds of understandings that are required for teaching elementary and secondary mathematics would be tremendously useful. (See, e.g., Ma, 1999.)

5.1.4 Aspects of Teaching Practice and Institutions that Affect Learning

What views of learning do tertiary mathematics teachers have and how do these affect their practice? Does a teacher's pedagogical knowledge closely resemble the kind of automated procedural knowledge that might be called upon in actual teaching practice? That is, can one predict, and ultimately change, moment-to-moment pedagogical strategies? (Schoenfeld, 1998.) In athletics, knowing how a game should be played is rather different from being able to play it.

Could some research be directed towards generating pedagogical content knowledge, e.g., how to teach the Chain Rule or an explanation of why some university students persist in adding fractions incorrectly? Such knowledge can be a major part of the pre-service teacher curriculum, but there is a dearth of it at the tertiary level. Perhaps some mathematicians would be interested in discovering and analyzing pedagogical content knowledge by conducting small teaching experiments, thereby making a contribution without having to delve deeply into the theoretical aspects of mathematics education research.

5.1.5 Philosophical, Theoretical, and Pragmatic Questions

Views of mathematics arising from the current philosophical climate tend to treat mathematics as a social or mental construct, and sometimes equate objectivity with social agreement (Ernest, 1998). This appears inconsistent with the ideas of many mathematicians who often see themselves as approaching some kind of abstract knowledge that is independent of time and place. Is there a synthesis of these two apparently contradictory positions that is compatible with both?

Various approaches, e.g., APOS theory (Asiala, et al, 1996), Schoenfeld's (1985) framework for problem solving, and Pirie and Kieren's (1994) theory of the growth of mathematical understanding, have elucidated individual learning and cognition. Can these be pushed further, both theoretically and empirically? In mathematics education in general, more attention is being paid to 'the social.' Can progress be made toward a theoretical melding of the cognitive and social perspectives?

As technology permeates the curriculum of engineering and other students, questions of which mathematics to teach, as well as how to teach it, come to the fore. Also, students can often carry out tasks using software imbued with mathematics they haven't yet learned. Thus learning to critique the solutions generated (e.g., recognizing the effect of varying parameters) is important. (See Kent and Noss, this volume, pp. 395-404.) Research questions include: How does the design of technology-based service courses in mathematics shape students understandings of mathematics? What kinds of messages about mathematics do students receive in such courses?

5.2 Professional Research Organizations and Journals

In order for a thriving community of tertiary mathematics education researchers to develop and prosper, there need to be adequate opportunities for the presentation and publication of research results. Currently, the following journals and refereed publications will accept tertiary mathematics education research articles, but many publish research mainly at the school level:

- *Journal for Research in Mathematics Education*
- *Educational Studies in Mathematics*
- *Journal of Mathematical Behavior*
- *Proceedings of PME and PME-NA*
- *Recherches en Didactiques des Mathématiques*
- *Focus on Learning Problems in Mathematics*
- *For the Learning of Mathematics*
- *International Journal of Computers for Mathematical Learning*
- *Mathematical Thinking and Learning.*

In addition, the *Research in Collegiate Mathematics Education* volumes specialize in refereed tertiary mathematics education research papers and function much like a journal. These appear in the Conference Board of the Mathematical Sciences series, Issues in Mathematics Education, published by the American Mathematical Society.

Furthermore, tertiary mathematics education researchers need professional organizations devoted to their interests. Some promising developments along this line are occurring. In January 1999, the Association for Research in Undergraduate Mathematics Education (ARUME) was formed in affiliation with the MAA; it is devoted to research in undergraduate mathematics education and its applications. This organization is now a special interest group of the MAA known as SIGMAA on RUME. Meeting in conjunction with the MAA provides an opportunity to interact with, and possibly influence, mathematicians. (See http://www.maa.org/sigmaa/arume.) In February 1998, the Advanced Mathematical Thinking (AMT) Working Group was established at the British Society for Research in Learning Mathematics (BSRLM). Its main aims include continuing to develop a psychology of advanced mathematical thinking, understanding how mathematicians and students think about advanced mathematics, and to provide better ways of teaching students. (See http://www.soton.ac.uk/~amt.)

REFERENCES:

Artigue, M. (1991). Chapter 11: Analysis. In D. Tall (Ed.), *Advanced Mathematical Thinking*, pp. 167-198. Dordrecht: Kluwer Academic Publishers.

Artigue, M. (1992). The importance and limits of epistemological work in didactics. In W. Geeslin and K. Graham (Eds.), *Proceedings of the Sixteenth Conference of the International Group for the*

Psychology of Mathematics Education, Vol. III pp. 3-195 to 3-216. Durham, NH: University of New Hampshire.

Artigue, M. (1998). Research in mathematics education through the eyes of mathematicians. In A. Sierpinska and J. Kilpatrick (Eds.), *Mathematics Education as a Research Domain: A Search for Identity*, ICMI Study, Vol. 2, pp. 477-489. Dordrecht: Kluwer Academic Publishers.

Asiala, M., Brown, A., DeVries, D. J., Dubinsky, E., Mathews, D. and Thomas, K. (1996). A framework for research and curriculum development in undergraduate mathematics education. In J. Kaput, A. H. Schoenfeld, and E. Dubinsky (Eds.), *Research in Collegiate Mathematics Education*, II, pp. 1-32. CBMS Issues in Mathematics Education, Vol. 6. Providence, Rhode Island: American Mathematical Society.

Bass, H. (1998). Research on University Level-Mathematics Education: (Some of) What is Needed, and Why? *Pre-proceedings of the ICMI Study Conference*, Singapore.

Brousseau, G. (1983). Les obstacles épistémologiques et les problèmes en mathématiques. *Recherches en Didactiques des Mathématiques*, Vol. 4.2, 164-168.

Brousseau, G. (1997). *Theory of Didactical Situations in Mathematics*. Edited and translated by N. Balacheff, M. Cooper, R. Southland and V. Warfield, Dordrecht: Kluwer Academic Publishers.

Charles, R. I. and Silver, E. A. (1989). *The Teaching and Assessing of Mathematical Problem Solving, Research Agenda for Mathematics Education, Vol. 3*. Reston, VA: Lawrence Erlbaum and NCTM.

Chevallard, Y. (1991). *La transposition didactique*, 2nd ed., Grenoble, France: La Pensée Sauvage.

Cobb, P. (1994). Where is the mind? Constructivist and sociocultural perspectives on mathematical development. *Educational Researcher*, 23(7), 13-20.

Dahlberg, R. P. and Housman, D. L. (1997). Facilitating learning events through example generation. *Educational Studies in Mathematics*, 33, 283-299.

Davis, R. B. (1990). The knowledge of cats: Epistemological foundations of mathematics education. In G. Booker, P. Cobb and T. N. de Mendicuti (Eds.), *Proceedings of the Fourteenth Conference of the International Group for the Psychology of Mathematics Education*, Vol. I., pp. PI.1 - PI.24. Mexico: CINVESTAV.

Dreyfus, T. (1990). Advanced mathematical thinking. In P. Nesher and J. Kilpatrick (Eds.), *Mathematics and Cognition: A Research Synthesis by the International Group for the Psychology of Mathematics Education*, pp. 113-134. ICMI Study Series. Cambridge: Cambridge University Press.

Dubinsky, E. and Leron, U. (1994). *Learning Abstract Algebra with ISETL*. New York: Springer.

Ernest, P. (1998). *Social Constructivism as a Philosophy of Mathematics*, SUNY Press.

Forbes, D. R. (1980). *The Texas System: R. L. Moore's Original Edition*, Ph.D. Dissertation, University of Wisconsin, University Microfilms, Inc., Ann Arbor, Michigan.

Gibson, D. (1998). Students' use of diagrams to develop proofs in an introductory analysis course. In A. H. Schoenfeld, J. Kaput and E. Dubinsky (Eds.), *Research in Collegiate Mathematics Education*, III, pp. 284-307. CBMS Issues in Mathematics Education, Vol. 7. Providence, Rhode Island: American Mathematical Society.

Gravemeijer, K. (1998). Developmental research as a research method. In A. Sierpinska and J. Kilpatrick (Eds.), *Mathematics Education as a Research Domain: A Search for Identity*, Vol. 2, pp. 277-295. An ICMI Study. Dordrecht: Kluwer Academic Publishers.

Grouws, D. A. (Ed.) (1992). *Handbook of Research on Mathematics Teaching and Learning*, New York: Macmillan.

Hanna, G. (1998). Evaluating research papers in mathematics education. In A. Sierpinska and J. Kilpatrick (Eds.), *Mathematics Education as a Research Domain: A Search for Identity*, Vol. 2, pp. 399-407. An ICMI Study. Dordrecht: Kluwer Academic Publishers.

Hiebert, J. (1999). Relationships between research and the NCTM Standards. *Journal for Research in Mathematics Education*, 30(1), 3-19.

Jones, F. B. (1977). The Moore Method. *American Mathematical Monthly*, 84 (4), 273-278.

Legrand, M. (1993). Débat scientifique en cours de mathématiques. *Repères IREM*, no. 10, Topiques Editions.

Leron, U. and Dubinsky, E. (1995). An Abstract Algebra Story. *American Mathematical Monthly*, 102(3), 227-242.

Lester, F. K., Jr. and Lambdin, D. V. (1998). The ship of Theseus and other metaphors for thinking about what we value in mathematics education research. In A. Sierpinska and J. Kilpatrick (Eds.), *Mathematics Education as a Research Domain: A Search for Identity*, Vol. 2, pp. 415-425. An ICMI Study. Dordrecht: Kluwer Academic Publishers.

Ma, Liping. (1999). *Knowing and Teaching Elementary Mathematics*. Mahwah, NJ: Lawrence Erlbaum Associates.

Noss, R. and Hoyles, C. (1996). The visibility of meanings: Modeling the mathematics of banking. *International Journal of Computers for Mathematical Learning*, (1)1, 3-31.

Pirie, S. and Kieren, T. (1994). Growth in mathematical understanding: how can we characterise it and how can we represent it? *Educational Studies in Mathematics*, 23, 505-528.

Pozzi, S., Noss, R. and Hoyles, C. (1998). Tools in practice, mathematics in use. *Educational Studies in Mathematics*, 36(2), 105-122.

Rasmussen, C. L. and King, K. (2000). Locating starting points in differential equations: A realistic mathematics education approach. *International Journal of Mathematical Education in Science and Technology*, 31(2), 161-172.

Schoenfeld, A. H. (1985). *Mathematical Problem Solving*. New York, NY: Academic Press.

Schoenfeld, A. H. (1994). Some notes on the enterprise (research in collegiate mathematics education, that is). *Research in Collegiate Mathematics Education*. I, pp. 1-19. CBMS *Issues in Mathematics Education* Vol. 4. Providence, RI: American Mathematical Society.

Schoenfeld, A. H. (1999). The core, the canon, and the development of research skills: Issues in the preparation of education researchers. In. E. Lagemann and L. Shulman (Eds.), *Issues in Education Research: Problems and Possibilities*, pp. 166-202. New York: Jossey-Bass.

Schoenfeld, A. H. (1998).On theory and models: The case of teaching-in-context. In S. B., Berenson, K. R., Dawkins, M., Blanton, W. N., Coulombe, J., Kolb, K. Norwood, and L. Stiff, (Eds.), *Proceedings of the Twentieth Annual Meeting of the North American Chapter of the International Group for the Psychology of Mathematics Education*, Vol. 1, pp. 27-38. Columbus, OH: ERIC.

Selden, A. and Selden, J. (1993). Collegiate mathematics education research: What would that be like?, *The College Mathematics Journal*, 24 , 431-445.

Selden, A. and Selden, J. (1995). Unpacking the logic of mathematical statements. *Educational Studies in Mathematics*, 29, 123-151.

Sfard, A. (1991). On the dual nature of mathematical conceptions: Reflections on processes and objects on different sides of the same coin. *Educational Studies in Mathematics*, 22, 1-36.

Sierpinska, A. (1985). Obstacles épistémologiques relatifs à la notion de limite. *Recherches en Didactiques des Mathématiques*, 6.1, 5-67.

Sierpinska, A. (1992). On understanding the notion of function. In G. Harel and E. Dubinsky (Eds.), *The Concept of Function: Aspects of Epistemology and Pedagogy*, pp. 25-58. MAA Notes, Vol.25. Washington, DC: Mathematical Association of America.

Smith, E., Haarer, S. and Confrey, J. (1997). Seeking diversity in mathematics education: Mathematical modeling in the practice of biologists and mathematicians. *Science and Education*, (6)5, 441-472.

Tall, D. and Vinner, S. (1981). Concept image and concept definition with particular reference to limits and continuity. *Educational Studies in Mathematics*, 12, 151-169.

Tall, D. (Ed.) (1991). *Advanced Mathematical Thinking*. Dordrecht: Kluwer Academic Publishers.

Tall, D. (1992). The transition to advanced mathematical thinking: Functions, limits, infinity and proof. In Douglas A. Grouws (Ed.), *Handbook of Research on Mathematics Teaching and Learning*, pp. 495-511. New York: Macmillan.

Tucker, T. W. (Chair) (1990). *Priming the Calculus Pump: Innovations and Resources*. MAA Notes Vol. 17. Washington, D.C: Mathematical Association of America.

Wilson, P. S. (Ed.) (1993). *Research Ideas for the Classroom: High School Mathematics*. New York: Macmillan. (See NCTM's Research Interpretation Project.)

Annie Selden,
Tennessee Technological University, Tennessee, USA
selden@tntech.edu

John Selden,
Mathematics Education Resources Company, Tennessee, USA
js9484@usit.net

JEAN-LUC DORIER AND ANNA SIERPINSKA

RESEARCH INTO THE TEACHING AND LEARNING OF LINEAR ALGEBRA

1. INTRODUCTION

It is commonly claimed in the discussions about the teaching and learning of linear algebra that linear algebra courses are badly designed and badly taught, and that no matter how it is taught, linear algebra remains a cognitively and conceptually difficult subject. This leads to (a) curricular reform actions, (b) analyzing the sources of students' difficulties and their nature, as well as (c) research based and controlled teaching experiments. This chapter is organized along these three axes of action and research. The chapter does not claim the title of an exhaustive review of research on the teaching and learning of linear algebra at the undergraduate level. All omissions should be blamed on the authors' ignorance rather than their bad will.[1]

2. LINEAR ALGEBRA CURRICULUM REFORM MOVEMENT

In many countries, the problems related to the teaching and learning of linear algebra have received much attention within the mathematical community in the past decade.

The 1999 winter meeting of the Canadian Mathematical Society in Montreal had a plenary lecture devoted to this topic (by David C. Lay, the author of the much praised textbook, 'Linear Algebra and its Applications', Addison-Wesley, 1994), and a whole two-day working section with as many as seven 45 minute presentations, followed by a discussion forum. The presentations provided the audience with accounts of many reform efforts across the country.

In the United Stated, the plans for a reform of the teaching of linear algebra at the undergraduate level started to germinate around 1989 and they have been presented and discussed at a number of meetings and workshops, some of them sponsored by the National Science Foundation. In 1990, a group of mathematicians started a study group, the 'Linear Algebra Curriculum Study Group', or LACSG (David C. Lay was one of its first organizers), which proposed a set of recommendations grounded both in the practice of teaching and in research based on

---

1 In several cases, we only gave the most recent and accessible references when different papers have been published by the same author or group of authors. We also avoided references to doctoral dissertations or other type of not easily accessible literature. A more complete bibliography and chronology of the various works is given in (Dorier 2000).

Derek Holton (Ed.), The Teaching and Learning of Mathematics at University Level: An ICMI Study, 255—273.
© 2001 *Kluwer Academic Publishers. Printed in the Netherlands.*

algebra in the United States has become visible through its publications such as the Special Issue on linear algebra in January 1993 of the *College Mathematics Journal*, papers in many of its subsequent issues, papers in the *American Mathematical Monthly*, and a volume of the MAA Notes (No. 42), entitled *'Resources for the Teaching of Linear Algebra'* (1997). Publications of the LACSG contain many concrete ideas about how these recommendations could be implemented in practice, many challenging problems and project ideas, and some accounts of real life implementations of a new linear algebra curriculum in mathematics departments. Harel is the main 'theoretician' and professional researcher in mathematics education of the LACSG. There will be more about his research later in this paper. Other efforts in the direction of an identification and explanation of students' difficulties on the basis of a theory of learning were made, in the United States, by Dubinsky and his collaborators (see Dubinsky and MacDonald, this volume, 273-280).

Without being based in any recognized organizational forms, like the LACSG in the US, curricular changes have been an ongoing process in many countries. In countries like France, Poland or Morocco, where linear algebra has been traditionally taught in a very theoretical way, this teaching tends now to be more orientated towards numerical computations, even if the structural approach remains important for the minority of students majoring in mathematics. The research works we will present now were a source of reflection that nourished these curricular changes.

3. LINEAR ALGEBRA AS A COGNITIVELY AND CONCEPTUALLY DIFFICULT SUBJECT

Attempts at adapting the curriculum to the students' interests and learning processes notwithstanding, one still has to accept the fact that linear algebra is and will remain a difficult subject for most students. We turn now to describing some research that has been done into the sources of this difficulty. For the organization of this section, we would wish to make a distinction between two kinds of sources of students' difficulties: the nature of linear algebra itself (conceptual difficulties), and the kind of thinking required for the understanding of linear algebra (cognitive difficulties). It has to be understood, however, that these two aspects are often inseparable in the real processes of learning and knowing. It is therefore rather difficult to find research in mathematics education that deals with one aspect and not with the other. The separation of the discussion of research works into a section on 'the nature of the beast' (to use an expression of Hillel), and a section on 'the kinds of thinking required for the understanding of linear algebra' is thus a little artificial.

4. THE 'NATURE OF THE BEAST'

4.1 Epistemological and historical analyses

A first attempt at revealing the sources of students' difficulties with linear algebra through a historical and epistemological analysis can be found in Robinet (1986). Work in this direction was then pursued by Dorier (1995a; 1996a; 1997 and 2000, Part 1). This research served not only as a reference for a better understanding of students' mistakes and difficulties, but it has also been used as an inspiration in the design of activities for students.

One particularly illuminating result of this research was concerned with the last phase of the genesis of the theory of vector spaces. The roots of this final step can be found in the late nineteenth century, but it really started only after 1930. It corresponds to the axiomatization of linear algebra, that is to say a theoretical reconstruction of the methods of solving linear problems, using the concepts and tools of a new axiomatic central theory. These methods were operational but they were not explicitly theorized or unified. It is important to realize that this axiomatization did not, in itself, allow mathematicians to solve new problems; rather, it gave them a more universal approach and language to be used in a variety of contexts (functional analysis, quadratic forms, arithmetic, geometry, etc.). The axiomatic approach was not an absolute necessity, except for problems in non-denumerable infinite dimension, but it became a universal way of thinking and organizing linear algebra. Therefore, the success of axiomatization did not come from the possibility of reaching a solution to unsolved mathematical problems, but from its power of generalization and unification and, consequently, of simplification in the search for methods for solving problems in mathematics. This approach marked a new level in abstraction, the concept of vector space being an abstraction from a domain of already abstract objects like geometrical vectors, n-tuples, polynomials, series or functions. It represents a shift of perspective, which induces a sophisticated change of level in mental operations. Indeed, one can distinguish two stages in the construction of a unifying and generalizing concept (which correspond to two mental processes in learning):

- recognition of similarities between objects, tools and methods brings the unifying and generalizing concept into being;
- making the unifying and generalizing concept explicit as an object induces a reorganization of old competencies and elements of knowledge.

As a consequence, one of the most noticeable difficulties encountered in the teaching of unifying and generalizing concepts are associated with the pre-existing, related elements of knowledge or competencies of lower level. Indeed, these need to be integrated within a process of abstraction, which means that they have to be looked at critically, and their common characteristics have to be identified, and then generalized and unified. From a didactic point of view, the problem is that any linear problem within the reach of a first year university student can be solved without using the axiomatic theory. The gain in terms of unification, generalization and simplification brought by the use of the formal theory is only visible to the expert.

One solution would be to give up teaching the formal theory of vector spaces. This is what Hillel (2000) advocates, questioning the sense of introducing the general theory of vector spaces in an undergraduate linear algebra course, only to show the isomorphism between any n-dimensional vector space and the R^n spaces, and then work with finite dimensional spaces most of the time.

However, many people find it important that students starting university mathematics and science studies get some idea about the axiomatic algebraic structures of which vector space is one of the most fundamental. In order to reach this goal, the question of formalism cannot be avoided. Therefore, students have to be introduced to a certain type of reflection on the use of their previous elements of knowledge and competencies in relation with new formal concepts. This led Dorier, Robert, Robinet and Rogalski to introduce what they called 'meta level activities'. These activities are introduced and maintained by an explicit discourse on the part of the teacher about the significance of the introduced concepts for the general theory, their generalizing and unifying character, the change of point of view or a theoretical detour that they offer, the types of general methods they lead to, etc. The students are then engaged in debates on such 'meta issues', and they have to answer on a 'meta level' in written tests. Indeed, this 'meta' dimension is also present in the evaluation of the students' learning. It hinges on the general attitude of the teacher who induces a constant underlying meta-questioning concerning new possibilities or conceptual gains provided by the use of linear algebra concepts, tools and methods.

This does not mean that the students should spend their time reflecting on actions that they would never have time to carry out. On the contrary, it is an absolute priority that any meta discussion or reflection be embedded in a concrete mathematical context, with a problem to be solved. Moreover, the choice of the mathematical problem is fundamental if the aim is for the students to regard the meta-questions as their own and not just a satisfaction of the teacher's or curricular requirements. For examples and evaluation see (Dorier 1995b; Dorier 1997 and 2000, Part II, Chapter 4; Robert and Robinet 1996; Robert 1998).

4.2   Analyses of the language of linear algebra

No matter how hard we try to help the students 'concretize' the abstract linear algebra objects, they have to concretize these abstract objects and not the models of these objects in mental isolation from each other. If they don't, they have a lot of trouble distinguishing between the objects and their representations. Linear algebra proposes and studies a range of 'languages', each of which throws a different light on (or constructs different concepts of) the objects.

The formal language. In France, the first diagnostic work of students' difficulties in linear algebra courses was conducted by Robert and Robinet, and Rogalski in the late 80s (see Dorier 1997 and 2000, Part II, Chapter 1) Responding to a questionnaire, students voiced their concern with the excessive use of formalism, the overwhelming number of new definitions and theorems, and the lack of connection with what they already knew in mathematics. It was quite clear that these students had the feeling of landing on a new planet and were not able to find their way in this new world. On the other hand, teachers complained about their students' erratic use of the basic tools of logic or set theory, as well as lack of skills in elementary Cartesian geometry. The latter was blamed for the failure of the students to use their intuition in building geometrical representations of the basic concepts of the theory of vector spaces. These complaints have a certain validity, but the few attempts at remedying this state of affairs – with the teaching of Cartesian geometry or/and logic and set theory prior to the linear algebra course – did not seem to improve the situation substantially. Students' difficulties with the formal aspect of the theory of vector spaces are not just a general problem with formalism but mostly a difficulty of understanding the specific use of formalism within the theory of vector spaces and the interpretation of the formal concepts in relation with more intuitive contexts like geometry or systems of linear equations, in which they historically emerged. Various diagnostic studies conducted by Dorier, Robert, Robinet and Rogalski between 1987 and 1994 pointed to a single massive obstacle appearing for all successive generations of students and for nearly all modes of teaching, namely, what these authors termed the *obstacle of formalism* (Dorier 1997 and 2000 Part II, Chapter 1).

The 'geometric', the 'algebraic' and the 'abstract' languages of linear algebra. Hillel (2000) distinguished three basic languages used in linear algebra: the ('abstract') language of the general abstract theory including vector spaces, dimension, linear transformations of vector spaces, the general eigenvalue theory, etc.; the ('algebraic') language of the R^n theory, including n-tuples, matrices, rank, solution of systems of linear equations, etc.; and the ('geometric') language of the two- and three-dimensional spaces including directed line segments, points, lines, planes, transformations of geometric figures. Hillel showed how misleading the geometric language could be for students who would take it too literally.

The 'opaqueness' of the representations seems to be ignored by lecturers. Hillel referred to an investigation in which five experienced teachers were videotaped when dealing with the topic of eigenvalues and eigenvectors in their courses. These tapes illustrated how the lecturers were constantly shifting the notation and modes of description. Moreover, these shifts were usually made without a pause and without any attempt to alert the students in any explicit way. Nor was it always clear when the geometric description was used to illustrate a specific case. By far, the most confusing case for students is the shift from the abstract to the algebraic representation when the underlying vector space is R^n. In this case, an n-tuple is represented as another n-tuple relative to a basis. Thus the object and its representation are the same 'animal' if not exactly the same list of values. The same is true for a matrix transformation A, first considered as a linear operator, and then having a (possibly different) matrix representation relative to a given basis. This confusion leads to persistent mistakes in students' solutions related to reading the values of a linear transformation given by a matrix in a basis (Hillel and Sierpinska 1994).

The 'graphical', the 'tabular' and the 'symbolic' registers of the language of linear algebra. Duval (1995) defined *semiotic representations* as productions made by the use of signs belonging to a system of representation which has its own constraints of meaning and functioning. Semiotic representations are, according to him, absolutely necessary in mathematical activity, because its objects cannot be directly perceived and must, therefore, be represented. Moreover, semiotic representations are not only a means to externalize mental representations in order to communicate, but they are also essential for the cognitive activity of thinking. In fact, they play a role in developing mental representations, in accomplishing different cognitive functions (objectification, calculation, etc.), as well as in producing knowledge. Duval stressed the distinction between *semiosis* or the comprehension or production of a representation by a sign, from *noesis* or the conceptual comprehension of an object, while claiming that the two acts cannot be separated in actual cognitive processes. In cognitive activities linked to *semiosis*, he distinguished three types of activities: the formation of a representation which can be identified as belonging to a given register; the processing and transformation of a representation within the register where it was created; and finally the *conversion*, i.e., the transformation of a semiotic representation from one register to another. He stressed the importance of the third activity by describing it as a necessary passage for coordinating registers attached to the same concept and claimed that, while the first two activities seem to be taken into account in mathematics teaching, the third one is usually ignored.

In her work, Pavlopoulou (see Dorier 2000, pp. 247-252) applied and tested Duval's theory in the context of linear algebra. She distinguished between three registers of semiotic representation of vectors: the graphical register (arrows), the table register (columns of coordinates), and the symbolic register (axiomatic theory of vector spaces). Through several studies, she has shown that the question of registers, especially as regards conversion, is not usually taken into account either in teaching or in textbooks. She also identified a number of students' mistakes that could be interpreted as a confusion between an object and its representation (especially a vector and its geometrical representation) or as a difficulty in converting from one register to another. She analyzed in detail what is at stake when converting from one register to another, pointing out the various difficulties. Conversions involving the symbolic register turned out to be always more difficult for the students, but conversions *into* the symbolic register were more difficult than *from* it. Pavlopoulou designed and experimented with a way of teaching linear algebra that explicitly took into account the question of registers and especially the conversions from one register to another. Her results showed, not only that students could be successfully trained in order to perform correct conversions but also that this training improved their linear algebra grades. This seemed to validate Duval's hypothesis regarding the importance of the ability to manage the conversion of registers in understanding a concept (the importance of semiosis for noesis). However, the fact that the experimental group performed conversions from one register to another better than the other group was only due to a specific training. Morever, apart from this type of tasks, the comparative test involved only a few tasks of a rather low level of difficulty. Thus, even if this work was essential, the conclusions it led to needed to be confirmed by further research questioning the importance of the semiotic aspects in more varied and difficult tasks in linear algebra. This was partly achieved by the works we will present now.

4.3 Characteristics of thinking required for the understanding of linear algebra

Cognitive flexibility. The research of Alves-Dias (see Alves-Dias and Artigue 1995, and Dorier 2000, pp. 252-256) aimed at extending the work of Pavlopoulou. Alves-Dias generalized the necessity of conversions from one semiotic register to another for the understanding of linear algebra to the necessity of 'cognitive flexibility': Linear algebra being a model for various mathematical contexts, it deals with several frameworks (in Douady's (1986) sense), registers of semiotic representation, and viewpoints at the same time, and requires the learner to be able to move freely between them. Moreover, she focused her study on the question of articulation between the Cartesian and parametric representations of vector subspaces, which is not a mere question of change of register, but deals with more complex cognitive processes involving the use of concepts like rank and duality.

Indeed, a Cartesian representation of a subspace V of R^n is given by a set of p linear equations in n variables. A parametric representation of V is a translation, in terms of equations on the coordinates, of the fact that any vector in V is a linear combination of a set of generators. When looking for a parametric representation of V, given a Cartesian representation, once a set of generators is known, it is therefore only a question of change of register. Thus, the main question is to find a set of generators of V, which is not just a change of register, nor is it an elementary cognitive process. In fact, any vector whose coordinates satisfy the equations of the Cartesian representation is in V. But it is usually difficult to prove directly that a set of vectors spans the subspace, unless the dimension 'd' of the subset is known. Indeed, in this case any set of d independent vectors constitutes a basis, and therefore a spanning set. Basic results on duality show that the dimension of V is $n-r$, r being the rank of the Cartesian equations representing V. Thus the reasoning and the calculations involved in the task consisting in finding a parametric representation of V are such that competencies with regard to the concept of rank and duality are indispensable. Moreover, in order to avoid easy mistakes in calculations or reasoning, it is necessary to be able to have some control over the results obtained. For instance, in the preceding examples, it is important to realize, from the beginning, that $dim\ V \geq n-p$ which means that the parametric representation will use at least $n-p$ parameters.

Alves Dias showed that in textbooks and classes, in general, the tasks offered to students are very limited in terms of flexibility. She developed a series of exercises that required the student to mobilize more changes of settings or registers and to exert explicit control via the concepts of rank and duality. She experimented with these exercises with students both in France and in Brazil. Her analysis of the students' responses demonstrated a variety of difficulties for the students. For instance, students often identified one type of representation exclusively through semiotic characteristics (a representation with x's and y's would be considered as obviously Cartesian) without questioning the meaning of the representation. Concerning the means of control over the validity of the statements by the students and anticipation of results or answers to problems, she found that a theorem like: $dim\ E = dim\ Ker\ f + dim\ Im\ f$, is known and used correctly by many students, but it is very seldom used for those purposes even in cases in which it would immediately bring up a contradiction with the result obtained, or in cases in which it offered valuable information in order to anticipate the correct answer.

Trans-object level of thinking. Hillel and Sierpinska (1994), stressed that a linear algebra course which is theoretically rather than computationally framed requires a level of thinking that is based on what has been termed by Piaget and Garcia as the 'trans-object level of analysis' which consists in the building of conceptual structures out of what, at previous levels, were individual objects, actions on these objects, and transformations of both the objects and actions (Piaget and Garcia 1983, p. 28). A similar claim was made by Harel (2000), in his assertions that a substantial range of mental processes must be encapsulated into conceptual objects by the time students get to study linear algebra. In particular, functions must not be just rules for producing numbers from other numbers but objects in themselves, that can be added, multiplied by scalars, and combined (see the 'Concreteness Principle' in section 3.1).

The difficulty of thinking at the trans-object level leads some students to develop 'defense mechanisms' (to 'survive' the course), consisting in trying to produce a written discourse formally similar to that of the textbook or of the lecture but without grasping the meaning of the symbols and the terminology. This appeared as a major problem for Sierpinska, Dreyfus and Hillel, and the team set out to design an entry into linear algebra that would make this behaviour or attitude less likely to appear in students (Sierpinska, Dreyfus and Hillel 1999). The designed teaching-learning situations were set in a dynamic geometry environment (Cabri-geometry II) extended by several macro-constructions for the purposes of representing a two-dimensional vector space and its transformations (Sierpinska, Trgalova, Dreyfus and Hillel 1999; Sierpinska 2000). It turned out that, while, indeed, in this environment, most students were learning the linear algebra concepts 'with understanding' and the production of nonsensical formal discourse was rare (but was not completely eliminated), this understanding was too often at odds with the meanings intended by the theory. The phenomenon was blamed, partly, on the Cabri environment whose very concrete graphical and dynamic representations may have influenced the students' thinking. But the persistence of some 'misconceptions' even in the more varied environment in an experimental course suggested that the dynamic geometry environment could not bear the entire responsibility for the phenomenon.

Theoretical as opposed to practical thinking. Further analysis of the students' behaviour in the experimented situations led Sierpinska to postulate certain features of their thinking that could be held partly responsible for their erroneous understandings and difficulties in dealing with certain problems (especially the problem of extending a transformation of a basis to a linear transformation of the whole plane). She proposed that these features be termed "a tendency to think in 'practical' rather than 'theoretical' ways" (Sierpinska 2000).

The distinction between these two ways of thinking was inspired by the Vygotskian notion of scientific, as opposed to spontaneous or everyday concepts. In particular, it was assumed that

- Theoretical thinking (TT) is a specialized mental activity in itself, whereas practical thinking (PT) is an auxiliary activity that accompanies and guides other activities. TT expresses itself in the written word or texts; PT expresses itself in direct action upon the environment. A theoretical thinker is aware of her being engaged in this kind of thinking and often reflects on and is able to speak about her ways of thinking.
- In TT language and other sign systems are not 'taken for granted' as they are in PT; TT problematizes the existing semiotic means of representation of knowledge and seeks new and better performing notational systems.
- TT is based on systems of concepts rather than aggregates of ideas, whence the sensitivity of TT to contradiction in thoughts.
- In TT reasoning is based on logical and semantic connections between concepts within a system rather than on the 'logic of action', contingent upon empirical, functional and contextual associations between objects or events. In particular, definitions of concepts, comparison between concepts and their differentiation are based on their relation to more general concepts rather than on the basis of their most 'typical' examples.

The behaviour of students who were encountering difficulties in the experimentations suggested that their ways of thinking had the features of PT rather than TT as described in the above characterization. In particular they had trouble going beyond the appearance of the graphical and dynamic representations in Cabri that they were observing and manipulating: their relation to these representations was 'phenomenological' rather than 'analytic'. For example, some of them started thinking of vectors as elastic arrows whose length and direction could be changed at will, and of linear transformations as dynamic pairs of such arrows, one of which could be dragged independently while the other would follow the movement in some special way.

Students' relative insensitivity to the systemic character of mathematical knowledge (and thus, to contradiction) was also held responsible for their difficulties. In linear algebra, the number of new concepts that students are introduced to grows very quickly and the chains of deduction become longer and longer. But many students in the experimentation were either unable or unwilling to perform longer chains of reasoning, always wanting to restrict their field of thought to the problem and context at hand. Some of them also did not seem interested in checking their assertions, even if this could be done using the facilities of Cabri. This often led them to contradictions.

By far the most blatant feature of the students' practical thinking was their tendency to base their understanding of an abstract concept on 'prototypical examples' rather than on its definition. For example, linear transformations were understood as 'rotations, dilations, shears and combinations of these'. This way of understanding made it very difficult for them to see how a linear transformation could be determined by its value on a basis, and consequently, their notion of the matrix of a linear transformation remained at the level of procedure only.

The above conjecture, about the students' tendency to practical thinking as a source of difficulty in linear algebra, has been refined through a more recent (on-going and yet unpublished) research on ways of thinking in high achieving linear algebra students who had studied linear algebra in a paper and pencil environment only.

The analytic-arithmetic and analytic-structural modes of thinking in linear algebra. Parallel with the three languages 'spoken' in linear algebra identified by Hillel, the 'geometric', the 'algebraic' and the 'abstract', there are three modes of thinking that have led to the development of these languages and that are necessary for an understanding of the domain. Sierpinska et al. called them the 'synthetic-geometric', 'analytic-arithmetic', and 'analytic-structural' (Sierpinska, Defence, Khatcherian and Saldanha 1997). The difference between the 'synthetic' and the 'analytic' modes of thinking is that, in the *synthetic* mode, an object of thought appears, in a way, as *directly accessible* to the mind that then tries to describe it. In the *analytic* mode, an object of thought is seen as *conceived* or *constructed* in a language and a conceptual system; it can be given indirectly by a definition, or a set of properties. The further distinction between analytic-arithmetic and analytic-structural modes was made via a reference to the historical development of linear algebra. It referred to "two large steps, related to two processes. One was the *arithmetization of space*, as it took place in the passage from synthetic geometry to analytic geometry (and the theory of linear equations). The other was the de-arithmetization of space or its *structuralization*, whereby vectors lost the coordinates that anchored them to the domain of numbers and became abstract elements whose behaviour is defined by a system of properties or axioms" (Sierpinska et al. 1997). Through examples of students' solutions of linear algebra problems and exercises, the article showed how each of these modes of thinking can present its specific difficulties for the students, suggesting that by far the hardest thing is for the students to use each of them in a conscious manner where appropriate and move flexibly between them (which brings us back to the postulate of cognitive flexibility).

5. THE TEACHING OF LINEAR ALGEBRA

5.1 *Postulated principles of teaching linear algebra and their violation in practice*

Harel posits three 'principles' for the teaching of linear algebra, inspired by Piaget's psychological theory of concept development: the Concreteness Principle, the Necessity Principle and the Generalizability Principle.

According to Piaget, the development of abstract concepts requires the cognitive operation of 'reflective abstraction', i.e. a mental operation on mental representations of some physical actions which leads to an identification of these mental representations as objects in themselves. Based on this idea, the Concreteness Principle states, "For students to abstract a mathematical structure from a given model of that structure, the elements of that model must be conceptual entities in the student's eyes; that is to say, the student has mental procedures that can take these objects as inputs" (Harel 2000, p.180). This principle is violated whenever the general concept of vector space is taught as a generalization from structures such as R^n, spaces of polynomials of restricted degree, spaces of matrices, continuous functions, etc., to students who have not (yet) constructed the elements of these structures as mental entities on which other mental operations can be performed.

Starting from the premise that "students build their understanding of a concept in a context that is concrete to them", Harel then proceeded to investigate which one of the two contexts, algebraic or geometric, would lead students to a better understanding of the vector space concept. His experiment led him to conclude that a sustained emphasis on a geometric embodiment of abstract linear algebra concepts appeared to produce a quite solid basis for students' understanding. He insisted, however, that it would be incorrect to conclude that a linear algebra course should start with geometry and build the algebraic concepts through some kind of generalization from geometry. A teaching experiment built on this premise allowed Harel to observe that "when geometry is introduced before the algebraic concepts have been formed, many students view the geometry as the raw material to be studied. As a result, they remain in the restricted world of geometric vectors, and do not move up to the general case." Harel's warning was that "we need to consider carefully the way geometry is introduced and used. We, as teachers, see how the geometric situation is isomorphic to the algebraic one and so we believe that the geometric concept can be a corridor to the more abstract algebraic concept. Unfortunately, many students do not share this important insight" (Harel, in press).

The Necessity Principle – "For students to learn, they must see an [intellectual, as opposed to social or economic] need for what they are intended to be taught" - is based on the Piagetian assumption (which has also been adopted by the Theory of Didactic Situations elaborated by Brousseau (1997)) that knowledge develops as a solution to a problem. Learning is assumed as not reduced to the result of a transmission of information from teacher to students. Learning is understood more as sense making of situations in a milieu, and developing ways of coping with them. Teaching of a knowledge K consists in organizing the didactic milieu in such a way that knowledge K becomes necessary for the student to survive in it. If the situations in a mathematics class are such that a certain type of social behaviour is sufficient for survival in them, without any use of mathematical knowledge, then it is the social behaviour, not the mathematical knowledge that the students will learn. If the teacher solves the problems for the students and only asks them to reproduce the solutions, they will learn how to reproduce teacher's solutions, not how to solve problems. As an example of a violation of the necessity principle, Harel quotes the "deriving the definition of vector space from a presentation of the properties of R^n that correspond to the vector space axioms" for a student unfamiliar with the axiomatic approach and the economy of thought it provides for mathematical theory building. "For this student", says Harel, "properties of R^n are self evident; they do not warrant the attention they get" (Harel 2000). A student with this mind-set will not see the point of proving such things as "for any vector **v** in a vector space, $(-1)\mathbf{v}=-\mathbf{v}$", or "there is only one zero vector in a vector space" (see also Sierpinska 1995).

The last, Generalizability Principle postulated by Harel, is concerned more with didactic decisions regarding the choice of teaching material than with the process of learning itself. "When instruction is concerned with a 'concrete' model, that is a model that satisfies the Concreteness Principle, the instructional activities within this model should allow and encourage the generalizability of concepts." This principle would be violated if the models used for the sake of concretization were so specific as to have little in common with the general concepts they were aimed at. For example, the notion of linear dependence introduced in a geometric context defined through collinearity or co-planarity is not easily generalizable to abstract vector spaces.

5.2 Consistency and limitation of a geometry based linear algebra course

Robert, Robinet and Tenaud (1987) designed and experimented with a geometric entry into linear algebra. The aim was to overcome the obstacle of formalism by giving a more 'concrete' meaning to linear algebra concepts, in particular, through geometrical figures that could be used as metaphors for general linear situations in more elaborate vector spaces. However, as in Harel's study mentioned above, the connection with geometry proved to be problematic. Firstly, geometry is limited to three dimensions and therefore some concepts, like rank, for instance, or even linear dependence, have a quite limited field of representation in the geometric context. Moreover, it is not rare that students refer to affine subspaces instead of vector subspaces when working on geometrical examples within linear algebra.

A historical analysis shows that the relation between geometry and linear algebra is less 'natural' than it may appear at first sight. Indeed, the basic notions of the theory of vector spaces were created in an algebraic context (linear equations or field theory) more than in a geometric one. The connection with geometry was more indirect; indeed it was related to more sophisticated tools, mainly transformations and their classification. However, the role of geometry in the genesis of the theory of vector spaces had been quite decisive at the beginning of the 20th century in the early stage of 'functional analysis'. Schmidt's and Riesz's use of geometric language in the study of functional equations made it possible to use a 'synthetic'[2] approach that contrasted with 'analytic'[3] approaches developed especially by Hilbert. In the analytic approach, a function was represented by an infinite series of coefficients (e.g. Fourier's coefficients), and the functional equation was then expressed as an infinite system of linear equations. By using a geometrical metaphor (especially concerning topological relations: distance, norm and orthogonality), Schmidt and Riesz introduced the concept of functional vector space in which the function

---

[2] 'analytic-structural' in Sierpinska et al.'s terminology (1997)

[3] 'analytic-arithmetic' in Sierpinska et al.'s terminology (ibid.)

appeared as a single object denoted by a letter. In this approach the function was independent of its representation by infinite series, just as geometric vectors could be conceived of as independent from their representation in a coordinate system (Dorier 1997 and 2000, Part I, and Moore 1995).

In her work, Chartier (see Dorier 2000, pp. 262-264)[4] conducted an epistemological study of the connection between geometry and linear algebra, using the evidence from both historical and modern texts. One of her main goals was to characterize what was meant by geometrical intuition, in relation with linear algebra, by various authors. In the second part of her work, she was interested in the didactic aspect of the question. She analyzed several textbooks from different countries and different periods; she also designed questionnaires for teachers and students at various university levels.

She found that the necessity of geometric intuition was very often postulated by textbooks or teachers of linear algebra. However, in reality, the use of geometry was most often very superficial. Some students would use geometrical representations or references in linear algebra, without this always being to their advantage, however. Indeed, as already mentioned, some of them could not distinguish the affine space from the vector space structure; they also often could not imagine a linear transformation that would not be a geometric transformation. In other words, the geometrical reference acted as an obstacle to the understanding of general linear algebra. On the other hand, some very good students were found to use geometric references very rarely. They could operate on the formal level without using geometrical representations. It seems that the use of geometrical representations or language is very likely to be a positive factor, but it has to be controlled and used in a context where the connection is made explicit.

5.3 Study of long-term teaching experiments.

Most of the research conducted in France on the teaching and learning of linear algebra has been more or less directly connected with an experimental course implemented by Rogalski (1991)[5]. This course was built on several interwoven and long-term strategies, using meta level activities as well as changes of settings (including intra-mathematical changes of settings), changes of registers and points of view, in order to obtain a substantial improvement in a sufficient number of students.

'Long-term strategy' (Robert 1992) refers to a type of teaching that cannot be divided into separate and independent modules. The long-term aspect is vital because the mathematical preparation and the changes in the 'didactic contract' (Brousseau 1997) have to operate over a period which is long enough to be efficient for the students, in particular as regards assessment. Moreover, the long-term strategy refers to the necessity of taking into account the non-linearity of the

---

4 Chartier defended her PhD thesis, supervised by Dorier, in November 2000: Chartier, G. (2000). Rôle du géométrique dans l'enseignement et l'apprentissage de l'algèbre linéaire, PhD Thesis, équipe DDM, Laboratoire Leibniz, Grenoble.

5 See also (Dorier 1997 and 2000, Part II, Chapter 3).

teaching due to the use of change in points of view, implying that a subject is (re)visited several times in the course of the year.

The teaching design had the following main characteristics:

- In order to take into account the specific epistemological nature of the concepts, some activities are introduced, at a favourable and precise time of the teaching, in order to induce a reflection on a 'meta' level.
- A fairly long preliminary phase precedes the actual teaching of elementary concepts of linear algebra. It prepares the students to understand, through 'meta' activities, the unifying role of these concepts.
- As much as possible, changes of settings and points of view are used explicitly and are discussed.
- Finally, the concept of rank is given a central position in this teaching.

For a long-term teaching design, it is difficult to choose the time suitable for its evaluation, as interference may occur due to students' own organization of their time and work, in a way that cannot be kept under control. Thus, phenomena of maturing, depending on students' level of involvement (which varies during the year) are difficult to take into account in the evaluation of the teaching.

Moreover, such a global teaching design cannot be evaluated by usual comparative analyses, because the differences with the standard course are too important. However, internal evaluations have been conducted, showing several positive effects, even if some questions remain open.

The effects of the long-term strategy have also been analyzed through another type of work by Behaj (see Behaj and Arsac 1998, and Dorier 2000, pp. 259-262). Behaj is interested in what he calls the 'structuration du savoir' (structuring of knowledge) for teachers as well as for students. The experimental part of this work dealt with the following set of concepts: vector (sub)space, linear (in)dependence, generators, basis, rank and dimension. Each of these concepts could be defined in many different ways and, moreover, they could be presented in various different orders. Teachers were asked to describe how they usually present these notions to their students as well as to give their opinion on two different proposals of presentation. They were also asked to explain the relative importance they awarded examples, exercises and proofs in their teaching. Students, from second to fifth years of university mathematics and science programmes in France and Morocco were interviewed in pairs. They were asked to write in common a plan of a lesson presenting the concepts for first year students and their discussions were recorded.

Concerning the teachers, the results of the experiment showed a great variability. Moreover, it suggested that teachers may have different priorities at different times of their teaching. In the long-term strategy, local choices made by a teacher would have to be evaluated as a whole, as a dynamical system with complementarity but also contradictions and side effects.

Concerning the students, Behaj's experiment was interesting in that it gave an idea of what remained one, two, three or four years after the course. Having to build a course for other students made them suddenly realize that they had finally understood things that were obscure at the beginning. It appeared as well, that several students explained that they had to wait until the very end of the first year, while preparing the exam, or even until they were in their second year, to overcome their difficulties with formalism and finally understand something in linear algebra.

These results are important as regards the question of the evaluation of long-term teaching. Maturity is a necessity; it may take more time for some students than for others. These results show that the question of the evaluation of long-term teaching design taking into account students' individual work and institutional constraints, is a real theoretical challenge for research in mathematics education.

5.4 Study of the relationship between tutor-student-text interactions and the type of knowledge learned

One can change the curriculum, make the instructors more knowledgeable about the students' learning processes, and prepare the students, already at the high school and college level, for theoretical thinking, but one may still fail to obtain a satisfactory understanding of linear algebra in students. The above-mentioned study, conducted by Behaj, points to the links between the way a teacher 'structures' knowledge and the knowledge that the students learn. In another long-term study, based on observations of students learning linear algebra from textbooks with the help of tutors over a period of 13 weeks, it was suggested that the mathematical meanings constructed by the students depend, to a large extent, on the 'formatting' of the way they use the text, and of the way they work with it as students and as mathematicians (Sierpinska 1997). This 'formatting' is a function of both the text's strategy of defining it's 'model reader' (the constraints it imposes on its possible interpretations) and of the interactions about the reading of the text between the tutor and the student. Sierpinska found it useful to distinguish between the process of formatting and it's result, the format. In her research, she focused not so much on the social aspects of interactions as on their mathematical content, aiming at identifying the mechanisms of stabilization of mathematical meanings through various kinds of formatting, and at analyzing thus established formats or 'standards' of understanding in mathematics. She gave examples that illustrated how different these standards could be in relation with different formatting of the use of a textbook and with different formatting of the student's mathematical behaviour.

6. CONCLUSIONS

The products of the research into the teaching and learning of linear algebra are not, in general, statements of the type, "Such and such a way of teaching linear algebra will result in a better understanding, by all students, of the basic concepts of the theory and a better performance on standard examination questions", supplemented by classroom materials, textbooks, software, and the like. Such statements and the steady supply, on the market, of linear algebra textbooks, tutorials and software, are not, usually, based on research. In fact, the most reliable results of the research are, in a sense, 'negative' in nature. They explain, to some extent, why so many students fail in linear algebra courses. They show why some innovative approaches to teaching linear algebra can fail. These results do lead to some well founded 'recommendations' or 'good advice' concerning the practice of teaching, but the advice is not to be confused with an infallible recipe; in fact, the recommendations are but conjectures that are still open to questioning.

This paper attempted to pinpoint some reasons *why many students find linear algebra so difficult to learn*. These are:

1. The axiomatic approach to linear algebra appears gratuitous to many students. All linear problems within the reach of first year university students can be solved without using the theory of vector spaces. Thus this theory has little chance of being perceived as an intellectual necessity for these students. The students may 'acknowledge' its existence, or even 'appreciate' it as significant information, but the theory is unlikely to become part of the 'cognitive furnishing' of the students' minds. It will be considered as superfluous and meaningless by many students.
2. Linear algebra is an 'explosive compound' of languages and systems of representation. There is the geometric language of lines and planes, the algebraic language of linear equations, n-tuples and matrices, the 'abstract' language of vector spaces and linear transformations. There are the 'graphical', the 'tabular' and the 'symbolic' registers of the languages of linear algebra. There are also the 'Cartesian' and the 'parametric' representations of subspaces. Teachers and texts constantly move between these languages, registers and modes of representation without allowing for the time necessary to make the conversions and discuss their validity. They appear to assume that these conversions are 'natural' and obvious and don't need any conceptual work at all. Linguistic and epistemological studies, as well as observations of students demonstrate, however, how deceiving these assumptions can be.
3. Linear algebra is highly demanding from the cognitive point of view. On the most general level, an understanding of linear algebra requires a fair amount of 'cognitive flexibility' in moving between the various languages (e.g. the language of matrix theory and the language of the vector space theory), viewpoints (Cartesian and parametric) and semiotic registers. Understanding of linear algebra requires also that the students encapsulate, into identifiable conceptual structures, a substantial range of what were, previously, individual objects and actions on these objects. (For example, functions have to be viewed as objects in themselves, elements of a vector space rather than procedures of assigning numbers to other numbers). Moreover, understanding of linear algebra requires the ability to resort to 'theoretical thinking'. Indeed, it is indispensable to have a means of control over context dependent way of thinking, like 'intuitions' and mental imagery, characteristic of 'practical thinking'. 'Practical thinking' is, however, necessary, to avoid a situation where linear algebra is no more than a foreign, cryptic and formal language that can be written but not used to think with.

We also asked the question, *what are the implications of research for the teaching of linear algebra*. We feel that:

1. It must be recognized that a very large majority of students taking the undergraduate linear algebra courses are not preparing to specialize in mathematics, but in a variety of other domains and disciplines. The axiomatic approach in these courses is, therefore, highly questionable.
2. If an axiomatic approach is used in the teaching of linear algebra, then even the most motivated and able students must be given an opportunity to reflect on and discuss 'meta-level' issues such as the generalizing and unifying character of the linear algebra concepts.
3. The structuring of the subject matter by textbooks and teachers should take into account what is known about how people learn and do mathematics. Starting from an intellectually engaging question, the process then evolves through finding and doing examples, making conjectures, solving problems, proving theorems and communicating with others in talking and writing. The choice of such 'engaging questions' which trigger the process of learning is of primary importance for the effectiveness of teaching.
4. The success rate in linear algebra courses cannot be improved by obliging students to take courses in logic and set theory as prerequisites. In linear algebra, students have difficulty not with formalism in general, but with the interpretation of the formal concepts of vector space theory in the more intuitive contexts of geometry and systems of linear equations.
5. Even a 'matrix oriented' linear algebra course is cognitively demanding; students must be allowed the time to start to view vectors and matrices as objects, or conceptual entities in themselves.
6. Although geometric embodiments may help the students in understanding the more abstract concepts, it is not a good idea to start a linear algebra course with vector geometry and build the algebraic concepts as a generalization from geometry. The relation between linear algebra and geometry is, epistemologically, less natural than it may appear. In some cases, geometry may act as an obstacle to students' understanding of linear algebra.
7. The didactic procedure of deriving the axioms of the general vector space from the properties of R^n, in a kind of bottom-up or 'inductive' generalization is ineffective in terms of the students' understanding. Research suggests that the axiomatics of vector space are better understood from the more general perspective of an algebraic system and notions such as that of a neutral element with respect to an operation, opposite or inverse element, etc.
8. The learning of linear algebra is a long process, requiring a maturation of thought and an evolution of the points of view. Short-term teaching experiments are likely to pass over some of the important factors of the learning of this subject.
9. The number of undergraduate linear algebra texts is overwhelming. Some authors attempt to write teacher-proof texts, heavily annotated with didactic advice and guidance for the students. But research suggests that the same text may teach the student very different knowledge depending on how she herself or the teacher 'formats' her use of that text. Teachers need suggestions regarding the structuring

of the knowledge they teach and a supply of good examples, questions, exercises, and problems, and they appreciate texts which provide them with that. But they don't need and cannot be replaced by strongly didactic texts that appear to ignore the role teachers play in the process of learning mathematics.

REFERENCES

Alves Dias, M. and Artigue, M. (1995). Articulation Problems Between Different Systems of Symbolic Representations in Linear Algebra. In *The Proceedings of PME19*, Volume 2, pp. 34-41 Universidade Federal de Pernambuco, Recife, Brazil..

Behaj, A. and Arsac, G. (1998). La Conception d'un Cours d'Algèbre Linéaire. *Recherches en Didactique des Mathématiques*, 18 (3), 333-370.

Brousseau, G. (1997). *Theory of Didactic Situations in Mathematics*. Dordrecht: Kluwer Academic Publishers.

Dorier, J.-L. (1995a). A General Outline of the Genesis of Vector Space Theory. *Historia Mathematica*, 22 (3), 227-261.

Dorier, J-L. (1995b). Meta Level in the Teaching of Unifying and Generalizing Concepts in Mathematics. *Educational Studies in Mathematics*, 29 (2), 175-197.

Dorier, J.-L. (1996). Basis and Dimension: From Grassmann to van der Waerden. In G. Schubring (Ed.), *Hermann Günther Grassmann (1809-1877): Visionary mathematician, Scientist and Neohumanist Scholar*, pp. 175-196, Dordrecht: Kluwer Academic Publishers.

Dorier, J.-L. (Ed.) (1997). *L'enseignement de l'Algèbre Linéaire en Question*. Grenoble: La Penséee Sauvage éditions.

Dorier, J.-L. (1998). The Role of Formalism in the Teaching of the Theory of Vector Spaces. *Linear Algebra and its Applications (275)*, 1 (4), 141-160.

Dorier, J.-L. (Ed.) (2000). *On the Teaching of Linear Algebra*. Dordrecht: Kluwer Academic Publishers.

Dorier, J.-L., Robert A., Robinet J. and Rogalski M. (2000). On a Research Program about the Teaching and Learning of Linear Algebra in First Year of French Science University. *International Journal of Mathematical Education in Sciences and Technology*, 31 (1), 27-35.

Douady, R. (1986). Jeu de Cadres et Dialectique Outil-Objet. *Recherches en Didactique des Mathématiques*, 7 (2), 5-31.

Duval, R. (1995). *Semiosis et Pensée Humaine. Registres Sémiotiques et Apprentissages Intellectuels*. Bern: Peter Lang.

Harel, G. (2000). Principles of Learning and Teaching Mathematics, With Particular Reference to the Learning and Teaching of Linear Algebra: Old and New Observations. In J-L. Dorier (Ed.), *On the Teaching of Linear Algebra*, pp. 177-189. Dordrecht: Kluwer Academic Publishers.

Hillel, J. and Sierpinska, A. (1994). On One Persistent Mistake in Linear Algebra. In *The Proceedings PME 18*, pp. 65-72, University of Lisbon, Portugal.

Hillel, J. (2000). Modes of Description and the Problem of Representation in Linear Algebra. In J-L Dorier (Ed.), *On the Teaching of Linear Algebra*, pp. 191-207. Dordrecht: Kluwer Academic Publishers.

Moore, G. (1995). The Axiomatization of Linear Algebra. *Historia Mathematica*, 22 (3), 262-303.

Piaget, J. and Garcia, R. (1983) *Psychogenesis and the History of Science*. New York: Columbia University Press.

Robert, A., Robinet, J. and Tenaud, I. (1987). De la Géométrie à l'Algèbre Linéaire. *Brochure, 72*, IREM de Paris VII.

Robert, A. (1992). Projets Longs et Ingénierie pour l'Enseignement Universitaire: Questions de Problématique et de Méthodologie. Un exemple: un Enseignement Annuel de Licence en Formation Continue. *Recherches en Didactique des Mathématiques*, 12 (2/3), 181-220.

Robert, A, and Robinet, J. (1996). Prise en Compte du Méta en Didactique des Mathématiques. *Recherches en Didactique des Mathématiques*, 16 (2), 145-176.

Robert, A. (1998). Outil d'Analyse des Contenus Mathématiques Enseignés au Lycée et à l'Université. *Recherches en Didactique des Mathématiques*, 18 (2), 191-230.

Robinet, J. (1986). Esquisse d'une Genèse des Concepts d'Algèbre Linéaire. *Cahier de Didactique des Mathématiques*, 29, IREM de Paris VII.

Rogalski, M. (1991). Un enseignement de l'algèbre linéaire en DEUG A première année. *Cahier de Didatique des* Mathématiques, n°53, IREM de Paris 7.

Rogalski, M. (1994). L'enseignement de l'Algèbre Linéaire en Première Année de DEUG A. *La Gazette des Mathématiciens*, 60, 39-62.

Rogalski, M. (1996). Teaching Linear Algebra: Role and Nature of Knowledge in Logic and Set Theory which Deal with Some Linear Problems. In *The Proceedings PME 20*, Volume 4, pp. 211-218, Valencia Universidad,. Spain.

Sierpinska, A. (1995). Mathematics: 'in context', 'pure', or 'with applications'?. *For the Learning of Mathematics*, 15 (1), 2-15.

Sierpinska, A. (1997). Formats of Interaction and Model Readers. *For the Learning of Mathematics*, 17 (2), 3-12.

Sierpinska, A., Defence, A., Khatcherian, T., Saldanha, L. (1997). A propos de trois modes de raisonnement en algèbre linéaire. In J.-L. Dorier (Ed.), *L'Enseignement d'Algèbre Linéaire en Question*, pp. 249-268. Grenoble: La Pensée Sauvage éditions.

Sierpinska, A., Dreyfus, T. and Hillel, J. (1999). Evaluation of a Teaching Design in Linear Algebra: The Case of Linear Transformations. *Recherches en Didactique des Mathématiques*, 19 (1), 7-40.

Sierpinska, A., Trgalova, J., Hillel, J., Dreyfus, T. (1999). Teaching and Learning Linear Algebra with Cabri. Research Forum paper. In *The Proceedings of PME 23*, Volume I, 119-134, Haifa University, Israel.

Sierpinska, A. (2000). On Some Aspects of Students' Thinking in Linear Algebra. In J-L. Dorier (Ed.), *On the Teaching of Linear Algebra*, pp. 209-246. Dordrecht: Kluwer Academic Publishers.

Jean-Luc Dorier
Equipe DDM – Laboratoire Leibniz, Grenoble, and IUFM de Lyon,
France
Jean-Luc.Dorier@imag.fr

Anna Sierpinska
Concordia University, Montréal, Canada
sierp@vax2.concordia.ca

ED DUBINSKY AND MICHAEL A. MCDONALD

APOS: A CONSTRUCTIVIST THEORY OF LEARNING IN UNDERGRADUATE MATHEMATICS EDUCATION RESEARCH

1. INTRODUCTION

The work reported in this paper is based on the principle that research in mathematics education is strengthened in several ways when based on a theoretical perspective. Development of a theory in mathematics education should be, in our view, part of an attempt to understand how mathematics can be learned and what an educational programme can do to help in this learning. We do not think that a theory of learning is a statement of truth and although it may or may not be an approximation to what is really happening when an individual tries to learn one or another concept in mathematics, this is not our focus. Rather we concentrate on how a theory of learning mathematics can help us understand the learning process by providing explanations of phenomena that we can observe in students who are trying to construct their understandings of mathematical concepts and by suggesting directions for pedagogy that can help in this learning process.

Theories in mathematics education can

- support prediction,
- have explanatory power,
- be applicable to a broad range of phenomena,
- help organize one's thinking about complex, interrelated phenomena,
- serve as a tool for analyzing data, and
- provide a language for communication of ideas about learning that go beyond superficial descriptions.

We would like to offer these six features, the first three of which are given by Alan Schoenfeld in Schoenfeld (1998), both as ways in which a theory can contribute to research and as criteria for evaluating a theory. In this paper, we describe one such perspective, APOS Theory, in the context of undergraduate mathematics education. We explain the extent to which it has the above characteristics, discuss the role that this theory plays in a research and curriculum development programme and how such a programme can contribute to the development of the theory, describe briefly how working with this particular theory has provided a vehicle for building a community of researchers in undergraduate

Derek Holton (Ed.), The Teaching and Learning of Mathematics at University Level: An ICMI Study, 275—282.
© 2001 *Kluwer Academic Publishers. Printed in the Netherlands.*

studies, both by researchers who are developing it as well as others not connected with its development. We have constructed, in connection with this paper, an annotated bibliography of research reports that involve this theory. It can be found on the Web (McDonald, 2000).

2. APOS THEORY

The theory we present begins with the hypothesis that mathematical knowledge consists in an individual's tendency to deal, in a social context, with perceived mathematical problem situations by constructing mental actions, processes, and objects and organizing them in schemas to make sense of the situations and solve the problems. We refer to these mental constructions as *APOS Theory*. The ideas arise from our attempts to extend the work of Piaget on reflective abstraction in children's learning, to the level of collegiate mathematics. APOS Theory is discussed in detail in Asiala, Brown, et al. (1996). We will argue that this theoretical perspective possesses, at least to some extent, the characteristics listed above and, moreover, has been very useful in attempting to understand students' learning of a broad range of topics in calculus, abstract algebra, statistics, discrete mathematics, and other areas of undergraduate mathematics. Here is a brief summary of the essential components of the theory.

An *action* is a transformation of objects perceived by the individual as essentially external and as requiring, either explicitly or from memory, step-by-step instructions on how to perform the operation. For example, an individual with an action conception of left coset would be restricted to working with a concrete group such as Z_{20} and could construct subgroups, such as $H = \{0,4,8,12,16\}$ by forming the multiples of 4. Then the individual could write the left coset of 5 as the set $5 + H = \{1,5,9,13,17\}$ consisting of the elements that have remainder 1 when divided by 4.

When an action is repeated and the individual reflects upon it, he or she can make an internal mental construction called a *process* which the individual can think of as performing the same kind of action, but no longer with the need of external stimuli. An individual can think of performing a process without actually doing it, and therefore can think about reversing it and composing it with other processes. An individual cannot use the action conception of left coset described above very effectively for groups such as S_4, the group of permutations of four objects and the subgroup H corresponding to the 8 rigid motions of a square, and not at all for groups S_n for large values of n. In such cases, the individual must think of the left coset of a permutation p as the set of all products ph, where h is an element of H. Thinking about forming this set is a process conception of coset.

An *object* is constructed from a process when the individual becomes aware of the process as a totality and realizes that transformations can act on it. For example, an individual understands cosets as objects when he or she can think about the number of cosets of a particular subgroup, can imagine comparing two cosets for equality or for their cardinalities, or can apply a binary operation to the set of all cosets of a subgroup.

Finally, a *schema* for a certain mathematical concept is an individual's collection of actions, processes, objects, and other schemas which are linked by some general principles to form a framework in the individual's mind that may be brought to bear upon a problem situation involving that concept. This framework must be coherent in the sense that it gives, explicitly or implicitly, means of determining which phenomena are in the scope of the schema and which are not. Because this theory considers that all mathematical entities can be represented in terms of actions, processes, objects, and schemas, the idea of schema is very similar to the concept image which Tall and Vinner (1981) have introduced. Our requirement of coherence, however, distinguishes the two notions.

The four components, action, process, object, and schema have been presented here in a hierarchical, ordered list. This is a useful way of talking about these constructions and, in some sense, each conception in the list must be constructed before the next step is possible. In reality, however, when an individual is developing her or his understanding of a concept, the constructions are not actually made in such a linear manner. With an action conception of function, for example, an individual may be limited to thinking about formulas involving letters which can be manipulated or replaced by numbers and with which calculations can be done. We think of this notion as preceding a process conception, in which a function is thought of as an input-output machine. What actually happens, however, is that an individual will begin by being restricted to certain specific kinds of formulas, reflect on calculations and start thinking about a process, go back to an action interpretation, perhaps with more sophisticated formulas, further develop a process conception and so on. In other words, the construction of these various conceptions of a particular mathematical idea is more of a dialectic than a linear sequence.

APOS Theory can be used directly in the analysis of data by a researcher. In very fine grained analyses, the researcher can compare the success or failure of students on a mathematical task with the specific mental constructions they may or may not have made. If there appear two students who agree in their performance up to a very specific mathematical point and then one student can take a further step while the other cannot, the researcher tries to explain the difference by pointing to mental constructions of actions, processes, objects and/or schemas that the former student appears to have made but the other has not. The theory then makes testable predictions that if a particular collection of actions, processes, objects and schemas are constructed in a certain manner by a student, then this individual will likely be successful using certain mathematical concepts and in certain problem situations. Detailed descriptions, referred to as genetic decompositions, of schemas in terms of these mental constructions are a way of organizing hypotheses about how learning mathematical concepts can take place. These descriptions also provide a language for talking about such hypotheses.

3. DEVELOPMENT OF APOS THEORY

APOS Theory arose out of an attempt to understand the mechanism of *reflective abstraction*, introduced by Piaget to describe the development of logical thinking in children, and extend this idea to more advanced mathematical concepts (Dubinsky, 1991). This work has been carried on by a small group of researchers called a Research in Undergraduate Mathematics Education Community (RUMEC) who have been collaborating on specific research projects using APOS Theory within a broader research and curriculum development framework. The framework consists of essentially three components: a theoretical analysis of a certain mathematical concept, the development and implementation of instructional treatments (using several non-standard pedagogical strategies such as cooperative learning and constructing mathematical concepts on a computer) based on this theoretical analysis, and the collection and analysis of data to test and refine both the initial theoretical analysis and the instruction. This cycle is repeated as often as necessary to understand the epistemology of the concept and to obtain effective pedagogical strategies for helping students learn it.

The theoretical analysis is based initially on the general APOS theory and the researcher's understanding of the mathematical concept in question. After one or more repetitions of the cycle and revisions, it is also based on the fine-grained analyses described above of data obtained from students who are trying to learn or who have learned the concept. The theoretical analysis proposes, in the form of a genetic decomposition, a set of mental constructions that a student might make in order to understand the mathematical concept being studied. Thus, in the case of the concept of cosets as described above, the analysis proposes that the student should work with very explicit examples to construct an action conception of coset; then he or she can interiorize these actions to form processes in which a (left) coset gH of an element g of a group G is imagined as being formed by the process of iterating through the elements h of H, forming the products gh, and collecting them in a set called gH; and finally, as a result of applying actions and processes to examples of cosets, the student encapsulates the process of coset formation to think of cosets as objects. For a more detailed description of the application of this approach to cosets and related concepts, see Asiala, Dubinsky, et al. (1997).

Pedagogy is then designed to help the students make these mental constructions and relate them to the mathematical concept of coset. In our work, we have used cooperative learning and implementing mathematical concepts on the computer in a programming language that supports many mathematical constructs in a syntax very similar to standard mathematical notation. Thus students, working in groups, will express simple examples of cosets on the computer as follows.

$$Z20 := \{0..19\}$$
$$op := |(x,y) \rightarrow x + y \ (\mathrm{mod}\ 20)|;$$
$$H := \{0,4,8,12,16\};$$
$$H5 := \{1,5,9,13,17\};$$

To interiorize the actions represented by this computer code, the students will construct more complicated examples of cosets, such as those appearing in groups of symmetries.

$$Sn := \{[a,b,c,d] : a,b,c,d \ \text{in} \ \{1,2,3,4\} \mid \#\{a,b,c,d\} = 4\};$$
$$op := |(p,q) \rightarrow [p(q(i)) : i \ \text{in} \ [1..4]]|;$$
$$H := \{[1,2,3,4], [2,1,3,4], [3,4,1,2], [4,3,2,1]\};$$

p   := [4,3,2,1];

pH   :={p .op q : q in H};

The last step, to encapsulate this process conception of cosets to think of them as objects, can be very difficult for many students. Computer activities to help them may include forming the set of all cosets of a subgroup, counting them, and picking two cosets to compare their cardinalities and find their intersections. These actions are done with code such as the following.

SnModH := { {p .op q : q in H} : p in Sn};

#SnModH;

L := arb(SnModH); K := arb(SnModH); #L = #K; L inter K;

Finally, the students write a computer program that converts the binary operation op from an operation on elements of the group to subsets of the group. This structure allows them to construct a binary operation (coset product) on the set of all cosets of a subgroup and begin to investigate quotient groups. It is important to note that in this pedagogical approach, almost all of the programs are written by the students. One hypothesis that the research investigates is that, whether completely successful or not, the task of writing appropriate code leads students to make the mental constructions of actions, processes, objects, and schemas proposed by the theory. The computer work is accompanied by classroom discussions that give the students an opportunity to reflect on what they have done in the computer lab and relate their activities to mathematical concepts and their properties and relationships. Once the concepts are in place in their minds, the students are assigned (in class, homework and examinations) many standard exercises and problems related to cosets.

After the students have been through such an instructional treatment, quantitative and qualitative instruments are designed to determine the mental concepts they may have constructed and the mathematics they may have learned. The theoretical analysis points to questions researchers may ask in the process of data analysis and the results of this data analysis indicates both the extent to which the instruction has been effective and possible revisions in the genetic decomposition.

This way of doing research and curriculum development simultaneously emphasizes both theory and applications to teaching practice.

4. REFINING THE THEORY

As noted above, the theory helps us analyze data and our attempt to use the theory to explain the data can lead to changes in the theory. These changes can be of two kinds. Usually, the genetic decomposition in the original theoretical analysis is revised and refined as a result of the data. In rare cases, it may be necessary to enhance the overall theory. An important example of such a revision is the incorporation of the triad concept of Piaget and Garcia (1989) which is leading to a better understanding of the construction of schemas. This enhancement to the theory was introduced in Clark, et al. (1997) where they report on students' understanding of the chain rule, and is further elaborated upon in three recent studies: sequences of numbers (McDonald, et al., 2000); the chain rule and its relation to composition of functions (Cottrill, 1999); and the relations between the graph of a function and properties of its first and second derivatives (Baker, et al., 2000). In each of these studies, the understanding of schemas as described above was not adequate to provide a satisfactory explanation of the data and the introduction of the triad helped to elaborate a deeper understanding of schemas and provide better explanations of the data.

The triad mechanism consists in three stages, referred to as Intra, Inter, and Trans, in the development of the connections an individual can make between particular constructs within the schema, as well as the coherence of these connections. The *Intra* stage of schema development is characterized by a focus on individual actions, processes, and objects in isolation from other cognitive items of a similar nature. For example, in the function concept, an individual at the Intra level, would tend to focus on a single function and the various activities that he or she could perform with it. The *Inter* stage is characterized by the construction of relationships and transformations among these cognitive entities. At this stage, an individual may begin to group items together and even call them by the same name. In the case of functions, the individual might think about adding functions, composing them, etc. and even begin to think of all of these individual operations as instances of the same sort of activity: transformation of functions. Finally, at the *Trans* stage the individual constructs an implicit or explicit underlying structure through which the relationships developed in the Inter stage are understood and which gives the schema a coherence by which the individual can decide what is in the scope of the schema and what is not. For example, an individual at the Trans stage for the function concept could construct various systems of transformations of functions such as rings of functions, infinite dimensional vector spaces of functions, together with the operations included in such mathematical structures.

5. APPLYING THE APOS THEORY

Research utilizing APOS theory has focussed on mathematical concepts such as: functions; various topics in abstract algebra including binary operations, groups, subgroups, cosets, normality and quotient groups; topics in discrete mathematics such as mathematical induction, permutations, symmetries, existential and universal quantifiers; topics in calculus including limits, the chain rule, graphical understanding of the derivative and infinite sequences of numbers; topics in statistics such as mean, standard deviation and the central limit theorem; elementary number theory topics such as place value in base n numbers, divisibility, multiples and conversion of numbers from one base to another; and fractions. In most of this work, the context for the studies are collegiate level mathematics topics and undergraduate students. In the case of the number theory studies, the researchers examine the understanding of pre-college mathematics concepts by college students preparing to be teachers. Finally, some studies such as that of fractions, show that APOS Theory, developed for 'advanced' mathematical thinking, is also a useful tool in studying younger students' understanding of more basic mathematical concepts.

The totality of this body of work, much of it done by RUMEC members involved in developing the theory, but an increasing amount done by individual researchers having no connection with RUMEC or the construction of the theory, suggests that APOS Theory is a tool that can be used objectively to explain student difficulties with a broad range of mathematical concepts and to suggest ways that students can learn these concepts. APOS Theory can point us towards pedagogical strategies that lead to marked improvement in student learning of complex or abstract mathematical concepts and in students' use of these concepts to prove theorems, provide examples, and solve problems. Data supporting this assertion can be found in the papers listed in the annotated bibliography (McDonald, 2000).

6. USING APOS THEORY TO DEVELOP A COMMUNITY OF RESEARCHERS

At this stage in the development of research in undergraduate mathematics education, there is neither a sufficiently large number of researchers nor enough graduate school programmes to train new researchers. Other approaches, such as experienced and novice researchers working together in teams on specific research problems, need to be employed at least on a temporary basis. RUMEC is one example of a research community that has utilized this approach in training new researchers.

In addition, a specific theory can be used to unify and focus the work of such groups. The initial group of researchers in RUMEC, about 30 total, made a decision to focus their research work around the APOS Theory. This was not for the purpose of establishing dogma or creating a closed research community, but rather it was a decision based on current interests and needs of the group of researchers.

RUMEC was formed by a combination of established and beginning researchers in mathematics education. Thus one important role of RUMEC was the mentoring of these new researchers. Having a single theoretical perspective in which the work of RUMEC was initially grounded was beneficial for those just beginning in this area. At the meetings of RUMEC, discussions could focus not only on the details of the individual projects as they developed, but also on the general theory underlying all of the work. In addition, the group's general interest in this theory and frequent discussions about it in the context of active research projects has led to growth in the theory itself. This was the case, for example, in the development of the triad as a tool for understanding schemas. As the work of this group matures, individuals are beginning to use other theoretical perspectives and other modes of doing research.

7. SUMMARY

In this paper, we have mentioned six ways in which a theory can contribute to research and we suggest that this list can be used as criteria for evaluating a theory. We have described how one such perspective, APOS Theory, is being used in an organized way by members of RUMEC and others to conduct research and develop curriculum. We have shown how observing students' success in making or not making mental constructions proposed by the theory and using such observations to analyze data can organize our thinking about learning mathematical concepts, provide explanations of student difficulties and predict success or failure in understanding a mathematical concept. There is a wide range of mathematical concepts to which APOS Theory can and has been applied and this theory is used as a language for communication of ideas about learning. We have also seen how the theory is grounded in data, and has been used as a vehicle for building a community of researchers. Yet its use is not restricted to members of that community. Finally, we point to an annotated bibliography (McDonald, 2000), which presents further details about this theory and its use in research in undergraduate mathematics education.

REFERENCE

Asiala, M., Brown, A., DeVries, D., Dubinsky, E., Mathews, D. and Thomas, K. (1996). A framework for research and curriculum development in undergraduate mathematics education. *Research in Collegiate Mathematics Education II, CBMS Issues in Mathematics Education*, 6, 1-32.

Asiala, M., Dubinsky, E., Mathews, D., Morics, S. and Oktac, A. (1997). Student understanding of cosets, normality and quotient groups. *Journal of Mathematical Behavior*, 16(3), 241-309.

Baker, B., Cooley, L. and Trigueros, M. (2000). A Calculus graphing schema. *Journal for Research in Mathematics Education*, 31(5), 557-578.

Clark, J., Cordero, F., Cottrill, J., Czarnocha, B., DeVries, D. J., St. John, D., Tolias G. and Vidakovic, D. (1997). Constructing a schema: The case of the chain rule. *Journal of Mathematical Behavior*, 16(4), 345-364.

Cottrill, J. (1999). Students' understanding of the concept of chain rule in first year calculus and the relation to their understanding of composition of functions. Doctoral dissertation, Purdue University.

Dubinsky, E. (1991). Reflective abstraction in advanced mathematical thinking. In D. Tall (Ed.), *Advanced Mathematical Thinking*, pp. 95-126. Dordrecht: Kluwer Academic Publishers.

McDonald, M. A. (2000). http://galois.oxy.edu/mickey/APOSbib.html.

McDonald, M. A., Mathews, D. and Strobel, K. (2000). Understanding sequences: A tale of two objects. *Research in Collegiate Mathematics Education IV, CBMS Issues in Mathematics Education*, 8, 77-102.

Schoenfeld, A. H. (1998). Toward a theory of teaching-in-context. *Issues in Education*, 4 (1), 1-94.

Piaget, J. and Garcia, R. (1989). *Psychogenesis and the History of Science*, New York: Columbia University Press.

Tall, D. and Vinner, S. (1981). Concept image and concept definition in mathematics with particular reference to limits and continuity. *Educational Studies in Mathematics*, 12, 151-169.

Ed Dubinsky,
Unaffiliated
edd@mcs.kent.edu

Michael A. McDonald,
Occidental College, California, USA
mickey@oxy.edu

ALINE ROBERT AND NATASHA SPEER

RESEARCH ON THE TEACHING AND LEARNING OF CALCULUS/ELEMENTARY ANALYSIS

1. INTRODUCTION

Learning calculus (a term we shall use interchangeably with 'elementary analysis') is difficult for students, no matter what country they live in. Interestingly, however, researchers in different countries take vastly different approaches to the study of the teaching and learning of calculus. It is challenging to report on these differences, given the different traditions of research and instruction. This paper is one effort to examine these issues.

The paper is divided into two main parts. In the first part we discuss approaches based on research in which the research is grounded, at least partly, in content and epistemological analyses of the target knowledge. This work tends to be European or Israeli. In much of this work, applications to teaching are limited and sometimes not even considered. The second part of this paper, which contrasts with the first, focuses on work done in the United States. Much American research on calculus learning was motivated by a pragmatic programme of 'calculus reform.' Several calculus reform projects and studies associated with them (often focusing on the impact of instruction) will be discussed.

Although somewhat complementary, the European/Israeli and the American approaches are related in a variety of ways. Both approaches stress the importance of student conceptions of the reals, functions, and limits; both recognize the complexity of the field and difficulties in student learning; and both tie what is taught before calculus to the actual introduction to calculus.

2. THEORY-DRIVEN RESEARCH

By way of preface, we begin by listing the main areas covered in this survey: functions of one and several variables, considered both locally and globally, limits, continuity, derivatives, sequences, definite and indefinite integrals, and differential equations. The reals are fundamental to the study of analysis, but the reals are not always studied on their own terms with regard to calculus. Generally, the main intellectual constructs and difficulties that students encounter are concerned, whether explicitly or implicitly, with the concepts of approximation and convergence.

Derek Holton (Ed.), The Teaching and Learning of Mathematics at University Level: An ICMI Study, 283—299.
© 2001 *Kluwer Academic Publishers. Printed in the Netherlands.*

pathways for students intending mathematical careers, scientific careers, or neither. Of course, any curriculum or syllabus may be taught with different emphases or pacing. Broadly speaking, most of the research reported has been conducted with university students (recall that, world-wide, calculus is frequently taught at the secondary level); also, there are relatively few studies concerning 'non-scientific' students (Kent and Noss, this volume, pp. 395-404).

There are two main parts to this section of this paper. The first part describes some 'basic' (i.e., widely known) results in the study of calculus/elementary analysis. Given the space constraints, we can only provide a brief description of major trends in this foundational work. For more detail see *Advanced Mathematical Thinking*, (Tall, 1991); *The concept of function: Aspects of epistemology and pedagogy* (Harel and Dubinsky, 1992); *International Handbook of Mathematics Education* (Bishop, Clements, Keitel and Kilpatrick, 1996), and Artigue's (1996) comprehensive synthesis of research on the teaching and learning of analysis. Generally speaking, the contents first examined were related to limits and functions; then studies related to integrals, derivatives, and area appeared. There is, today, also a significant amount of research related to the use of technology.

Much contemporary research returns to the study of the basic concepts listed above, often with new insights or new perspectives. The second part of this section discusses studies emblematic of current work. Specifically, it focuses on Praslon's (1999) study of discontinuities that occur between the secondary and tertiary levels in student understanding of the derivative, and Project AHA's attempts to overcome students' epistemological obstacles in a course in calculus. Technology is also briefly discussed.

2.1 Foundational Work in the study of Calculus/elementary Analysis

Three dimensions will organize this review: the epistemological, cognitive, and pedagogical dimensions. Of course they are not independent with (at least) the last cutting across the first two.

2.1.1 Epistemological dimension and obstacles

Studies of the history of analysis, cited in Tall (1991), influenced many later works on teaching or learning analysis. Epistemological obstacles related to infinity and limits were studied first (Cornu, 1991; Hauchart, 1985; Sierpinska, 1985). Later others were identified, related for instance to the multiple ways one can understand a tangent, or various ways to express a function (e.g., with a graph, as a subset set of a Cartesian product, using an explicit formula). In typical work, Sierpinska (1992) noted that it may be difficult for "an individual whose experience of functions is in terms of formulae and computation to accept a definition which does not involve these attributes." Other work also involves obstacles about integrals, areas and volumes computations (Schneider, 1991, see below).

Recent French historical analyses (Robert, 1998), suggest that when the formalization of a mathematical concept occurs a long time after its first empirical use by mathematicians, that concept will be especially difficult to teach and to learn. This is related to the fact that the formalization of the notion typically generalizes and unifies many previous aspects of it. This was, for example, the case with regard to the convergence of sequences. The different uses of the same notion according to the degree of formalism correspond to the different conceptualization levels of the students (Robert, 2000). More abstract and general notions are more difficult for students to understand.

The history of calculus has also been looked at from a socio-epistemological perspective by other research groups, for instance the research group on advanced mathematics at CINVESTAV in Mexico. This group starts with the assumption that the present structure of theoretical mathematical discourse in analysis obscures the essential empirical sources of the development of the field. Thus, looking at historical development provides alternative ways to introduce and develop knowledge in the field. This is especially necessary if one has in mind not the training of future mathematicians but the training of scientific students and engineers. Such a perspective has been used by CINVESTAV in order to study the learning and teaching of variation, from high school to studies in advanced engineering. They have coordinated mathematical and historical analyses, the socio-analysis of the way variation is dealt with in different professional and social contexts, the cognitive analysis of learning processes, and didactic engineering designs (see e.g., Cantoral and Farfán, 1998; Cordero, 1994; Farfán, 1997; Ortega, 2000).

2.1.2 Cognitive dimension, conceptions and learning theories

The core studies referenced above influenced studies that followed. A great deal of subsequent research consisted of diagnoses of student learning in light of a specific (teaching and learning) theory, or a proposed (teaching) theory to overcome students' learning difficulties. Early studies of students' conceptions (of limits, for example) confirmed both the power of the 'epistemological obstacles' point of view and the importance of taking into consideration the fine-grained study of student understanding.

In a recent study, Pian (2000) has explored the relationship between the nature of the tasks given to students and the students' success. Pian defined three levels of tasks. We illustrate these levels using tasks from analysis typical of those given to French undergraduate mathematics majors.

The 'technical' level refers to situations where students are asked to apply definitions, properties and theorems directly. Tasks at this level also entail the use of formal language to the degree appropriate to the student's stage of schooling. A typical analysis task at the technical level is:

For $p \in N$, $p \geq 1$ and $x \geq 0$, let $u_p(x) = xe^{-px}$.
Show that $\sum u_p(x)$ is point-wise convergent for all $x \in [0, +\infty)$.

The 'mobilizable' level refers to a broader way of applying mathematical knowledge. The task is not a direct application: several steps are needed, or something has to be transformed or recognized to apply the property or theorem involved. Here is an example of a task involving mobilizable level:

In the previous exercise, show that, $\forall a > 0$, $\sum u_p(x)$ converges uniformly on $[a, +\infty)$.

The 'available' level refers to the ability to solve problems without hints or context, to give counterexamples, to change methods, to use knowledge from different mathematical fields. For example:

Study (without using the Lebesgue integral) $\displaystyle\lim_{n \to \infty} \int_0^1 (2/x + 1)^n dx$.

Of course these levels are relative: they depend on the level of study and also on students' experience. Pian's findings were that if some adaptation of an analysis task is necessary to solve it (that is, if the task is presented at the mobilizable or available level instead of the technical level), more than half of the students will fail to solve it. This is consistent with a general perception on the part of mathematics faculty that students are often capable of solving 'imitative' exercises' but not problems that require transformation and reformulation.

Subsequent research, grounded in the kinds of basic research described above, provided more organized, global theories. In his recent synthesis of research on function and calculus, Tall (1996) describes Sfard's (1992) general thesis that an operational view of mathematics precedes a structural view in terms of objects and formal definitions. This has implications for teaching theory. Tall (1996) reprises the difference between *concept definition* and the wider notion of *concept image*. This distinction allowed some authors (Tall, Vinner, Dreyfus 1981, 1989) to explain some failures of student understanding. This is related to theories of learning – cf. Selden and Selden, this volume, pp. 237-254).

Dubinsky and McDonald (this volume, pp. 275-282) and Artigue (this volume, pp. 207-220), discuss how APOS theory is an adaptation of the Piagetian theory of reflective abstraction (see also Tall, 1996, Dubinsky, 1995). This has implications for a teaching theory, implemented with the use of computers: the principle of proposing "actions which become repeatable processes, which are encapsulated as objects and then related in a wider schema is realized by programming activities which place visualization at the end of the line as a visual representation of a function constructing through encapsulating a programming process as an object" (Dubinsky, 1995).

Tall and Gray (1994) use the notion of procept (a combination of process and concept), claiming it is particularly adapted to the study of calculus and beginning analysis. For them, function, derivative, integral and the fundamental limit notions are all examples of procepts. Note, for example, that the standard notation for limits,

lim both denotes a process and the object which is produced by this process. They
$n \to \infty$
claim that the theory of function and calculus can be summarized in outline as the
transition between process and objects, what they call "study of the 'doing' and
'undoing' of the processes involved."

In some French research (e.g., Dorier and Sierpinska, this volume, pp.255-274;
Robert, 1998) one finds similar ideas expressed in a different manner. These authors
emphasize the importance of changes in settings or registers (semiotic forms) in
learning notions such as functions or real numbers. Seeing concepts in multiple
settings and from multiple perspectives lets students better organize their
knowledge. This is considered as a necessary cognitive condition to learning at this
level. This can be seen as an aspect of the general idea of flexibility discussed in
Dreyfus (1996).

2.1.3 Didactical dimension

Of importance is what teachers (and institutions) expect from students, both in
terms of topics and levels of difficulty. Thus research from this perspective often
begins with a detailed analysis of the curriculum and syllabus under investigation.
Other, more synthetic studies have described general expectations in analysis, by
comparing these aspects at the secondary and tertiary levels. For example, Guzman,
Hodgson, Robert, and Villani (1998) show that significant differences occur in
proofs and use of formalism. At the tertiary level, non-trivial proofs are expected
and one must use some formalism correctly to succeed. Moreover, recall that, at this
level, students are expected to be able to solve problems that are not merely copies
of problems they have been shown.

Other works (especially French ones) propose what is called 'ingénieries' –
studies of didactic engineering – in which carefully specified instructional
interventions are assessed. For example, such a study was conducted on teaching a
qualitative approach to differential equations using computers. We quote some
results, taken from Artigue (1992). We focus here on the relations between
didactical and cognitive points of view.

> The constructing of inter-setting relations about functions, necessary for a first approach to the
> qualitative solution of differential equations is accessible to beginning students although it is a
> high level task. Students encounter various cognitive difficulties [tied to] interpretation and
> prediction registers; in the proof register, cognitive difficulties are more constraining and some
> kind of geometrical proof has to be legitimized in order to cope with them.
>
> [These results] confirm the initial hypothesis of coexistence between cognitive difficulties and
> didactical ones in the construction of these inter-setting relations and help, it seems, to better
> understand how these two kinds of difficulties are interwoven and to evaluate the influence of
> this interweaving on their obstinacy. In particular, they show that this interweaving is not
> independent of the registers in which the interaction is called on to function. It arises here, above
> all, in the register of proofs. In this register, difficulties of a didactic nature are especially
> difficult to overcome and the problems encountered in the progressive adaptation of the
> experimental device clearly show the strength of these difficulties. Beliefs and habits about the
> status and role of the graphic setting act as didactic obstacles and they have to be explicitly
> questioned in order to obtain the necessary epistemological changes both in teachers and
> students. (Artigue, 1992. p.132).

Another attempt to teach analysis (which is applicable to other areas of mathematics) is developed in Legrand's work. This approach, called 'scientific debate' (see Artigue 2001; Legrand, 2001) is based on a specific form of discussion among students regarding the validity of theorems. In structured arguments about mathematical content, students develop deeper understandings of fundamental concepts.

New didactical issues arise when dealing with computers to help teach analysis. These issues are discussed in the section of this volume that deals with the uses of technology in instruction.

2.2   Some new aspects of research

2.2.1   The secondary/tertiary transition

One new direction of study involves the transition between secondary and tertiary levels. This has become increasingly important because of large increases in university enrolments (the 'democratizing' of higher education: see the work of the secondary/tertiary transition group, Wood, this volume, pp. 87-98). As more students enrol in university mathematics courses, it becomes increasingly critical to examine transitions between secondary and tertiary levels. Some recent research studies attempt to describe the differences. The manner in which teachers or institutions (by the way of programmes and curricula) expect students to use basic concepts is a good indicator of the differences.

One might consider continuity, or derivatives, or sequences as typical topics. Many of the differences in teaching (and learning) seem to depend upon whether students are typically asked to work on or with specific functions or mathematical objects, or whether the functions and objects are described generically. Are students asked to explore the properties of a particular function f (say $f(x) = 7x^3 + 5x^2 - 2x + 4$), or the properties of an arbitrary cubic? Are they asked to explore the properties of a particular sequence, or those of sequences with particular properties? It has been shown that there is a real discontinuity between the two levels when it comes to the use of particular examples instead of general objects, see Robert (1998).

More generally, as Artigue (this volume, pp. 207-220) notes, differences between the two levels may be seen by examining which notions appear in problems. Formalism and new modes of proofs appear much more frequently at the tertiary level. In what follows we describe one study in some detail, because we judge it emblematic of this type of research.

In his thesis, Praslon (2000) examines the concept of derivative as a case study of discontinuities in the secondary/university transition. Praslon's study of the way institutions at each level deal with the derivative concept (and students' resulting understandings thereof) reveals the existence of two very distinct cultures. These differences can be seen as many little discontinuities, often related to tasks proposed (or not!) to students, more than one or two big 'breakdowns.' Results attest to "the attempts made by students in order to adapt their mathematical resources to the problematic and complex situations proposed to them. These attempts most often lead to oversimplifications, favoured by their reduced field of experience." To summarize the results, their experience with secondary mathematics does not prepare many students with an understanding of the derivative that will be expected at the tertiary level. This is often because they have only been asked to work isolated, simple, exercises that require no adjustment in order to be solved.

2.2.2 Technology in instruction

Another new direction occurs with the adoption of computer-based technology in teaching. (For details, see the section of this volume on Technology, pp. 347-404.) We will observe, briefly, that two kinds of studies may occur. One has been already seen in the description of Dubinsky's (1995) or Tall's (1996) papers: the use of computers to help students think in a certain manner, or to show them powerful technologically-dependent information (e.g., graphics, or very fast computation). There is now another direction of work, in which formal calculus is to be used by students to solve problems. Here, the theoretical and pragmatic issues involve the manner of integrating those technological tools in learning. There is, for example, the possibility of devising 'ingénieries' making integral use of technology. Trouche's (1996) thesis and subsequent research in France, are examples of this second approach, within the framework of developing carefully specified instructional practices in mathematics.

2.3 Extended analyses of long-term projects

Finally, in some recent research, the authors have conducted extended studies of a coherent body of instruction. The work is often complex and supports multiple lines of inquiry as part of its development. We will discuss below, the AHA project as a representative example of those projects. Such research-intensive development programmes contrast with some of the 'development first' approaches to calculus reform seen in the second half of this chapter.

Project AHA, Heuristic Approach to Analysis, (Hauchart and Schneider, 1996) takes place in the two years at the end of high school in Belgium. This is comparable to the collegiate level in the U.S. One of its main ideas is to help students overcome epistemological obstacles in beginning analysis.

Many obstacles (more than 60!) are enumerated. However, one can 'easily' find them because they generate errors in students' procedures. Not surprisingly, many of those obstacles are related to limits and infinity. Others are related to the area and volume computations. For example, there is a well-known obstacle on convergence of sequences: many students believe that in order for a sequence of positive terms to converge to 0, the sequence must be decreasing. Or there is the conviction that 0.9999... ≠ 1. Students also find it very difficult to relate the slope of the tangent to the limit of the rate of increment of the function (that is, to see the tangent line as the limit of secant lines, visually and analytically).

At the macro-didactic level, the authors have chosen a 'three-components' approach, that is an heuristic approach, taking place in a constructivist perspective, and with a broad historic inspiration. Combining a Piagetian view with heuristic and historic approaches leads to the selection of particular kinds of problems. These problems which help give some meaning to the notions to be taught, are given to students when they first encounter the concepts. If possible, in those problems different points of views are involved (e.g., different settings, different registers). Then teachers add some cognitive 'mediation' – discussions of relevant mathematics or methods or big ideas. They have students try to overcome the previously identified obstacles by working and discussing these carefully chosen problems. Finally, this process leads to the formalization of concepts that the students can use themselves when writing proofs.

For instance, functions appear first in many contexts: as tools for modelling, in different registers, and in some real situations (numerical, or geometrical, or dynamic situations, tied to velocity for instance). It is from those situations that teachers choose initial problems. Graphs, or arrays, or analytic expressions may be used to represent functions at the beginning, depending on the situations explored. Some novel problems are proposed to students, such as finding the expression of a function when you are given its graph or a description of how it changes. After this initial work, students are then invited to study not only isolated functions but classes of functions (e.g., polynomials of third degree) and they work on creating descriptions that characterize all members of the class.

Another choice illustrates the authors' intentions to make connections: they try to have the concept of derivative emerge from the connection between velocity in calculus and affine approximations. Another example of design with regard to epistemological obstacles: the authors first propose computing areas and volumes using Cavalieri's principles, in order to avoid and correct weak intuition.

Finally, and very importantly: formalization is not given at once. For example, limits are studied many times in the project, but they are used in an approximate way for a long time. Limits are progressively formalized – and only in the last three chapters of the whole project. The authors invoke the 'proof-generated concept' of Lakatos: students will have to transform instrumental notions into formal ones when they need them. This may be done when students work on difficult proofs, where the need for formalization appears. For example, students cannot prove assigned theorems without quantifiers and inequalities, even intuitive ideas suggest why the theorems should be true.

We want now to emphasize the emblematic character of such a project for present research. This is a long project, with goals and means overlapping. It involves more than one source (historic, heuristic, cognitive, didactic). In consequence, it is very difficult to analyze and assess the whole project. When does one test student knowledge? How can one relate results on learning and teaching? How can one assess homework? These are all issues to be confronted. This is very complex and difficult work (for example, understanding Cavalieri's principles is not simple!). Teachers have to be trained to teach in such a different manner than usual, both with regard to the mathematics content and with regard to the management of classroom dynamics. Particularly interesting is that this mix of characteristics (or a similar if not identical mix) is now appearing with some frequency in current research.

A final note. Issues related to effective teaching practices seem to occupy many authors these days. This, of course, is because any project –no matter how good it is – will be affected by what occurs with students (whether or not the teacher is aware of what is causing those effects).

3. DEVELOPMENT AND RESEARCH IN THE UNITED STATES

The past decade and a half have brought significant change to college calculus instruction in the United States. The calculus reform movement is an effort to improve student learning of calculus by modifying both what students are taught and the manner in which it is taught to them. The initial motivation for change came from perceptions of poor student understanding and low retention rates in course sequences. The proposed remedies were to revise the standard course syllabus, introduce new, more active teaching practices and incorporate technology into instruction.

The motivation for reform and the manner in which it has taken place are both very pragmatic. Students were doing poorly and faculty sought to improve the situation. In the U.S. there was no established field of college-level educational research and few theories of teaching and learning were used to guide the initial development and direction of calculus reform. This is unlike the situation in Europe, which had longer traditions of research and theory building at the college level. More recently, a field of 'research in collegiate mathematics education' has emerged in the U.S., but significant contributions in this field followed most design and implementation of calculus reform.

This section of the paper describes the history of U.S. calculus reform, the growth of the field of collegiate mathematics education research, and how these developments relate to reform of and research in pre-college mathematics education.

3.1 History of calculus reform in U.S.

The calculus reform movement was the mathematical community's response to low student achievement and retention rates. Students were not emerging from the courses with a strong understanding of calculus concepts and they were also not choosing to pursue further mathematics courses. One catalyst for reform was a conference held at Tulane University (Sloan Conference) in 1986. The conference had two main themes: calculus courses should address fewer topics in more depth and students should learn through active engagement with the material. These central tenets were echoed in an *MAA Notes* volume from the conference entitled *A Lean and Lively Calculus* (Douglas, 1986) as well as other publications that appeared shortly thereafter.

The Tulane conference created a focus on curriculum reform and consciousness-raising about the state of student understanding. Early on much of the effort was devoted to the examination of existing curricula, development of new curricula and exploration of uses of technology. The U.S. National Science Foundation funded projects to develop and implement new curricula and instructional practices. These projects each carried out the mission of 'lean' syllabi and 'lively' learning opportunities in very different ways and these differences did much to shape the directions the movement would take.

3.2 NSF projects and themes

Although they shared some overarching goals, the initial projects differed considerably. Both the specific instructional goals and the means of achieving them varied substantially from project to project. Below we give short descriptions of some projects to provide the reader with a sense of the character of the movement and the different approaches to reform chosen by different projects. This is far from a comprehensive review. There is, of course, more to each of the projects than is described below: the descriptions given highlight one or two features of each project that are illustrative of themes found in the overall movement (see also Keynes and Olson, this volume, pp. 113-126 and Smith, this volume, pp 167-178).

Project CALC. Project CALC reflects three themes of the movement: investigation of 'real world' problems, intensive use of technology and emphasis on communication. In Project CALC courses, students investigate questions, formulate problems and communicate their solutions to others. Students use technology as an integral part of the problem solving process to explore real-world problems with real data. They also write up their results and conclusions in reports to be shared with others.

Calculus&Mathematica. Calclulus&Mathematica (C&M) illustrates several other themes: use of software packages and maths 'labs.' C&M utilizes Mathematica (a commercially available software package) to allow students to do extensive experimentation with mathematical ideas. Class time is also utilized in a manner similar to labs that are a common feature of courses in the physical or biological sciences. C&M's use of Mathematica allows students to work and rework many examples using different numbers and functions.

Harvard Consortium. The Harvard calculus project represents two other themes: emphasis on multiple representations and the creation of a new syllabus. The Harvard project made a very explicit focus on the use of multiple representations. The 'Rule of Three' stipulated that concepts should be taught and understood from an algebraic, numerical and graphical perspective. This was later broadened to include written representatives (or communication more generally). This significant use of multiple representatives was a shift away from the algebraic focus of traditional curricula. Second, the project re-examined the traditional syllabus. Many projects had trimmed the traditional syllabus by eliminating topics. In contrast, the Harvard project started from scratch and built a new syllabus reflecting their views of what students should learn in calculus.

Calculus, Concepts, Computers and Cooperative Learning (C4L, Purdue University). The C4L project represents two other themes: constructivism and cooperative learning. Essentially, the constructivist perspective says that students must make sense of and 'construct' their own understanding of the ideas and, as a result, teaching cannot consist merely of telling. C4L's use of constructivism was made more explicit than in many other projects. They created laboratory activities where students work with examples and 'discover' the main conceptual ideas for themselves prior to participating in a lecture or discussion about the concepts.

C4L also made extensive use of cooperative learning. In the simplest of terms, cooperative learning is the umbrella term used to describe a variety of ways in which pairs or groups of students work on mathematical problems. Cooperative learning has become a common feature of many reform projects. In C4L, after discovering the mathematical ideas, students work in groups to further examine those ideas by doing problems.

One or more of these themes of technology, cooperative learning, syllabus revision, constructivism, 'real world' problems and communication can be found in most reform projects. Major projects reflecting one or more of these themes were also launched at New Mexico State University, University of Michigan, St. Olaf College and the Five Colleges consortium. For information about these and other projects, see Solow (1994), Tucker and Leitzel (1995) and Schoenfeld (1997).

3.3 Relationship to K-12 (pre-collegiate) reform

At approximately the same time that reform was initiated in calculus, a movement to change K-12 mathematics instruction in the U.S. began. The pre-collegiate mathematics reform movement has led to. change in the mathematical content that students experience as well as the ways in which they experience it. The goals of K-12 reform (as represented by the National Council of Teachers of Mathematics' Curriculum and Evaluation Standards (NCTM, 1989) and Professional Standards for Teaching Mathematics (NCTM, 1991) reflect many of the same ideals as those of calculus reform. These include increased conceptual understanding, use of active approaches to teaching and learning, and incorporation of technology. The goal is for students to come to a deeper, more conceptually-oriented understanding of mathematical ideas through revision of the curricula and use of new instructional practices by teachers.

Although many of the issues and the motivation were the same, reform at the college and K-12 levels have remained relatively isolated movements in the U.S. One main difference between calculus reform and reform at the K-12 level is the extent to which the movements are connected to educational research. When K-12 maths reform began, there was an established field of K-12 mathematics education research that examined issues related to curricular design, learning and teaching. Concurrent with curricular and instructional changes, K-12 mathematics education research experienced considerable growth. These developments have involved both the nature of the questions being asked and the methods used to investigate those questions. Where the field was once dominated by laboratory-based, quantitative studies of memory and skill acquisition, there is now a vast array of studies ranging from large-scale quantitative surveys to highly detailed qualitative case studies. These advances in mathematics education research and the timing of them with the reform movement have made the examination of the impact of pre-college educational reform possible in ways it would not have been several decades ago.

Although the K-12 reform created a need for evaluation of its impact on students, there was also an existing cadre of educational researchers who were interested in the study of complex questions related to teaching and learning. Theory and methods in pre-college educational research had developed to an extent that made it possible to examine what learning was happening and how that learning was happening.

Unlike the situation with pre-college reforms, the field of college mathematics education research was not especially prepared to address the issues related to reform when calculus reform began. The existing work in college-level maths education research was sparse and the field was not poised to experience rapid and significant growth. Although the changes in curriculum materials, instructional strategies, course goals, employed technologies and fundamental premise of what it means to understand calculus, all provide very fertile opportunities for educational research, the circumstances were not right at the time for significant advances to be made. In addition, despite the fact that many of the issues were quite similar, the pre-college research and literature was rarely used as a resource for work in calculus reform. As a result, even though there has been an expansion in the amount of collegiate mathematics education research being undertaken, the process of implementing the college reforms has been out of step with the available educational research.

3.4 Publications: descriptions, opinions, and reports

Initial efforts in calculus reform grew from pragmatic concerns and the design of solutions was pragmatic as well. This approach was reflected in the publications about reform. Having designed and implemented a project, the people involved then communicated their ideas and shared materials with others. This dissemination of information was one of the ways in which the movement spread from the small collection of original NSF sites to other schools. People also sought out opportunities to express their opinions about specific projects or issues of reform more generally. To date, the vast majority of publications related to calculus reform are opinions or descriptions of projects, curricula and software where the objective is to inform others of what was done and how. Relevant examples can be found in Solow (1994), Douglas (1995), and Schoenfeld (1995).

People were also interested in the impact that the reform projects were having on students. This led to the design and implementation of evaluation studies. The emphasis of these evaluations tended to be very pragmatic. In essence, frequently the main question asked was "Has it worked?" The specifics of how this question was pursued differed from project to project, but the experimental design of comparing students from traditional classes with students from reform classes was predominant.

In many instances, the evaluations used experimental designs borrowed from physical or biological sciences (or early educational research). They used 'experimental' and 'control' groups, typically comparing the test scores of classes that used the reform approach and those that did not. Sometimes the tests were designed specifically to assess particular types of student performance such as proficiency with traditional algorithms or ability on conceptual problems. (For examples, see Armstrong, Garner and Wynn, 1994; Bookman and Friedman, 1994.) The limitations of such comparative studies is well known (see, e.g., Schoenfeld, 1994). Comparative studies of this sort are poorly-suited to the study of phenomena as complex as teaching and learning. These studies often did a reasonable job of answering a very specific question for a particular course or programme, however, instructional situations are complex and differ so dramatically from project to project that answering a project-specific question usually cannot contribute much to our understanding of teaching and learning as a whole.

Even if comparative methods were appropriate and we had ideal control and experimental groups, knowing that one group out-performed the other does not tell us what brought about the differences. Because of this, many of these evaluation studies are best described (in K-12 research) as a 'pilot' study. In many cases, the research identified what may be interesting phenomena but the results and conclusions fall short of contributing to our understanding of the processes of teaching and learning. In most cases, having identified an interesting occurrence (perhaps differential performance on exam questions by students in different courses), a study utilizing a different experimental design could be carried out. For example, problem solving interviews and/or observation of students' in-class behaviour might generate results with explanatory power for the differences in performance. In a small number of cases, those methods have been used to examine research questions of a non-comparative nature. For examples of non-comparative studies, see Palmiter (1991); Selden, Selden and Mason (1994); Bonsangue and Drew (1995); Park and Travers (1996). Unfortunately, studies that yield results with strong explanatory power are scarce and the comparative studies undertaken in connection with calculus reform have not significantly advanced our understanding of how students learn calculus and how different instructional circumstances influence that learning.

3.5 Research

Some research on calculus has gone beyond comparative studies and evaluation of particular courses. Such work has not been primarily about calculus *reform*, but instead looked at student understanding of calculus concepts. This work examined student understanding in ways that have explanatory power and utility – the aim was not to answer some yes/no question but to explore some of the underlying mechanisms through which learning occurs. The researchers often made connections between existing research and their current research and the nature of the work also permits future researchers to build upon it.

Research of this sort examined how student understanding of particular concepts interacts with their understanding of calculus. In particular, researchers examined how calculus understanding is influenced by student understanding of variables, functions and limits. It is accepted in much of the educational research community that students' understanding of one concept influences their learning and understanding of related concepts. In terms of calculus learning, this means that the understandings of the concept of a variable, functions and limits will influence the development of their understanding of derivatives and other calculus concepts. Research has revealed that what may appear to be weaknesses in students' understanding of calculus concepts can really be just manifestations of their pre-existing understanding of a related concept. For example, students may understand the concept of function in ways that served them well in certain contexts but that are incompatible with, or do not support the development of, a robust understanding of derivative. Examples of this research can be found in Monk (1987), Williams (1991), Tall (1992), Ferrini-Mundy and Graham (1994), and White and Mitchelmore (1996).

3.6 Looking to the future

Although the research associated with calculus reform has, for the most part, been limited to comparative studies, that may be starting to change. At least two recent developments have the potential to strengthen research on calculus reform. First, some researchers have begun to examine issues of teaching and learning at the college level with methods that show promise for generating informative, explanatory results. At the present time, most of this work has been focused on the influence that student understanding of concepts such as function and limit has on learning calculus, but research in this area could be extended to concepts students encounter later in the curriculum as well as to issues more specifically connected with reform.

Second, there is a growing interest in collegiate mathematics education research in the U.S. This has led to the formation of organizations that support and promote research of teaching and learning at the college level through conferences and publications. Through the exchange of information and ideas, the members of these groups have the potential to build on existing research and move beyond pilot-stage research into issues of calculus reform. In addition to communication within or among these groups, some communication between the K-12 maths education research and collegiate maths education research fields has begun. This has the potential to be beneficial for both groups. K-12 researchers can learn more about the teaching and learning that they may be preparing their students for and collegiate researchers can learn about research designs that allow for the investigation of deeper, more complex questions that will help move the field forward.

At the start of the movement, the type of communication that people engaged in was primarily limited to expressions of opinion and descriptions of projects. As the movement progressed, people began to examine and reflect on the impact of the reforms. Now what remains is to move beyond these comparative studies motivated by local, pragmatic concerns and to begin to look closely at the complex teaching and learning issues involved. The challenge will be to channel the existing enthusiasm and efforts toward research questions and methods that will lead to increasingly substantive understanding of teaching and learning. If the existing enthusiasm for reform is sustained, armed with a deeper understanding of the issues, the community will have a greater chance of designing, refining and implementing programmes that are sustainable and that have a lasting impact on calculus education.

4. CONCLUSION

As noted in the introduction, there have been two very different traditions of research on calculus/introductory analysis. These traditions might almost be called 'theory-driven,' as reflected in section 2; and 'practice-driven' as described in section 3. Interestingly, there appears to be a move toward convergence of the two types. On the one hand, the theoretical work described in section 2 has given rise to some studies of 'didactic engineering.' On the other hand, now that various efforts at reform have been developed and stabilized, as described in section 3, such courses provide excellent sites for basic research. Ultimately, the field will make progress on effective teaching and learning only if it deals meaningfully with theoretical and pragmatic issues simultaneously. This paper reflects movement in that direction. All the articles cited – some of which focus on theoretical considerations, some on reform, and some on both theory and reform – are part of the foundations on which we build.

REFERENCES

Armstrong, G., Garner, L. and Wynn, J. (1994). Our Experience With Two Reformed Calculus Programs, *Primus*, Vol IV (4), 301-311.
Artigue, M. (1992). Functions from an algebraic and graphic point of view: cognitive difficulties and teaching practices. In G. Harel and E. Dubinsky, *The Concept of Function : Aspects of Epistemology and Pedagogy*, pp. 109-132. MAA Notes vol. 28. Washington, DC: Mathematical Association of America.
Artigue, M. (1996). Teaching and learning elementary analysis. In C. Alsina et al, 8th international congress on mathematics education, selected lectures, pp. 15-30. Sevilla: SAEM Thales.
Artigue, M. (2001). What Can We Learn From Educational Research At The University Level? this volume, pp. 207-220.
Bonsangue, M. V. and Drew, D. E. (1995). Increasing Minority Students' Success in Calculus. *New Directions for Teaching and Learning*, 61, 23-33.
Bookman, J. and Friedman, C. P. (1994). A Comparison of the Problem Solving Performance of Students in Lab Based and Traditional Calculus. In E. Dubinsky, A. H. Schoenfeld and J. Kaput (Eds.), *Research in Collegiate Mathematics Education I* ,Vol. 4, pp. 101-116. Providence, RI: American Mathematical Society.
Cantoral, R. y Farfán, R. (1998). Pensamiento y lenguaje variacional en la introducción al análisis. *Epsilon*, No. 42, 353 - 369.
Cordero, F. 1994. Cognición de la integral y la construcción de sus significados (un estudio del discurso matemático escolar). Tesis doctoral, Cinvestav- IPN.
Cornu, B. (1991). Limits. In D. Tall (Ed.), *Advanced Mathematical Thinking*, pp. 153-166. Dordrecht: Kluwer Academic Publishers.
Dorier, J-L. and Sierpinska, A. (2001). Research Into the Teaching and Learning of Linear Algebra, this volume, pp. 255-274.
Douglas, R. G. (Ed.). (1986). *Toward a Lean and Lively Calculus*. (Vol. 6). Washington, DC: Mathematical Association of America.

Douglas, R. G. (1995). The First Decade of Calculus Reform. *UME Trends*, 6 (6), 1-2.

Dreyfus, T. and Eisenberg, T. (1996). On different facets of mathematical thinking. In R.J. Sternberg et T. Ben-zeev (Eds.), *The Nature of Mathematical Thinking*, pp. 253-284. Mahwah, NJ: Lawrence Erlbaum Associates, Inc.

Dubinsky, E. (1991). Reflective abstraction in advanced mathematical thinking. In D. Tall (Ed.) *Advanced Mathematical Thinking*, pp. 95-126. Dordrecht: Kluwer Academic Publishers.

Dubinsky, E. (1995). A programming language for learning mathematics. *Communications in Pure and applied Mathematics*, vol. 48, 1-25.

Dubinsky, E. and McDonald, M. (2001). APOS: A Constructivist Theory of Learning in Undergraduate Mathematics Education Research, this volume, pp. 275-282.

Farfán, R. (1997). Ingeniería Didáctica. Un estudio de la variación y el cambio. México: Grupo Editorial Iberoamérica.

Ferrini-Mundy, J. and Graham, K. (1994). Research in Calculus Learning: Understanding of Limits, Derivatives, and Integrals. In J. J. Kaput and E. Dubinsky (Eds.), *Research Issues in Undergraduate Mathematics Learning: Preliminary Analysis and Results*, Vol. 33, pp. 29-45. Washington, DC: The Mathematical Association of America.

Gray, E.M. and Tall, D. (1994) Duality Ambiguity and Flexibility: A prospectal view of simple arithmetic. *Journal of Research in Mathematics Education*, vol. 26, 115-141.

Guzman, M., Hodgson, B.R., Robert, A. and Villani, V.(1998). Difficulties on the passage from secondary to tertiary education. *Proceedings of the International Congress of Mathematicians, Berlin*, 1998, vol. III, 747-762.

Hauchard, C. (1985). Sur l'appropriation des concepts de suite et de limite de suite. Dissertation doctorale, Université de Louvain.

Hauchard, C. and Schneider, M. (1996). Une approche heuristique de l'analyse. *Repères IREM* no. 25, 35-62.

Kent and Noss, Finding a Role for Technology in Service Mathematics for Engineers and Scientists, this volume, pp 395-404.

Keynes, H. and Olson, A. (2001). Professional Development for Changing Undergraduate Mathematics Instruction, this volume, pp. 113-126.

Legrand, M. (1993). Débat scientifique en cours de mathématiques et spécificités de l'analyse. *Repères IREM* no. 10, 123-158.

Monk, S. and Nemirovsky, R. (1994). The Case of Dan: Student Construction of a Functional Situation through Visual Attributes. In E. Dubinsky, A. H. Schoenfeld and J. Kaput (Eds.), *Research in Collegiate Mathematics Education I*,Vol. 4, pp. 139-168. Providence, RI: American Mathematical Society.

National Council of Teachers of Mathematics (1989). *Curriculum and Evaluation Standards for School Mathematics*. Reston, VA: The National Council of Teachers of Mathematics.

National Council of Teachers of Mathematics (1991). *Professional Standards for Teaching Mathematics*. Reston, VA: The Nation Council of Teachers of Mathematics.

Ortega, G. M. (2000). Elemento de enlace entre lo conceptual y lo algoritmico en el Calculo integral. *Relime*, vol. 3.2, 131-170.

Palmiter, J. R. (1991). Effects of Computer Algebra Systems on Concept and Skill Acquisition in Calculus. *Journal for Research in Mathematics Education*, 22(2), 151-156.

Park, K. and Travers, K. J. (1996). A Comparative Study of a Computer-Based and a Standard College First-Year Calculus Course. In J. Kaput, A. H. Schoenfeld and E. Dubinsky (Eds.), *Research in Collegiate Mathematics Education II* Vol. 6, pp. 155-176. Providence, RI: American Mathematical Society.

Praslon, F. (1999). Discontinuities regarding the secondary/university transition: the notion of derivative, as a special case. In O. Zavlavsky (Ed.), Proceedings of the 23rd Conference of the International Group for the Psychology of Mathematics, tome 4, pp. 73-80. Haifa: Technion Printing Centre.

Robert, A. (1998). Outils d'analyse des contenus à enseigner à l'université. *Recherches en Didactique des Mathématiques*, vol. 18.2, 139-190.

Schneider, M. (1991). Un obstacle épistémologique soulevé par des découpages infinis de surfaces et de volumes. *Recherches en Didactique des Mathématiques*, vol. 11.2/3, 241-294.

Schoenfeld, A. H. (1994). Some Notes on the Enterprise (Research in Collegiate Mathematics Education, That Is). In E. Dubinsky, A. H. Schoenfeld and J. Kaput (Eds.), *Research in Collegiate Mathematics Education I*, Vol. 4, 1-20. Providence, RI: American Mathematical Society.

Schoenfeld, A. H. (1995). A Brief Biography of Calculus Reform, *UME Trends*, 6(6), 3-5.

Schoenfeld, A. H. (Ed.), (1997). *Student assessment in calculus : a report of the NSF Working Group on Assessment in Calculus.* MAA Notes. Washington, DC, Mathematical Association of America.

Selden, A. and Selden, J. (2001). Tertiary Mathematics Education Research and its Future, this volume, pp. 237-254.

Selden, J., Selden, A., and Mason, A. (1994). Even Good Calculus Students Can't Solve Nonroutine Problems. In J. J. Kaput and E. Dubinsky (Eds.), *Research Issues in Undergraduate Mathematics Learning: Preliminary Analysis and Results*, Vol. 33, pp. 17-26. Washington, DC: The Mathematical Association of America.

Sfard, A. (1992). On the dual nature of mathematical conceptions: reflections, on processes and objects as different sides of the same coin. *Educational Studies in Mathematics*, vol. 22, 1-36.

Sierpinska, A. (1985). Obstacles épistémologiques relatifs à la notion de limite. *Recherches en Didactique des Mathématiques*, vol. 6.1, 5-68.

Sierpinska, A. (1992). Theoretical perspectives for development of the function concept. In G. Harel and E. Dubinsky (Eds.), *The concept of function : Aspects of Epistemology and Pedagogy*, MAA no. 25, pp. 23-58. Washington, DC: The Mathematical Association of America.

Sloan Conference (1986). Conference/workshop to develop alternative curriculum and teaching methods for calculus at the college level. Tulane University, January 2-6.

Smith, D. (2001). The Active/Interactive Classroom, this volume, pp. 167-178.

Solow, A. (Ed.). (1994). *Preparing for a New Calculus: Conference Proceedings*. MAA Notes No. 36, The Mathematical Association of America, Washington, DC.

Tall D. and Vinner, S. (1981). Concept image and concept definition in mathematics with particular reference to limits and continuity. *Educational Studies in Mathematics*, vol. 12/2, 151-169.

Tall, D. (Ed.), (1991). *Advanced Mathematical Thinking*. Dordrecht: Kluwer Academic Publishers.

Tall, D. (1991). Intuition and rigor : the role of visualization in the calculus. In S. Cunningham and W.S. Zimmermann (Eds.), *Visualization in Teaching and Learning Mathematics*, MAA no. 19, pp. 105-119. The Mathematical Association of America, Washington, DC.

Tall, D. (1992). The transition to advanced mathematical thinking: Functions, limits, infinity and proof. In D. A. Grouws (Ed.), *Handbook of Research on Mathematics Teaching and Learning: A project of the National Council of Teachers of Mathematics*, pp. 495-511. New York: Macmillan Publishing Co, Inc.

Tall, D. (1996). Functions and calculus. In A.J. Bishop, K. Clements, J. Kilpatrick and C. Laborde (Eds.), *International Handbook of Mathematics Education*, pp. 289-325. Dordrecht: Kluwer Academic Publishers.

Trouche, L. (1996). A propos de l'apprentissage des limites de fonctions dans un 'environnement calculatrice' : étude des rapports entre processus de conceptualisation et processus d'instrumentation. Doctoral thesis, Université de Montpellier 2.

Tucker, A. C. (Ed.), (1995). *Models That Work: Case Studies in Effective Undergraduate Mathematics Programs*, Vol. 38. Washington, DC: The Mathematical Association of America.

Vinner, S. and Dreyfus, T. (1989). Images and definitions in the concept of function. *Journal of Research in Mathematics Education*, vol. 20/4, 356-366.

White, P. and Mitchelmore, M. (1996). Conceptual Knowledge in Introductory Calculus. *Journal for Research in Mathematics Education*, 27(1), 79-95.

Williams, S. R. (1991). Models of Limit Held by College Calculus Students. *Journal for Research in Mathematics Education*, 22(3), 219-236.

Wood, L. (2001). The Secondary-Tertiary Interface, this volume, pp. 87-98.

Aline Robert,
IUFM de Versailles, France
robert@math.uvsq.fr

Natasha Speer,
University of California at Berkeley, USA
nmspeer@socrates.berkeley.edu

SECTION 4

MATHEMATICS AND OTHER DISCIPLINES

Edited by Urs Kirchgraber and Mogens Niss

LYNN ARTHUR STEEN

REVOLUTION BY STEALTH: REDEFINING UNIVERSITY MATHEMATICS

1. INTRODUCTION

Change, growth, and accountability dominate higher education at the dawn of the twenty-first century. According to delegates at UNESCO's recent World Conference on Higher Education, change is unrelenting, 'civilizational in scope,' affecting everything from the nature of work to the customs of society, from the role of government to the functioning of the economy. Growth in higher education has been equally dramatic, with worldwide enrolment rising from 13 million in 1960 to near 90 million today (World Conference on Higher Education, 1998).

Concomitant with change and growth is pressure from governments everywhere for greater accountability from professors and leaders of higher education, for evidence that, in a world of rapid change, universities are working effectively to address pressing needs of society. Autonomy, the prized possession of universities, presupposes accountability. The escalating pressures of global change, growth, and accountability will create, according to UNESCO Director Federico Mayor, "a radical transformation of the higher education landscape not only more but *different* learning opportunities" (Mayor, 1998).

Much of the upheaval in society and employment is a consequence of the truly revolutionary expansion of worldwide telecommunication, as is the stunning increase in demand for higher education. But this demand is predicated on the belief that universities properly anticipate signals from the changing world of work and create optimal linkages between students' studies and expectations of employers. Unfortunately, few universities have taken up this challenge, at least not unless pushed by external forces.

As the world changes rapidly and higher education grows explosively, universities evolve leisurely. Courses, curricula, and examinations remain steeped in tradition, some centuries old, while autonomy and academic freedom rule in the classroom. Few institutions of higher education readily embrace the culture of assessment that is required to ensure relevance and effectiveness of their curriculum. Under these circumstances it is only natural for political leaders to demand stronger connections between the classroom and the community, between the ivory tower and the industrial park.

303

Derek Holton (Ed.), The Teaching and Learning of Mathematics at University Level: An ICMI Study, 303—312.
© 2001 *Kluwer Academic Publishers. Printed in the Netherlands.*

2. UNDERGRADUATE MATHEMATICS

Demands for relevance and accountability are no strangers to undergraduate mathematics. Indeed, post-secondary mathematics can be viewed as higher education in microcosm. Growth in course enrolments has been enormous, paralleling the unprecedented penetration of mathematical methods into new areas of application. These new areas—ranging from biology to finance, from agriculture to neuroscience—have changed profoundly the profile of mathematical practice (see Odom, 1998). Yet for the most part these changes are invisible in the undergraduate mathematics curriculum, which still marches to the drumbeat of topics first developed in the eighteenth and nineteenth centuries.

It is, therefore, not at all surprising that the three themes identified at the UNESCO conference are presaged in the Discussion Document for this ICMI Study: the rapid growth in the number of students at the tertiary level; unprecedented changes in secondary school curricula, in teaching methods, and in technology; and increasing demand for public accountability (see Discussion Document, 1997). Worldwide demands for radical transformation of higher education bear on mathematics as much as on any other discipline.

Post-secondary students study mathematics for many different reasons. Some pursue clear professional goals in careers such as engineering or business where advanced mathematical thinking is directly useful. Some enrol in specialized mathematics courses that are required in programmes that prepare skilled workers such as nurses, automobile mechanics, or electronics technicians. Some study mathematics in order to teach mathematics to children, while others, far more numerous, study mathematics for much the same reason that students study literature or history: for critical thinking, for culture, and for intellectual breadth. Still others enrol in post-secondary courses designed to help older students master parts of secondary mathematics (especially algebra) that they never studied, never learned, or just forgot. (This latter group is especially numerous in countries such as the United States that provide relatively open access to tertiary education, see Phipps, 1998.)

In today's world, the majority of students who enrol in post-secondary education study some type of mathematics. Tomorrow, virtually all will. In the information age, mathematical competence is as essential for self-fulfilment as literacy has been in earlier eras. Both employment and citizenship now require that adults be comfortable with central mathematical notions such as numbers and symbols, graphs and geometry, formulas and equations, measurement and estimation, risks and data. More important, literate adults must be prepared to recognize and interpret mathematics embedded in different contexts, to think mathematically as naturally as they think in their native language (see Steen, 1997).

Since not all of this learning can possibly be accomplished in secondary education, much of it will take place in post-secondary contexts, either in traditional institutions of higher education (such as universities, four- and two-year colleges, polytechnics, or technical institutes) or, increasingly, in non-traditional settings such as the internet, corporate training centres, weekend short-courses, and for-profit universities. This profusion of post-secondary mathematics programmes at the end of the twentieth century contrasts sharply with the very limited forms of university mathematics education at the beginning of this century. The variety of forms, purposes, durations, degrees, and delivery systems of post-secondary mathematics reflects the changing character of society, of careers, and of student needs. Proliferation of choices is without doubt the most significant change that has taken place in tertiary mathematics education in the last one hundred years.

3. MATHEMATICAL PRACTICE

The primary purpose of mathematics programmes in higher education is to help students learn whatever mathematics they need, both for their immediate career goals and as preparation for life-long learning. Today's students expect institutions of higher education to offer mathematics courses that support a full range of educational and career goals, including:

• Agriculture	• Law
• Biological Sciences	• Mathematical Research
• Business	• Management
• Computing	• Medical Technology
• Elementary Education	• Physical Science
• Electronics	• Quantitative Literacy
• Economics	• Remedial Mathematics
• Engineering	• Secondary Education
• Finance	• Social Sciences
• Geography	• Statistics
• General Education	• Technical Mathematics
• Health Sciences	• Telecommunications

Even without exploring details of specific curricula or programmes, it should be obvious that the multiplicity of student career interests requires, if you will, multiple mathematics. Consider a few examples of how relatively simple mathematics is used in today's world of high performance work:

- Precision farming relies on satellite imaging data supplemented by soil samples to create terrain maps that reflect soil chemistry and moisture levels. These methods depend on geographic information systems that blend spreadsheet organization with a variety of algorithms for geometric projections (e.g., for rendering onto flat maps oblique satellite images of earth's curved surface).
- Technicians in semiconductor manufacturing plants, analyze real-time data from production processes in order to detect patterns of change that might signal an impending reduction in quality before it actually happens. These methods involve measurement strategies, graphical analyses, and tools of statistical quality control.
- Teams that design new commercial airplanes now engage designers, manufacturing personnel, maintenance workers, and operation managers in joint

planning with the goal of minimizing total costs of construction, maintenance, and operation over the life of the plane. This enterprise involves teamwork among individuals of quite different mathematical training as well as innovative methods of optimization.

- Emergency medical personnel need to interpret quickly and accurately dynamic graphs of heart action that record electrical potential, blood pressure, and other data. With experience, they learn to recognize both regular patterns and common pathologies. With understanding, they can also interpret uncommon signals.

These examples are not primarily about the relation of mathematical theory to applications—the traditional poles of curricular debate—but about something quite different: mathematical practice (see Denning, 1997). Behind each of these situations lurks much good mathematics (e.g., projection operators, optimization algorithms, fluid dynamics, statistical inference) that can be applied in these and many other circumstances. However, most students are not primarily motivated to learn this mathematics, but rather to increase crop yield, minimize manufacturing defects, reduce airplane costs, or stabilize heart patients. Although a mathematician will recognize these as mathematical goals—to increase, minimize, reduce, stabilize—neither students nor their teachers in agriculture, manufacturing, engineering, or medicine would recognize or describe their work in this way. To these individuals, the overwhelming majority of clients of post-secondary mathematics, mathematical methods are merely part of the routine practice of their profession.

Indeed, mathematics in the workplace is often so well hidden as to be invisible to everyone except a discerning observer. In the United States different industries have created skill standards for entry-level employees (e.g., electronics (American Electronics Association, 1994), photonics (Center for Occupational Research and Development, 1995), health care (FarWest Laboratory, 1995), and National Skills Standards Board, 1998). Virtually all of these standards include substantial uses of mathematics, but most such applications are embedded in routine job requirements without any visible hint of the underlying mathematics. Although mathematics is now ubiquitous in business and industry, the mathematics found there is often somewhat different from what students learn in school or college (see Davis, 1996 and Packer, 1997). Similarly, in the wider world of public policy, the gradual incursion of statistics and probability in measuring (and sometimes controlling) personal health, societal habits, and national economies has created whole new territories for students and professors to explore (see Bernstein, 1996, Porter, 1995 and Wise, 1995).

In sharp contrast to this profligate flowering of practical mathematics in diverse post-secondary settings, university mathematics—what mathematicians tend to think of as 'real mathematics'—matured in the last century as a tightly *disciplined* discipline led by professors of world-wide renown who held major chairs in leading universities and research institutes. However, this university mathematics, 'real mathematics' as practiced in real universities, now constitutes only a tiny fraction of post-secondary mathematics. One data point: in the United States, fewer than 15%

of traditional undergraduate mathematics enrolments are in courses above the level of calculus (Loftsgaarden, Rung, and Watkins, 1997). And this does not count non-traditional enrolments, where the variety of offerings is even greater. A realist might well argue that 'real mathematics' is found not in the traditional curriculum inherited from the past but in today's widely dispersed courses, where a multitude of students learn a cornucopia of mathematics in diverse situations for a plethora of purposes.

4. LEARNING MATHEMATICS

Where do students learn mathematics? Some take traditional mathematics courses such as calculus, geometry, and statistics. Some take courses specifically designed for certain professions—mathematics for nurses, statistics for lawyers, calculus for engineers—that are offered either by mathematics departments or by the professional programmes themselves. But many, perhaps even most, pick up mathematics invisibly and indirectly as they take regular courses and internships in their professional fields (e.g., in physiology, geographic information systems, or aircraft design).

Any university dean knows that statistics is more often taught outside of statistics programmes than inside them. The same is true of mathematics, but is not as widely recognized. Every professional programme, from one- and two-year certificates to four- and five-year engineering degrees, offers courses that provide students with mathematics (or statistics) in the context of specific professional practice. This is entirely natural, since most students find that they learn mathematics more readily, and are more likely to be able to use it when needed, if it is taught in a context that fits their career goals and in which the examples resonate with those that appear in their other professional courses.

The appeal of context-based mathematics is no surprise, nor is its widespread presence in university curricula. But what is somewhat new—and growing rapidly—is the extent to which good mathematics is unobtrusively embedded in routine courses in other subjects. Anywhere spreadsheets are used (which is almost everywhere) mathematics is learned. It is also learned in courses that deal with such diverse topics as image processing, environmental policy, and computer-aided manufacturing. From technicians to doctors, from managers to investors, most of the mathematics people use is learned not in a course called mathematics but in the actual practice of their craft. And in today's competitive world, where quantitative skills really count, embedded mathematical tools are often as sophisticated as the techniques of more traditional mathematics.

So tertiary mathematics now appears in three forms: as traditional mathematics courses (both pure and applied) taught primarily in departments of mathematics; as context-based mathematics courses taught in other departments; and as courses in other disciplines that employ significant (albeit often hidden) mathematical methods. I have no data to quantify the 'biomass' of mathematics taught through these three means, but to a first approximation I would conjecture that they are approximately equal.

If that is true, then I can with some confidence suggest that the biomass of mathematics that is learned, remembered, and still useful five years later is decidedly tilted in favor of the hidden curriculum. Most of what is taught in the traditional mathematics curriculum is forgotten by most students shortly after they finish their mathematical studies, but most of what students learn in the context of use is remembered much longer, especially if students practice in the field for which they prepared.

Thus this paradox: as widespread use makes post-secondary mathematics increasingly essential for all students, mathematics departments find themselves playing a diminishing role in the mathematical education of post-secondary students. The obvious corollary poses another paradox: to exercise their responsibility to mathematics, mathematicians must think not primarily about their own department but about the whole university. Mathematics pervades not only life and work, but also all parts of higher education.

Many people in higher education can think of nothing more dangerous than to have mathematicians take seriously their leadership responsibility for university-wide mathematics. I daresay that most faculty and students think that mathematics is too important to be left to mathematicians. After all, mathematicians tend to think that they alone know what mathematics is and how it should be taught. Courses that do not fit their standards for real mathematics—for example, mathematics for elementary school teachers, or a reprise of secondary school algebra—are often marginalized, if not ignored. Changes in the secondary school programmes are resisted or deflected under the pretense that they should not apply to university-bound students. (In fact these changes have a major impact on university mathematics because so many students—in some countries, the majority—now take some form of post-secondary education.) Also ignored, but for different reasons, is the vast array of mathematics taught in other departments.

5. RESPONDING TO NEEDS

The growth of higher education, the pervasiveness of mathematics, and the pace of change in society fuel legitimate public demand for tertiary programmes in mathematics that are effective, flexible, and responsive to current conditions. Mathematicians know perfectly well how to make systems that are effective, flexible, and responsive: they require feedback loops. But in their own completed work, mathematicians disdain feedback loops. Feedback smacks of approximation, of trial and error, of processes that overshoot and undershoot. Good mathematics hits the nail on the head. It provides a clean argument based on solid principles and strict reasoning. Who ever heard of a completed proof that relied on a feedback loop to correct its errors?

As mathematicians, we tend to design curricula the same way that we design proofs. We select our goal, we think carefully and systematically about all relevant factors, and then we write down all the steps that are logically necessary to reach that goal. What we do not do very often is to listen carefully for feedback, from students, from other faculty, and especially from professionals who employ mathematical methods in their daily work. If we listen carefully, here is some of what we might hear:

- the panic of an architecture student who took several years of secondary mathematics but is still mystified by simple equations and graphs;
- the fear of a minority-culture student who is afraid of mathematics but wants to return to his or her home community to teach young children;
- the frustration of an employer over a mathematics tradition that favours esoteric concepts over mundane examples;
- the anger of college-educated adults who despite their education are unable to understand the economic implications of changes in interest rates;
- the confusion of consumers trying to decide among competing offers for mortgages or cellular telephone service;
- the embarrassment of recent university graduates who find themselves inexperienced with common mathematical, statistical, and graphical software.

Rarely, if ever, would we hear the lament of a former student who felt disabled for lack of understanding the Lebesgue integral or the Sylow theorems.

It turns out that many business leaders, politicians, and innovative educators are listening more attentively than are mathematicians. Having failed to persuade mathematicians to expand their horizons, these outsiders are busy setting up alternatives to university mathematics, indeed, alternatives to the university itself. The Open University in Great Britain was perhaps the first major alternative of this type. A more recent example is the for-profit University of Phoenix which, in less than a decade and with a faculty that is over ninety percent part-time, has become one of the ten largest universities in the United States. It is fully accredited, intensely evaluated, and traded on a major stock exchange (see Fischetti et al., 1998, Leatherman, 1998, Shea, 1998 and Traub, 1997).

Recently several internet-based 'virtual universities' have begun. The State of California sponsors one of these. Another one is sponsored by a dozen states in the western half of the U.S. where students in need of education are often geographically isolated and employed in jobs that prevent them from studying full time on a traditional campus (see Hungry Minds, 2000). As satellites enable wireless worldwide telecommunications, courses and credits from these alternative universities will soon be available almost everywhere (Marchese, 1998, Slaughter and Leslie, 1997 and Strosnider, 1998). Economic viability will depend on feedback from students, now called 'consumers.' Convenience learning will transcend both institutional and even national boundaries. In this networked future the choice of what mathematics to study, under whom, and at what time will be made not by professors, but by students.

6. THE NEXT CENTURY

As we approach the twenty-first century we are greeted by an astonishingly rich landscape of undergraduate mathematics that extends vast distances in several different directions:

- in terms of *level*, it ranges from elementary topics ordinarily taught in the primary and secondary schools to advanced courses at the interface with graduate education;
- in terms of *content* it ranges from classical topics of analysis and algebra to modern interdisciplinary subjects such as neural networks, image processing, and the pricing of financial derivatives;
- in terms of *context*, it ranges from traditional mathematics courses offering basic theory and selected applications, to authentic scientific or work-related situations with embedded mathematics that is learned through use but rarely studied separately;
- in terms of *setting*, it ranges from scheduled classes in traditional universities and corporate short-courses on weekends and after working hours, to on-line courses designed for professional development or pre-professional qualification and for-profit educational corporations that constantly re-invent the means of delivery.

One corner of this landscape shelters a tiny village somewhat isolated from the new superhighways of post-secondary education. This village is world-famous for its continued commitment to mathematics courses and degrees using traditions inherited from the nineteenth century. However, many of today's young people disdain these traditions as largely irrelevant for the twenty-first century. Others learn these traditions and come to respect them, but nonetheless choose to follow a different path. Only a few follow the tradition faithfully, preserving the past and advancing its accomplishments.

Outside this village, across the vast landscape of post-secondary education, a more contemporary and flexible form of university-level mathematics thrives. This mathematics pervades life and work, is required of virtually every profession and career, is taught both explicitly and implicitly in courses throughout higher education, and is available at any time and in any place—in colleges, at work, after work, on-line, and at home. The reality is clear to anyone overlooking this terrain: post-secondary mathematics is no longer a craft practiced only in the village of university mathematicians.

Nor should it be. The critics are right: mathematics *is* too important to be left to mathematicians. The forces of change, growth, and accountability have spread both innovation and mathematics across the post-secondary landscape, freeing mathematics from the confining traditions of university departments and opening it to innovative content and pedagogy of other fields. Abetted by these external forces, mathematics the discipline—if not mathematics the department—is thriving. As observers of this vast landscape, we become witnesses to a revolution by stealth, a virtual takeover of undergraduate mathematics by its entrepreneurial clients who now set the agenda. As mathematicians, we should welcome our new colleagues and thank them for propelling mathematics into the twenty-first century.

REFERENCES

American Electronics Association (1994). *Setting the Standard: A Handbook on Skill Standards for the High-Tech Industry*. Santa Clara, CA: Amercan Electronics Association.

Bernstein, Peter L (1996). *Against the Gods: The Remarkable Story of Risk*. New York, NY: John Wiley.

Center for Occupational Research and Development (1995). *National Photonics Skill Standards for Technicians*. Waco, TX: Center for Occupational Research and Development.

Davis, P. W. (1996). *Mathematics in Industry*. Philadelphia, PA: Society for Industrial and Applied Mathematics.

Denning, Peter J. (1997). Quantitative Practices. In Lynn Arthur Steen (Ed.), *Why Numbers Count: Quantitative Literacy for Tomorrow's America*, pp. 106-117. New York, NY: The College Board.

Discussion Document (1997). ICMI Study on the Teaching and Learning of Mathematics at the University Level. *Bulletin of the International Commission on Mathematical Instruction* 43, December, 3-13. URL: http://elib.zib.de/IMU/ICMI/bulletin/43/Study.html.

FarWest Laboratory (1995). *National Health Care Skill Standards*. San Francisco, CA: FarWest Laboratory.

Fischetti, Mark, Anderson, John, Watrous, Malena, Tanz, Jason, and Gwynne Peter (1998). Education? The University of Phoenix is Just Around the Corner, *University Business*, (March/April) 45-51.

Higher Education in the Twenty-First Century (1998). *World Conference on Higher Education*, 5-9 October. Paris: UNESCO, 1998. URL: http://www.unesco. org/education/educprog/wche/principal/principal.htm.

Hungry Minds (2000). *The On-Line Learning Marketplace*. URL: http://www.hungryminds.com.

Leatherman, Courtney (1998). University of Phoenix's Faculty Members Insist They Offer High-Quality Education. *The Chronicle of Higher Education*, (October 16) A14-A16. URL: http://chronicle.com/weekly/v45/i08/08a01401.htm.

Loftsgaarden, Don O., Rung, Donald C., and Watkins, Ann E. (1997). *Statistical Abstract of Undergraduate Programmes in the Mathematical Sciences in the United States*. Washington, DC: Mathematical Association of America.

Marchese, Ted (1998). Not-so-Distant Competitors: How New Providers are Remaking the Post-secondary Marketplace, *Bulletin of the American Association of Higher Education (AAHE)* 50, 9, May, 3-11.

Mayor, Frederico (1998).. World Conference on Higher Education, 5-9 October 1998. Paris: UNESCO. URL: http://www.unesco.org/education/educprog/wche/world.htm.

National Skills Standards Board (1998). *Occupational Skills Standards Projects*. Washington, DC: National Skills Standards Board. URL: http://www.nssb.org/.

Odom, William E. (1998). *Report of the Senior Assessment Panel for the International Assessment of the U.S. Mathematical Sciences*. Washington, DC: National Science Foundation. URL: http://www.nsf.gov/cgi-bin/getpub?nsf9895.

Packer, Arnold (1997). Mathematical Competencies that Employers Expect. In Lynn Arthur Steen (Ed.), *Why Numbers Count: Quantitative Literacy for Tomorrow's America*, pp. 137-154. New York, NY: The College Board.

Phipps, Ronald (1998). *College Remediation: What it Is, What it Costs, What's at Stake*. Washington, DC: The Institute for Higher Education Policy, December. URL: http://www.ihep.com/PUB.htm.

Porter, Theodore M. (1995). *Trust in Numbers: The Pursuit of Objectivity in Science and Public Life*. Princeton, NJ: Princeton University Press.

Shea, Christopher (1998). Visionary or 'Operator'? Jorge Klor de Alva and his Unusual Intellectual Journey. *The Chronicle of Higher Education*, July 3, A8-A10. URL: http://chronicle.com/chedata/articles.dir/art44.dir/issue43.dir/43a00101.htm.

Slaughter, Sheila and Leslie, Larry L. (1997). *Academic Capitalism: Politics, Policies, and the Entrepreneurial University*. Baltimore, MD: Johns Hopkins Univ. Press.

Steen, Lynn Arthur (Ed.) (1997). *Why Numbers Count: Quantitative Literacy for Tomorrow's America*. New York, NY: The College Board.

Strosnider, Kim (1998). For-Profit Higher Education Sees Booming Enrollments and Revenues, *The Chronicle of Higher Education*, Jan 23, A36-A38. URL: http://chronicle.com/che-date/articles.dir/art-44.dir/issue-20.dir/20a03601.htm.
Traub, James (1997). Drive-Thru U. Higher Education for People who Mean Business, *The New Yorker*, October 20 and 27, 114-123.
Wise, Norton M. (1995). *The Values of Precision*. Princeton, NJ: Princeton University Press.

Lynn Arthur Steen
St. Olaf College, Minnesota, USA
steen@stolaf.edu

JEAN-PIERRE BOURGUIGNON

MATHEMATICS AND OTHER SUBJECTS

1. INTRODUCTION

In all its history, mathematics has evolved under a double influence: an internal one, often driven by the search for solutions of outstanding problems and by esthetic considerations, and an external one, stimulated by other disciplines or challenges coming from problems or needs of other subjects.

Nowadays, the training of students at the university level requires exposure to different disciplines and among them mathematics. This trend has been much reinforced by the use and prospective need of mathematics in many more areas of employment than before. This poses the question of how it relates to other subjects in this context. This was certainly the basis on which ICMI made its decision to have a working group devoted to the theme 'Mathematics and other Subjects' in the Singapore conference. The present article, based on the preliminary note produced by the author prior to the conference, has been enriched through the discussions that took place in the special session held in Singapore.

The expression 'other subjects' is quite undefined, and forces us to be ready to face real diversity. Indeed, it depends on many things, in particular the institutional context in which the teaching of mathematics takes place and the backgrounds of teachers involved in the process.

The fact that people involved in the discussions and their preparation were only mathematicians or specialists of the didactics of mathematics is *a priori* a serious methodological weakness. Indeed, in order to discuss all aspects of this topic in depth, confrontation with and testimonies of students and specialists from other disciplines, in particular those who are the furthest away from mathematics, seem compulsory. Such exogeneous contacts did not take place. This fact must be kept in mind, in particular since mathematicians have too often the tendency of reconstructing the world from their own peculiar viewpoints, and as a result sometimes overlook the fact that other people may take other radically different approaches which have at least equal value.

Three tendencies force us to seriously consider the possibility of advocating major changes for the future in relation with the theme discussed:

- There is a large and growing demand for mathematics from other fields because of the now widespread use of mathematical models. The demand is very diverse: models used in applied endeavours can either be rather simple from the mathematical point of view or pose high mathematical demands; new subjects (e.g. connection with Biology, or the telecommunications industry) may require new mathematics and/or bring users of mathematics having other types of training.

313

Derek Holton (Ed.), The Teaching and Learning of Mathematics at University Level: An ICMI Study, 313—320.
© 2001 *Kluwer Academic Publishers. Printed in the Netherlands.*

- The emergence of new knowledge-based jobs requires mathematical ability in connection with other disciplines. It may even require more in the future with the development of the e-economy, and other aspects of virtual reality. Hence a stronger need for an adapted training to share mathematical knowledge.

- The generalization of the use of software packages greatly contributes to broadening the scope of mathematical modelling. Training students to their use cannot any more be delayed by mathematicians, or systematically delegated. A basic understanding of the strengths and limits of mathematical technology has become compulsory.

2. A WIDE ARRAY OF SITUATIONS

The situation differs extremely from one country to another, and also from one institution to another. At the same time the willingness among mathematicians to work on this front and opinions on how to deal with such problems vary enormously. We will successively consider the diversity of situations from the point of view of students, from that of other disciplines and finally more globally.

2.1   What students are interested in relating mathematics with other subjects?

In many countries, during the first years of university study, curricula are multidisciplinary, so that students who are taught mathematics are not only students who will eventually major in mathematics. A healthy by-product there is that mathematicians-to-be are exposed to the foundations of other disciplines.

It has become traditional to claim that engineering students cannot do without being introduced to or even made comfortable with mathematical skills. In many universities they represent an important (and most of the time very motivated) population attending maths classes. Some of the courses to be taught in these areas are basic, some others are rather advanced and specialized.

Other important groups of students need to be educated in mathematics, with proper relations to their main subject of study. Among them, one finds very different profiles, from economists to biologists, from physicians to architects.

2.2   What are the other subjects concerned?

As a result, subjects to be put in relation with mathematics are very diverse. Some sciences, such as physics or mechanics, develop in a framework that is mathematical in nature. The apparatus of calculus is embodied in physical and mechanical theories. They cannot be understood without a working knowledge of mathematical concepts and techniques. Therefore, basic skills required from practitioners of these areas are not substantially different from those indispensable to students who want to become mathematicians. Vice-versa one can say that exposure to physics and mechanics can help students grasp the power of mathematical theories and also understand how they grow and become transformed.

The case of computer science is another instance of a theory whose foundations are very close to those of mathematics. Mathematical concepts involved there are more algebraic or number theoretical in general, except for specialists of image processing who are interested in (sometimes sophisticated) geometrical theories. Points of view may differ but concepts are identical. In such an area, one deals more often with discrete than with continuous mathematics.

The study of economics involves general ideas about mathematics, but also very special branches (often connected to non-smooth objects) such as optimization, stochastic models, and more recently neural networks.

In disciplines, such as biology and medicine, where the framework is less theoretical, other branches of mathematics, such as dynamical systems or statistics, are at the top of the list of requirements.

In humanities, needs are centred around discrete mathematics, and statistical models. With philosophy we are confronted with another fact concerning mathematics, namely that it gets its value from a subtle interaction between an array of concepts and techniques allowing it to deal with many different examples. Students in art and architecture are often interested in mathematical concepts underlying the concept of space, i.e. they want to know more about geometry.

2.3 What are the different learning environments?

The learning environments can also be very diverse. Quite often mathematics courses are part of the curriculum, and students from different origins are mixed, either deliberately or because of the size of the institution. Then the connection with courses in other disciplines taken by students usually remains minimal.

In some other instances, they are given to a homogeneous group of students. This typically happens in engineering schools. In such situations there are instances where courses are delivered by non-mathematicians, who want to be sure that the emphasis is put on the context. The further remote from mathematics the subject is, the greater is the risk that mathematics is presented in a narrow perspective.

3. THE EMERGENCE OF NEW TYPES OF KNOWLEDGE-BASED JOBS

The presence of a mathematical component in a larger number of jobs is intimately related to the emergence of new types of knowledge-based jobs. Many factors contribute to this. If one is to look into the future, it is certainly important to analyze the main sectors where jobs involving mathematics appear.

3.1 The birth of a virtual world

Defining and controlling a virtual world usually requires both analytical and geometrical models of this world. This can only be achieved with the proper use of mathematical concepts and skills mediated by computers.

These automata are likely to proliferate in many everyday situations, opening new types of jobs. This also means new training needs.

3.2 Consequences of a product-based economy

Modern economy is based on the definition, the production and the maintenance, of products whose life and death are programmed. This fact prevails in many areas of economic life, from house appliances to cars, from food industry to medical prescriptions. These products are optimized in many different ways: in their contents, in their shape, in their production process, in their packaging (if needed), in their modular structure for adequate maintenance, etc. This opens a number of doors for mathematics to enter because the various optimization steps require more than the knowledge particular to the domain to which the product belongs. They require genuine mathematics.

3.3 A new relation to mathematics

The preceding development probably requires a broader view, namely to claim that modern societies establish a new relation to mathematics [1]. It is a fact that the analysis of complex systems (e.g., climate changes, health systems, social conditions, transportation systems) has a major mathematical component, which, because it is not present in the first naive approach that one can develop, is often overlooked. To get a feeling for it cannot be achieved without having a sufficiently open training in mathematics. This is a typical situation where relations of mathematics with other subjects are relevant.

New uses of mathematics appear in many areas: e.g., in telecommunications – sophisticated cryptographic procedures, in medicine – image processing, computer aided surgery, in banking – stochastic models for the pricing of derivatives, etc. This requires a much wider mathematical knowledge than previously expected from students mastering basic calculus techniques and a few standard algorithms.

The use of sophisticated software, which could *a priori* imply less mathematical knowledge, in fact requires more general ideas about mathematics.

4. THE VARIETIES OF THE MATHEMATICAL CONTENTS PRESENTED

It is difficult to actually present the different contents of mathematical courses in connection with other subjects. This would require a cumbersome inventory. Here we prefer to consider the type of contents rather than the actual contents themselves.

4.1 Skills

For a number of activities involving a mathematical background, it is compulsory that the training aims at a real mastery of mathematical skills. These can be more or less sophisticated, but students have to be drilled on well-calibrated

---

[1] See, Bourguignon, Jean-Pierre (2000). A Basis for a New Interaction Between Mathematics and Society. In B. Engquist and W. Schmid (Eds.) *Mathematics Unlimited – 2001 and Beyond*, pp. 171-188. Heidelberg, New York: Springer.

exercises, which are likely to be given in context. Most of the time foundational courses in mathematics strongly reflect the inherent nature of our discipline, but should not be purely inward looking. In some countries unfortunately, beginner courses contain rather few genuine applications outside of mathematics itself.

That the control of the efficiency with which students perform mathematical exercises be in line with future professional use is another side of the same coin.

4.2 Concepts

In some other cases, transmitting a real understanding of mathematical concepts, ensuring a deeper acquisition of their knowledge, is the main issue. In that case the training will certainly have a much more leisurely tempo since personal involvement is required.

In the case of domains whose mathematization is either recent or even in the making, the search for the right mathematical counterpart of a concept that has been properly identified in its context may require the involvement of mathematical researchers. This is a typical situation where teaching should be delivered by mathematicians actively engaged in research.

The truth is that people tend to impose a given technique on a problem, with insufficient consideration of its appropriateness. For instance linearity may be imposed because linear mathematics is simple, yet applying non-linear techniques may result in immense improvements in the practical device or product being designed. It is important to be honest about the limitations and the reasons why a particular technique is being used. This requires a lot of thinking when choosing concepts to be incorporated in curricula which are only partly mathematical.

4.3 The pervasiveness of mathematical models

'Modelling' is a broad term, and the development of mathematical models has historically played an important role in many sciences. In recent years one has witnessed an explosion of mathematical models, which granted mathematics access to many new areas. This, of course, results in increased demand for mathematical training.

Some models are very sophisticated, some others are rather simplistic. In all cases, having an oversight of the mathematical context in which they are developed increases the possible impact a great deal.

As usual when one speaks of models, the value that one gets out of them does not depend only on the quality of the mathematics used but also of the adequacy of the model itself to reality. This part of the process is not mathematical and requires real exchanges between mathematicians and non-mathematicians.

5. INVOLVEMENT OF AND CONSTRAINTS ON TEACHING PERSONNEL

It is well known that the teaching-learning process cannot be reduced to solely considering the students or the curricula. Teachers as persons play a crucial role. It is therefore paramount to examine how their profiles, their attitudes and the conditions under which they work, affect the process.

5.1 Competence

When the teaching of mathematics is not undertaken by mathematicians, the question of competence may be raised. This view is too simplistic though. A symmetric concern with competence must also be raised. Mathematicians who participate in the training of students majoring in other subjects must also have a basic knowledge of these topics, at least to illustrate their teaching with stimulating examples.

In case one sets up more involved courses, requirements are even higher. This can only be achieved through a serious investment in the other subject. It is certainly fruitful to have a partnership between the mathematician and the appropriate expert in the applications area. This can facilitate the choosing of suitable modelling problems for incorporation into the mathematics course. Mathematicians need to be ready to learn from their colleagues in the disciplines of application.

The lack of recognition and reward for mathematicians who undertake the onerous task of redesigning courses can certainly be cited as a constraint. A vicious circle has been established in some institutions where promotion and tenure depend almost solely on research, teaching efforts being given only side consideration.

In some small departments of course, the problem is of another nature, and solely comes from the size constraint.

5.2 Teamwork

In order to produce the mixing of competencies required, it seems *a priori* that teamwork is the most appropriate tool. Since most higher education institutions are organized by disciplinary departments, such work is not so natural, and sometimes difficult to set up.

Another reason why teamwork could be useful is the example that it sets for students. Seeing this type of collaboration at work can be considered as part of the training that students ought to receive.

5.3 Relations to research work

At most higher education institutions, teaching and research are intimately connected. The cooperation of mathematicians and non-mathematicians in teaching can induce collaborative research work.

Since many higher education institutions involve students very early in the research process, this can also be an excellent framework to prepare students to confront their knowledge (both mathematical and from other domains) of multi-disciplinary situations.

5.4 Training for the use of software and computers

The advent of powerful computers and mathematical software is challenging teachers and researchers to reconsider the aims and scope of mathematics courses at the university level for both specialists and students from other disciplines. Having technology available means, in particular, more realistic modelling at hand.

Employers often use sophisticated computational tools that students must understand. Mathematicians have therefore a responsibility to open up the use and understanding of computers and software, and to work to increase their general availability. Even further, mathematicians may need to see how the practitioners use packages, and learn new approaches.

The main issue is to know whether the time necessary for this new training will be taken away from more traditional training or whether new more active forms of pedagogy need to be developed. This question cannot be answered without taking into account the technical constraints connected with the use of machines, and the presence of technicians to provide the appropriate support.

6. SOME RECOMMENDATIONS

After all the emphasis put on the diversity of situations and the constraints that teachers endure, one cannot expect that 'magic' solutions can be made to radically and definitely deal with the questions raised in this article. Here you will find gathered some recommendations that emerged from the discussions of the working group [2].

In several universities in different countries, the pressure towards applied areas has sometimes resulted in taking away positions from mathematics departments, and also moving the responsibility of some mathematics courses to other departments. This should convince mathematicians to adopt a more aggressive attitude.

6.1 The necessary evolution of curricula

It is important that modelling, i.e. applying appropriate mathematics to practical problems, be embedded as a key part of the curriculum already in elementary and secondary schools. It should not be left to begin at university level since what is required for a successful study is rather a different attitude, and the sooner students are exposed to it, the better.

Realistic applications of mathematics to the relevant fields of interest of students need to be incorporated into university courses, including first year. The number may have to be restricted, but there should be a discussion of the appropriateness of

---

[2] The author thanks the participants for their lively and constructive contributions.

the assumptions underlying the mathematical formulation, and how (realistic) changes in those assumptions would affect the conclusions.

This being said, one must acknowledge that mathematics curricula are under increasing pressure. Some teaching areas may therefore be eliminated or de-emphasised. Above all, the essence of mathematics, its logical structure and its repertoire of beautiful and powerful results must be preserved, not only for their own sake, but, interestingly enough, because ultimately they form the bedrock for worthwhile applications. This dimension must not be overlooked. This choice has to be made. It is often a difficult one that can challenge the departments' cohesion.

6.2 Towards multidisciplinary teaching teams?

The need to make the proper connection with colleagues of other disciplines pleads for a close contact at the personal level. Multi-disciplinary teaching teams could best achieve it but such solutions can be put in place only when the student population is relatively homogeneous, or the institution large enough.

Such solutions have a better chance of success when multidisciplinary work is already common practice at the research level. This shows one more time the possible impact of research activities of a department on its teaching. In fact this also goes in the opposite direction since contacts at the teaching level, in particular on the occasion of the supervision of small projects for students, can be a good opportunity to identify some research questions coming from outside mathematics that are worthy of further investigations.

6.3 How to favour the involvement of teachers?

The continuing vitality, even long-term survival, of mathematics departments may depend on their ability to implement curriculum reform. This will not be achieved if efforts in this direction by faculty members are not rewarded. In many departments, this will require a major change in promotion and hiring policies.

Another issue of a general nature is the teaching load of professors. This is particularly so if more active pedagogical practices such as small projects for students (or groups of students), are put in place. They can represent major steps in improving the efficiency of the teaching, and the perception by students of the linkage between mathematics and other fields, but they can be extremely time consuming. Such innovative practices have to be properly taken into consideration when estimating the faculty load. This too may require changing present practices.

BURKHARD KÜMMERER

TRYING THE IMPOSSIBLE

Teaching Mathematics to Physicists and Engineers

1. UNSOLVABLE PROBLEMS

In high school Andrea became fascinated by theoretical physics. She was captivated by the ideas of the relativity of time and the big bang. She could see herself working with other physicists on how to build quantum computers. So she decided to enter the local university with the intention of majoring in physics and unlocking some basic secrets of the universe.

Her classmate, Rolf, had more practical interests. He had constructed a radio receiver and installed an alarm system in his parent's house. He had also successfully changed a hard disk in his PC, and he knew that a large company nearby was offering high salaries for electrical engineers. So he decided to enter the local university to major in electrical engineering.

Their high hopes were damped, however, by the requirements for both of these majors. First of all, neither of them found courses on the ideas that interested them. Instead, Andrea found that she had to take large introductory lecture courses on topics like mechanics, optics, and the theory of heat. The required course on chemistry was even farther from what she thought that she would be studying. Rolf feels the same way about the introductory course in physics. And worst of all, both found that they had to spend a third of their time on courses on mathematics.

This is not what they had expected. To make things worse older students let them know that mathematics is really hard and many students fail because of mathematics. But surviving this hell is a kind of consecration for becoming a genuine engineer or physicist. When I, as a mathematician, talk to some engineers, it takes only a few minutes before they will inevitably report on their experiences in the mathematics courses that they 'enjoyed' as students. Their attitude towards mathematics expresses a good deal of respect – even though they believe that for them most of mathematics is useless. Indeed, to students such a course must appear fairly esoteric: While learning about real numbers and convergence in mathematics, students are already struggling with differential equations and multiple integrals in their other courses.

Derek Holton (Ed.), The Teaching and Learning of Mathematics at University Level: An ICMI Study, 321—334.
© 2001 *Kluwer Academic Publishers. Printed in the Netherlands.*

on top of all this, is useless. Is there any alternative to this scenario? My answer is: No! Because there are unsolvable problems to face.

Physicists and many engineers

- need a wider knowledge in mathematics than an average mathematics student will ever acquire;
- are not primarily interested in mathematics;
- should already have large parts of mathematics at their disposal before they attend the first course on their main subject.

In the following text I first continue with analysing this situation. Then follows a discussion of some general principles which I consider to be important for teaching mathematics to physicists and engineers. The second half of this text discusses some concrete measures for the daily routine of teaching.

2. SOLUTIONS TO UNSOLVABLE PROBLEMS?

These problems are felt by everybody who is involved in teaching mathematics to students of science and engineering, and one can distinguish different approaches.

The workman approach. Many engineers view mathematics as a kind of machine which, when one follows the rules, automatically produces the right answer to a problem. They feel no need for a deeper understanding of this machine which is kept running by the mathematicians. Therefore, some traditional courses on higher mathematics are like training camps for computing the value of an infinite series, for solving integrals by tricky substitutions, for classifying quadratic forms, for solving ordinary differential equations by completely unexpected transformations, and so on. Everything has to be done against a stopwatch, since these skills are tested in a final written examination. In such a course 'mathematics' means 'use the right formula at the right time'.

The gentle approach. Another popular and modern approach to teach mathematics to 'users' is 'to make things user-friendly'. A typical preface to a book following this approach promises to present only the absolutely necessary material. Again, the intention is not to teach understanding. Instead, after some heuristic explanations the results are presented, frequently in coloured boxes, where the colours may even be chosen according to the importance of the result. In such a box, types of conic sections might be listed with an elaborate enumeration. In a book just lying on my desk, there is a box defining the product of two complex numbers of the form $a + ib$ and shortly after there is another box defining (!) the quotient of two such numbers, a third box gives the rule for the n-th power (fortunately, this is formulated as a rule, not as a definition). Since complex numbers are supposed to be difficult, they are introduced only very late. First, all analysis in one real variable is done, including power series. Next, linear algebra and even Fourier series are discussed in this book before complex numbers are introduced. Everything is done in order not to trouble students with difficult mathematics.

There is a danger in trying to make things too easy, which is frequently overlooked: Oversimplification troubles the good students. They sense when something is not correct, but often they are not yet able to recognize what is wrong. They wonder why one should define the quotient of two complex numbers when this follows from the product, but they accept it. They think that they do not understand and become frustrated.

The preparatory approach. Sometimes I am confronted with the idea that all mathematics should be taught *before* students start their first physics or engineering courses. Then a lecturer in physics, say, can refer to all the material whenever necessary. This has the disadvantage that the whole motivation for a student is based on the promise of future applicability; no applications can be discussed in the mathematics course because of the students' lack of knowledge in physics. Such a course fails to relate mathematics to the intended applications. I would rather consider it as an important part of a course in mathematics to demonstrate how mathematics leads to a deeper understanding in other areas.

The unifying approach. At some universities, students of physics join the mathematicians in their first courses in calculus and linear algebra. Again there is the danger that no relation between mathematics and applications becomes apparent, as many students of mathematics don't want to be bothered with more than elementary physics. Indeed, this problem has actually increased in recent years as nowadays most students of mathematics are attracted by computer science rather than by physics. In such a situation an additional course on modelling and applications for students of physics might be helpful.

The optimal approach? The optimum might be to develop an integrated course that teaches mathematics in combination with the intended applications. It should be a fascinating task to do so. However, in most cases developing such a course is so far from reality that I will concentrate on discussing some suggestions that can be realized within the limits of a traditional course.

3. SHAPE AND CONTENTS OF A TYPICAL COURSE

A traditional course might be a course on mathematics for physicists and/or engineers, which runs parallel to their other introductory courses. This type of course will commonly take some 250 to 300 hours of lecturing accompanied by tutorials with exercises. In order to close the time gap between the systematic approach of mathematics and the immediate needs of applications, it can be helpful to offer a few weeks' crash course beforehand. In such a course students can learn the meaning of a partial derivative or of a multiple integral, and other frequently used parts of the mathematical vocabulary. This can provide a provisional basis for following the other courses.

The present discussion grew out of some of my own experiences. I taught some courses on higher mathematics attended by several hundred students of physics and engineering. In addition, I gave a number of more specialized courses for non-mathematicians on subjects like probability theory, mathematical foundations of quantum mechanics, Fourier analysis, and differential equations. The following considerations, however, primarily concern introductory courses on higher mathematics.

When preparing such a course one is faced with a frightening list of topics in mathematics that a physicist or engineer should master. These include calculus in one and several variables, linear algebra, Fourier analysis and integral transforms, ordinary differential equations and dynamical systems, partial differential equations, distributions, special functions, some functional analysis, in particular Hilbert spaces (for quantum mechanics but also for linear systems and filtering), vector analysis, complex functions, variational calculus, optimization, differential geometry, probability and statistics, and finally numerical methods for all the foregoing. In their other courses the students will be confronted with all this difficult material. The list is still not complete and some students will need even more advanced topics like Lie groups and topology (for physics) or algebra and logic (for electrical engineering). On the other hand, nowadays workmanlike skills for solving integrals and differential equations are not as important as they used to be. Instead students need to know how to use (and control) computer algebra systems and numerical software packages for doing these jobs.

In addition, students should be trained in logical thinking. It is my experience that it always takes considerable time before students are able, for example, to formulate the negation of a composed statement (like 'every computer is out of order for at least one day a year'). Students should also get acquainted with the mathematical vocabulary in order to enhance their ability to look up subjects in a book. Moreover, they should develop a sense of structural clarity that enables them to deal with complex situations.

4. GUIDING PRINCIPLES

Learning as many facts as possible on the above subjects seems to be the wrong strategy. Given the limited time, it will be impossible to acquire enough information. Instead, contrary to some of the approaches outlined above, I want to ground the discussion on the following thesis:

The most efficient way to learn mathematics is to understand mathematics.

Understanding mathematics is demanding, especially for students, who never wanted to study mathematics. However, in the long term, instead of trying to collect mathematical facts piece by piece it is more efficient to acquire a basic understanding of mathematics. On this basis additional knowledge can be added in the future according to individual needs. Therefore, as a primary goal a course on higher mathematics should enable students to handle mathematics on their own.

There are many more good reasons, educational as well as cultural, to teach understanding of mathematics. But these reasons are frequently countered by arguments of limited time and the different needs of the students. I want to emphasize that even when there is limited time and differing needs, mathematics is best taught with understanding.

A note in the margin, it is interesting to observe that since the beginnings of modern mathematical teaching around 1800 at the École Polytechnique at Paris there have been demands for strengthening the understanding of mathematics instead of training technical skills. And since the beginnings, the practice of teaching remains far behind all theory.

4.1 Concrete and applied versus abstract and pure?

Various approaches to teaching mathematics to students of engineering and science seem to rely on the assumption that in such a course one has to teach concrete mathematics (and not abstract mathematics) and one has to teach applied mathematics (and not pure mathematics). However, making these distinctions should be questioned. The universal applicability of mathematics relies on its abstractness: Going from the concrete to the abstract means at the same time to increase the number of situations where structural parallels can be recognized. It thus increases the range of possible applications. As the other side of the same coin, abstraction decreases the number of facts to be handled separately and thus forms an important part of what is called 'understanding'. Therefore, it is of special importance to let students experience the power of abstraction.

Referring to a frequently used example, mathematics provides a unified approach to seemingly different things like interest rates, growth of strains of bacteria, or the charge of a capacitor. Extracting the exponential function from these different situations is an act of abstraction that should be emphasized in a course. In the same spirit: when, after having discussed diagonalization of matrices and their Jordan form, one nevertheless solves an n-th order linear differential equation and after that a linear system of first order differential equations (both with constant coefficients) by assuming a certain exponential form of its solution, then one has missed an important point. (Conversely, to improve the understanding of these solutions is enough reason to discuss the Jordan form of a matrix.) It is the interplay between abstract and concrete that brings mathematics to life. Making this clear, very explicitly, is an important part of teaching mathematics, in particular to students of science and engineering.

What is applied mathematics? I don't know: Most of mathematics emerged directly or indirectly from applications. For example, Cantor was led to his set theory when studying points of convergence of Fourier series. There are parts of mathematics which traditionally play a role in many applications like partial differential equations and special functions. Other parts of mathematics became important for applications more recently, like martingale theory for financial engineering, modern topology for quantum field theory and solid state physics, or number theory for coding of information and cryptography. Many real world problems require a sort of mathematics that is still too difficult for the mathematicians. There are people who apply mathematics and others who don't; but mathematics itself seems to resist any division into an applied part and a pure part. It follows that students have to be taught mathematics *and* how to apply it.

4.2   What makes it easier

So far our discussion seems to indicate that many students of science and engineering need to spend an unrealistic amount of effort on mathematics. Fortunately, the following principles can lighten their burden:

- understanding foundations and relations is prior to knowing many details;
- being able to use results is prior to being able to prove results;
- being able to check a computation is prior to being able to do all computations by oneself.

Before being misunderstood let me explain in more detail what I mean.

Understanding foundations. There are some basic mathematical definitions, concepts, and results which everybody who uses mathematics should have at their disposal: on the level of calculus I would include the concepts of convergence, continuity and differentiability. In fact, as far as these basic concepts are concerned I would be very firm. I really mean that the correct mathematical definitions, including their understanding, should be reproducible at any time. Similarly, basic criteria of convergence of infinite sums or the form of the exponential series must be known (by heart). To argue that these things can be looked up when needed is like arguing that one could play chess without knowing the rules since they can be looked up. On the other hand, students should be able to immediately derive criteria for maxima of differentiable functions or the formula of the radius of convergence of a power series from their basic understanding (that is not the same as knowing it by heart). But I would not expect them to know Cauchy's generalized mean value theorem or Cauchy's condensation test for series if they are not obvious from an advanced level of understanding.

Using results. Students need to know what a proof is and this has to be trained. Moreover, they should have understood some basic arguments like the derivation of criteria for convergent series from comparing them with the geometric series. However, I would not expect them to know the complete proof of Taylor's formula. But they have to know how to use it. They must know that not every function can be expanded into a Taylor series, that there is a radius of convergence and how to compute it (in many cases looking for the singularities of the complex extension of a function is easier than to use the root criterion), and that Taylor approximation is the best approximation by polynomials of a function and its derivatives at one point but not the best uniform approximation on an interval (here I would include a hint to Chebyshev polynomials and to splines). Finally, I emphasize the fact that determining a Taylor series by computing the derivatives of a function can be the wrong way. In many cases it is easier to derive its form from a few known power series expansions by using uniqueness of the power series (still, in a recent written examination some students preferred to compute the Taylor series of $\dfrac{1}{1-x^2}$ around zero by computing the derivatives).

Checking a computation. Modern computer algebra programs are able to determine more integrals and solve more differential equations explicitly than students could ever do. Thus it is important to know and to train the basic principles of such computations, but it is no longer necessary to acquire great experience in finding tricky substitutions. Instead, one has to be able to check, whether the result is reasonable. For example, one of the standard computer algebra programs 'computes' $\int_{-1/2}^{1/2} \frac{\arcsin\sqrt{1-x^2}}{\sqrt{1-x^2}} dx = 0$. A student should immediately question such a result on the basis of symmetry arguments.

5. TOWARDS THE PRACTICE OF TEACHING

Although it is easy to formulate general principles and guidelines for lecturing, it is much harder to bring them to life. Here are some of my ideas and suggestions for daily teaching. Of course, all teaching depends to a great deal on the personality of the teacher; there is no set of rules that automatically leads to good teaching.

5.1 Taking time for the foundations

One can spend more time on the foundations at the beginning and speed up later. When faced with the enormous amount of material to be covered in such a course, one is tempted to start as quickly as possible. Doing the contrary takes a lot of nerve. But in the long term it pays off when students know the difference between axioms, definitions and theorems, or when they know about relations, in particular equivalence relations. It is not so easy for the students to understand what it means when several conditions are equivalent and lead, when being fulfilled, to a new definition. For example, for a function the ε-δ criterion and the sequence criterion (for continuity) are equivalent and if they are fulfilled, the function is called continuous. Taking enough time for the foundations at the beginning does not mean to start gently. On the contrary, foundations are difficult. As an additional advantage, such a start gives the opportunity to start off with some material that is new to the students: Instead of simply repeating what students already know it is better to begin with material that is genuinely new to them. Otherwise, there is a great danger that students will not realize the point at which they are entering new territory and get lost.

5.2 Teaching paradigm cases

One can teach the material through paradigm cases. For example, instead of discussing all cases of what a continuous function on an open set can do near the boundary, I discuss the case of a bounded open interval and indicate how the other cases are only variations of this. I do the same with one-sided differentiability and improper integrals. Contrary to this approach some books for engineers discuss all cases in a detailed list and treat them all on the same level, without pointing out that they are only variations on one theme. Another good example concerns elementary methods for solving ordinary differential equations of the first order. Many applied books feel obliged to discuss all methods in great detail. In my opinion, separation of variables, linear and exact differential equations, and a few typical examples of substitutions are enough. I would, however, mention names of other frequently used differential equations just to make students feel more comfortable when they are confronted with such equations. A remark in passing, in order to give students a good understanding of exact differential equations, it is essential to have discussed the theorem on implicit functions in detail. Again this is something that is frequently supposed to be too difficult for students in applied sciences, but it finally pays off when they have truly understood.

It is obvious that in all such cases understanding increases the efficiency of learning.

5.3   Glimpsing at further topics

One can take advantage of opportunities to look ahead as early as possible to more advanced parts of mathematics. When determining extrema of functions by using differential calculus, this is the obvious place to say a few words on the calculus of variations. When discussing the second derivative and convexity, one can include some remarks on convex sets and convex functions and give hints to convex optimization, barycentres, and entropy. When introducing differentiability, I like to show a picture of the path of a Brownian particle. It shows that nowhere-differentiable functions can occur 'in nature', and students have heard of Brownian motion and white noise. It takes only a little bit of imagination to add many more examples. Since it is much easier to show such prospects in a lecture than in a book, there are not many books that are helpful in this respect. The aim of these hints could be that at the end of the course students should be able to read the titles of most 'Introductions' in a mathematical library. In addition, such hints can stimulate good students.

5.4   Using the advantages of an integrated course

One can use the unique opportunity of an integrated course to present mathematics as a unity, not as a collection of various seemingly unrelated subjects. Most courses in mathematics for physicists and engineers are integrated courses, i.e., a single course covers everything from analysis and linear algebra to more advanced subjects. This gives the opportunity to emphasize the relations between different parts of the course. Obviously, after having treated diagonalization of linear maps one can continue with linear differential equations. It should become clear that Fourier and Laplace transforms 'diagonalize' the derivative and that this explains why they are so useful. In addition, this point of view makes most formulas and rules for these transformations obvious. They don't have to be learned. Instead they emanate from understanding, since they are obvious for exponential functions. In addition, in an integrated course one can try to work out a few basic ideas that keep recurring. The concept of a vector field can be used in order to present vector analysis, theory of dynamical systems, and theory of complex functions from a unified point of view. The idea of linearization (via the derivative) can serve as a guideline through large parts of calculus. Theorems on inverse and implicit functions, on transformation of integrals or on volume preserving dynamical systems, should be clear for the linear functions before the general case is treated. Also, when prepared suitably, students can recognize some type of argument such as Banach's fixed point principle on many occasions.

5.5 *Getting different points of view on the same object*

One idea that I like to stress during such a course is that it is crucial to develop different ways of looking at the same mathematical object. The usual way of introducing the concept of continuity is to define continuity by, say, the ε-δ criterion, to discuss the concept for a while, and then to prove that there is also the sequence criterion. I think, it is clearer to start by showing that both criteria are equivalent and then to define continuity through these equivalent criteria. They describe different points of view of the same object. Similarly, when introducing holomorphic functions, one usually starts with a definition, say by using Cauchy-Riemann, and afterwards one proves many other properties like complex differentiability, Cauchy's theorem or analyticity. I prefer to formulate a big main theorem that for a function on an open set of the complex numbers with certain differentiability conditions (depending on the level of the course), complex differentiability, validity of Cauchy-Riemann, existence of a local primitive, representability by a power series, validity of Cauchy's integral formula (and perhaps some more), are all equivalent and lead to the definition of a holomorphic function. I follow these points with a discussion about which point of view one can use for which kind of problem. My experience is that such an approach increases lucidity, since important facts are kept together. It also enables the weaker students to keep them in mind. Moreover, students should learn that it is always a good strategy to get different points of view of a problem. In particular, when dealing with mathematical models – and that is what many of them will do – this is an important experience. Wonderful examples from physics are the local and the global formulations of Maxwell's equations or Schrödinger's and Heisenberg's approach to quantum mechanics. The Fourier transform of a function (the 'frequency picture') gives another good example.

5.6 *Presenting the whole truth*

I try to present the whole truth. For instance, the general notion of a vector space is not easy to understand. Mathematicians have been using it for only a hundred years. Therefore, some authors decide to develop large parts of linear algebra for R^n only. But the notion of a vector space is so fundamental that it is worth pointing out from the beginning that a vector space can be formed by the points in R^n as well as by continuous functions, linear mappings, or linear functionals like the integral or the electric field. Again, to do this takes some time since it is difficult to get acquainted with such a general concept. But after having done this, it can improve clarity to point out that the integral is a linear functional and the derivative is a linear map on a vector space of functions. Moreover, having started with the notion of an abstract vector space, it is easier to introduce the representation of a vector by its coordinates with respect to a basis and to change the basis according to needs. Most physicists find tensors difficult since usually they are not told that tensors are elements in a certain vector space (of multilinear forms) which can be represented by its coordinates, when a basis is given. Again this can be the subject of a quick glimpse at an early stage of linear algebra. Another example is complex numbers. I would introduce them at the very beginning of such a course and do everything with complex numbers, whenever it makes sense. It would not be the whole truth to discuss polynomials or the eigenvalue problem without complex numbers.

5.7 Teaching on different levels

Some of the above suggestions should be helpful for all students, some for the average student, some only for the gifted ones. Thus it is important to keep in mind, that teaching mathematics – as all teaching – should be done on at least three levels: Students should be able to recognize the material that is absolutely necessary (this forms a miminum level). About 95 percent of the time should be spent addressing the average student (the medium level), interspersed with some additional information, hints and prospects for the gifted (the upper level), preferably inconspicuously enough so as not to trouble the others. When preparing a lecture, I always try to keep track of the particular level of the material.

6. MOTIVATING STUDENTS

It is obvious from the foregoing that for physicists or engineers, learning mathematics requires an enormous effort. Therefore, it is essential to motivate students to invest a lot of work in such a course. The best way of doing this seems to be to convince them that mathematics can help them in understanding their own subject. It easily happens that there is a huge gap between the approach in the mathematics course and the way they are using the material in the other courses. Students need some help to find the bridge between the two worlds. Usually such help is not given in other courses, therefore it has to be given in the mathematics course. The following suggestions might be helpful.

6.1 Keeping in touch with parallel courses

I always try to be informed about what students are currently doing in the other courses. From time to time there is the opportunity to copy a page from the current script in experimental physics or electrical circuits and to demonstrate to the students that they can now (hopefully) understand better what is written there. In most cases a translation into the other language is necessary before students can recognize that one is discussing the same subject. For example, the representation of oscillations by complex numbers and the usage of complex amplitudes is not always easy to understand from a physics textbook. It often appears to the students as a mystery that this representation works when they use it according to certain rules, but they cannot understand its meaning. For example, one can start with explaining how any linear time invariant system reacts on an input of a complex harmonic and how this reads, when it is decomposed into its real and imaginary part.

6.2 What is a differential???

A permanent source of troubles is the usage of differentials like dx or df, which are usually interpreted as infinitesimals (for example when used in the chain rule). A clean introduction of the derivative, also in higher dimensions, and its interpretation as a linear map that approximates (up to a constant) a given function near a point, can help a lot. As above, it is primarily the good students who are puzzled by the misuse of differentials. If left alone, they get accustomed to living in two different worlds, where these symbols are handled according to different rules. One should take a few pages of a physics book and have an exemplary discussion on how to read them. A chapter on phenomenological thermodynamics is a bottomless source for such a discussion.

The aim of such a discussion should not be to ridicule 'faulty mathematics'! To a physicist's eye a differential means something different than to a mathematician's eye. Thinking in terms of small increments can be a good approach for a discussion of some physics. Thinking in terms of linear maps helps to understand the structure of the situation and thus can help to make a complicated situation easier to understand. It is important to use both languages and to know how they are related to each other. To make this clear should be a goal of such a discussion. Moreover, one can use this discussion as another opportunity to point out that there are different ways of 'seeing' a derivative.

6.3 Being open to students' questions on other courses

Being open to students' questions on how to understand some mathematical discussion in another course is always a good idea. It helps to get a feeling for the problems that students have in closing the gap between the mathematics course and the usage of mathematics in the other courses. If it is obvious that many students have the same problem, it even makes sense to interrupt your own course for a moment in order to answer such a question for the whole audience. Such a discussion can further help to convince students of the value of mathematics. For instance, when I was in the middle of linear algebra I was told that students had tremendous problems with Green's functions in some other course. I interrupted my course for about half an hour, reminded them of the fact that differential operators are linear mappings (I had already pointed out this at the beginning of linear algebra), made it plausible that 'δ-functions' are something like a canonical basis for vector spaces of functions (this was the right time for a hint on distributions), and told them that the inverse of a differential operator can sometimes be written as an integral operator, i.e., as a kind of a continuous matrix with respect to the basis of 'δ-functions'. (For this discussion it was very helpful that I always emphasized that there are abstract vectors and abstract linear mappings, which can be represented by coordinates when a certain basis is given. Taking time for foundations and fundamental principles, even if they are 'abstract', pays off!) The effect was that good students could understand what I said (on the level I did it), many students got an idea of what was going on, and all students got the feeling that I was taking care of them and that mathematics is not as unrelated to their other courses as it seems. The time was well spent.

6.4 Helping to understand physics by mathematics

Sometimes mathematics can help students to understand physics even better. (The following example is meant as a hint for the good students.) In three-dimensional space, points, linear forms and two-forms form three-dimensional real vector spaces whose elements are not distinguished in most physics books. However, an electrical field is, by its physical definition, measured by the work done on a probe when moved along a path, so it is a one-form, i.e., the vector of an electrical field represents a linear functional on the physical space. Similarly, whenever a quantity is physically defined as a flux per area, such as the magnetic induction B, it is a two-form. Thus, the vector of the magnetic field acts as a linear functional on orientated areas. In order to further convince students of this distinction it is very instructive to let them observe in a physics book which quantities are acted upon by the operators *rot* and *div*, respectively. Such a hint can be given in a few sentences.

To give another example, to many physics students it seems difficult to understand the meaning of a canonical transformation in classical mechanics. But it is easy to help them. As soon as one has discussed the gradient field coming from a scalar function, one can include a remark on how to obtain a Hamiltonian vector field from it (in even dimensions), and how a canonical transformation commutes with forming the Hamiltonian vector field. To explain what this means (and how it is written out) takes ten to twenty minutes but it saves the physics students much more time. And once more it convinces them of the value of a systematic mathematical approach.

7. ACTIVATING THE STUDENTS

So far, we have discussed what a lecturer can do. However, the students must also become active. The principle 'learning by doing' certainly applies to mathematics. Here are some suggestions of how to make students become more active.

I am convinced that physics students and students of engineering have to deliver **written solutions to exercises**, which are then corrected by a tutor and discussed in a small group. It is important to learn how to formulate a simple mathematical argument and how to do a complete computation. According to my experience it takes a year of weekly exercises, before the solutions improve considerably. Among the exercises there should always be examples which show how the material can be applied to real world problems. Again, if one refers in some problems to their other courses one can thereby increase the students' motivation.

Students should learn to solve problems in a **team**. For example, we formed teams of up to three students and encouraged them to solve problems in this team. In many cases, years later the students prepared their final exam still within the same team.

In addition, students should learn to **talk about mathematics**. For example, they can present their written solutions to a group of other students. The tutors are asked not only to check the mathematics, but also to help with the presentation.

From time to time one can include problems that force students to go to the **library** and to find something in a book. For example, the instructor can point out a definition that was deliberately omitted in the lecture or a theorem that is not contained in every standard text. Most students find it difficult to use a mathematical library, but being able to do so could be crucial in their future life.

Another valuable type of exercise is: 'Browse through your book or script on, say, physics, and **find examples in your book** of the use of determinants/total derivatives/contour integrals...', depending on the subject currently discussed. Again, this can help to relate mathematics to the other courses.

Recently we started a new experiment: Each week different teams had to prepare a 15 to 20 minutes **report** on a given subject, concerning the material of the last few hours of the course. The idea was to let the students think about what was really important and to repeat this. Typical subjects were 'Changing the basis: What is important and how can I keep it in mind?', 'What do I have to keep in mind on the diagonalization of matrices?', 'Why are continuous functions important?', 'Illustrate the total derivative of f: $R^m \to R^n$ by examples for some interesting values of m and n'. After some resistance, students started to like this opportunity to recapitulate the most important parts of the course.

To bring these ideas to life it is essential to have good tutors. They should be familiar not only with the mathematical material but also with the philosophy behind such a course. On one occasion I organized a course for my tutors to discuss goals and ways of teaching before we started with the course on higher mathematics. This turned out to be very useful and did improve the tutors' teaching.

8. THE STUDENTS' REACTION

As indicated above, I have given several courses on higher mathematics, where I tried to bring the above ideas to life. Preparing such a course is very time consuming and I succeeded only in part. However, efforts pay off.

At the beginning, students expected a very different course, and I first had to resist many of their complaints. When comparing my course with other similar courses I seemed to be slower and I spent more time with mathematical arguments.

In addition, I did not refer to one complete accompanying book or script (but to many references). Therefore they had to attend the lecture and could not defer learning mathematics to some indefinite time in the future. Moreover, through written solutions to exercises, preparation of their own small lectures, and intermediate tests, they had to keep up with the material. Altogether, they had to spend more time on mathematics in my course than they would have spent in another course. But as time went on I could convince more and more of them of the value of this approach. Having an enthusiastic team of tutors and assistants was fundamental to the final success of these courses. If tutors are not convinced of the value of such an approach, there is little chance of convincing the students.

Students finally noticed that they were really learning something useful. In the end I got very good evaluations from the students for these courses and I was even awarded prizes for good teaching. A typical comment from the evaluations was: "Mathematics was the hardest of my courses but it was the course I liked most". Even students who were really struggling with it emphasized later that it was most useful for them to have such a sound basis in mathematics. Mathematical language in other courses no longer distracted their attention from the main topics. The final examinations confirmed my impression that many students had made major advances in their understanding of the material.

9. CONCLUSION

In my experience there are encouraging signs that it is possible to develop successful courses in mathematics for scientists and engineers. These courses do not necessarily solve the unsolvable curricula problems, but they are valuable for the students. There is no need to give up the idea of teaching the understanding of mathematics. The students will experience it as demanding but also as motivating. I have given some suggestions that might be helpful in designing a course of this type. There are, however, many other approaches. I thus wish all lecturers the courage to develop their own visions and the strength to carry them through.

ACKNOWLEDGMENTS

Writing this article was an adventure. I benefited from many people's encouragement, advice and opinions. I would like to mention N. Gebhardt (Stuttgart), D. Holton (Dunedin), D. and J. Hoy (Santa Cruz), F. Karsten (Stuttgart), K. Kirchgässner (Stuttgart), U. Kirchgraber (Zürich), R. König (Reutlingen), C. Köstler (Stuttgart, Kingston/Ontario), and H. Maassen (Nijmegen). Several generations of students were exposed to my attempts to teach, based on the method of trial and error; their reactions were fundamental for the present text. It is a pleasure to thank all of them.

Burkhard Kümmerer
Mathematisches Institut A, Universität Stuttgart, Germany
kuemmerer@mathematik.uni-stuttgart.de

JOHNNY T. OTTESEN

DO NOT ASK WHAT MATHEMATICS CAN DO FOR MODELLING

Ask what modelling can do for mathematics!

1. INTRODUCTION

Frequently people discuss which parts of mathematics students should learn to become good 'scientists' in various disciplines. However, this is not the principal question one should pose. A more fundamental question is which competencies related to mathematics are of benefit to these disciplines. More specifically, since mathematics is applied to other disciplines through mathematical modelling, the question should be which competencies related to mathematical *modelling* are of benefit to these disciplines. This is not an easy question to answer. Indeed, to obtain an answer one has to be familiar with which competencies the teaching of mathematics, and especially of mathematical modelling, may serve to develop, and to possess a deep insight into the various disciplines that may benefit from such competencies.

However, to ask what mathematics can do for modelling in some domain is certainly interesting, but it is not the only interesting question related to the interaction between mathematics and the real world. In this paper, emphasis is on the reverse question, what can modelling do for mathematics and its teaching and learning? The first answer may be that modelling may motivate some students. Of course, this is true, but as we shall argue below, it is not the whole truth, nor the most important truth.

In section 2, the first question "what can mathematics do for modelling?" is discussed rather briefly. The second question "what can modelling do for mathematics?" is discussed in section 3. The discussion of the first question serves to put the second and, in this paper, the main question into perspective. In section 4 a summary is given. Of course the programme outlined here is very ambitious, thus this paper should be seen as an attempt to initiate a debate rather than as an exhaustive analysis in its own right.

Before continuing, it is important to comment on some central concepts used throughout this presentation. The terms model and modelling are used in the sense of *mathematical* models and *mathematical* modelling, which implies that mathematics and mathematical competencies are involved to some degree. Another keyword is 'competency', which is used to indicate a set of intellectual and other mental powers related to abilities and skills, that may well involve subject matter

335

Derek Holton (Ed.), *The Teaching and Learning of Mathematics at University Level: An ICMI Study*, 335—346.
© 2001 *Kluwer Academic Publishers. Printed in the Netherlands.*

knowledge but involves much more than that. The notion of competencies is used in contrast to a usual mathematical syllabus, i.e. a list of mathematical topics and methods known from most mathematical textbooks. A further specification of this term will appear in section 3 but until then the definition just given will suffice.

2.   WHAT CAN MATHEMATICS DO FOR MODELLING?

As an associate professor in mathematics I became professionally involved in mathematical modelling in physiology several years ago. During this involvement I became more and more interested in examining precisely how mathematics does contribute to furthering knowledge and development in other disciplines such as physiology. In my opinion, by and large the primary goal of university education is to develop the students' intellectual powers to (or above) the level of most academic researchers. Thus, the analysis of what mathematics can do for modelling may be applied directly, essentially without changes, to analyse what the teaching of mathematics can do for other disciplines that apply modelling at the university level. Hence, I will solely address the former issue here. To examine how mathematics contributes to the furthering of knowledge and development in physiology is indeed more important than ever. While theoretical physics has long enjoyed the status of a well-established discipline, mathematical biology, for example, is far from having gained such universal esteem. Despite this relative lack of status, important and meaningful biological experimentation can rarely be performed without theory and concepts that permit hypotheses to be formulated by means of experimentally testable mathematical models. Nonetheless, most biologists still believe that mathematical models are of no use in their disciplines. Non-mathematicians applying mathematics in their field have, in general, different perceptions as to how mathematics interacts with their discipline from mathematicians. More specifically, they do not necessarily have a single perception, but the ones they have surely deviate from those of mathematicians. These perceptions influence their conceptions of how mathematics should be taught, and thus of which competencies students should learn. In my experience, it is rather common to view disciplines other than one's own, as disciplines of service, which should be taught rapidly and by rote as a technique, assuming that it could be obtained by students in some intravenous way.

However, recently some interest has been shown in the development of theory and mathematical modelling by scientists in biology and medicine, especially in physiology. Leading biological journals have actually begun to publish papers in which the identification and study of mechanisms that govern the functioning and the dynamics of biological systems are based on rigorous mathematical modelling, see P. Tracqui, J. Demongeot, P. Auger and M. Thellier (2000). The reason for this growing interest is not just the rapid spread of fast computers which make access to simulations of complex systems easier. What has also contributed to the increasing interest in modelling has been the increasing availability of continuously sampled clinical data. From this, new insight into the dynamics of physiological systems (and not only into their steady state behaviour pattern), can be gained.

Yet another and even more important factor is the attempt to focus on precise definitions of physiological concepts in order to avoid ambiguity, confusion and misunderstandings, see Ottesen and Danielsen (2000). When building a model, a physiologist is often forced to make specific unambiguous statements about relationships that previously might have been only loosely collected thoughts. This challenge sharpens the thinking, often requires a thorough review of the literature, and reveals areas in which quantitative information is missing. Once a mathematical model has been constructed, its predictions can be used to identify and design more meaningful laboratory protocols and to avoid protocols that have little overall value. Furthermore, mathematics may provide a tool for structuring ideas and thoughts, a point that is gaining increasing attention these years.

Frequently mathematical models of physiological hypotheses are preferred to verbal descriptions, and not only in order to clarify thoughts as stated above. For example, physiological homeostasis usually involves many different biological control and feedback processes that occur simultaneously and often at various time scales. The sequential nature of ordinary language is far inferior to a mathematical description when it comes to this simultaneity, and to the course of biological events that are aggregate results of many parallel and simultaneously occurring processes, see Ottesen (2000a). Many of the most important attributes of biological systems, such as non-linearity, redundancy, and hysteresis, cannot be properly described in verbal terms. In contrast, a mathematical description of concepts such as these is often both compact and precise, see Ottesen, (1997a, 1997b).

In addition to its scientific significance, the importance of the interdisciplinary field of mathematical physiology, which is based on mathematical models, has received growing recognition as an applied mathematical activity in modern industry, see Ottesen (1997c, 2000b).

An extended analysis of the reflections just presented, which serve as a basis for the discussion in this section, will appear elsewhere, Ottesen (in preparation). Let us summarise the elementary yet non-trivial findings from this analysis in a condensed form.

- Models may be used to falsify hypotheses. By means of models the range of concepts and underlying assumptions may be tested.
- The use of models may serve to reveal *ad-hoc* elements, assumptions, and vaguely or even ambiguously defined concepts.
- Models may improve the structuring of ideas and thoughts, and increase the understanding of the system being modelled. Moreover, the influence of various parameters on the model system may be investigated along with a corresponding interpretation thereof with respect to the system being modelled.
- The use of models may resolve conter-intuitive observations and paradoxes, and give insight into the causal functioning of the system modelled, with particular regard to cause and effect relationships.
- Models provide the possibility of asking not only new questions but also *new types* of questions that could not be stated without the involvement of models.

Furthermore, new useful concepts may be proposed and defined, and ideas and thoughts may be tested.

Caution has to be taken when applying mathematics to the world outside mathematics. Not everything can be put into a mathematical framework in a meaningful manner. Terrible examples appear in many fields, for example in psychiatry, see Rodin (2000). A very common mistake is to use descriptive models involving many parameters that often cannot be estimated in a sensible way, or do not have a well-defined interpretation in the reality domain at issue. These flaws are serious, especially when models are applied in the making of decisions with a great impact on people or society. Some examples are given in Niss, Blum and Huntly (1991). However, the misuse of models is beyond the scope of this presentation and will not be pursued further.

In my opinion, the above points, mainly referring to the research process itself, may essentially be transferred without change to what students should be exposed to during their education. The knowledge of what researchers from other disciplines in principle (should) demand – more or less consciously – from the teaching of mathematics may serve as a background and a perspective for analysing what modelling can do for the teaching of mathematics itself. It follows from this and the next section that the points of views of mathematicians and non-mathematicians applying mathematics, need not be mutually contradictory but may well complement each other in some symbiotic way.

3. WHAT CAN MODELLING DO FOR MATHEMATICS?

A few years ago I became involved in designing and developing a special first-year course (study unit), called BASE (which stands for Basic Analysis, Modelling and Simulations), for the multidisciplinary two-year basic programme in the sciences at my university. Participants in this course are mainly students whose primary interests are not in mathematics or physics. Two thirds of the science and maths students at our university are supposed to take the course. A faculty member is solely responsible for a single class consisting of a maximum of 25 students. The faculty takes care of the teaching as well as of evaluating the students' homework etc. In class, emphasis is put on discussing the goals of the course as a whole and of each component in particular. Similarly, a lot of emphasis is spent on guiding the students individually regarding which competencies they should seek to improve and how this may be done. Hereby the faculty gets in rather close contact with the students. It should be emphasised that in this presentation the BASE course serves merely as an example, from which more general conclusions will be drawn. Likewise, my judgement of the students' cognitive levels and development is not a result of systematic investigations but is, rather, based on concrete experiences in teaching this course (as well as numerous other courses). Instead of basing the course on a fixed mathematical syllabus, a list of competencies, necessary for someone who is to apply mathematics to another scientific field, such as biology, chemistry or geography, was produced. As mentioned in the introduction, the term 'competency' is used to indicate intellectual and other mental powers related to abilities and skills. Based on this list of competencies, a huge number of smaller real-life problems and about five two-week projects using realistic data were developed for the course. Here the term 'two-week projects' (not to be confused with the larger one-semester projects at our university, described by Mogens Niss this volume, pp. 153-166) denotes the investigation of a partly open but limited problem. Most of the smaller problems deal with the mathematization part of the modelling process (i.e. the translation of verbal descriptions into mathematical representations), together with an analysis of the resulting mathematical system, and a discussion of both. The two-week projects are on modelling at large. Hereby, the students are supposed to develop motivation, realising that they will benefit from such a course by obtaining an idea of how and why they themselves can and should use mathematics. Lecture notes give the necessary mathematical background, thus making the course material largely self-contained.

The substance of the competencies mentioned above is drawn from an expanded list due to Mogens Niss, among others, see Niss (1999) (in Danish only). In compressed and translated form it reads: mathematical way of thinking; problem posing and solving; modelling; mathematical representation; mathematical formalism and symbols; reasoning; communication; and IT-competencies.

Since I am mainly dealing with modelling competencies here, I shall restrict myself to elaborating on this point. The modelling competency may be subdivided into components. They consist in the ability to:

- analyse the foundation and quality of given models, and estimate the validity and scope of such models;
- structure the field or situation to be modelled;
- mathematise, i.e. translate extra-mathematically given problems, objects and relations into mathematical representations;
- handle models, including solving the mathematical problems arising in models;
- validate models, internally as well as externally;
- reflect on, analyse and exert a critique of models both in their own right and with respect to possible alternative models;
- interpret, i.e. translate from mathematical to extra-mathematical statements, and decode elements of models and results in relation to the field or situation under consideration;
- communicate about models and results based on models;
- control the entire modelling process, i.e. to master all of the above elements actively on the basis of an insight into the interrelationships between the various elements involved.

Of course, not all of the individual elements of the previously mentioned extensive and ambitious list of competencies enjoy the same degree of attention in the BASE course all the time. Besides, it seems naïve to believe that the students can fully achieve those competencies or that these will all be achieved at a uniform level in just one year. Nonetheless, all competencies are frequently discussed in the course. However, because of the emphasis on problems and two-week projects, students have a tendency to focus mainly on those competencies closely related to modelling.

Selected experiences from the competencies-based BASE course that are hoped to be of general interest are described in the examples below. In these examples many if not all of the general competencies mentioned above are involved. However, those through which students experience a deeper understanding of mathematics or of the strength of mathematics attract the greatest attention.

Example 1. Interpreting the concept of derivative. In high school, students learn about the concept of the derivative of a function of a real variable, usually based on the familiar analytic and geometric constructions. For various reasons, some students were not able to understand this concept sufficiently well. So, in the course, we redo this introduction with emphasis on a discussion of the interrelations between the analytic and the geometric approaches. Hereby students become better able to find the derivatives of elementary functions in the usual operative manner. Before the students 'are allowed' to go on and use this concept, various possible ways to read 'verbal meaning' into the concept are discussed. While students often do not have severe difficulties in grasping the individual verbal statements as such, they can be expected to have a hard time understanding the limitations in the *interpretations*, i.e. various translations from a mathematical to an extra-mathematical form. Surprisingly, most students in the BASE course have a very hard time understanding the interpretation of mathematical statements. For example, given the derivative, $y'(t) = kt^2$, of the function y, where k is a constant and t the time, the following interpretations may be relevant:

- The function y *changes* at any given time t with a *speed* proportional to the square of that time.
- At any $t = t_1$, the *difference quotient* of the function y tends to k times t_1 squared.
- The *derivative of the function* y is equal to k times the function $f(t) = t^2$.
- The *slope of the tangent* to the graph of the function y at any point is k times the square of the time.
- *Locally* near a point $(t_1, y(t_1))$ the graph of the function y may be *approximated* with the straight line of slope $y'(t_1)$ passing through the point $(t_1, y(t_1))$.

The given class exercises, reveal the students' lack of making such active interpretations in various cases. Even more troubles arise when the students have to translate *between* the various interpretations. This seems to be very provocative and usually brings the students into a state of 'constructive astonishment.' Those students who do overcome this interpretation obstacle seem to use the statements to read sense into the mathematical concept and its formulation. After a while these students develop an understanding of the mathematical concept itself, apparently without referring to the interpretations anymore. They somehow use the interpretations as a 'tin opener' to the mathematical concept. My conclusion from the course is that students who have not gained the competency of interpreting do not understand the fundamental idea of the concept of derivative, and so they will not be able to translate between this mathematical concept and corresponding aspects of the world not formulated in mathematical terms. Focussing for a while on those competencies which relate to interpretation may help remedy, to some extent, the incompleteness of their understanding.

Example 2. Understanding the concept of derivative. When re-introducing the derivative of a function of a real variable, in the manner outlined in example 1, the aim is to help the students to acquire a deeper understanding of the concept. In that way, students are expected to understand the concept so as to be able to use it within, as well as outside of, mathematics. However, in the BASE course most students, to begin with, learn the concept of derivative by heart. Thus, they are not quite able to employ the concept beyond the types of examples studied in class. By means of algebraic operations the students can find the derivative of a function, such as $f(x) = 3x^2 + 7x - 13$, and can relate the results to a geometric point of view. The derivative, $f'(x) = 6x + 7$, is generally obtained by the students by means of rules and procedures, and totally without further reflection. Severe problems then arise for the students when they are to use the concept in mathematization and modelling. In one of the problems, students are told that arteries of the human body can be thought of as elastic tubes filled with blood. Furthermore, the volume (V) of a segment of such an artery is often assumed to be proportional to the pressure difference (P) across the wall of that artery, $V = cP$, where c denotes the constant of proportionality. Also they are told that the flow out of the right end of the segment is supposed to be proportional to the pressure inside that segment, assuming that the flow out of the segment goes into a domain of approximately zero pressure. The situation of no flow in or out of the left end of the segment is considered. The students are then guided to find an expression for the time-derivative, $V'(t)$, of V as a function of time. The students usually solve this part of the exercise (albeit not easily for this category of students) as follows: the change of V per time unit is given by the flow into the segment minus the flow out of the segment. Thus for some constant k_1, $V'(t) = -k_1P(t)$, due to the assumption made. Thereafter, it seems to be intellectually very difficult for most students to get the idea of substituting P by $\frac{1}{c}v$ on the right hand side to obtain $V'(t) = -kV(t)$, for some constant k. Next, the students are asked to give an expression that describes the derivative of the pressure, P, in the segment. Indeed it is a point here that no strategy for dealing with this sub-question is shown in advance. Serious problems then begin to occur. The students take a lot of different approaches but until now none of them has been very efficient. It turns out that there are at least three obstacles. Firstly, the students are not able to deduce themselves that $V'(t) = cP'(t)$ from the explicitly given equation, $V = cP$. Secondly, they are not able to combine this expression with the earlier one, $V'(t) = -kV(t)$. The first part is very surprising, since the students do not have any problems in finding the derivative of functions like $f(x) = cx$, $f(x) = cx^2$, or even $f(x) = cg(x)$, where $g(x)$ is another function. Thus this suggests that differentiation is done procedurally by heart and is not based on any understanding of the derivative as a mathematical object. The second part is supposed to simply be due to an insufficient breadth of view and to inexperience. Of course, this may also be a factor concerning the first part of the obstacle. Thirdly, an insufficient insight into the meaning and the idea of an equation is without doubt present as well. This third part of the obstacle is closely related to the competencies of mathematical representations and that of mathematical formalism and symbols. So the students are not able to fully understand symbolic representations or to manipulate them in fairly elementary situations. After having realised this, students are generally willing to work to get rid of these obstacles in a conscious way, a precondition for gaining a deeper understanding. Finally, it should be emphasised that students capable of taking an alternative approach are encouraged to do so if possible. For example, one could imagine that some students were able to solve the differential equation $V'(t) = -kV(t)$ and afterwards use the equation $V = cP$ to obtain a solution for $P(t)$. However, this happens only rarely with this group of students.

Example 3. Enrichment of the concept of steady state solution. In introducing a single ordinary differential equation, compartmental terminology is used. The idea is to keep account of a certain quantity under study over time, say $V(t)$. The time-derivative, $V'(t)$, of the quantity may be interpreted as the change of the function $V(t)$ per time unit. If $I(t)$ denotes the amount of the quantity flowing into the compartment (account) per time unit at time t and $U(t)$ the amount flowing out per time unit at time t, the net flow into the compartment per time unit at time t equals the change of the quantity $V(t)$ at time t, $V'(t) = I(t) - U(t)$.

There exists a constant solution, V_0, to this differential equation if and only if $I(t) = U(t)$. This does not usually cause trouble for students' understanding of the model. However, when allowing inflow, I, and outflow, U, to be functions of $V(t)$ troubles do appear. A steady state solution $V(t) = V_0$ of the differential equation exists if and only if $I(V_0) = U(V_0)$. One might say that with respect to a steady state, explicit time having been left out of the problem, the relevant dependency is how $I - U$ behaves as a function of V. Some students are not ready to climb over this obstacle when it is presented to them for the first time. Thus they are able neither to discuss stability of steady state solutions nor to carry the analysis further. The following example may open their minds. Consider the average volume over time (for example during the last 24 hours) of liquid in the human body. Then it seems reasonable to assume that U and I are continuous functions of how much liquid, V, the body contains. After a little thinking it becomes meaningful to state that V does not change in time if and only if $I = U$. If the students do this reasoning while drawing graphs representing I and U the general idea gradually becomes clear. In this way some students verbally interpret, read sense into, or visualise whole mathematical statements or ensembles of concepts. This interpretation differs from the earlier interpretation, discussed in example 1 and 2, due to the higher level of complexity of the subject under study. Hence, these students are able to overcome some of their conceptual obstacles by means of modelling, not primarily because they were motivated but because an applied point of view opened their minds to the meaning and significance of mathematical concepts. Other students who, in the beginning, thought they understood the concepts of steady state solution had to revise their perception based on the very same example. So, the point is not that modelling is a shortcut to understanding mathematics, on the contrary, it is a hard yet fruitful way for many students if not for all to achieve a deeper understanding. The understanding of the concept of steady state solution for a single differential equation is crucial later in the course when the qualitative analysis of systems of differential equations is put on the agenda.

Example 4. Mathematics as a key to knowledge. In the last two-week project, at the end of the course, students have to set up a model describing the situation of anaesthetising a hospitalised patient. Prior to the project the students are given three pages in which the situation is presented. In compressed form it reads:

> "During anaesthesia patients have to be given different drugs. Some drugs, for example Pancuronium, serve to relax the muscles. Since the heart is a muscle as well, the level of this drug has to be below a certain level. On the other hand, the level has to be above a certain minimum to prevent patient twitches. Pancuronium affects the musculature locally. In summary, the concentration of Pancuronium in the muscles is desired to be kept between two values, c_{min} and c_{max}. Furthermore, the anaesthesiologist often divides the human body into two compartments, the bloodstream and the rest of the body. When a drug is injected directly into the bloodstream by means of a drop, it will be almost uniformly distributed in the bloodstream in less that one minute. The amount of a drug in the bloodstream is easily measured. The other compartment consists of all the various organs, fat, tissue and muscles. It is usually not possible to measure the amount of a drug outside of the bloodstream directly."

The first half of the project is to put up and analyse a model describing an anaesthetised patient. The second half deals with the problem of whether the amount of a drug in the muscles can be controlled solely on the basis of knowledge of the amount of the drug in the bloodstream, and, if so, how can it be done? As we shall see, the second half of the project will be answered by use of mathematics. Furthermore, it can only be answered by use of mathematics. It is even more important to emphasise that the mathematical formulation and analysis related to the first half of the project gives rise to the statement of very important questions that could not be stated without the use of mathematics.

Based on some assumptions, students are able to put up a model relatively easily. It turns out that the concentration of a drug in the bloodstream, c_b, and in the muscles (i.e. the extra-bloodstream compartment), c_m, fulfil a system of differential equations such as

$$c_b'(t) = -(a_1 + a_3)c_b(t) + I(t)$$
$$c_m'(t) = a_1c_b(t) - a_2c_m(t)$$

Here t denotes the time, a_1 describe the rate of the drug flowing from the bloodstream into the muscles, a_2 the reversed rate, and a_3 the elimination of the drug from the body due to urination etc. The amount of drug injected into the bloodstream per minute is described by the function $I(t)$. In addition to putting up the model, students have to perform a qualitative analysis of the system for the special case when the patient gets a certain amount of the drug to start with, while $I(t) = 0$ thereafter. Based on such a mathematical model and the related analysis, it becomes natural to ask the question, is it possible to estimate the rate parameters from measurements and, if so, how could it be done? In fact, this question can only be stated by means of a mathematical formulation, and it is not possible in any meaningful way to state it without access to such a formulation. (By the way, the answer to the question is "yes" and it turns out that it can be done from a single time-series of the concentration of Pancuronium in the blood!) Thus students begin to see that mathematics provides them with a complementary and deeper insight into the subject modelled. Furthermore, they recognise that this insight can be gained only by means of mathematics. Frequently such insight is surprising or even counter-intuitive to students and it may throw them into a state of fascination. Moreover, most students realise that theoretical mathematical concepts and definitions, such as eigenvalues, are meaningful and relevant. Hereby these students also obtain a better overall understanding of the mathematical structure of the problem. Finally, this modelling problem forces the students to relate different mathematical concepts, for example the derivative and a corresponding differential equation, a differential equation and the corresponding compartments diagram, a differential equation and the corresponding eigenvalues, etc. In this way, students develop a stronger understanding of the mathematical concepts involved.

In answering the second half of the problem "can c_m be suitable controlled?" students may be guided to look for the steady state solution $c_m = c_0$ where, for example,

$$c_0 = \frac{c_{min} + c_{max}}{2}.$$

Thus from the second differential equation one obtains

$$c_b(t) = \frac{a_2}{a_1} c_0.$$

Insertion into the first differential equation yields

$$I(t) = \frac{a_2 a_3}{a_1} c_0.$$

Once the steady state level is reached, it may be maintained by injecting the constant value found above. This answer is clearly based on the use of mathematics. However, only a few students are able to handle this part of the problem with confidence.

My experience is that through such a two-week project students become familiar with parts of the real world that are influenced by mathematics, learn to express themselves clearly in mathematical terms, and learn to analyse and solve problems involving mathematics. Hence students begin to see how important it is to possess competencies related to manipulating equations and symbolic representations. Without these competencies they are not able to build a model and subsequently are unable to draw relevant conclusions from it. Furthermore, students learn to ask certain types of questions that can only be *answered* by the use of mathematics, as well as types of questions that can only be *posed* by use of mathematics. Thus the students realise some of the strengths and depths of mathematical formalism and ways of thinking. Thereby their resistance to work with and to discuss fundamental mathematical structures and arguments is significantly lowered. Moreover, the students become able to reflect on the approaches taken. All of the competencies involved in working with this two-week project are indeed closely related to the competency of understanding and manipulating with symbolic representations. In fact, a substantial portion of the competencies mentioned earlier does indeed come into play during this project.

4. SUMMARY

At the beginning of the BASE course, most students tend to believe that mathematics has to be learnt by heart, not by such insight and understanding that generate the mastery of the competencies of mathematics. Often students do not understand the background to, or the basic idea of, the mathematical concepts they are supposed to learn. So they are not able to translate between purely mathematical concepts and parts of the extra-mathematical world not represented in mathematical terms. Focusing on these translation competencies for a while, students' insufficient understanding reveals itself to them, which is a necessary prerequisite for progress. Of course, the students do not become perfect masters of mathematics during this one-year course, but their minds become opened to the thinking and structures of mathematics and to how these may be used within mathematics as well as outside mathematics.

When some students realise that they are experiencing difficulties in modelling, they become able to work on reducing these difficulties in a conscious way, and from here a deeper understanding is often gained. In other words, these students may be able to overcome some of their difficulties through modelling, not only because they are motivated but also because different points of view have opened their minds. In a first approach, these students use the reality being modelled as a 'tin opener' for grasping mathematical concepts. Continuing working with various interpretations and translations seems to result in an understanding of a mathematical concept that goes beyond the original, more superficial one. In roughly the same way, after some time, mathematical statements of higher complexity or ensembles of concepts can be grasped by a model-based interpretation leading to independent mathematical understanding. Furthermore, some students, who initially thought they understood a given concept, have to revise their perception on the basis of modelling examples. The point is not that modelling is a shortcut to understanding mathematics, on the contrary, it is a hard yet fruitful way for students in general to achieve a deeper understanding.

As pointed out earlier, through projects students become familiar with parts of the real world which are or can be influenced by mathematics. A large number of the competencies mentioned earlier come into play during such projects. Students learn to ask certain types of questions that can only be answered by means of mathematics, as well as types of questions that can only be posed by means of mathematics. Also they learn to express themselves clearly in mathematical terms, to analyse and solve problems involving mathematics. Moreover, these competencies are not only essential to students whose main interests do not lie with mathematics. Every student, including those mainly interested in mathematics, should acquire these competencies. Hereby a greater understanding of the role of mathematics in culture and society and of the development of mathematics as a discipline is gained. Hopefully this will also contribute to preventing too narrow-minded mathematicians to graduate from our programme.

In summary, modelling should be taught not only to motivate students, but also because it may help grasping mathematical concepts and statements, as well as serving as an enrichment of the comprehension of mathematical concepts. However, it should be emphasised that this kind of situated learning, on the one side, and learning of well known ordinary structured and conventionally organised mathematics, on the other side, in my opinion should be somehow equally represented in the educational system at all levels. Neither of these approaches should be omitted in the teaching of mathematics. The ordinary, structured and conventionally organised mathematics represents supplementary comprehension, in addition to being suitable for guiding students into various domains and perspectives that they could hardly have achieved otherwise. The two points of views mentioned, that mathematics can do something for modelling and that modelling can do something for (the learning of) mathematics, are not at all mutually contradictory. If both points of view are taken seriously they complement each other in a symbiotic way, thus giving rise to a more complete understanding of mathematics.

4.1 Acknowledgement

I would like to thank Mogens Niss for suggesting to me that I contribute to this book, for his constructive suggestions, and for many fruitful discussions.

REFERENCES

Rodin, E.Y. (Ed.) (2000). *Mathematical and Computer modelling*, vol. 31, no 4-5, pp. 1-225. Special issue: Guest editor Györi, I. Proceedings of the conference on dynamical systems in biology and medicine.

Niss, M. (1999). Kompetencer og uddannelsesbeskrivelse. *Uddannelse*, no. 9, Novenber. København: Undervisningsministeriet.

Niss, M. (2001). University Mathematics Based on Problem-Oriented Student Projects: 25 Years of Experience with the Roskilde Model, this volume, pp.153-166.

Niss, M., Blum, W. and Huntley, I. (Eds.) (1991). *Teaching of mathematical modelling and applications*. New York: Ellis Horwood.

Ottesen, J.T. (1997a). Nonlinearity of baroreceptor nerves. *Surv. Math. Ind.*, 7, 187-201.

Ottesen, J.T. (1997b). Modelling of the baroreflex-feedback mechanism with time-delay. *J. Math. Biol.*, 36, 41-63.

Ottesen, J.T. (Ed.) (1997c). *Surveys on mathematics for industry*, vol. 7, no. 3.

Ottesen, J.T. (2000a). General compartmental models of the cardiovascular system. In J.T. Ottesen and M. Danielsen (Eds.), *Mathematical modelling in medicine*, Studies in health technology and informatics series, vol. 71 pp. 121-138. Amsterdam: IOS press.

Ottesen, J.T. (2000b). Modelling the dynamical baroreflex-feedback control. *Mathematical and Computer Modelling*, vol. 31, no. 4-5, pp. 167-173.

Ottesen, J.T. and Danielsen, M. (Eds.) (2000). *Mathematical modelling in medicine*, vol. 71 in Studies in health technology and informatics series. Amsterdam: IOS press.

Ottesen, J.T. (in preparation). Matematiske modellers betydning i kredsløbsfysiologi – set i et historisk perspektiv.

Tracqui, P., Demongeot, J., Auger, P. and Thellier, M. (Eds.) (2000). *Sciences de la vie, Life sciences*. Tome 323, No 1, pp. 1-144, Janvier.

Johnny T. Ottesen
Roskilde Universty, Denmark
johnny@mmf.ruc.dk

SECTION 5

TECHNOLOGY

Edited by Michèle Artigue and Joel Hillel

KAREN KING, JOEL HILLEL AND MICHÈLE ARTIQUE

TECHNOLOGY

A working group report

1. INTRODUCTION

The technology working group focused on the various ways in which technology can impact upon the teaching and learning of mathematics. As was already underlined in the Discussion Document for this ICMI Study, "Worldwide, increasing use is being made of computers and calculators in mathematics instruction. Much mathematical software and many teaching packages are available for a range of curriculum topics. This, of course, raises such issues as what such software and packages offer to the teaching and learning of the subject, and what potential problems for understanding and reasoning they might generate." The Discussion Document proposed to identify and analyze innovative projects and research that are particularly fruitful for advancing our thinking in this domain.

Reflecting on the impact of information technologies on the teaching of mathematics is not new for an ICMI Study – ICMI had already launched a study in 1985 entitled "The influence of Computers and Informatics on Mathematics and its Teaching". That ICMI Study touched all levels of instruction and underlined primarily the impact of computers on several areas, including:

- on mathematics itself; computers have prompted the revisiting of familiar notions such as number and elementary functions, the revitalizing of old problems, and the emergence of new domains. They have extended the range of applications of mathematics, and have blurred the boundaries between pure and applied mathematics;
- on the notion of proof in view of computer-assisted proofs;
- on the practice of mathematicians; computers have led to an increase in experimentation and the use of simulations. They afford new means of communication and accessing information that affect the way mathematicians carry on their professional lives.

This previous ICMI Study also recognized that despite an abundance of interesting experiences, the impact of technology on teaching was still globally weak, and that the introduction of computers in the classroom had not necessarily

349

Derek Holton (Ed.), *The Teaching and Learning of Mathematics at University Level: An ICMI Study, 349—356.*
© 2001 *Kluwer Academic Publishers. Printed in the Netherlands.*

led to any discernible improvements. The working group discussion focused on the present-day role of technology in teaching at the post-secondary level, on the perspectives envisaged for the future and on broader research questions that are affected by the use of technology. It centred mostly on the use of technological tools for supporting students' learning, particularly via visualization; computation, and programming. But, it also recognized the role of such tools for: demonstration by the teacher; presentation of lessons via distance learning; student assessment; and student drill.

2. TECHNOLOGY AS A MEANS FOR SUPPORTING STUDENTS' LEARNING

At the university level in general, and at the collegial level in particular, the introduction of technologies was seen as a means to renew pedagogical practices and to circumvent a style of teaching that was too formal or too algorithmic. It was intended to create better coherence between teaching practice and the constructivist approach to learning. Celia Hoyles, in her description of the potential contribution to post-secondary education of researches carried in the secondary level, has emphasized that:

"There is considerable evidence of the computer's potential to:

- foster more active learning using experimental approaches along with the possibility of helping students to forge connections between different forms of expression, e.g. visual, symbolic ;
- provoke constructionist approaches to learning mathematics where students learn by building, debugging and reflection, with the result that the structure of mathematics and the ways the pieces fit together are open to inspection ;
- motivate explanations in the face of "surprising" feedback : that is, start a process of argumentation which can (with due attention) be connected to formal proof ;
- foster cooperative work, encouraging discussion of different solutions and strategies ; computer work is more visible and more easily "conveyed" between lecturer and students ;
- open a window on to student thought processes : students hold different conceptions of mathematical ideas which are hard to access, even in the case of articulate adults. How students interact with the computer and respond to feedback can give insight into their conceptions and their beliefs about mathematics and the role of computers."

Hoyles hastened to add that a successful integration of computers necessitates the rethinking of "the content and sequence of the mathematics courses given that students and mathematics have (or should have) changed in the light of the new technology [...] teaching approaches to take into account the broad range of response inevitable in interacting with computers [...] and the relationship of 'computer maths' to paper and pencil maths" (Hoyles, 1999).

The question of what constitutes 'successful integration' of technology to the teaching and learning process was central to the working group discussion. Several presentations by participants on the way in which they have used technology to teach mathematics at the undergraduate level, helped to focus the discussion. These included presentations by: Karen King, on teaching differential equations; Ed Dubinsky, on programming using ISETL; Joel Hillel, on using Maple in teaching linear algebra; and, Rosalind Phang, on using statistical software.

2.1 Changes in the Nature of the Mathematics Taught

King's example illustrated the nature of the changes in teaching differential equations made possible by using a technology that graphs slope fields and direction fields. These enable students to engage in qualitative analyses of previously inaccessible differential equations rather than use traditional analytic techniques. Thus, the focus of a differential equations' course could shift from just finding the solution functions, to graphically organizing the space of solution functions using slope fields and bifurcation diagrams, and to examining the nature of the solution functions (see Rasmussen, 1999).

If one considers, for example, the differential equation

$$dy/dt = 0.3y(1-y/8)(y/3-1),$$

one could attempt to solve this using separation of variables but would not deduce a closed-form general solution. However, with a slope field as shown in Figure 1 derived from a TI-92 program written by King, a student can examine the types of solution functions and their general behaviours, given different initial conditions.

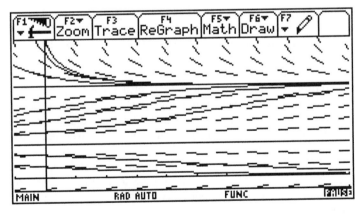

Figure 1 slope field program with slopes and several approximations
dy/dt=0.3y(1-y/8)(y/3-1), y(0)=1, 2, 3, 4, 5, 6, 8, 9, 11, 12

This example provides an instance where changes to an entire course can be made, including the order in which topics are taught and the mathematics with which the students engage (see Artigue, 1992, and Rasmussen and King, 1999). Such changes, in turn, lead to other changes in the curriculum. For example, the study of dynamical systems has been greatly impacted by the availability of computing technology and has resulted in an early focus on systems of differential equations in many courses. This is but one example where a particular mathematical discipline is changed by technology which, in turns, affects changes in the nature of how it is taught.

2.2 The role of professional tools

Secondary schools tend, by and large, to use software products and calculators that have been specifically conceived for teaching. In contrast, universities mostly tend to use professional tools be they general symbolic manipulators (e.g. Maple, Mathematica, MuPad, Matlab, SciLab) or tools for specific domains such as Statistics (APSS, SASS), though some specific educational software such as Geometer's Sketchpad and Cabri are also relevant for instruction at the tertiary level. Faculty members are familiar with these professional tools since they use them in their own mathematical work and, consequently, they tend to be widely available on campus. There is an ever increasing number of texts that integrate the use of a software package, for example, "Calculus and Mathematica" (Uhl, 1999) or "Ordinary Differential Equations using MATLAB" (Polking, 1995). Individual universities have also written primers that bridge between the particular program or technology that they use and the mathematics texts in use in the department (see Colgan, 1999 for a discussion of such a primer from Australia).

Professional software tools are particularly powerful and, at first sight, seem to take full charge of what traditionally has been the mathematics work expected of students. They embody a tremendous amount of mathematical knowledge that, nevertheless, remains invisible and inaccessible to the users. The availability of these powerful tools raised the inevitable question in the working group regarding the necessary mathematical knowledge of users if they are to become efficient and in reasonable control of such tools. These tools also force us to both question and redefine the content of mathematical training, notably in sectors where mathematics is a service course. It prompts us to ask under what conditions can they become means for students to construct mathematical knowledge, over and above their role as powerful computational tools.

In response to the question regarding the necessary skills/concepts that students must possess before they can use a powerful CAS tool, Hillel suggested seizing the 'black box' feature as a learning opportunity. He presented an example on teaching the Cayley-Hamilton Theorem in linear algebra, where students are first asked to use Maple to build inductive evidence for that theorem. By using the software, students can compute the characteristic polynomial $f(x)$ of a given matrix A and then compute $f(A)$. Among other things, it becomes apparent that the result of the computation is a square matrix, not a number, to which students must therefore attend and about which they must be explicit. Students also can explore these computations for several matrices, focusing the process of computing $f(x)$ rather than on the actual computations. Such activities can take place prior to introducing the theorem and its proof in class. This is a pedagogical choice, a kind of 'didactical inversion' which is made within the larger context of the instructor's course design (see Kent and Noss, 1999, and Noss, 1999).

In a slightly different vein, participants also recognized that there are computer applications designed for other purposes that could be mathematically exploited (e.g., Excel). This raised the questions of how one would characterize the difference (for instructors, for users, and for the types of tasks and interactions) in using educational mathematical tools, professional mathematical tools, and the mathematical usage of tools designed for other purposes.

2.3 The role of programming

In the analysis of the potential of computers for mathematics learning, programming has always played an important role. In the early days of computers when tools for scientific calculations were very different from those of today, programming was essential. But even if software packages have evolved, programming can be seen as a means to change students' relation to algorithmic work, so important in mathematics, by putting the accent on the construction of algorithms rather than on their execution. This shift is seen as a way to give sense to both the algorithms and to the underlying concepts.

Dubinsky presented to the working group the use of programming in a function-based program language (ISETL) to facilitate students' learning about functions (see Dubinsky, 1999). Instead of having students use conventional programs, the students write their own. His work illustrates particularly well the conceptual gains that students make when they have to write mathematical constructions as programs. His approach is built on a theoretical model that looks at learning in terms of actions, processes, and objects. ISETL is particularly well adapted for mathematics, since it favours transforming actions into processes and encapsulation of processes as mathematical objects (see Dubinsky and MacDonald, this volume. pp. 275-282).

Programming activities could also be implemented via scripting which is an automatic execution of an often used sequence of commands. Scripting capabilities are now built into many applications, such as Excel. Whether one uses a programming or scripting language, it is important to pay attention to the kinds of instructional tasks that fit well with the language. Tasks that are appropriate to a function-based language such as ISETL, would not be so in other languages that do not operate the same way.

Finally, it was noted that programming can also play a large role in introducing students to the world of algorithms and the concomitant notions of complexity, validity, and efficiency.

3. TECHNOLOGY AS COMMUNICATION

3.1 Old and new frontiers

Many instructors, particularly in the USA, use the overhead projector as a demonstration tool while teaching. Both graphing calculators and computers can be attached to panels, allowing the image to be projected on a large screen. These enable instructors to illustrate some difficult and sophisticated mathematical. (Though sometimes the temptation is also to include a lot of unnecessary distractions, as evident in many slick PowerPoint presentations.) This use of technology is still fairly traditional in the sense that it assumes an instructor in front of a class. But the advent of the Internet has now allowed ideas for distance learning to permeate educational discussions. Already, technology entrepreneurs are offering to fund virtual universities that bring the 'best' courses from prestigious universities to clients around the world. Many universities are also beginning to offer their own virtual programmes and degrees over the Internet, and departments are busy designing courses to present over the Internet.

More modest uses of distance learning opportunities exist. For example, students can explore geometry problems using a Java version of Cabri Geometry (http://www.cabri.net/cabrijava/). Students might also submit certain assessments using the Internet, ask questions through email messages, or receive help through Web sites and chat rooms organized by the instructor.

As technology revolutionizes the means of communication and information, changes in these domains could, in a very short time, exert even more radical forces on the education system. What is at stake is the whole organization of the pedagogical landscape, the role of the teacher, the types of resources available to the learners, and the forms of access of these resources. The text of Lynn Steen (this volume, pp. 303-312) touches on some of these issues. Since very little is known about the effects of these more recent developments, it is not surprising that the working group this raised such questions as: Is mathematics a suitable subject for learning through this medium? How can information, particularly symbolic and graphical representations, easily be transmitted through this medium? What will these 'classrooms' look like and to what types of materials will students be exposed (e.g., the book and all interactions on the Internet, hardcopy book with Internet interactions, etc.)? What we have today is mainly an explosion of activities, innovations, and texts that tend to be descriptive. What is lacking is scholarly work that reflects on the whole nature of the enterprise.

4. CONCERNS AND RESEARCH QUESTIONS

4.1 Concerns

One concern expressed in the working group was that "what students see is not always what is 'there'." The interface between the student and the workings of the computer or graphing calculator will lie at varying degrees of 'transparency' to the student and this transparency may become opaque (see Noss, 1999). Students' understanding of the inscriptions and the underlying processes that lead to 'answers' on the screen can develop reflexively. However, this means that software developers and instructors must pay specific attention to the instruction design and enactment of the curriculum in which these understandings will emerge. More generally, it is important to view the use of technology within the larger paradigm of learning and instruction, not in isolation. The focus is on the teaching and learning of mathematics and technology as a possible tool toward facilitating this end, but not the end itself.

Some concerns raised about writing programs or using professional tools in the classroom were related to the issue of learning of new syntax. Since the learning curve for the syntax is often high, how does having to teach it impinge on other mathematical activities? Can the learning of syntax itself be problematized to present mathematical learning opportunities? There was also a concern raised that unless one uses computer tools frequently, students' ability to use these tools in later studies will be quite shaky. It was noted that problems also arise when a department does not adopt a single software tool for a range of sequential courses, so students are faced with having to learn new syntax every time.

Some ethical concerns were touched upon very briefly. In particular, how do we make sure that technology does not amplify existing inequalities between the sexes, and social classes? How can we make sure that technology becomes a tool for democratization of teaching rather than favouring just those who can tap into the technological culture outside out of school? Some of these concerns are raised in the section by Zevenbergen (this volume, pp. 13-27).

4.2 Research questions

There has been substantial research on using technology in the teaching and learning of mathematics. However, much of this research has been of the form, "What is the effect of calculator/computer use on students' learning of X?" The working group felt that the research base needed to be broadened in order to incorporate many of the issues raised in the discussion. Of particular interest is the nature of the interactions that students have with the technology and the nature of classroom interactions in which technology is a part.

The following is a list of questions, some of which have been discussed in this report that members of the working group found worthy of attention.

- How can technology be used to teach theoretical concepts?
- The genesis of epistemological obstacles is cultural. Are epistemological obstacles going to change now that the culture has new tools?
- How does technology change mathematics (what is considered mathematics, how it is done)?
- Does the existing literature make convincing arguments for using technology?
- How do we characterize teacher-student interactions with technology (the Internet, calculators, computers)?

- Should we focus on the current curriculum and how to integrate technology into it or should we consider what the mathematics curriculum could be now we have technology?
- What do we need to know in order to use technology to teach mathematics well?
- How do we question mathematics in the light of technology?
- What strategies (e.g., starting with a black box and exploring) do we have for using technology to teach mathematics?
- How do we design technology and build it into the curriculum?

REFERENCES

Artigue, M. (1992). Functions from an algebraic and graphic point of view: Cognitive difficulties and teaching practices. In G. Harel and E. Dubinsky (Eds.), *The Concept of Function: Aspects of Epistemology and Pedagogy*, pp. 109-132. Mathematical Association of America: Washington, DC.

Colgan, L. (1999). MATLAB in First Year Engineering Mathematics. *International Journal of Mathematical Education in Science and Technology*, 31 (1), 15-25.

Dubinsky, E. (1999). Writing Programs to Learn Mathematics. Panel Discussion in Pre-Proceedings of the ICMI Study Conference 8-12 December 1998, Singapore, 37-38.

Hoyles, C. (1999). Panel Discussion: Technological Windows on Undergraduate Service Mathematics. Panel Discussion in Pre-Proceedings of the ICMI Study Conference 8-12 December, 1998, Singapore, 39-40.

Kent, P. & Noss, R. (1999). The Visibility of Models: Using Technology as a Bridge Between Mathematics and Engineering. *International Journal of Mathematical Education in Science and Technology*, 31 (1), 61-69.

Noss, R. (1999). Panel Discussion: Technological Windows on Undergraduate Service Mathematics. Panel Discussion in Pre-Proceedings of the ICMI Study Conference 8-12 December, 1998, Singapore, 41-45.

Polking, J. C. (1995). *Ordinary Differential Equations using MATLAB*. Prentice Hall: Englewood Cliffs, NJ.

Rasmussen, C. (1999, April). Symbolizing and unitizing in support of students' mathematical growth in differential equations. Paper presented at the Research Pre-session of the National Council of Teachers of Mathematics, San Francisco, CA, USA.

Rasmussen, C. L. and King, K. D., (1999). Undergraduate Mathematics from an RME Perspective: Differential Equations as an Example. Pre-proceedings of the ICMI Study Conference, 8-12 December 1998, Singapore, 222-225.

Uhl, J. (1999). *Calculus and Mathematica*. Champaign, IL: Wolfram Research Group.

Karen King
Michigan State University, Michigan, USA
kdking@math.msu.edu

Joel Hillel
Concordia University, Montreal, Canada
jhillel@alcor.concordia.ca

Michèle Artigue
Université Paris 7, France
artigue@math.jussieu.fr

JOAN GARFIELD, BETH CHANCE AND J. LAURIE SNELL

TECHNOLOGY IN COLLEGE STATISTICS COURSES

1. OVERVIEW

Although software has long been available for doing statistical analysis, the role of technology in teaching and learning statistics is still evolving. Computers and calculators reduce the computational burden and allow more extensive exploration of statistical concepts. The availability of powerful computing tools has also led to newer methods of analyzing data and graphically exploring data. During the past 20 years, since personal computers became available in homes and schools, the developments in educational technology have progressed at an accelerated pace (Ben-Zvi, 2000).

The types of technology that are now being used in statistics instruction fall into one or more of the following categories:

- statistical packages and spreadsheets for analyzing data and constructing visual representations of data;
- multimedia materials to teach, tutor, and/or test students' statistical knowledge and skills;
- Web or computer-based tools, including simulations, to demonstrate and visualize statistical concepts;
- graphing calculators for computation, graphing or simulation;
- programming languages that students can use to set up more complicated simulations or numerical analyses.

As the capabilities of technology have increased and more software tools have become available, it has become important to consider the most appropriate uses of technology in facilitating students' learning of statistics in different situations. Ben-Zvi describes how technological tools are now being designed to support statistics learning in the following ways (2000, p. 128):

1. Students' active construction of knowledge, by 'doing' and 'seeing' statistics.
2. Opportunities for students to reflect on observed phenomena.
3. The development of students' metacognitive capabilities, that is, knowledge about their own thought processes, self-regulation, and control.

Derek Holton (Ed.), The Teaching and Learning of Mathematics at University Level: An ICMI Study, 357—370.
© 2001 *Kluwer Academic Publishers. Printed in the Netherlands.*

Moore (1997) sees a reform of statistics instruction and curriculum based on strong synergies among content, pedagogy, and technology. He cautions statisticians to remember that we are "teaching our subject and not the tool" (p.135), and to choose appropriate technology for student learning, rather than use the software that statisticians use (which may not be pedagogical in nature).

This paper examines the ways technology is being used in a variety of college-level statistics courses: introductory statistics, probability, mathematical statistics, and intermediate statistics. Although there is some overlap in the types of technological resources being used in these different courses, an attempt is made to isolate the particular types of technology or software that are most appropriate or most used in each type of course. Because of the large number of Web sites referenced in this paper, a Web version of the paper with links to all the sites mentioned can be found at www.dartmouth.edu/~chance/technology.html.

2.   THE INTRODUCTORY NON-CALCULUS STATISTICS COURSE

Like the calculus reform movement in college mathematics programmes, a statistics reform movement has called for changes in the introductory statistics course. Changes are recommended in course content (e.g., more emphasis on real data, less emphasis on formal probability and mathematical formulae), pedagogy (use of active learning, cooperative groups, and real data) and student assessment (use of alternative approaches including student projects, reports, and writing assignments). In addition, instructors are urged to incorporate more technology in their courses, to represent, analyze, and explore data, as well as to illustrate abstract concepts (Cobb, 1992).

The 1990 Mathematics Undergraduate Program Survey (CBMS) showed that the enrolment of students in statistics courses taught in departments of mathematics in four-year colleges and universities and in two-year college mathematics programmes, more than doubled from 1970 to 1990 (from 76,000 to 179,000 students). Recently, many institutions have added the requirement of a data-oriented, or quantitative-literacy course in their core curriculum, increasing the statistics enrolment, so that now more sections of statistics courses are offered than calculus courses in colleges in the USA. The 1995 CBMS study reported 3530 sections of elementary statistics taught in mathematics departments, 820 in statistics departments, and 2566 sections taught in two-year colleges. Numerous elementary statistics courses are also taught in other disciplines, such as psychology, business, education, sociology, and economics.

In a recent survey of the introductory statistics course, Garfield (2000) found that most students in these courses are required to use some type of technology, and most of the faculty surveyed anticipated increasing or changing the use of technology in their courses. The type of technology is often different depending on the type of institution. Students in two-year college courses are more likely to use graphing calculators (for computations using small data sets). Students in four-year colleges or universities are more likely to use a statistical software or spreadsheet program.

2.1   Statistical Software Packages

A variety of different software packages are used in introductory statistics courses, several of which are reviewed by Lock (1993). The most widely used statistical software program in introductory statistics courses is Minitab (Joiner and Ryan, 2000), which is perceived as easy to learn and use, and is available on both Mac and PC platforms. Data are entered in a data window and pull-down menus allow students to easily graph or analyze data. In addition, 'macros' are small programmes that instructors or students can create to generate data or run simulations. Minitab is incorporated into over 250 textbooks directly or as a special supplement and has been used in over 2000 colleges and universities around the world (Shaughnessy, Garfield, and Greer, 1997).

Data Desk (Velleman, 1998) offers a strong, visual approach to data exploration and analysis, integrating dynamic graphics with tools to model and display data. The intent of this program is to allow students to explore data with an open mind, no preconceptions, and no formal hypotheses, in the spirit of detective work. Users are asked to consider the question: "Can I learn something useful?" Data Desk provides many unique tools that allow students to look for patterns, ask more detailed questions about the data, and 'talk' with the program about a particular set of data. Despite the many unique features of this software, it has not been widely used in introductory courses. Data Desk is sometimes difficult for students who are accustomed to a spreadsheet system for data entry, and some instructors do not like the non-standard method of manipulating data.

A newer software package, developed to help students explore and learn statistics, is Fathom. Originally called 'Dataspace', Fathom is described as a dynamic computer learning environment for data analysis and statistics based on dragging, visualization, simulation, and networked collaboration (Finzer and Erickson, 1998). Although originally aimed at middle school and high school level students, Fathom is beginning to be used in college level statistics classes as well.

Although not actually a statistics program, the Excel software, produced by Microsoft, is widely available in most computing labs and on many personal computers. This spreadsheet software has add-ons that allow the software to perform some statistical analyses. However, concerns have been expressed about using this software instead of more authentic statistics software (e.g., McCullogh & Wilson, 1999).

2.2   World Wide Web Resources

In addition to graphing calculators and statistical or spreadsheet software, technology resources available on the World Wide Web are increasing daily. Several Web sites have been established that make links to other sites to assist instructors of introductory statistics courses as well as their students. For example, Robin Lock's Web site organizes and links Web resources by categories and awards medals to the ones he finds most useful or of the highest quality. His categories include:

- On-line Course Materials (e.g., course syllabi and materials for particular instructors);
- On-line Textbooks (some are free and some are available for a fee);
- JAVA applets (a platform-independent Web programming language which is used to develop interactive demonstrations that can be accessed via any JAVA-capable browser). Examples include randomly generated scatterplots from which students are asked to guess correlations, dynamic regression-lines that change when points are added or removed, graphical visualizations illustrating the power of a statistical test when different parameters are varied, and histograms where students manipulate the bin width to see the effect on the overall pattern;
- Data sets or links to repositories of data (such as the Data and Story Library that contains data sets and accompanying stories along with analyses, and government agencies such as the Census bureau, that collect and store data);
- Miscellaneous sites (such as the electronic *Journal of Statistics Education*, sites that consist of links to sample test and quiz questions, or collections of links and resources for teaching an introductory level course).

The development of the World Wide Web has produced unprecedented means for instructors to easily share their ideas on ways to improve the teaching of statistics (Lock, in press). The links on Robin Lock's Web site represent just a sample of resources that are currently available. He encourages faculty not to be daunted by the volume of on-line materials but to search and try out resources that may greatly enhance students' learning.

An example of a statistics course that is based exclusively on the World Wide Web is The Visualizing Statistics project which has developed an on-line introductory statistics course, *CyberStats* (Kugashev, 2000). This course, consisting of over 50 modular units, includes several components: including text material, case studies, Java applets, self-assessment questions for students, exercises, and a glossary. This commercial product is designed to be flexible and can be adapted to different educational settings and courses. In some cases, using these materials has enabled instructors of even large statistics courses to spend more time in class on group work or computer activities (e.g., Harkness, 2000).

Another example of an integrated set of Web materials is the Rice Virtual Lab in Statistics, which is freely available to students. This Web site contains an online statistics textbook (HyperStat Online), links to Java applets that demonstrate statistical concepts, case studies that provide examples or data, and some basic statistical analysis tools. This is a dynamic set of resources, and instructors are invited to contribute links to appropriate pages of the Web site.

2.3　Multimedia Materials

Velleman and Moore (1996) define multimedia as a computer-based system that combines several components such as text, sound, video, animation, and graphics. Since their article appeared, which described the 'promises and pitfalls' of using multimedia in statistics courses, several types of multimedia materials have been developed and are being used in various ways in statistics courses.

ActivStats (Velleman, 2000), is a multimedia resource for students to learn basic statistics. It can be used by itself or along with a textbook. This CD ROM which runs on either Mac or PC platforms, contains videos of uses of statistics in the real world, mini-lectures accompanied by animation, tools similar to Java applets to illustrate concepts interactively, and a student version of Data Desk. Lessons on the CD ROM instruct students in how to use the software, and many homework exercises are included that have data sets formatted for analysis with Data Desk. Other versions of *ActivStats* may be used with software programmes such as Excel or SPSS.

Cumming and Thomason (1998) describe *StatPlay* as multimedia designed to help students develop a sound conceptual understanding of statistics and to overcome misconceptions. These materials consist of dynamically linked simulations and microworlds, 'play grounds' and estimation games, along with recorded mini-lectures or directions on analyzing data or using the software tools.

Another unique CD ROM is the *Electronic Companion to Statistics* (Cobb and Cryer, 1997). In contrast to *ActivStats* and *StatPlay*, this is a study guide for students that provides interactive illustrations and exercises to present examples of the different topics covered. It allows students to explore the relationships between statistical concepts using concept maps, and includes a variety of self assessment items and animations to help students review or better understand important concepts.

Many other multimedia resources are currently being developed around the world, several of which were described at the Fifth International Conference on Teaching Statistics (Pereira-Mendoza, 1998).

2.4 Stand-alone Simulation Software

Some concepts in statistics are particularly challenging for students to learn, such as the idea of a sampling distribution and the Central Limit Theorem. The Tools for Teaching and Assessing Statistical Inference project (Garfield, delMas and Chance, 2001), has developed some simulation programmes that are easy for students to use, and are accompanied by structured lab activities and assessment instruments. Although there are other simulation tools on the Internet or on CD ROM (such as ActivStats), this software is unique in that it provides more detail and flexibility. It allows students to manipulate a variety of populations (based on discrete or continuous data), and draw samples from these populations for different sample sizes. Templates are provided to compare the sampling distribution to the population and to a normal distribution, as well as to calculate probabilities. The tool can be used to illustrate confidence intervals and p-values as well.

The RESAMPLING STATS Software (Simon and Bruce, 1991) is a software package that offers an easy-to-use, powerful tool to conduct repeated simulations (including the bootstrap), calculate test statistics, and analyze and view the results.

2.5 Other technological resources

Two other types of technological resources that are used in introductory statistics classes are video and email. Moore (1993) discusses strengths and weaknesses of using videos in statistics classes. The *Against All Odds* video series produced by COMAP is an excellent set of videos that illustrate real-world applications of statistical topics. A shorter series of segments of these videos were produced and distributed as *Decisions Through Data*. Some of these videos are now included in the multimedia materials described above. Additional videos are described in the section on Quantitative Literacy Courses.

Email is a technological tool that is impacting all courses, including statistics courses. For example, Chance (1998) reported on the use of email in introductory statistics courses to foster outside-of-class communications between the student and instructor and between students. This type of support network appears to be especially relevant in a course such as statistics, which students often enter with much trepidation and unease. Hyde and Nicholson (1998) extend this communication to intercontinental collaboration by linking statistics classrooms in different countries, allowing students to collect and analyze data comparing themselves to their international peers.

3. THE ROLE OF TECHNOLOGY IN A PROBABILITY COURSE

As described earlier, the arrival of the computer in the classroom and the ability to analyze real data in the elementary statistics course has completely changed the way this course is taught in many colleges and universities. This has resulted also in a change in the way probability is taught in these statistics courses. The types of technology now used in these courses simulate simple experiments such as coin tossing, rolling dice, and choosing random samples from a population. Minitab has the ability to write macros to simulate any kind of chance experiment. The new statistical package Fathom can also simulate a wide range of experiments.

Statistics courses have taken advantage of this by having their students learn the basic probability ideas such as the Binomial distribution, the law of large numbers, and the Central Limit Theorem, through simulations. This means that the statistics teachers do not have to spend time on the mathematics of probability associated with the study of combinatorics, sample spaces and formal properties of probability measures. A separate probability course still develops the mathematics of probability but instructors in these courses also take advantage of simulation to help the students better understand theoretical results and to solve problems which do not lend themselves to mathematical solutions.

3.1 Software and programming languages

Teachers of probability courses are beginning to take advantage of powerful symbolic mathematical software such as Mathematica, Maple, and Matlab. However, those who have tried using these packages have had mixed success. Simulating a chance experiment or solving a combinatorial problem to calculate a probability often requires writing a procedure and students find this difficult. They may spend so much time trying to write the procedure, that they end up feeling that they did not learn enough from this effort to justify the time they put in.

One area where the students can appreciate the value of these packages is in the study of Markov chains. The ability to raise matrices to powers and to solve equations saves the students enormous amount of work and makes Markov Chains come to life. Of course the teacher can use these packages to write procedures to illustrate basic probability results. Elliot Tanis shows on his Web site how this can be done for the Central Limit Theorem.

Traditional probability courses rarely deal with real life data in their courses. Perhaps the success of the statistics reform will convince probabilists that it is very natural and rewarding to show students how their probability models fit real data. Technology facilitates the introduction of real data into probability models, helping students become more aware of the role of variation in the real world.

3.2 World Wide Web Resources

The World Wide Web should play an important role in the future of the probability course. The Web has been particularly successful in leading to the development and dissemination of interactive text materials. An interactive probability book by Siegrist at his 'Virtual Laboratories in Probability and Statistics' has applets to go with each topic. Several of the on-line interactive statistics books also have chapters on probability and make good use of applets (e.g., HyperStat by David Lane at Rice University). Individual applets that are particularly useful in teaching a probability course are: Binomial probabilities, Normal approximation to the binomial distribution, Central Limit Theorem, and Brownian motion.

Probability is full of surprises, and Susan Holmes has a Web site titled "Probability by Surprises." This site includes applets related to interesting probability paradoxes and problems such as: the birthday problem, the box-top problem, the hat-check problem and others. Alexander Bogomolny also provides a discussion of a number of puzzling probability problems illustrated with applets at his Web site. His list includes: Benford's law, the Buffon needle problem, Simpson's paradox, and Bertrand's paradox.

More traditional probability books are also available on the Web. *Introduction to Probability* by Grinstead and Snell (1997) is an example of such a book. This book includes applets as well as Mathematica and Maple programmes illustrating basic probability concepts. Waner and Costenoble have an on-line probability book 'Calculus Applied to Probability and Statistics for Liberal Arts and Business Majors' which deals exclusively with continuous probability.

One of the greatest strengths of the Web is the ability to find very current information on probability theory that cannot yet be found in textbooks but might be good to use in a probability course. The classic example of this is David Griffeath's 'Priomordial Soup Kitchen'. For the past ten years David has provided each week a new beautiful coloured picture showing how simple cellular automaton rules create fascinating structures from random initial states. Another good example of this is the 'Web Site for Perfectly Random Sampling with Markov Chains' which provides the latest information on the use of Markov Chain simulations which have recently found numerous applications in physics, statistics, and computer science. As in the case of Griffeath's site, the fact that, unlike publishers, the Web has no trouble with coloured pictures makes the Web a wonderful way to transmit the current progress in these fields.

Finally, Phil Pollett maintains a Web site called 'The Probability Web' that provides a comprehensive resource for links to various probability resources on the Web.

4. TECHNOLOGY IN QUANTITATIVE LITERACY COURSES

Over the past 10 years a new course is increasingly being offered in mathematics and statistics departments, often referred to as Quantitative Literacy or Statistical Literacy. This course typically attempts to help students develop an understanding of quantitative information used in the world around them, including basic concepts in statistics and probability. Gal (2000) describes what a course on statistical literacy should be and how it differs from a standard introductory statistics course. He argues that such a course should be aimed at consumers of statistics rather than producers of statistics. The basic statistical concepts taught are not different but the emphasis should be different. For example, a much broader discussion of types of experiments is essential to understanding reports in the news on medical experiments. Students need to understand different interpretations of probabilities (subjective and objective) and risk (relative and absolute) etc. than would normally be taught in a first statistics course. Gal has started a web site, "Adult Numeracy: Research and Development Exchange," which provides information on statistical literacy for adults and students.

The course called 'Chance' at Dartmouth, developed with several other colleges in 1992, is an example of this new type of quantitative literacy course. This course was designed to help students understand statistics used in the media and utilized articles from the current news to focus class discussions each day. The Chance Web site provides resources for teaching a Chance course or a standard course enriched with discussion of news items that include reference to ideas of chance. The Chance project produces a monthly electronic newsletter called Chance News that abstracts, and provides discussion questions for, current issues in the news that use probability or statistics concepts. This newsletter is sent out by email and archived on the Chance Web site.

Many major newspapers have Web versions and are a good source of articles to use in a quantitative literacy course. Search engines such as Lexis-Nexis may be used to locate articles on particular topics such as DNA fingerprinting, weather forecasting, polls and surveys, and lotteries. Articles in science and medical journals are also available on Web sites where the full text of the articles may be accessed.

Another source of information on how probability and statistics is used in the real world is provided by the videos of the lectures from the Chance Lecture Series available on the Chance Web site. In this series, experts in areas of probability and statistics that are mentioned regularly in the news provide lectures on their topic. Some of these are David Moore from the Gallup Organization on problems in polling, Arnold Barnett from MIT on estimating the risk of flying, Bruce Weir on the use of DNA fingerprinting in the news, etc. The lectures are also available on a CD ROM for those with slow Internet connections.

The Web site *Chance and Data in the News* maintained by Jane Watson is a useful resource for quantitative literacy courses. This is a collection of newspaper articles from an Australian newspaper grouped according to the five topics: Data Collection and Sampling, Data Representation, Chance and Basic Probability, Data Reduction and Inference. Each topic starts with general questions for articles related to this topic. In addition, each article has specific questions pertaining to it and references to related articles.

National Public Radio (NPR) also covers the major medical studies as well as other chance news. They keep all of their programmes in an archive that students can access and listen to with free RealAudio software. Their reports are usually in the form of questions and answers from the researcher who did the work or other experts in the field or both. These questions often anticipate questions readers of the newspaper report might have and so often make a significant contribution. NPR has a good search engine to look for discussions of older articles. The Chance Web site has a number of these interviews under *Video and Audio*.

5. TECHNOLOGY IN OTHER STATISTICS COURSES

This section focuses on more 'advanced' introductory statistics courses. By this we mean introductory courses with a calculus prerequisite (often serving students majoring in statistics, mathematics, science, or engineering) as well as second semester introductory courses. Similar to the statistics reform movement in the introductory course, a slower movement has addressed the content and pedagogy of these courses, with technology again playing a central role. While progress has been slower there is also tremendous potential for innovative uses of technology in these courses.

Web applets and computer packages now allow a greater emphasis on the conceptual ideas underlying the statistical methods. For example, students can use Minitab or a Java program to simulate the sampling distribution of the least squares regression slope or the chi-square statistic. This expands the types of simulations students use in other statistics courses to study more complicated techniques. Students can use Excel to graph and numerically estimate the maximum likelihood estimator, while easily changing parameter values to see how the maximum likelihood function updates. Students are therefore able to focus on the function and less on the calculus involved.

SPlus is another computer package that allows students to program routines and perform simulations. Similar to programming in the C language, students can be asked to write simple scripts for performing analyses. SPlus also offers exemplary graphical techniques for exploring more complicated data sets. A similar, but free, student version, R, can be downloaded from the Web.

These more advanced courses have also seen an increase in the number and size of data sets that can be analyzed, due to the availability of technological resources. For example, *The Statistical Sleuth* (Ramsey and Schafer, 1997) and *StatLabs* (Nolan and Speed, 2000) each follow a series of case studies to demonstrate the application of intermediate and mathematical statistics tools. These examples deal with real, and often messy, genuine data sets. Students must learn to deal with the messiness inherent in real data as well as better linking their knowledge to the practice of statistics.

Current technology allows students to take advantage of, and learn from, more recent computationally intensive statistical methods. For example, Jenny Baglivo is developing laboratory materials for use with Mathematica that incorporate simulation, permutation, and bootstrap methods throughout a math/stat course. These materials are available at her Web site. Tanis (1998) has been integrating computer algebra system modules into mathematical statistics courses.

Experience using technology and the need to be able to communicate statistical knowledge are considered important competencies for students who major in statistics. This is seen by an emergence of 'capstone' courses that require students to apply their knowledge and then use technology to produce integrated reports of their results and analysis discussion (e.g., Derr, 2000; Spurrier, 2000). Video is also used extensively by Derr to present students with examples of good and bad consulting sessions. This is a very effective instructional tool, enabling students to 'be there' and to see alternative examples of the same session. However, no one claims that such videos should completely replace personal experience.

6.   RESEARCH ON THE ROLE OF TECHNOLOGY IN PROBABILITY AND STATISTICS COURSES

Although it is apparent to statistics educators that technology has had a huge impact on the content of current courses as well as the types of experiences students have in these courses, there is little research to document the actual impact of technology on student learning. There is also lack of information evaluating the effectiveness of particular types of software or activities using technology. Biehler (1997) cautions that statistics educators need a system to critically evaluate existing software from the perspective of educating students, and to produce future software more adequate both for learning and doing statistics in introductory courses.

A round table conference sponsored by the International Association for Statistics Education was convened in 1996 to examine research on the role of technology in learning statistics. Although there was little empirical research to report at that time on the impact of technology on student learning, a main outcome of this conference was to identify important issues and research questions that had not yet been explored (Garfield and Burrill, 1997).

Two years later, at the Fifth International Conference on Teaching Statistics, there were 35 papers presented in seven different categories, under the topic heading "The Role of Technology in the Teaching of Statistics" (Pereira-Mendoza, 1998). Additional papers on the use of technology were in many of the other topic sessions as well. Most of these papers discussed ways technology can be used in courses, rather than offering data on learning outcomes. However, empirical studies (e.g., Finch and Cumming, 1998; Shaughnessy, 1998) provided valuable information on how technological tools can both improve student learning of particular concepts as well as raise new awareness of student misconceptions or difficulties (e.g., Batanero and Godino, 1998).

While controlled experiments are usually not possible in educational settings, qualitative studies are increasingly helpful in focusing on the development of concepts or the use of skills that technology is intended to facilitate. Biehler (1998) used videos and transcripts to explore students' thinking as they interacted with statistical software. Lee (2000) and Miller (2000) are examples of qualitative studies of how a course that integrates technological tools can support a student-centered, constructivist environment for statistics education. delMas, Garfield, and Chance (1999) provide a model of collaborative classroom research to study the impact of simulation software on students' understanding of sampling distributions.

7. SUMMARY, CONCLUSIONS, RECOMMENDATIONS

Thistead and Velleman (1992), in their summary of technology in teaching statistics, cite four obstacles that can cause difficulties when trying to incorporate technology into college statistics courses. These include equipment (e.g., adequate and updated computer labs), software (availability and costs), projection (of computer screens in classrooms), and obsolescence (of hardware, software, and projection technologies). Eight years later, we can see increased availability of computers and access to graphing calculators, updated and more widely available software, often via CD's bundled with textbooks or on the World Wide Web, and new methods of projecting computer screens such as interactive White Boards.

The ways that these technological resources appear to have changed the teaching of probability and statistics include:

- Less of a focus on computations. This frees students to spend more time focusing on and understanding the concepts. There is also less focus on manipulating numbers, or on exercises using only small and/or artificial data sets.
- Improved visualization of statistical concepts and processes. Students are better able to 'see' the statistical ideas, and teachers are better able to teach to students who are predominantly visual learners.

- Dynamic representations and analyses. Discussions or activities may focus on 'what if?' questions by changing data values or manipulating graphs and instantly seeing the results.
- Increased use of simulations. Simulations provide an alternative to using theoretical probability when teaching introductory statistics, better motivate probability theory when teaching probability, and offer better ways to convey ideas of long-run patterns.
- Empowering students as users of statistics. Students are able to solve real problems and use powerful statistical tools that they may be able to use in other courses or types of work. This allows them to better understand and experience the practice of statistics.
- Facilitating discussions about more interesting problems by using technological tools to explore interesting data sets (which may be large and complicated), often accessed from the Internet.
- Allowing students to do more learning on their own, outside of class, using Web-based or multimedia materials. This frees the instructor to have fewer lectures during class and to spend more time on data analysis activities and group discussions.
- Making the course relevant and connected to everyday life. Web resources make it easy to connect course material to real world applications and problems through data sets, media resources, and videos.

Despite the capabilities that technology offers, instructors should be careful about using sophisticated software packages that may result in the students spending more time learning to use the software than they do in applying it. Even in this advanced technological society, students are not always ready for the levels of technology used in courses. It is important that videos and simulation games do not become 'play time' for students and that computer visualizations do not just become a black box generating data. Rather than replace data-generating activities with computer simulations, educators may choose to use a hands-on activity with devices such as dice or M&Ms to begin an activity, and then move to the computer to simulate larger sets of data. In this way students may better understand the simulation process and what the data actually represent.

What is still lacking is knowledge on the best ways of integrating technology into statistics courses, and how to assess the impact of technology on student understanding of statistics. With an increased emphasis on statistics education at all educational levels, we hope to see more high quality research, incorporating a variety of methods and theoretical frameworks that will provide information on appropriate uses of technology.

Finally, probability and statistics are specialized subjects, and many colleges may not have a faculty member whose expertise is in these areas. The methods being developed for distance learning, which incorporate many innovative uses of technology, may allow schools to share resources and make a high quality probability or statistics class at one university available to others. However, with increased distance learning courses, it is also unclear as to how much of a course can be taught exclusively using technology, what the appropriate role of an instructor should be, and how much emphasis should still be placed on students generating calculations and graphs by hand.

REFERENCES

Batanero, C. and Godino, J. D. (1998). Understanding graphical and numerical representations of statistical association in a computer environment. In L. Pereira-Mendoza (Ed.), Proceedings of the Fifth International Conference on Teaching Statistics, pp. 1017-1023. Voorburg, The Netherlands: International Statistical Institute.

Ben-Zvi, D. (2000). Toward understanding the role of technological tools in statistical learning. Mathematical Thinking and Learning, 2,127-155.

Biehler, R. (1997). Software for learning and doing statistics. International Statistical Review, 65, 167-189.

Biehler, R. (1998). Students – statistical software – statistical tasks: A study of problems at the interfaces. In L. Pereira-Mendoza (Ed.), Proceedings of the Fifth International Conference on Teaching Statistics, pp. 1025-1031. Voorburg, The Netherlands: International Statistical Institute.

Chance, B. (1998). Incorporating a Listserve into Introductory Statistics Courses. 1998 Proceedings of the Section on Statistical Education, American Statistical Association.

Cobb, G.W. (1992). Teaching Statistics. In L. Steen (Ed.), Heeding the Call for Change: Suggestions for Curricular Action, pp. 3-43. Washington, DC: Mathematics Association of America.

Cobb, G. W. and Cryer, J. (1997). Electronic Companion to Statistics. New York: Cogito Learning Media.

Cumming, G. and Thomason, N. (1998). StatPlay: Multimedia for statistical understanding. In L. Pereira-Mendoza (Ed.), Proceedings of the Fifth International Conference on Teaching Statistics, pp. 947-952. Voorburg, The Netherlands: International Statistical Institute.

delMas, R., Garfield, J., and Chance, B. (1999). A model of classroom research in action: Developing simulation activities to improve students' statistical reasoning. *Journal of Statistics Education*, 7(3). (electronic journal, http://www.amstat.org/publications/jse/)

Derr, J. (2000). Statistical Consulting: A Guide to Effective Communication. Pacific Grove, CA: Duxbury Press.

Finch, S. and Cumming, G. (1998). Assessing conceptual change in learning statistics. In L. Pereira-Mendoza (Ed.), Proceedings of the Fifth International Conference on Teaching Statistics, pp. 897-904. Voorburg, The Netherlands: International Statistical Institute.

Finzer, B.and Erickson, T. (1998). DataSpace – A computer learning environment for data anlaysis and statistics based on dynamic dragging, visualization, simulation, and networked collaboration. In L. Pereira-Mendoza (Ed.), Proceedings of the Fifth International Conference on Teaching Statistics, pp. 825-830. Voorburg, The Netherlands: International Statistical Institute.

Gal, I. (2000). Statistical literacy: Conceptual and instructional issues.In D. Coben, J. O'Donoghue, & G. FitzSimons, (Eds.), Perspectives on adults learning mathematics: Research and practice (pp. 135-150). London: Kluwer.

Garfield, J. (2000). A snapshot of introductory statistics. Paper presented at Beyond the Formula conference, Rochester, NY.

Garfield, J. and Burrill, G. (Eds.), (1997). Research on the Role of Technology in Teaching and Learning Statistics. Voorburg, The Netherlands: International Statistical Institute.

Garfield, J., delMas, R. and Chance, B. (2001). Tools for teaching and assessing statistical inference. Paper presented at the Joint Mathematics Meetings, New Orleans, LA.

Grinstead, C. and Snell, J. L. (1997). Introduction to Probability. Washington, DC: The American Mathematical Society.

Harkness, W. (2000). Restructuring the elementary statistics class: The Penn State model. Paper presented at the Joint Statistical Meetings, Indianapolis, IN.

Hyde, H. and Nicholson, J. (1998). Sharing data via email at the secondary level. In L. Pereira-Mendoza (Ed.), Proceedings of the Fifth International Conference on Teaching Statistics, pp. 95-102. Voorburg, The Netherlands: International Statistical Institute.

Joiner, B.L., and Ryan, B.F. (2000). MINITAB(r) Handbook. Pacific Grove, CA: Brooks/Cole Publishing Co.

Kugashev, A. (2000). Statistical instruction in distance learning. Paper presented at the Joint Statistical Meetings, Indianapolis, IN.

Lee, C. (2000). 'Developing Student-Centered Learning Environments in the Technology Era - the Case of Introductory Statistics," presented at Joint Statistical Meetings, August, 2000.

Lock, R. H. (1993). A comparison of five student versions of statistics packages. The American Statistician, 47, 136-145.

Lock, R. H. (in press). A Sampler of WWW Resources for Teaching Statistics. In T. Moore (Ed.), Teaching Statistics: Resources for Undergraduate Instructors. Washington, D.C.: Mathematics Association of America.

McCullough, B.D. and Wilson, B. (1999). On the accuracy of statistical procedures in Microsoft Excel 97. Computational Statistics and Data Analysis, 31, 27-37.

Miller, J. (2000). The Quest for the Constructivist Statistics Classroom: Viewing Practice Through Constructivist Theory. Ph.D. Dissertation, Ohio State University.

Moore, D.S. (1993). The place of video in new styles of teaching and learning statistics. The American Statistician, 47, 172-176.

Moore, D. S. (1997). New pedagogy and new content: the case of statistics. International Statistical Review, 635, 123-165.

Nolan, D. and Speed, T. P. (2000). StatLabs: Mathematical Statistics Through Applications. New York: Springer.

Pereira-Mendoza, L. (Eds.), (1998). Proceedings of the Fifth International Conference on Teaching Statistics. Voorburg, The Netherlands: International Statistical Institute.

Ramsey, F. and Schafer, D. (1997). The Statistical Sleuth. Pacific Grove, CA: Duxbury Press.

Shaughnessy, J.M. (1998). Immersion in data handling: Using the chance-Plus software with introductory college students. In L. Pereira-Mendoza (Ed.), Proceedings of the Fifth International Conference on Teaching Statistics, pp. 913-920. Voorburg, The Netherlands: International Statistical Institute.

Shaughnessy, J. M., Garfield, J.B. and Greer, B. (1997). Data Handling. In A. J. Bishop, K. Clements, C. Keitel, J. Kilpatrick and C. Laborde (Eds.), International Handbook on Mathematics Education, pp. 205-237. Dordrecht: Kluwer Academic Publishers.

Simon, J. L., and P. Bruce. (1991). Resampling: A tool for everyday statistical work. Chance, 4(1), 22-32.

Spurrier, J.D. (2000). The Practice of Statistics: Putting the Pieces Together. Duxbury Press.

Tanis, E. (1998). Using Maple for instruction in undergraduate probability and statistics. In L. Pereira-Mendoza (Ed.), Proceedings of the Fifth International Conference on Teaching Statistics, pp. 199-204. Voorburg, The Netherlands: International Statistical Institute.

Thistead, R. A. and Velleman, P. F. (1992). Computers and modern statistics. In D. Hoaglin and D. Moore (Eds.), Perspectives on Contemporary Statistics, pp. 41-53. Washington, DC: Mathematics Association of America.

Velleman, P. (1998). Learning Data Analysis with Data Desk. Reading, MA: Addison-Wesley.

Velleman, P. (2000) ActivStats. Incorporating a Listserve into Introductory Statistics Courses, 1998 Proceedings of the Section on Statistical Education, American Statistical Association.

Velleman, P. and Moore, D.S. (1996). Multimedia for teaching statistics: promises and pitfalls. The American Statistician, 50, 217-225.

Joan Garfield
University of Minnesota, Minnesota, USA
jbg@tc.umn.edu

Beth Chance
California Polytechnic State University, California, USA
bchance@calpoly.edu

J. Laurie Snell
Dartmouth College, New Hampshire USA
jlsnell@dartmouth.edu

JOEL HILLEL

COMPUTER ALGEBRA SYSTEMS IN THE LEARNING AND TEACHING OF LINEAR ALGEBRA: SOME EXAMPLES

1. INTRODUCTION

This paper examines one point of view of the use of Computer Algebra Systems (CAS) in the teaching and learning of elementary linear algebra. Three examples are given, each with a different pedagogical purpose. These are followed by a description of an actual observation of students working in a CAS environment, as a way to point to different types of interactions that can occur within a session. Finally, the paper addresses the issue of trying to assess CAS as educational tools and the importance of articulating goals and a 'context' for CAS activities.

2. CAS: A BRIEF HISTORY

CAS have evolved dramatically from their humble beginning nearly 50 years ago, when the then existing computer programs were coaxed into performing some symbolic routines such as differentiation. In the course of their evolution, they have gone from being specialized tools for solving specific problems in fields such as mathematical physics and theoretical chemistry, to becoming all-purpose, autonomous mathematical tools, with their own language, syntax, and libraries of routines. While these powerful tools were first housed only on large mainframe computers and were not very accessible, they migrated to personal computers almost as soon as these become available. Nowadays, even hand-held calculators have some CAS capabilities.

It is not surprising that as far back as the late 1970s, when CAS became more accessible, questions about their potential use as teaching and learning tools began to be raised. For example, Wilf, discussing muMATH, which was one of the first CAS available for personal computers, correctly predicted that "the calculators-or-no-calculators dilemma that haunts the teaching of elementary school mathematics is heading in the direction of college mathematics" (Wilf, 1982). Though muMATH was rather primitive by today's standards (and without graphing capabilities), Wilf had already wondered how traditional calculus courses would have to be changed. By the mid-1980s the debate about the educational uses of CAS was raging. It was characterized by the rather acute polarization of the pro and con camps, each making dramatic predictions about the outcome of letting students have access to CAS. The

371

Derek Holton (Ed.), The Teaching and Learning of Mathematics at University Level: An ICMI Study, 371—380.
© 2001 *Kluwer Academic Publishers. Printed in the Netherlands.*

become a lot more familiar to mathematicians, whether they are using them as professional or as educational tools. It has become recognized that CAS could be used in instruction, to meet different goals, in different ways, and for different student populations, and that the modalities of using CAS are mitigated, in part, by the personal style and preferences of the individual instructors. Nowadays, there is also a more sober appraisal of what is possible with CAS, given the constraints of time (instructors' time, students' time, and course time), availability of equipment and labs, and availability of support, both instructional and technological. It is fair to say that in most educational uses, CAS are seen now as part of a 'system of means' which include lectures, text, as well as traditional paper-and-pencil problems.

3. CAS IN RELATION TO LINEAR ALGEBRA

Many examples of and ideas related to educational uses of CAS are now available, as well as both anecdotal and research-based assessments. It is not surprising that the teaching of calculus and differential equations are predominant among these examples, since with these, the graphical/symbolic/numerical capabilities of CAS can be fully exploited. One can also make a reasonable argument as to why using these different modes of representation helps students' conceptualization of fundamental ideas such as limits, tangent planes, or solution spaces of differential equations. In this respect, linear algebra is somewhat different. Students are known to experience difficulties with the structural aspect of the subject, since for most of them it is the first course in which the subject matter is built in a fairly systematic, and definition-based way. They also are often confused by the confluence of three languages of description (abstract, geometric, and algebraic) and knowing how these are related and which is appropriate for a given situation (Hillel, 2000, Dorier and Sierpinska, this volume, pp.255-274). While CAS afford a great tool for manipulating matrices and solving system of equations, they do not offer obvious means to help students' understanding of the abstract constructs of the general theory of vector spaces. But, if one stays within the confines of the concrete level of R^n (and many have argued that this is the appropriate level for the first course in the subject, see, for example Carlson, Johnson, Lay and Porter, 1993), CAS activities can be used in a variety of ways to enhance students' understanding and appreciation of the subject.

4. THREE EXAMPLES

The following are three prototypical activities that I have used with students (from mathematics and other different disciplines), who are in their first 1- or 2-semester course in linear algebra. They exemplify how CAS can be easily integrated in the teaching of the subject. Most of the students are not familiar with CAS prior to starting the course, so learning some basic syntax has to be done within the course itself.

Activity 1: Start with a matrix A and look at the effect of computing successive powers of A.

Activity 2: Solve a system of equations by reducing the augmented matrix to echelon form.

Activity 3: Start with a matrix A and a vector $\mathbf{v}$ and look at the effect of iterating the action of A on $\mathbf{v}$.

These are certainly rather innocent-looking activities; they certainly don't require knowledge of extensive CAS syntax. Yet each activity is initially chosen with a slightly different pedagogical purpose in mind. It can stand on its own as a one-time lab activity or can be expanded and linked to some important aspects of the underlying theory, as elaborated in the next sections.

4.1 Activity 1: Surprises

Ask any beginning student of linear algebra about matrices that satisfy, say, $A^2 = 0$ and their likely answer is $A = 0$. This, of course, should not come as a surprise as students' 'concept image' of a multiplicative structure is derived from their knowledge of the real number system. However, by the time they complete their first course, students will hopefully be more sophisticated and recognize that there are non-zero matrices that satisfy the equation. But their conception of such matrices is rather inaccurate. Most often they believe that such matrices must be sparse, and a matrix such as $\begin{bmatrix} 0 & 1 \\ 0 & 0 \end{bmatrix}$ serves as the prototype.

Activity 1 is one of a set of CAS activities whose main aim is to first and foremost create a **surprise**, in this specific case by asking students to compute successive powers of a judiciously chosen and 'large' matrix A such as

$$A = \begin{bmatrix} -18 & 24 & -42 & 0 & 6 \\ 0 & 15 & -33 & -27 & 24 \\ 6 & -2 & 8 & -6 & 4 \\ -12 & -2 & 44 & 102 & -86 \end{bmatrix}$$

The discovery that $A^5 = 0$ is initially met with incredulity. Sometimes, students will run through their computations a second time just to make sure that they haven't made a mistake. Several other examples of the same type should quickly convince them that this is a legitimate phenomenon.

The above is one of a whole range of activities involving taking successive powers of specially pre-selected matrices. The object of such activities is clearly not to do the calculations but rather to observe that complicated-looking matrices can have interesting properties, properties that later can be named (e.g. nilpotent, idempotent, etc.). If presented very early in a linear algebra course, these activities give students quite a different insight about behaviour of matrices.

Gaining greater familiarity with matrices is already a worthwhile goal. But, suppose some students' curiosity is piqued and they ask where do matrices such as

A above come from. Or, perhaps, they could be prompted to think about this question by being given the challenging task of constructing, say, a 4×4 nilpotent matrix with non-zero entries. Now the initial activity can be also linked to some important strands of the theory, for example, similarity and properties of matrices that are invariant under similarity.

One may add here that students' experience with certain operations on matrices is always one directional: modify a relative complex-looking matrix into something simpler (via similarity, row equivalence, etc.)[1]. The idea that the same operations allow one to 'complexify' a simple matrix, yet maintain some key property such as nilpotency, comes again as a surprise. They eventually can become privy to 'secrets of the trade', namely that instructors and textbook authors create non-trivial examples of matrices with specific properties by essentially working backwards from simple cases.

4.2 Activity 2: Clarifications

All linear algebra students learn to solve a system of equations $AX = B$ by reducing the augmented matrix to echelon form. This is such a routine procedure that students often are oblivious to the status of the intermediate matrices, obtained by elementary row operations. They are not conscious of the fact that they are replacing one system of equations by equivalent, albeit, simpler ones. In the worse case scenario, students obtain a solution X_0 from the echelon form but lose the connection of this vector to the original system of equations.

Activity 2 starts with a relatively large homogeneous system with non-trivial solutions and has students using CAS to perform elementary row operations on the coefficient matrix A_0. Students are also instructed to label successive matrices as A_1, A_2, A_3, ... till echelon form is reached, and to find a particular solution X_0. Students are then asked to calculate A_iX_0 for $i = 0$, 1,2, 3, ... The fact that all of these calculations produce a zero column vector brings to light that a solution is a solution of the original, final, and all the intermediate systems of equations.

The purpose of such activity is **clarification** of theory covered in class. Grasping the relation between elementary row operations and equivalent systems is the key notion, not the actual procedure for row-reducing matrices. Once understood, I see no reason why students should not be given free rein to use CAS and go directly to row reduced echelon form of a matrix without actually performing the row operations (and later, to go directly from a system of equations to a CAS-generated solution).

Having an immediate access to the reduced echelon form of a matrix opens the door to another activity: students are asked to ascertain whether several large $m \times n$ systems are equivalent. Hopefully, students would recognize that the answer resides completely in the reduced echelon forms of these systems, and that they would not

---

[1] This point was expressed succinctly by one student during a demonstration in class that row-equivalence is an equivalence relation when he asked "why bother doing the elementary row operations in the other way?"

bother to go through the actual row reduction. Once again, such activity presents an opportunity to probe a bit further: the fact that some of the systems in the example given are equivalent is obviously not just serendipity so one can raise the question of how these systems were constructed in the first place. As in Activity 1, the same idea is at play: start with a simple matrix (a reduced echelon form) and then 'complexify' via row equivalence to create other matrices.

4.3 Activity 3: Investigations

At one point in a more advanced linear algebra course, Jordan Canonical forms are discussed, and students have to work with spaces of generalized eigenvectors, i.e., vectors that are annihilated by some power of the matrix transformation $(A - \lambda I)$. Underlying the notion of generalized eigenvectors is the simple idea of iterating the action of a matrix on a vector. Again, students usually have very little experience with just playing around with iterates and noticing that some vectors have privileged properties *vis-à-vis* a matrix (e.g. $A^k \mathbf{v} = \mathbf{v}$, or $A^k \mathbf{v} = \mathbf{0}$, for some $k \geq 1$).

Students can gain access to these notions via activities that can be characterized as **investigations** that act as 'advanced organizers' for concepts that have not yet been covered in class. The following is an excerpt of a worksheet given to students taking a lab course on Maple and linear algebra: The ideas are introduced using non-technical language. They are a step up in sophistication compared to the previous examples because they require students to work through some definitions first.

The 'life span of $\mathbf{v}$ under A'.

Starting with a matrix A, a non-zero vector $\mathbf{v}$ may not 'survive' the first application of the matrix A in the sense that $A\mathbf{v} = \mathbf{0}$. It might survive the first application of A, so $A\mathbf{v} \neq \mathbf{0}$ but not the second, so we get that $A^2\mathbf{v} = \mathbf{0}$, and so on.

Let $\mathbf{v}_k$ be the kth iterate of A on $\mathbf{v}$, so $\mathbf{v}_k := A^k\mathbf{v}$. $\mathbf{v}_0 = \mathbf{v}$. If $\mathbf{v}_0 \neq \mathbf{0}$, but $\mathbf{v}_1 = \mathbf{0}$, we can say that '$\mathbf{v}$ has 1-life span under A'. If $\mathbf{v}_1 \neq \mathbf{0}$ but $\mathbf{v}_2 = \mathbf{0}$ then $\mathbf{v}$ has '2-life span under A', and so on. In general, the *life span of $\mathbf{v}$ under A* is k if $\mathbf{v}_{k-1} \neq \mathbf{0}$ but $\mathbf{v}_k = \mathbf{0}$. The zero vector is basically 'lifeless', so we can say that it has 0-life span under A.

We say that $\mathbf{v}$ has *finite-life* span under A if $\mathbf{v}_k = \mathbf{0}$ for some $k \geq 0$. Otherwise, $\mathbf{v}$ has *infinite-life span* under A.

The above text is accompanied by giving several examples of specific matrices and vectors and asking students to check the life spans. For example:

$$\text{Let } N := \begin{bmatrix} 8 & 8 & -5 & 4 & -3 & -2 \\ -5 & -5 & 8 & 4 & -3 & -2 \\ 3 & 3 & 3 & 8 & -6 & -4 \\ -2 & -2 & -2 & -1 & -4 & -6 \\ -1 & -1 & -1 & -7 & -11 & -3 \end{bmatrix}$$

a) What is the life span under N of the column vectors [1,2,1,2,-1,1], [-3,0,-3,2,-1,1], and [1,1,1,1,1,1]?

b) Pick several other vectors of your own choosing and find their life span under N.

This activity could be then followed by more theoretical questions that try to link this new notion to previous concepts that students have seen, for example, eigenvectors, nullspace, and nilpotent matrices. It can also lead to a subsequent work which addresses (informally) the concept of minimal polynomials of a vector under A.

5. OBSERVING STUDENTS AT WORK

Not too many mathematicians have the luxury nor, perhaps, the inclination, to carefully watch individuals or groups of students as they use CAS. Assessments of CAS as teaching and learning tools are often based on general impressions. In practice, CAS sessions do not necessarily follow the 'script' imagined by the designer of the activities; they may yield unanticipated outcomes that can be rewarding or disappointing, and sometimes the perspective of students and instructors diverges.

Dreyfus and Hillel (1998) had an occasion to analyze in detail a Maple lab session of four students working as a group on a problem involving inner products. The students were in their second linear algebra course in which Maple was introduced to the whole class in two lab sessions at the beginning of the course. After that, the instructor regularly demonstrated Maple generated solutions to class problems but Maple assignments were optional and only few students chose to attempt them. One of the topics covered in class was inner product space including orthonormal bases, projections on a subspace and the connection to least-square approximations. One example shown to students was that of continuous functions on an interval $[a,b]$ (with the inner product defined by $<f, g> := \int_a^b f(x)g(x)dx$) and least square approximation of some such functions by quadratic polynomials.

The students in the lab were working on the task of finding the least-square quadratic approximation of x^3 on the interval [0,1]. They already had, from class, an orthonormal basis of the subspace of polynomials of degree at most 2 on the interval [0,1]. The task was intended by the instructor to be rather routine. The use of Maple was envisaged to be simply as a calculator, enabling them to evaluate some integrals and to provide a graphical display of the initial function and its approximation (this is one case in which graphs do come into play in linear algebra).

The hour-long session eventually led to a correct solution. In the process, it enabled students to debug several misconceptions. These included:

- approximations should always be good, on the interval [0,1] and beyond;
- approximations should always share the same initial value as the function;
- regardless of the inner product, the norm of the constant $f(x)=1$ is always 1;

- the orthonormal basis obtained for the interval [0,1] will remain orthonormal when one changes the underlying interval (and hence the integral defining the inner product).

Evidently, this was a rather successful CAS session. But, the detailed analysis of tapes of the session clearly showed that there were several factors, aside from Maple, that contributed to its success. Furthermore, Maple had played different roles within the session. For one thing, the students needed substantial amount of time at the start of the session just to sort out what they had to do. The fact that the task was intended as a routine one was predicated on the assumption on the part of the instructor that students were up to date with the material covered in preceding lectures. We should, of course, know better. The students spent over 20 minutes discussing the problem, trying to remember terminology and meaning, erratically leafing through their notes and text, examining a similar Maple solution given in class and involving $sin(x)$, and recalling some things they thought they heard their teacher say in class. These exchanges among the students, though characterized by rather confused and very imprecise language, were eventually sufficient to get them started. Though Maple seemed to have been irrelevant at this phase of the solution process, Dreyfus and Hillel suggested that, implicitly, it acted as the animator for the discussion since the students were constantly looking at the screen with the Maple prompt and were anxious to start typing some commands. But in order to do so, they needed to first reach some consensus on what was to be done.

The analysis of the session also clearly demonstrated that, once Maple was used to perform the necessary calculations to compute the approximating quadratic function, the graphical feedback was critical for helping students refine their ideas about approximation. Since they arbitrarily chose to compare x^3 and its quadratic approximation on the interval [-3,3], they were extremely disappointed with the poor approximation that was given by the graph below:

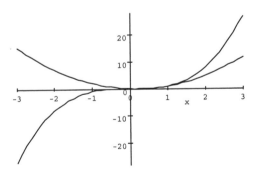

But the analysis also showed that two interventions by an observer were critical in unpacking some of the students' misconceptions. In the final part of the session, when the students became more aware about the role of the underlying interval, they conjectured that the quadratic approximation will be very good on the interval [0, 3]

if the inner product was redefined accordingly, using $\int_0^3$. Thus, they have turned Maple into an investigative tool. Their investigation was fraught with difficulties (because they assumed they still had the same orthonormal basis even though they changed the interval, hence the inner product), so the last part of the session was inconclusive and somewhat confusing. In fact, it was only several days after the session before the students were able to sort out the source of their problem.

This rather brief summary of a research finding is intended to highlight the fact that an analysis of what makes a particular CAS session 'work' or 'fail' shows that we are dealing with a dynamic process that is a result of several levels of interactions. These interactions can involve the student(s), the computer, the task, the instructor, as well as textbooks and students' notes. Each kind of interaction can be a source of feedback though the feedback is of somewhat different nature. In the above example, Maple was not the only 'agent' responsible for the resolution of the task and the clarification of some of the underlying notions; equally important were the collaborative nature of the activity and several key interventions by an observer.

6.   ASSESSING CAS AS EDUCATIONAL TOOLS

One often-heard challenge for those espousing the educational use of CAS is to provide evidence that they improve both teaching and learning. But without a context that takes into account questions such as, for whom, for what purpose, and how, what 'improvement' means and how it is assessed is not at all clear. Even when one has an articulated educational goal in mind, neither 'success' nor 'failure' can reside completely within the CAS. When successful learning outcomes with CAS are reported, we should examine whether there were other important agents in the success formula. This was clearly the case with the example on inner products discussed above. On the other hand, when failures are reported, we should ask whether the context was the right one for the intended goals in using CAS.

6.1 Describing a context

Assessing the pertinence of any CAS activities can only make sense if there is an articulation of the overall context in which they take place. We can imagine the following two very different scenarios for implementing these activities. In one scenario, they are part of students' normal assignment, to be done at home (if students have computers) or in a computer lab, which students can go to on their own time. Being part of an assignment, students then need to hand something tangible back, be it examples of their computations or answers to a set of accompanying questions. A very different scenario would have these CAS activities as an integral part of a class, so the whole class and the instructor are present at a lab at the same time. The students do not necessarily need to hand in their work but rather their observations become a basis for discussions, led and prompted by the instructor. It is likely that the same CAS-activities would follow different trajectories and outcomes within each scenario, so what constitutes a 'successful activity' has to take account of the setting in which it took place.

One can easily conceive of other scenarios and aspects that can vary even within each scenario. These pertain to: the extent to which an instructor uses and refers to CAS during his or her normal lectures; whether the CAS use in class is 'live' or primarily the presentation of already worked examples; and whether students are encouraged or even mandated to work in groups. Other aspects relate to whether the use CAS is occasional or nearly all the time, whether students need to know a large or a restricted set of CAS commands, and whether programming is part of the CAS activities. Finally, there are questions related to assessment: how are CAS activities assessed; is a functional knowledge of CAS part of the overall assessment of students; and are CAS allowed to be used by students during examinations.

In some universities, practical and institutional constraints actually leave very little choice. In other cases, the choice depends on personal preferences such as the instructor's style, commitment to the use of technology, and general attitude towards mathematics teaching and learning. (For a different example of CAS use, see Muller, this volume, pp. 381-394).

6.2 Pedagogical choices accompanying CAS activities

The three examples described in section 3, as well as most other uses of CAS in learning and teaching, are based on the premise that learners can better focus on ideas and concepts because the burden of doing the computations is taken over by CAS. But frequently, we make the pedagogical mistake of replacing computational burden by other equally distracting chores. For example, looking at Activity 2, one would certainly want to move away from the stereotypically 2×2 and 3×3 examples that permeate traditional linear algebra courses by having students work on larger-sized matrices and systems of equations. But, if one starts with, say, a 7×10 system of equations, it may take a substantial amount of time for students just to enter the data that defines the coefficient matrix A. It is easy to make some trivial mistakes along the way, mistakes that can render a session frustrating and unproductive for students and instructors alike.

A the introductory stages of CAS use, I would want to minimize aspects related to entering data and getting the right syntax, for example, by having the matrix of a large system of equations, already coded and available to the students. I would remind them of the syntax of the commands that need to be used. This certainly allows students to quickly get right what is intended as the core of the activity, and it frees up time for doing other related work during the same session.

Another practical concern relates to the fact that there are several modes of working with CAS which are quick and efficient if getting to the answer is the only goal. For example, it is easy with CAS to start with a matrix and perform elementary row operations without having to keep track of the successive matrices (not unlike someone working on the blackboard who just keeps erasing a row of a matrix and replacing it by another row). Similarly, one can iterate the application of A on **v** by repeated executing of a command to multiply A with the last-computed vector. This way of working does not keep track of intermediate steps in a calculation. However, something like Activity 2 requires keeping track and labelling intermediate matrices, and students need to be told explicitly to do so. In other words, CAS activities cannot be formulated just in terms of the mathematical content. In some cases, when the instructor has a particular goal in mind, students may need additional information such as suggestions on how to proceed, which calculations to label and save, useful commands to use, etc. After all, the students cannot guess beforehand the purpose of a given task.

These may seem like rather trivial issues when discussing the role of CAS as learning tools but, in real-life action, they are sometimes the most important factors in determining the course of a session. It is all too easy for us, as instructors, to get carried away with enthusiasm, and to create CAS activities that are exciting, mathematically speaking. But we need to be realistic about time constraints and not forget that novice users can 'waste' most of a session doing exactly those things that are of no mathematical significance.

7. CONCLUSION

What I have tried to point out in this paper with the examples of linear algebra work is that, as a mathematician, one does not need to be completely committed to using CAS nor drastically change teaching styles in order to devise some interesting CAS activities. Certainly, CAS are by and large black boxes which could easily be misused. Certainly, CAS work can entail a lot of fussing with syntax and other technical difficulties. But with thought, care, and experience, there are activities that can engage most students in a mathematical idea or confront them with unexpected results. In the process, students can become mathematically curious, and that is by itself a rewarding payoff.

REFERENCES

Carlson, D., Johnson C., Lay D. and Porter, A. (1993). The Linear Algebra Curriculum Study Group recommendations for the first course in linear algebra. *College Mathematics Journal*, 24.1, 41-46.

Dorier, J-L. and Sierpinska, A. Research into the Teaching and Learning of Linear Algebra, this volume, pp. 255-274.

Dreyfus, T. and Hillel, J. (1998). Reconstruction of Meaning for Function Approximation. *International Journal of Computers for Mathematical Learning*, 3, 93-112.

Hillel, J. (2000). Modes of description and the problem of representation in linear algebra. In J.-L. Dorier (Ed.), *On the teaching of linear algebra*, pp. 191-207. Dordrecht: Kluwer Academic Publishers.

Muller, E. (2001). Reflections on the sustained use of technology in undergraduate mathematics education, this volume, pp. 381-394.

Wilf, H. S. (1982). The Disk with a College Education. *The American Mathematical Monthly*, 89/1, 4-8.

Joel Hillel,
Concordia University, Montreal, Canada,
jhillel@alcor.concordia.ca

ERIC R. MULLER

REFLECTIONS ON THE SUSTAINED USE OF TECHNOLOGY IN UNDERGRADUATE MATHEMATICS EDUCATION

1. INTRODUCTION

David Tall (1991) in the Preface to the book Advanced Mathematical Thinking lists five cognitively appropriate approaches of presenting mathematics to learners. Some of these have demonstrated some empirical evidence of success in responding to "a wide variety of obstacles in students' mental imagery and often extremely limited conceptions of formal concepts." For the past thirty years, using developments in technology, the Department of Mathematics at Brock University has been addressing two of these approaches in a consistent and sustained manner. They are

- the building up of appropriate intuitive foundations of advanced mathematical concepts through an approach which balances cognitive growth and an appreciation of logical development, and
- the use of visualization, particularly utilizing a computer, to give the students an overall view of concepts and enabling more versatile methods of handling the information.

This paper does not provide a step by step review of how we have integrated technology into undergraduate mathematics education. For that the reader is directed to articles that we have previously written (see Muller, 1991a, Hodgson and Muller, 1992, and Ralph, 2000). Here we consider how technology has made it possible to provide a different route for students to build intuitive foundations, and present some of the issues that faculty continue to address. We begin with a brief history of the introduction of various technologies into our mathematics courses, and follow this with some discussion of issues arising from integrating technology into statistics courses and into all first year mathematics courses. We very briefly consider the implications of technology in the education of future teachers of mathematics and explore what the future may bring. Our task is assisted by the views of two mathematicians who have been immersed in the development of technology for the teaching and learning of mathematics, and of a third mathematician who has devoted much time to the development and running of a Web-based course. Their interviews, found in the Appendix to this chapter, make an appropriate conclusion to this paper on the use of technology in undergraduate mathematics education.

Derek Holton (Ed.), The Teaching and Learning of Mathematics at University Level: An ICMI Study, 381—394.
© 2001 *Kluwer Academic Publishers. Printed in the Netherlands.*

2. HISTORY

The Department of Mathematics at Brock University is recognized for the way in which it has integrated technology in undergraduate mathematics education and for its innovative programme for the education of middle school teachers. These two initiatives and emphases have fed off each other. The integration of technology was enhanced by faculty interest in issues of mathematics education and the education programme was enriched by their experience of technology in the teaching and learning of mathematics. It should be obvious, but it is worth stressing, that the integration of technology was driven first and foremost by considerations of students' mathematics learning in a developing technological environment. The objectives were not to duplicate more of the same with technology, but rather to use this medium to offer the students learning paths which were not available before. Integration of technology in mathematics courses was accomplished in small steps each planned to ensure that those courses that were working well would not be derailed. Each step was aimed at enhancing the experience of students while it retained a certain comfort level for the teaching faculty. In order to move beyond the transition and sustain the integration of technology, faculty support was viewed as being of primary importance. For more discussion on effecting curriculum change in undergraduate mathematics the reader is directed to Hodgson and Muller (1992, p. 99). The approach that we selected has its limitations. It does not, for example, allow for radically different courses in approach and content, in other words, courses that would put in question the traditional content and sequencing of mathematical ideas. Unfortunately such radical approaches have tended to have a short life span because they usually involve a single or a small group of faculty with limited departmental support. Our objective was to involve the majority of the faculty and to get support for the integration of technology through conviction rather than mandate.

By the end of the 1970s the Department of Mathematics had integrated technology into laboratory activities for introductory applied statistics courses. It was also using computer-generated data for each individual student in the large enrolment service courses (applied calculus, and applied statistics). The aim was to encourage students to work in small groups and to allow for more general assessment of student performance (Auer, Jenkins, Laywine, Mayberry and Muller, 1982). At that time some university senates were responding to student pressures that were looking for a broader set of assessment criteria than a final examination. By the end of the 1980s the Department of Mathematics was requiring laboratory activities using Minitab[1] in the introductory applied statistics courses and using Maple in the introductory applied calculus course. A number of factors made this possible. Mathematics and statistics software were becoming more user-friendly and could run on smaller, more affordable computer systems. An exploratory study of the use of Maple in a laboratory setting (Muller, 1991a) showed positive results especially in the area of retention rates. This study convinced the university administration to provide computer laboratory space for use by the Department of Mathematics in its large enrolment service courses. During the 1990s Brock

---

[1] The URLs of all of the software mentioned in this article can be found in the list of references.

University installed data projection devices in all its major classrooms, provided portable ones for smaller rooms, and developed a policy to update and expand its computer facilities for students. By the year 2000, the Department of Mathematics required the use of computer software by all students in its first year courses, irrespective of the students' choice of major disciplines. Maple is integrated in all sections and courses of calculus and linear algebra and it is site licensed for all university laboratories. Minitab is in use in all first year statistics courses. The integration of technology continues to be within laboratory settings and is used, to varying degrees, within lecture presentations. The Department is starting to integrate, into its calculus courses, "Journey Through Calculus", a CD ROM of calculus micro worlds developed by Ralph (2000). Some members of faculty provide course materials via the Web, and a course for future teachers uses the Web as a source of mathematics materials and integrates the use of Geometer's Sketchpad into its geometry component.

3. THE USE OF TECHNOLOGY IN THE TEACHING AND LEARNING OF STATISTICS

Although the discussion that follows is centered on applied statistics courses, the ideas and activities that are presented are used in mathematical statistics courses to support the development of the student's intuitive foundation of statistical concepts.

Applied statistics courses differ from mathematical statistics courses in a number of ways. Students take the course as a requirement for their programme which is not mathematics. Their preparation in mathematics is heterogeneous and generally weak. The time allocated to the course does not allow for an in-depth development of probability and the result is an inherent difficulty in bridging the gap between the mathematical probability and its use in the development of statistics. These courses play a fundamental role in the general education of a large number of individuals who should graduate with an understanding of the role of statistics in their society and who should be able to comprehend statistical arguments in their major discipline. These individuals should have developed a concept of statistical modelling that will enable them to make meaningful decisions in their world of work and to contribute to intelligent discussions with professional statisticians responsible for setting up procedures.

Students in first year applied statistics courses bring with them some beliefs, experiences and intuitive ideas of probability and some perception of measures of centrality. Many of their ideas in probability are naive and incorrect. They have little understanding of the role of variability, and they have a desire to set up causal schemes to explain random situations. In structuring opportunities for students to build up their 'concept image' (Tall and Vinner, 1981), one is aware that some of the strategies used in learning mathematics do not work for statistics. For example, in most statistical situations trial and error does not provide much insight, neither does the approach of breaking down examples into smaller ones. The feedback in statistics arises after a large number of trials, and perceivable patterns within randomness appear in the large and rarely in the small. It follows that computer technology with its power to simulate does provide a rich environment to help students explore statistical concepts and thereby build up their concept images (Bielher, 1991 and 1993).

Simulation, together with appropriate visual representations of results, offers a rich environment for the learner to develop intuitions about statistical concepts. From the visual displays of simulation results, students can explore the properties of probability distributions, look at the effects of sample size on the distributions of statistics (mean, proportion, etc.), explore the role of the significance level in confidence intervals and in hypothesis testing, etc. In a laboratory setting the technology allows the student to draw a sample of a particular size from a given probability distribution and to compare and discuss the differences in results with a student doing the identical activity at a neighbouring computer. Laboratories facilitate group work and support the discussion of information arising from their analysis of an individual data set or simulation. Time is spent on exploring data and not on data entry. Because the number of students in my lecture section is large, technology is used to analyze data generated by the class. During the very first class, students complete a questionnaire, and the information is loaded as a data file. It is used at various times during the year to explore properties of variables, to analyze different models and to explore effects of sampling from the class population. Students are generally interested in the data as they are part of it. This provides opportunities for the discussion of sampling procedures, errors and misunderstandings that they may have developed. In summary, computer simulation is used as a means to experience statistics, students obtain visual representations of the theory while they experience variability and receive reinforcement for the application of the theory.

The development and implementation of simulation with or without the use of technology can play a pedagogical role in probability situations. In a class for future teachers we discuss the counter-intuitive game show problem where the contestant is about to win a prize (see, for example, Barbeau, 1999). There are three doors, behind one is a car and behind each of the other two is a goat. We assume the contestant would prefer to win the car. The contestant selects one door, the game show host opens one of the two doors not selected by the contestant and exposes a goat. The contestant is invited to either stick with her first choice or to change her selection to the other unopened door. Based on this new information and using probabilities, should the player change her selection? The class is usually not in agreement on whether a change of door should or should not be made. Each section of the class is prepared to present convincing arguments to support their case. About one third of the class needs to be involved in a simulation to be spurred to look deeper into the assumptions on which they built their case. It is only after the class has collaborated to develop a simulation and has performed the simulation repeatedly in small groups, that everyone is able to come to a consensus. For some students the structuring and implementing of the simulation reveals characteristics of the situation that they had either not noticed or had dismissed as not important. Developing the simulation, in this situation, may play a similar role to programming, which some mathematicians (see, for example, Dubinsky and Schwingendorf, 1991) have used to foster the development of mental mathematical constructions. To conclude this discussion on simulation we note that its role has taken more significance in statistics because it is used with increased frequency in the development of new statistical procedures. Students majoring in statistics thus benefit from an early introduction to simulation concepts.

It is worth reflecting that the established statistical packages were developed by statisticians for statisticians. They have evolved with a view to making them more friendly for the knowledgeable statistician, one who is familiar with the terminology and who understands the concepts of statistics. The use of such software in statistics courses at Brock University has two objectives. The first and most important is to provide an environment where students can develop their statistical concepts. The second objective is to provide experience with a recognized statistical package that students will be able to apply in the analysis of data arising in their other courses. When such a software is used to develop statistical concepts and statistical thinking it is employed in a way for which it was not originally designed, and faculty must carefully consider how it is integrated into class and laboratory activities.

There is no doubt that technology has changed the scope and nature of statistics courses at Brock University. The lecture presentations stress concept development while students are much more engaged with the subject both in class and in laboratories.

4. SOME REFLECTIONS ON THE INTEGRATION OF MAPLE INTO UNDERGRADUATE MATHEMATICS EDUCATION

Maple belongs to one of a number of Symbolic Mathematical Systems that have evolved to include a huge array of mathematical techniques, both computational and algebraic, a capacity for scientific programming, stunning graphical representations that the user can manipulate, and substantial word processing capabilities. Maple provides a very different environment in which to do mathematics, and it also provides opportunities for the teaching and learning of mathematics that are not available with pencil and paper (Hodgson and Muller, 1992). From the transportation allegory we used in that paper, the car and the airplane provide differing opportunities for travel. The former provides occasions to stop and explore the local landscape, while the latter allows for rapid access to areas which would not normally be available if we insisted on going only by car. Maple provides a very different way of getting there, whether it be for exploration, development of knowledge, or as a tool for solving the well-structured mathematical problem. We can take our example of transportation a step further and compare the bicycle with the car. For the former we have to generate the energy for locomotion ourselves, whereas for the latter the engine provides the energy to take us to our desired destination. Computer-based materials are available that integrate mathematical activities with an automatic or on-request use of Maple as an engine. Examples of these are The Interactive Mathematics Dictionary (Borwein, Watters and Borowski), Journey Through Calculus (Ralph, 2000), and The Scientific Workbook.

A chapter by Tall (1996) in the International Handbook of Mathematics Education is a useful overview of the impact of technology on the curriculum development of functions and calculus. At Brock University, the use of Maple was first introduced in an applied calculus course in 1988. Since then it has become a requirement of all first year mathematics courses. Muller (1991a and 1991b) details the process that was followed in this implementation. Muller provides examples of student laboratory activities, looks at implications on course content and approach, and discusses some early assessment of the project. Certainly the agenda in the first few years was dominated by concerns of logistics, getting teaching faculty on board, developing appropriate activities, etc. As our experience grew we were able to work naturally within the new environment and pay much more attention to mathematics education issues such as the students' development of mental representations of mathematical concepts. In theory, one would have liked to have built the environment to respond to such issues. Unfortunately, other day-to-day concerns tended to dominate the agenda. Very early on we realized that Maple provided, with relative ease, different representations (numeric, algebraic, graphic), assisted in switching from one representation to another, and provided means to explore the shortcomings of one or other representation in a particular situation. However, these attributes are not evident because Maple was developed by mathematicians for mathematicians. Thus to integrate Maple into the learning and teaching of mathematics we realized that the students needed to be guided through activities in which the various representations were highlighted. Student activities were designed to make the processes explicit and teachers were encouraged to demonstrate these in their teaching. Maple makes some of the mathematics invisible, it is up to the teacher to decide how much of the mathematics should be exposed. Software, like Maple, that have the capability of having a number of representations simultaneously active, provide the students with an experience of a single set of related activities rather than a sequence of activities which may or may not form supporting links in their mind. We worked on the premise that mental representations of a concept are most useful when they link many aspects of that concept. As mathematicians, a multiplicity of links allows us to access various representations and to choose the one that appears most appropriate for the context in which the concept arises.

In 1991 we (Muller, 1991a) raised the question "What are the essential curriculum components of a calculus course using a Mathematics Computer Environment?". Lynn Steen (1992), in his article "Living with a New Mathematical Species", reflects on the broad implications of systems like Maple on mathematics education. He writes, "It is commonplace now to debate the wisdom of teaching skills (such as differentiation) that computers can do so well or better than humans. Is it really worth spending one month of every year teaching half of a country's 18-year-old students how to imitate a computer? ... It is not differentiation that our students need to learn, but the art of guessing. A month spent learning the rules of differentiation reinforces the student's ability to learn (and live by) the rules. In contrast, time spent making conjectures about derivatives will teach a student something about the art of mathematics and the science of order." How much of the techniques of differentiation should be covered in a calculus course now that one has access to Maple? Is it sufficient for the student to gain a good understanding, have a number of different mental representations of the derivative, and then rely on the technology to carry out the differentiation? We have wavered between more and less emphasis on techniques. It is clear that the majority of students in my courses have performed satisfactorily in their secondary school mathematics because they are able to carry out algorithms, and most of them get satisfaction from using the rules of differentiation to find the correct derivative. Aware of this and because my goal is for them to have a concept of differentiation, I use Maple in my class to compute derivatives from the limit definition. Their in-class work is to check Maple's result using the rules of differentiation.   From my point of view the techniques of differentiation provide the student with a means of checking the results of anti-differentiation performed by the technology. Students should develop an urge to check. Not just to check the work performed by someone else or by technology but by using a different (in this case reverse) process. We have decided to rely on the technology for anti-differentiation preferring to provide more time for the students to explore the concepts of anti-differentiation, the fundamental theorem of calculus, integration, and the solutions of differential equations. The aim is to get the students to develop mental images of these mathematical concepts and, as mentioned before, to then use their knowledge of differentiation to check the results produced by the technology. The debate continues among us who are involved with calculus courses and we appear to be no closer to answering the question we raised in 1991. In response, however, we have paid more attention to the role of Maple in curriculum development in calculus and other courses, aiming "to present the student with contexts in which cognitive growth is possible, leading ultimately to meaningful mathematical thinking in which the formalism plays an appropriate part" (Tall, 1991, p. 20). As a further development the department has started to integrate "Journey Through Calculus" (Ralph, 2000), in addition to Maple, into its calculus courses both in lecture presentation and in laboratory activities. This software differs from Maple in that it was developed by a mathematician for students. Its focus is very much the teaching and learning of calculus, it was developed as a curriculum directed environment while Maple was not. I will let William Ralph, the author of "Journey Through Calculus" speak for himself in an interview included in the Appendix. The Department is now using technologies that meet the two objectives it had set out to attain for its calculus courses. One is to provide for the learning of calculus, the other is to expose students to software used by mathematicians, a software that can be useful in other mathematics courses, and a tool that is not subject to the partitioning of mathematical concepts imposed by our undergraduate course administrative structure. Using the laboratory model, the department has integrated Maple into its first year linear algebra course. Examples of student activities can be found in Auer (1991) and Auer and Muller (1992). Some of the benefits of using Maple in learning linear algebra are the availability of rational arithmetic, the fact that operations within Maple have the same names as the standard mathematical operations, and that the symbolic manipulation allows students to examine, without drudgery, numerous examples of general properties and theorems without the frustration caused by algebraic errors. Our use of technology in the teaching and learning of linear algebra has been rather limited and we need to consider the issues that are raised by Dorier and Sierpinska (this volume, pp. 253-272).

The impact of Maple on the learning and teaching of mathematics at Brock University has been substantial. It has become an integral part of first year mathematics courses and faculty interest and use is now widespread. Maple is a very powerful mathematical system and instructors struggle to provide activities that integrate the learners at their individual stage of development. Learners use Maple with varying degrees of enthusiasm and success. The great majority can list something for which Maple has been particularly helpful to them in their learning of mathematics.

5. TECHNOLOGY AND THE PREPARATION OF TEACHERS OF MATHEMATICS

In Canada, education is a Provincial responsibility and in Ontario the education of primary and secondary teachers is based on a consecutive model. Teachers first complete an undergraduate degree in disciplines of their choice, and then apply to a Faculty of Education for a year of in-service professional development. Departments of Mathematics are therefore responsible for the mathematics education of future teachers who choose to take mathematics courses. The Mathematics Education Forum of the Fields Institute for Research in the Mathematical Sciences', located in Toronto, is developing recommendations for increasing the number of future elementary and secondary teachers who have some university courses in mathematics and is making recommendations to university mathematics departments about appropriate courses, both in terms of their content and in terms of their pedagogy.

Recent changes in Ontario school curricula have introduced technology at all levels. This includes the mandated use of calculators with increasing functionality in the higher grades and the use of Geometer's Sketchpad licensed for use in every Ontario school. Teachers of mathematics are now faced with implementing technology in their mathematics classes when their own mathematics education has been without it. From our own experience we expect that technology implementation concerns in the schools will first focus on logistic ones. To progress beyond that, mathematics education in-service with technology will have to be available. Furthermore, mathematics departments must now ensure that future teachers graduate with experience in the use of technology and have the opportunity to reflect on the role of technology in the learning and teaching of mathematics. How is this to be achieved? At Brock, students have experience with technology in their first year. Students in the special middle school mathematics and education programme take a course in their third year that explores mathematics learning and teaching issues, one of which is technology. With the integration of Geometer's Sketchpad into the school mathematics programme, how can this software be integrated into the appropriate geometry courses for teachers? We are just starting to address this and we have sought the views of Nicholas Jackiw, the developer of Geometer's Sketchpad. These views are found in an interview that is recorded in the Appendix.

6. LOOKING TO THE FUTURE

The implementation of technology for all students in first year mathematics classes has had some impact on upper year courses, but its use is very much a function of the instructor. Although the department received majority support for the use of technology in the first year, some faculty see no role for technology in their upper year courses, some faculty use it extensively and the rest suggest students use it in their assignments and projects. Recently appointed faculty are heavily involved with technology both in their research and in their teaching. One can therefore expect more intensive use of technology in all aspects of mathematics at Brock University.

Technologies aimed at the teaching and learning of mathematics are available and we can expect that mathematicians will develop new ones to provide for the learner of mathematics. For those of us not involved with the development of technologies how should we select them for use in our mathematics courses? We can take our cue from Noss and Hoyles (1996, p. 54) "Software which fails to provide the learner with a means of expressing mathematical ideas also fails to open any windows into the processes of mathematical learning. A student working with even the very best simulation, is intent on grasping what the simulation is demonstrating, rather than attempting to articulate the relationships involved. It is this articulation which offers some purchase on what the learner is thinking, and it is in this process of articulation that a learner can create mathematics and simultaneously reveal this creation to an observer." Another view is that the introduction of technology in undergraduate mathematics poses the same challenges as the use of manipulatives in school mathematics classes. The focus of these activities must be on helping the student to make the transition from the technology to the construction of the mental mathematical object. When using software that has been developed by mathematicians for mathematicians the instructor must provide activities which make the mathematics "visible" and "understandable" to the student. Visible in the sense that the student will develop a mental image of the mathematical object and understandable in the sense that this image will have links to other experiences and other mathematical objects already available to the student. Recent software developed specifically for the learning of mathematics include transitional activities where students communicate their mathematical understanding and are asked to make the links and explain their mental image of the mathematical concept. These software provide a quite different environment in which to learn mathematics. The mathematics can be very focused on specific topics as in "Journey Through Calculus", or can be much more exploratory as in Geometer's Sketchpad. We can expect that some students will select these environments as their preferred means for developing their mathematical knowledge. It is clear that some of the better students who like to work with Maple would be far more engaged in a mathematics programme that provided more incentive for exploration and for the making of conjectures.

What is the future role of the Web in mathematics education? For to be effective the Web must provide for efficient responsive communication with and of mathematics between the learner, the system and the teacher. Jorgenson (2000) explores a range of Web functionalities that can be expected to contribute to the learning of mathematics, and indicates that technological innovations are still required to ensure ease of communication with and of mathematics. The more mathematicians and mathematics educators work with this evolving technology, the more likely it will develop in ways that meet the needs of the learner and the instructor. Keith Taylor is a mathematician who has developed and run a mathematics course on the Web. He provides some insight into this experience in the Appendix.

Jorgenson (2000) has drawn a futuristic sketch of mathematics education in a fully integrated technological environment. The reader is encouraged to look at the implications of this scenario on undergraduate mathematics education.

REFERENCES

Auer, J.W. (1991). *Linear Algebra with Applications* and the accompanying Maple Solutions Manual. Scarborough: Prentice-Hall Canada Inc.

Auer, J.W., Jenkyns T.A., Laywine C.F., Mayberry J.P. and Muller E.R. (1982). Motivating non-mathematics majors through discipline-oriented problems and individualized data for each student. *Int. J. Math. Educ. Sci. Technol.*, 13, 221-225.

Auer, J.W. and Muller, E.R. (1992). Some examples of the use of CAS in teaching Linear Algebra. In Zaven A. Karian (Ed.), *Symbolic Computation in Undergraduate Mathematics Education*, pp. 101-108. Mathematical Association of America Notes 24. Washington, DC: Mathematical Association of America.

Barbeau E. (1999). *Mathematical Fallacies, Flaws, and Flimflam*. Spectrum Series, pp. 86-90. Washington, DC: Mathematical Association of America.

Bielher R. (1991). Computers in Probability Education. In R. Kapadia and M. Borovcnik (Eds.), *Chance Encounters: Probability in Education*, pp. 169-212. Dordrecht: Kluwer Academic Publishers.

Bielher R. (1993). Software Tools and Mathematics Education: The Case of Statistics. In C. Keitel and K. Ruthven (Eds.), *Learning from Computers: Mathematics Education and Technology*, pp. 68-100. Berlin: Springer-Verlag.

Borwein, J., Watters, C., and Borowski, E. *The interactive dictionary*. Halifax: MathResources Inc. URL: http://www.mathresources.com.

Dorier, J-L. and Sierpinska, A. (2001). Research into the Teaching and Learning of Linear Algebra, this volume, pp. 273-280.

Dubinsky E. and Schwingendorf K. (1991). Constructing Calculus Concepts. In L.C. Leinbach, J.R. Hundhausen, A.M. Ostebee, L.J.Senechal and D.B Small. (Eds.), *The laboratory approach to teaching calculus*, pp. 47-79. Mathematical Association of America Notes 20. Washington, DC: Mathematical Association of America.

Geometer's Sketchpad. URL: http://www.keypress.com.

Hodgson, B.R. and Muller, E.R. (1992). The impact of symbolic mathematical systems on mathematics education. In B. Cornu and A. Ralston (Eds.), *The influence of computers and informatics on mathematics and its teaching*, pp. 93-107. UNESCO Document Series 44. Paris: UNESCO.

JavaSketchpad. URL: http://www.keypress.com/sketchpad.

Jorgenson, L. (2000). Possible futures for mathematics on the Web. In B. Cload and T.A Jenkyns (Eds.), *Proceedings of the Conference on Technology in Mathematics Education at the Secondary and Tertiary Levels*, pp. 37-52. St. Catharines, Ontario: Brock University.

Minitab Inc. URL: http://www.minitab.com.

Muller, E.R. (1991a). Maple laboratory in a service calculus course. In L.C. Leinbach, J.R. Hundhausen, A.M. Ostebee, L.J.Senechal and D.B Small (Eds.), *The laboratory approach to teaching calculus*, pp. 111-118. Mathematical Association of America Notes 20. Washington, DC: Mathematical Association of America.

Muller, E.R. (1991b). Symbolic mathematics and statistics software use in calculus and statistics education. *ZDM*, 91 (5), 192-198.

Noss, R., and Hoyles, C. (1996). *Windows on Mathematical Meanings: Learning Cultures and Computers*. Dordrecht: Kluwer Academic Publishers.

Ralph, W. (2000). *Journey Through Calculus*. CD-ROM. California, USA: Brooks/Cole.

Scientific Workbook. URL: http://scinotebook.tcisoft.com.

Steen, L.A. (1992). Living with a new mathematical species. In B. Cornu and A. Ralston (Eds.), *The influence of computers and informatics on mathematics and its teaching*, pp. 33-38. UNESCO Document Series 44. Paris: UNESCO.

Tall, D. (1991). The psychology of advanced mathematical thinking. In D. Tall (Ed.), *Advanced Mathematical Thinking*, pp. 3-21. Dordrecht: Kluwer Academic Publishers.

Tall, D. (1996). Functions and Calculus. In A.J. Bishop K. Clements, C. Keitel, J Kirkpatrick and C. Laborde (Eds.), *International Handbook of Mathematics Education*, pp. 289-325. Dordrecht: Kluwer Academic Publishers.

Tall, D.O. and Vinner S. (1981). Concept image and concept definition in mathematics with particular reference to limits and continuity. *Educational Studies in Mathematics*, 12 (2), 151-169.

Technology, Computers, and Learning. URL: http://www.crito.uci.edu/tlc/html/findings.html.

The Math Forum. URL: http://www.mathforum.com.

Waterloo Maple Inc. URL: http://www.maplesoft.com.

APPENDIX

AN INTERVIEW WITH BILL RALPH

Bill Ralph teaches at Brock University. He is an accomplished mathematician, pianist and artist. His most recent computer art generated through the application of dynamical systems has attracted much attention.

(ERM) As a mathematician and a mathematics educator, why was it important for you to devote three years of your life to develop "Journey Through Calculus"?

(WR) I wrote JTC because I believe that multimedia is often better than traditional methods at teaching calculus concepts. I also believe that within 20 years the majority of mathematics teaching up to the first year university level will be handled by multimedia. Of course, there will be considerable resistance as there was when printed books dramatically altered oral traditions of teaching. I felt it would be a worthwhile undertaking if I could bring students a rich experience of calculus that helped them learn interactively, test themselves anonymously and enjoy the discovery of ideas. Since most traditional texts are designed to meet the needs of professors, it's not surprising that we don't expect students to learn calculus by reading the textbook. I specifically wrote JTC to give students an alternative to traditional resources that they could realistically be expected to use by themselves.

(ERM) Did your view of teaching and learning calculus change as you worked through this project?

(WR) Halfway through the development of JTC, I wanted to apologize to all of my students for all of the assumptions I'd made in 20 years of teaching them. Multimedia forced me to understand once again that calculus is hard and its concepts are not obvious. One of my students said that he liked JTC because it made so few assumptions and that was certainly one of my goals. When I first started working on JTC, I had trouble thinking outside of the pattern of lectures, textbooks and homework. In most lectures, the instructor is in charge and the students follow along. But the goal of JTC was to put the students in charge and keep them challenged and interested enough to move through the material by themselves. The positive response of my students to JTC has convinced me that, as educators, we are on the right track with multimedia.

(ERM) How would you like students to use JTC?

(WR) This last year, students watched me use JTC almost every day in class to demonstrate concepts and they used it in their weekly labs. In addition, students are given sections of JTC to work through at home to refine their understanding of specific concepts. My informal surveys say that students have found it very friendly and use it constantly. I can't ask for more than that.

(ERM) How would you like calculus instructors to use JTC?

(WR) Each screen in JTC is designed to be projected in front of a classroom. I use the simulations, games and modelling programs constantly in front of students to illustrate concepts. Lectures are so much more interesting when you can change simulations on the fly and then talk about what happened and why. With JTC, lectures are more motivated because they can be a response to something students have observed in a demonstration or simulation. In labs, I hope that instructors will make use of JTC's many programs for gathering data and modelling. It's so important for students to work with actual data so that they understand the practical applications of calculus and get a feeling for what quantities like derivatives really mean. Finally, instructors should be able to make good use of the nearly infinite number of algorithmically generated problems in JTC, many of which are purely visual.

(ERM) In the teaching and learning of mathematics, what opportunities will technology offer to future students?

(WR) Technology can give students greater insight into many concepts. For example, there's a game in JTC based on anti-derivatives in which you can't get a high score unless you understand the graphical relationship between a function and its anti-derivative. I saw students competing to get high scores in this game and there is no question that they understood the concept better than my students before JTC. Another example is modelling. In JTC, they work with real data and a simple program that allows them to visually fit their own personal model to the data. Students who use programs like JTC will be more comfortable generating and analyzing models they have fitted to data.

(ERM) What other questions would you have liked me to ask, and what would have been your responses?

(WR) I might have asked why there is so much resistance to introducing technology in the classroom. But I don't know the answer to that. Hopefully, people will see the possibilities in JTC.

AN INTERVIEW WITH KEITH TAYLOR

Keith Taylor is a Professor of Mathematics at the University of Saskatchewan and a Vice-President of the Canadian Mathematical Society.

(ERM) As a mathematician and mathematics educator, what attracted you to devote so much time to developing a Web-based course?

(KT) It happened incrementally. Our Extension Division wanted a pre-calculus course I had developed and converted to a correspondence version. As we looked for ways to get active learning built into this course, I became increasingly aware of the possibilities of JAVA for creating nice learning environments. Once I had a little bit of success, there were strong pressures to do more: there was a technical crew I wanted to keep employed; there were obvious improvements needed in our first course; there were potential students in remote communities who needed other courses; and there were organizations that wanted to provide financial resources. Each year, it took up more time.

(ERM) As you developed this project and as you worked with students in the course, has your view of teaching and learning mathematics changed?

(KT) A great deal. Somehow when we lecture to 100 students in the standard manner and 20 drop out, 25 fail, 20 get a final grade between 50 and 60%, and 20 are scoring over 80%, we consider that the natural course of things. We are not bothered by the fact that 65% of the students did not engage with us as we presented the material. In developing Web courses, I began to think a lot about the different ways in which someone can come to understand a mathematical concept, about the difference between technical facility and a more profound understanding, and about how we can teach the student how to write mathematics. These thoughts have begun to change my lecturing style.

(ERM) How would you like students to use your Web-based course?

(KT) My courses are aimed at upgrading and are not for credit. I would like students to be able to carry out that upgrading for its own sake - to build their level and confidence before they get into credit programmes.

(ERM) How would you like instructors to use your Web-based course, what is their role?

(KT) Our experience has been that students do much better with our courses if they have local support in the form of a proctor/tutor to keep them motivated and get them over tough spots. I don't think our courses are better than a full-time regular instructor can provide if one is available. They may find ways to use some of the tools for demonstrations. In fact, we are setting up a format that will make such usage more convenient.

(ERM) Can you visualize what learning mathematics will be like for students who decide to learn most of their mathematics through Web-based courses, either within a university setting or by distance education?

(KT) I have some ideas that might be completely wrong. I can see that it is now possible to create courses that are substantially better for many kinds of students than the traditional method. However, there are natural barriers that we may not cross for a long time. Such a course requires a skilled team (which includes an experienced mathematics professor) working full-time for at least two years. So the expense is too great for any university at this time unless they are creating a course that will be used widely. What may happen first is that single modules (say a module on Newton's method) gets developed which does a first rate job of the topic and includes features that adapt to the individual student's learning style. Instructors could assign that module to be completed in the next two weeks by students getting on the Web at their convenience with a rigorous evaluation procedure included - maybe with a laboratory style report required. We are currently working on such a module for conic sections to be used by high school teachers. Examples developed by others are beginning to emerge.

AN INTERVIEW WITH NICHOLAS JACKIW

Nicholas Jackiw is the developer of The Geometer's Sketchpad, one of today's leading examples of educational mathematics software.

(ERM) How would you characterize the environment that you have created in The Geometer's Sketchpad? What distinguishes this environment from other mathematics learning environments such as, pencil and paper, chalkboard, geometric models, etc.?

(NJ) Sketchpad is an open-ended workspace (both an environment and a set of tools) for exploring one's own geometric curiosity and understanding. At its heart, the software unfolds around a single principle, "Dynamic Geometry", which states that a visual, mathematical model should be as flexible, as transformable and permutable, and as mathematically general as is possible while maintaining precisely only the mathematical invariants (definitions, relationships, and constraints) which you specified explicitly when constructing the model. Explorations guided by this principle can be extremely useful within mathematics education (because they allow us to see "through" the specific to the general, and to readily identify invariance, relationships, and patterns of change). From a learner's perspective, such explorations wind up being engrossing, and even frequently entertaining (if I can use that word). This is because of the ways in which they simultaneously engage not only our capacity for logical deduction but our visual and spatial reasoning facilities, and even (due to the immediacy of the software's interaction paradigm) our haptic sense and potential for kinesthetic understanding. In a way, for the uninitiated, comparing Sketchpad to paper and pencil, models and chalk, is as interesting as contrasting it. Unlike lots of mathematical software, Sketchpad has little mathematics it "wants" to teach you; it has no curriculum to pursue. Like chalk and paper, Sketchpad is a medium as much for self-expression and inquiry as for executing assignments or responding to fixed questions. As an electronic medium, Sketchpad goes beyond the mechanical limits of paper and chalk or physical manipulatives, because it can electronically populate a mathematical realm somewhere on the border of the physical world and the learner's imagination, a world where triangles dance, lines really do go on forever, and physics need not apply.

(ERM) How do you visualize teachers using Geometer's Sketchpad in their teaching?

(NJ) Let me split my reply between the two perspectives of the "educational technologist." From an educator's perspective, I urge teachers to use Sketchpad first and foremost as learners would: as a tool for inquiry, for expression, and for communication. Experiencing new technologies from this perspective is unquestionably the best way to gauge their merits and their relevance to teachers' necessarily individual teaching situations and will provide a far more useful backdrop against which to frame, model, and assess student's interactions and achievements with a technology than any words its designer or other evangelist can give you. Second, as a technologist, I urge teachers to reflect critically on Sketchpad, as on all software, in terms of its contributions or impact (positive or negative) on their classrooms. Where teachers can expect conventional textbooks and other curricular materials to change astonishingly little over their professional career, software and more advanced technologies change every day. As professionals, teachers need experience not just in using particular pieces of software successfully, but in thinking about, understanding, and prioritizing the benefits of the many future potential software tools they'll encounter in their work. At the present moment, Sketchpad is regarded as something of an "exemplar" of educational technology. In a nationwide survey in the States last year "Technology, Computers, and Learning", maths teachers chose Sketchpad as the "most valuable software for students" by a wide margin. So I would hope teachers regard Sketchpad in two very separate lights: both as a practical and specific tool that can benefit their mathematics classroom and as a much more general example of how emerging technologies can change, and benefit, education.

(ERM) How do you visualize students using Geometer's Sketchpad?

(NJ) This sort of question cuts right to the heart of software design. Who is the student? What do I imagine her, or him, doing? In truth, I rarely have specific students or particular tasks in mind, other than imagining creative learners engaged in some pursuit of mathematical or graphical questions. In all its complexity, mathematics itself, and especially geometry, is marvellously self-organizing, from relatively simple axioms and a small number of conceptual tools. By providing such tools, and fertile mechanisms for their reapplication and reuse, Sketchpad attempts to provide multiple points of entry appropriate to various learners, ranging from middle school students to research mathematicians.

(ERM) What's the role of the Web in providing additional opportunities with Geometer's Sketchpad?

(NJ) For the past several years, I've been working on a variant of Sketchpad that operates within a Web browser. Hence activity and curriculum developers can "publish" dynamic geometry activities on the Web, or embed dynamic geometry illustrations and examples within the context of larger mathematical activities and sites devoted to them. A strange trajectory for most interesting, general purpose, mathematics software is that over time, they tend to evolve into "authoring environments" for student, teacher, and scholarly "publication" of materials. Thus you have Mathematica notebooks, MathCad documents, Maple environments, and so forth, all of which "wrap" software's innovative mathematics in some sort of publishable framework. In the past several years, however, the Web has emerged as the publishing medium par excellence. Not only can anyone write and illustrate Web pages, but one can embed any form of illustration or interactive content, and link to resources anywhere in the world. The Java Sketchpad variant is an attempt to integrate dynamic geometry, which is the most interesting part of Sketchpad, into that "authoring" model. But universal student access and fast internet connections are still not classroom realities, so I tend to think of this effort as more of a research prototype than a practical educational solution, except in established networked "distance learning" environments. (You can try Java Sketchpad yourself without needing access to any software beyond your Internet browser.) Separately, of course, the Web has proven itself a huge resource to teachers (and to a lesser degree, students) interested in sharing materials, ideas, activities, and student work. There are over 4,000 Web sites with some form of Sketchpad material on them! With that much information, it's important to develop sources of recommendation and reference to sites you find relevant, rather than just relying on search engines' ability to enumerate many of them. The Java Sketchpad site has a partial but by no means definitive list of useful world-wide resources, as does The Math Forum. Finally, the MSTE Sketchpad Teachers' Support Initiative at Queen's University is developing a site of resources recommended particularly for Ontario's teachers and provincial curriculum.

Eric R Muller
Brock University, Ontario, Canada
emuller@spartan.ac.brocku.ca

PHILLIP KENT AND RICHARD NOSS

FINDING A ROLE FOR TECHNOLOGY IN SERVICE MATHEMATICS FOR ENGINEERS AND SCIENTISTS

1. INTRODUCTION

The teaching and learning of 'service' mathematics — mathematics for non-specialists — is often considered quite separately from the didactical and epistemological problems facing specialist students. Compared to the corpus of work on undergraduate mathematics as a whole, the teaching of service mathematics remains relatively unexplored, and many of its fundamental assumptions (What is its purpose? What are the fundamental objects and relationships of study?) remain unexamined. In this paper, we will look, in particular, at service mathematics for engineers and scientists, but our main argument is intended to reflect more generally on the use and understanding of mathematics by non-mathematicians.

To begin with an obvious question: What does the word 'service' mean in the term service mathematics? Does it imply that the user needs only to know when to call for service, not how the service works or why? While there are some attractive implications of the notion that only professional mathematicians need to understand mathematics, they presume the extent to which the structure of mathematics must be formally learned and understood in order for it to be used effectively. And, while the idea that students can and should be taught mathematics by being shown what to do has become discredited within the mathematics education community at large, it is still the dominant paradigm of advanced mathematical education, particularly in the service domain. So it is worthwhile to consider approaches which throw light on the relationship between use and understanding, and to regard this relationship as a fundamental research issue rather than an axiom.

It *is* axiomatic that mathematical knowledge and techniques provide an essential 'service' to the engineering and scientific disciplines, but 'service mathematics' is not, according to many accounts over the last ten years or so, providing the right training (see, for example, Howson et al 1988, Sutherland and Pozzi 1995, IMA 1999). Service mathematics as a subject was stable in the United Kingdom for many decades, but is now affected by two major trends: firstly, curriculum change in schools means that too many students are not sufficiently prepared to tackle the traditional service curriculum. Secondly, the revolution in professional engineering and scientific practice brought about by computer technology means that the traditional curriculum, with its strong basis in pen-and-paper calculation techniques, has become of questionable relevance to professional practice.

Derek Holton (Ed.), The Teaching and Learning of Mathematics at University Level: An ICMI Study, 395—404.
© 2001 *Kluwer Academic Publishers. Printed in the Netherlands.*

mathematics by itself and a great deal more time working on, say, realistic engineering design projects, in which they will rely on computer software to handle calculations where necessary (indeed, the pen-and-paper calculation methods that are taught in the traditional service mathematics curriculum are often irrelevant *for the purposes of calculation* in such a context). Sutherland and Pozzi conducted a major opinion survey of UK engineering academics, and report that:

> There were some differences in opinion about the relative balance of more analytic mathematical skills and engineering design within engineering degrees. A view implicit in many of the responses is that analytic skills and engineering design are in opposition. (Sutherland and Pozzi 1995, p. 7)

They also report that "engineering design was ranked as being more important nowadays than mathematics by the majority of respondents" (p. 21) with pressure on mathematics teaching time as the content of engineering degree programmes shifts towards design, communication, languages, business and finance. It is remarkable that mathematics, sometimes championed as the 'language of technical communication', was not considered relevant to communication and design by this sample of engineers.

On the other hand, there are plenty of engineers and scientists who see virtues in the traditional service curriculum, and whilst recognising the need to update the service mathematics approach with a technological component, insist that students should still develop an appreciation for mathematics by doing mathematics *per se*. For example, a civil engineer colleague of ours emphasises the importance for his students in engaging with the 'logical' thinking processes of mathematics, compared with the heuristic and experiential nature of much else that they study in an engineering degree course, and in the culture of engineering practice. This view is echoed by Howson et al. (1988, p. 9): "It is a striking fact that non-mathematicians — even more than most mathematicians — insist on the power and value of a mathematical mode of thought. The idea is expressed equally forcefully by biologists ("never mind what you teach: teach students to reason well") and by engineers".

It seems to us that there is a need to call into question what is meant by mathematical knowledge from the viewpoints of different groups of people. Although there is apparently agreement on what 'mathematics' is, our perception, based on discussions with mathematicians, scientists and engineers, is that mathematics is quite a different thing from their different points of view. These differences are revealed, for example, in the way that 'mathematical' software packages have been designed to appeal to the ways of thinking of different types of users, with some targeted on the needs and interests of mathematicians and others on the needs and interests of engineers and applied scientists (see Kent and Noss 2000).

Another example of the difference was provided for us when discussing the mathematics used in structural analysis with a civil engineer. When a beam is supported at different points along its length, and then loaded with a variable weight (W), the theoretical result for the small deflections (y) of the beam is conventionally written by engineers in the form:

$$EIy = \frac{5}{3}(4w - 22275)x + (33000 - 4W)\frac{x^3}{60} - \frac{250}{3}(x - 3)^4$$
$$+ \frac{250}{3}(x - 6)^4 + 27000 + 14W\frac{(x - 10)^3}{60}$$

(this expression is calculated for a particular beam configuration — see Kent and Noss 2000 for details; *EI* is a large constant that is determined by the material strength and geometry of the beam, and for a beam of length 10 metres, so $0 \le x \le 10$, the deflection is of order millimetres). Now, there is a curious feature of the expression, in that the polynomial terms $(x - 3)^4$, $(x - 6)^4$ and $(x - 10)^3$ are all implicit *ramp functions*, functions defined to be zero when $x < 3$ and to be $(x - 3)^4$ when $x \ge 3$, etc. (Figure 1).

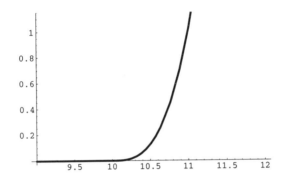

Figure 1. The ramp function '$(x - 10)^3$'.

For our engineer colleague there was nothing strange when he explained to us, "*of course*, these terms here are ramp functions ...", but we were surprised to be discovering something new about polynomials: basic and boring mathematical objects, but when looked at in this particular engineer's way, they *are* 'ramps'. It would be wrong to overstate the importance of a tiny episode, but it did bring home to us the fact that whenever people make use of mathematics, they attach new meanings, relevant to their own context, even to the most basic of mathematical objects.

In this paper, we explore the problem of service mathematics from this, essentially epistemological, angle. We contend that it is only by taking account of how mathematical knowledge is constructed in the 'host' domain, that we can engage with the difficulties that students have in this field.

2. TRENDS IN SERVICE MATHEMATICS IN THE U.K.

In the United Kingdom the last 10 years have witnessed major changes in university education in general, and mathematics in particular. In general, we have today a vastly enlarged student cohort, as government policy dictates that more and more young people should benefit from university-level education. As a result, it is no longer possible for most university departments to select only the most highly-qualified students from schools. And in mathematics, there has been major curriculum restructuring in school mathematics, with Euclidean geometry and algebra no longer being kingpin ideas in the secondary curriculum, and a much-reduced attention to differential and integral calculus in the 'advanced level' (ages 16-18) curriculum (see, for example, Sutherland and Pozzi 1995, Engineering Council 2000).

All these changes have undone the expectations of UK universities about incoming students, who are increasingly regarded as under-prepared for traditional mathematics courses. Indeed, this so-called 'mathematics problem' is a growing concern amongst academics and graduate employers. It has been the subject of a series of reports commissioned by various UK professional mathematical and engineering institutions (IMA et al. 1995; LMS 1995; Sutherland and Pozzi 1995; IMA 1999; Engineering Council 2000), as well as a number of conferences (eg. Mustoe and Hibberd 1995, Hibberd and Mustoe 1997).

As we have pointed out, the problem is not only the changed mathematical backgrounds of students. There have been significant shifts in the actual practice of mathematical work, as working scientists and engineers make use of more and more sophisticated 'invisible mathematics' embedded in computational tools such as CAD systems (where the most advanced numerical techniques for solving complex systems of equations are made available at the press of a button).

As a result of this ubiquitous technological presence, one remedy for the 'mathematics problem' has been simply to reduce the mathematical content in science and engineering courses, and to have students rely on computer-based tools to do much of the mathematical computations that they need to do. Sutherland and Pozzi (1995, p. 27) quote the following (rather extreme) view from an engineer: "Engineers use computers to design, build and run bridges, road systems, chemical plants etc. The clever people who write the packages need to understand calculus, but the engineer doesn't, in fact she/he is probably not aware of the complexity of analysis". This kind of reliance on technology is a pragmatic solution in the face of rapid change in universities, however its long-term viability is questionable. For one thing, we are seeing in the European Union the gradual harmonisation of professional qualifications; already the UK differs sharply from the rest of the EU in the lack of depth of mathematical preparation of its professional engineers (both at school and university levels), and to achieve harmonisation the service mathematics curriculum needs to be if anything significantly strengthened in content. (The acuteness of the problem in the UK is in some part due to the fact that there is not a two-tier — technician vs. professional — system of engineering qualifications, which is common in other countries; instead, all engineering students in the UK are required to aspire to the professional level. A two-tier system, consisting of 'incorporated', i.e. technician, and 'chartered' status, is currently under development — cf. IMA 1999.)

More fundamentally, though, there is the issue of use versus understanding. There is, undoubtedly, an argument that significant kinds of engineering and science can be done with mathematics which has already been done by someone else, and wrapped up into computational tools which are 'ready to use'. In many situations, this use of tools works fine. But there is a widespread feeling that this kind of use is not enough: situations can and do arise which are beyond the capability of the software, and which necessitate more than mere use — they require understanding. This point is well-expressed in IMA et al. (1995):

> Chartered Engineers not only need to have the ability to apply their knowledge but also must have a good understanding of the fundamental ideas and relevant techniques so as to enable them to adapt and update their knowledge to keep pace with future developments. Fluency in the execution of mathematics coupled with mathematical literacy is required. This concept is important and is analogous to language literacy, which is well understood. An educated user of English should be able to communicate clearly ideas on anything using written English. This does not mean that one is expert on all aspects but that one understands the way to express ideas. 'Literacy' is the intellectual ability that enables the process of good communication and requires a range of experience for its acquisition. 'Mathematical literacy' involves a similar process and the need for Chartered Engineers to possess it is confirmed by even a glance at the engineering literature. (*ibid*, p. 16.)

This problem is by no means restricted to university-level mathematics. In fact, we have located much the same need for this kind of mathematical 'literacy' (or 'habits of mind', to use the term proposed by Cuoco et al 1996) in a range of professional situations including nursing, commercial aeroplane piloting, and investment banking (see, for example, Noss and Hoyles, 1996; Noss, Pozzi and Hoyles, 1999; Hoyles, Noss and Pozzi, in press).

It is standard practice for software companies to incorporate more and more 'features' into their software in response to the demands of users, thus making them ever more unwieldy and complex. But the problem for the user does not lie within the software: it is about having sufficient mathematical literacy to understand what the software is doing, and being able to open up the 'black box' and see how it works. And the challenge for educators, with the responsibility for developing mathematical literacy in their students, is to decide which aspects of mathematics can be left boxed-up in professional software packages, and which will need to be opened up and explained. And crucially, what does mathematical literacy consist of (cognitively-speaking), and how does it develop?

As we expressed it in a previous paper (Kent and Noss 2000), these are questions about the 'visibility' of the mathematical models used in science and engineering, that is, the extent to which the mathematical 'mechanisms' which drive the model are explicit and evident to the user of the model. In fact, a key issue is not only whether or not to design for visibility, but to attempt to answer the question of what mathematics to make visible, and in what form. This is clearly an extremely complex question; we shall only outline part of an answer via the example in the next section.

3. TECHNOLOGY AND MATHEMATICS IN CONTEXT: THE MATHEMATICS LABORATORY

The 'application' metaphor underpins the mathematical training of undergraduate students in the UK: in service mathematics courses, they are taught 'the mathematics' that they will afterwards learn to apply to their subject. If it is the case that, as we have suggested, mathematical meanings are highly dependent on the application context in which mathematics is being used, then the pedagogical separation of mathematics from context can only be problematic.

In the following, we offer an example of mathematics learning 'in context', developed by one of us (P.K.) and his colleagues in the METRIC Project[1] at Imperial College in the University of London. Since 1994, the Chemistry Department at Imperial College has been developing, with METRIC's help, a completely new undergraduate mathematics course, the 'Maths Lab', which is a radical replacement for traditional service mathematics teaching.

The Maths Lab is technology-based (on the computer algebra system, Mathematica) and is closely aligned with the courses in physical chemistry (for detailed background, see Templer et al 1998). Half of the Chemistry graduates go on to do research in academia or industry, so this course has a strong 'research' flavour:

> We are trying to introduce mathematics as a natural, integral part of chemistry; Mathematica's power gives the students the opportunity to do this in an authentic, uncontrived way, with something of the flavour of real research. ...

> As one might expect, students experience a mixture of emotions when presented with this sort of challenge. One of these is certainly shock. We have deliberately set out to stretch students, but not in a way that they would expect. Their anticipation is that they will learn the techniques and tricks which most perceive as being the proper realm of mathematical experience. Instead we present them with the sort of chemical problems found in research, and with the help of a powerful mathematical toolbox, ask them to investigate the chemistry using those tools. That is not to say that they are not learning any mathematics, far from it. The problems are designed in such a way that they have to understand the mathematical processes that are being performed and the limitations that these may impart on their results. (Ramsden and Templer 1998)

And there is a very strong emphasis on the mathematical models used in chemistry:

> What sort of mathematician does the chemist of the 21st century need to be? Any lack of fluency in the techniques of traditional 'maths methods' will possibly hamper her less than it would have done her predecessors: most modern problems don't yield to such techniques anyway. But in any case, that kind of fluency can sometimes mask a more serious problem, namely a failure to understand and appreciate the mathematical models that lie *beneath* calculations (and, crucially, to appreciate their limitations). In a world where such models are ubiquitous, that kind of failure is potentially more serious than ever; the use of inappropriate models can utterly invalidate a piece of research, especially if (as often happens) the problem fails to show up until it's too late! ... [W]e needed [therefore] to include a substantial element based on the analysis, construction and critique of *underlying mathematical models*. (*ibid.*)

---

[1] That is, Mathematics Education Technology Research at Imperial College. See the project Web site at http://metric.ma.ic.ac.uk.

To give an example, one of the Maths Lab assignments is concerned with modelling the dynamics of ion collisions. The students study the different possible interactions between three ions, according a Newtonian mechanical model, where the ions are treated as charged point particles, and the equations of motion are solved numerically. Although this problem is somewhat simplified from a 'real' research problem (it is restricted to one dimension, so the particles are moving on a straight line) the students are explicitly working with a sixth-order, non-linear differential equation system. The solution of the equations is treated as a 'black box' operation by Mathematica (it would be very difficult to do otherwise, and fortunately it turned out that Mathematica's code is highly reliable). However, the students are required to check on the reliability of the numerical solutions, and hence the validity of the Newtonian model, by showing that the total energy of the ion system is (or is not) conserved.

It is important to note what mathematics has been hidden or not hidden in this energy check task. In fact, there is a *didactical inversion* in the design: that is, using the capability of the technology to allow students to carry out some task using mathematics which the students don't know yet (i.e. solving non-linear differential equations numerically) in order that they can focus on some conceptual points which the mathematics makes accessible (i.e. energy conservation in a non-trivial chemical interaction). The task in also noteworthy (and typical of the Maths Lab approach) in that the students must decide for themselves how they want to 'represent' the energy variations over time (most do use conventional energy-time graphs, but some opt for non-graphical, more statistical methods). This process of designing representations is not conventionally recognised as mathematical work, at least not for non-expert, beginning students. But, we think that it reflects a kind of mathematical thinking that has a great deal to do with having a well-developed mathematical literacy. That is, knowing how to adapt mathematical models to new situations, and not being confined to using the models only in the way that someone else has provided.

Kent and Stevenson (1999) carried out an evaluation of the ion collision task, with a particular interest in how the students' calculus knowledge was applied to the problem of determining the total energy. To do this, the students need to interconnect mathematics knowledge (integral calculus) with fundamental physics knowledge (the concepts of work and potential energy), in the context of a particular chemical problem. What we found (on the basis of detailed observation of the students at work, plus follow-up interviews) is that our sample of students all found this difficult. They knew all the components of the task but had to work hard at interconnecting them, and making them precise enough for expression in a computer program. As they worked, the students were focussed on solving the given problem (is energy conserved or not?), and were not necessarily thinking about mathematics, physics or chemistry with any clear separation. And for all the students that we interviewed (most with a very strong school mathematics background) any formal notion of integral calculus was a half-forgotten relic of school mathematics.

Is there a problem in that? If integration is to become a robust part of the student scientist's 'mathematical toolkit', what types of mathematical approach, allied to the subject context, *are* appropriate? Our view is that the formal notion of integral is backgrounded in this particular modelling task, but might not be in others. But these questions of detail are less important for first-year students at the beginning of their mathematical training, compared with getting them to develop good habits in mathematical thinking right from the start. The hypothesis is that students properly armed with the right habits of mind will be able to deepen their mathematical (and physical) knowledge where necessary later. Anyway, only a fraction of the students will specialise in mathematically-oriented (i.e. physical) chemistry.

But what happens later? Will the students ever 'properly' understand mathematics? From a mathematician's point of view, the answer is most likely 'no'. As one of the reviewers of (Kent and Stevenson 1999) put it: "Are the authors just developing a cook-book course using modern technology in which underlying concepts aren't even superficially understood, as they are at least in traditional courses?" The response from our Chemistry colleagues on this point is unequivocal: 'superficial' understanding isn't acceptable, even if it's 'proper' mathematics:

> This may not be the way that we were taught mathematics, nor the way a mathematician would describe mathematical training, but it is perhaps a more relevant and practical approach to training a chemist how to deal with the mathematics that may confront her in the laboratory. (Ramsden and Templer 1998)

4. CONCLUSION: MAKING CONNECTIONS

In this paper we have analysed the problems of separating mathematics from its context of 'application', the increasing crisis (in the UK at least) in service mathematics teaching (with its dominant metaphor of application), and the challenge of developing alternative approaches that recognise the contextualised roles and meanings of mathematics.

When choosing to teach mathematics 'in context', advice from experts in that context is essential, as we found out in the METRIC project on many occasions. Thus, this kind of curriculum development work necessitates not only making connections between knowledge domains, but making contacts between people working in the different domains. A form of contact which service mathematics tends to rule out of hand by enforcing a separation of domains.

Michael Clayton (1998), a mathematician working in the multidisciplinary environment of the telecommunications industry, has pointed to the connecting effects of technology on the relationships between mathematicians and engineers in industrial practice, by overturning the traditional roles of mathematicians as the makers of models, and other people in the design and production process as the consumers of models:

> General-purpose IT tools such as spreadsheets, and mathematically based environments and workbenches such as Mathcad and Matlab have made it easier for engineers, dealers, salesmen, managers and others to construct their own models and refine them for specific applications. When time is of the essence, the value of these tools lies in the rapid prototyping they allow: initial modelling ideas can be investigated by the potential users, and the resulting interaction often leads to an improved match between the model and the users' requirements. Modern graphical user interfaces ... can be designed to make even the most sophisticated special-purpose models accessible to the people who

need to use them, helping to remove the 'ivory tower' and 'back room' images that have sometimes been attached to mathematicians in the past. (Clayton 1998, p. 25.)

We see no reason why Clayton's observations should not carry over into university service mathematics teaching. Debates in universities about the future of service mathematics very often come down to an either/or situation: either mathematics should be taught by mathematicians (who understand the subject, but not the application) or by engineers and scientists (who know how to use mathematics in application). This kind of debate looks increasingly anachronistic in the light of technological change, and contemporary industrial working practice. We think it is more instructive to examine the potential of mathematical technology to change the relationships between mathematicians and those who 'apply' mathematics, and to find new ways to connect people and the knowledge domains in which they work.

REFERENCES

Clayton, M. (1998). Industrial applied mathematics is changing as technology advances: What skills does mathematics education need to provide?. In C. Hoyles, C. Morgan and G. Woodhouse (Eds.), *Rethinking the Mathematics Curriculum*, pp. 22-28. London: Falmer Press.

Cuoco, A.A., Goldenberg, E.P., and Mark, J. (1996). Habits of mind: An organizing principle for mathematics curriculum. *Journal of Mathematical Behavior*, 15, 4, 375-402.

Engineering Council (2000). *Measuring the Mathematics Problem*. London: The Engineering Council.

Hibberd, S. and Mustoe, L.R. (Eds.) (1997), *Proceedings of the 2nd IMA Conference on the Mathematical Education of Engineers*. Southend-on-Sea: Institute of Mathematics and its Applications.

Howson, A.G., Kahane, J.-P., Lauginie, P. and de Turckheim, E. (1988). *Mathematics as a Service Subject*. Cambridge: Cambridge University Press (ICMI Study Series).

Hoyles, C., Noss, R. and Pozzi, S. (in press) Proportional reasoning in nursing practice. *Journal for Research in Mathematics Education*.

I.M.A. et al (1995). *Mathematics Matters in Engineering*. Southend-on-Sea: Institute of Mathematics and its Applications.

I.M.A. (1999). *Engineering Mathematics Matters: Curriculum proposals to meet SARTOR 3 requirements for Chartered Engineers and Incorporated Engineers*. Southend-on-Sea: Institute of Mathematics and its Applications.

Kent, P. and Noss, R. (2000). The visibility of models: Using technology as a bridge between mathematics and engineering. *International Journal of Mathematical Education in Science and Technology*, 31, 1, 61-69.

Kent, P. and Stevenson, I. (1999). Calculus in context: A study of undergraduate chemistry students' perceptions of integration. *Proceedings of the 23rd Psychology of Mathematics Education Conference*, 3, 137-144.

L.M.S. (1995). *Tackling the Mathematics Problem* (A joint report of the LMS, the Institute of Mathematics and its Applications, and the Royal Statistical Society). London: London Mathematical Society.

Mustoe, L.R. and Hibberd, S. (Eds.) (1995). *Mathematical Education of Engineers (Proceedings of the 1st IMA Conference)*. Oxford: Oxford University Press.

Noss, R. and Hoyles, C. (1996). The visibility of meanings: Modelling the mathematics of banking. *International Journal of Computers for Mathematical Learning*, 1, 1, 3-31.

Noss R., Pozzi, S. and Hoyles, C. (1999) Touching epistemologies: Statistics in practice. *Educational Studies in Mathematics*, 40, 25-51

Ramsden, P. and Templer, R. (1998). A new approach to mathematical training for chemists. Unpublished article. [Download from http://metric.ma.ic.ac.uk/articles/New-approach.pdf]

Sutherland, R. and Pozzi, S. (1995). *The Changing Mathematical Background of Undergraduate Engineers: A review of the issues*. London: The Engineering Council.

Templer, R., Klug, D., Gould, I., Kent, P., Ramsden, P. and James, M. (1998). Mathematics laboratories for science undergraduates. In C. Hoyles, C. Morgan and G. Woodhouse (Eds.), *Rethinking the Mathematics Curriculum*, pp 140-154. London: Falmer Press.

Phillip Kent and Richard Noss,
University of London Institute of Education, United Kingdom
p.kent@mail.com
rnoss@ioe.co.uk

SECTION 6

ASSESSMENT

Edited by Mogens Niss

KEN HOUSTON

ASSESSING UNDERGRADUATE MATHEMATICS STUDENTS

1. INTRODUCTION

Any discussion of assessment must necessarily include a discussion of the curriculum, how it is designed and organised, and what it contains. It must examine the aims of the course that students are taking, and the objectives set for that course and the individual modules that comprise the course. (Here I am using terminology common in the UK. The 'course' students take is 'the whole thing', the 'programme'. A course in this sense consists of 'modules' or 'units', commonly called 'courses' in the USA, so beware of confusion!) The discussion must consider who is doing the assessing, why they are doing it, what they are doing and how it is being done. It must consider how assessors become assessors and how those assessed are prepared for assessment. And it must consider if the assessment is valid and consistent, and if it is seen to be so.

It might also be useful at this stage to define what we mean by a 'mathematician'. There is a real sense in which almost everyone could be described as a mathematician in that they make use of some aspect of mathematics – be it only arithmetic or other things learnt at primary/elementary school. The term could be used of those who have taken a first degree in mathematics and who use it in their employment. Or it could be reserved only for those who have a PhD and who are doing research in pure mathematics or an application of mathematics. We will use the middle of the road term. In other words, a mathematician will be one who has studied the subject at least to bachelors degree standard (and of course that varies across the world!), and who is using some aspect of advanced mathematics in their work. Such people could join a professional or learned society such as the UK based Institute of Mathematics and its Applications. So we are primarily concerned with the higher education of these people who can rightly be considered to be professional mathematicians. But also there are many disciplines wherein mathematics is an extensive and substantial component of study. Examples are physics or electronic engineering. The mathematical education of professionals in such fields as these could also come under the remit of this article in that many of the suggestions made could enhance the teaching, learning and assessment of students in these fields.

Traditionally assessment in higher education was solely summative and consisted of one or more time-constrained, unseen, written examination papers per module. A typical, and in some places predominant, purpose of assessment was to put students in what was believed to be rank order of ability. Students were, perhaps,

407

Derek Holton (Ed.), The Teaching and Learning of Mathematics at University Level: An ICMI Study, 407—422.
© 2001 *Kluwer Academic Publishers. Printed in the Netherlands.*

asked to prove a theorem or to apply a result, or to see if they could solve some previously unseen problem. Generally this method succeeded in putting students in a rank order and in labelling them excellent, above average, below average or fail. But was it rank order of ability in mathematics or rank order of ability to perform well in time-constrained, unseen, written examination papers? Sadly it was the latter, and while the two may coincide, this is not guaranteed. Taking time-constrained, unseen, written examination papers is a rite of passage, which students will never have to do again after graduation and which bears little relationship to the ways in which mathematicians work. While it is true that working mathematicians are sometimes under pressure to produce results to a deadline, the whole concept of time-constrained, unseen, written examinations is somewhat artificial and unrelated to working life.

It is in this context that people started to think about change, change in the way courses are designed and organised, change in the way course and module objectives are specified and change in the way students are assessed and in the way the outcomes of assessment are reported. It is usually the case that 'what you assess is what you get', that is, the assessment instruments used determine the nature of the teaching and the nature of the learning. Learning mathematics for the principal purpose of passing examinations often leads to surface learning, to memory learning alone, to learning that can only see small parts and not the whole of a subject, to learning wherein many of the skills and much of the knowledge required to be a working mathematician are overlooked. In time-constrained, unseen, written examinations no problem can be set that takes longer to solve than the time available for the examination. There are no opportunities for discussion, for research, for reflection or for using computer technology. Since these are important aspects of the working mathematician's life, it seems a pity to ignore them. And it seems a pity to leave out the possibilities for deep learning of the subject, that is, learning which is consolidated, learning which will be retained because it connects with previous learning, learning which develops curiosity and a thirst for more, learning which is demonstrably useful in working life.

This is, of course, a caricature of 'traditional' assessment, but it is not too far from the truth, and it brings out the reasons why some people in some societies became unhappy with university and college education. Consequently those who educate students now pay attention to stating aims and objectives, to redesigning curricula and structures and to devising assessment methods which promote the learning we want to happen and which measure the extent to which it has happened. And they pay attention to the need to convince students and funding bodies that they are getting good value for their investment of time and money.

The discussion on course design and assessment is also tied up with the discussion on 'graduateness'. What is it that characterises college or university graduates and distinguishes them from those who are not? Is it just superior knowledge of a particular topic, or is it more than that? It is, of course, more than that. It is not easy to define or even to describe, but it has to do with an outlook on life, a way of dealing with problems and situations, and a way of interacting with other people. (This is not to denigrate the learning that non-college graduates get from 'the university of life', nor to suggest that they are inferior as people. It is to do with considering the 'added value' of college or university education.) Traditionally graduateness was absorbed, simply through the university experience, but now that we have systems of mass education in many countries of the world, we need to pay attention to the development of graduate attributes in students so that they do, indeed, get value for money. In many instances, and mathematics is no exception, it is the 'more than' that is important when it comes to finding and keeping employment. Subject knowledge is important but so also are personal attributes. It is highly desirable that students develop what have come to be known as 'key skills' while they are undergraduates, and not just because employers are saying that the graduates they employ are weak in this area. Innovative mathematics curricula seek to do this by embedding the development of key skills in their teaching and learning structures. (Key skills are often described as employability skills or transferable skills. They include such skills as written, oral and visual communication, time management, group-work and team-work, critical reflection and self assessment, and computer and IT, and aural skills.)

Who are the stakeholders in an undergraduate's education? First and foremost are the students themselves. They are investing time and effort and they want to know that they are getting a return on this investment. Most of them realise that it is not enough for them to be given a grade; they know that they have to earn it. So they need to know what performance standards are required and they need to be able to recognise within themselves whether they have achieved these standards or not. This raises the question of self-assessment and ways of promoting self-assessment. Giving 'grades that count' is one way of encouraging students to carry out tasks.

The next stakeholder to consider are the teachers. It is their job to enable learning and so they need to know what learning has taken place. Financial sponsors of students are also stakeholders. They, too, want to know if they are getting a good return on their investment. Finally, in the stakeholder debate, there is a demand from society, students themselves, universities, prospective employers, that students be summatively assessed, ranked and labelled in such a way that they may be measured, not just against what they are supposed to have learned, but also against their peers across the world.

This chapter will consider all of these features, but will focus on assessment, as that is its theme. It will look at the purposes and principles of assessment and then it will move on to consider the aims and objectives of courses and modules. Innovative methods of assessment will be reviewed and discussed, and this will be the biggest part of the chapter. Ways of disseminating information about new assessment practices will be discussed, as will obstacles to change. Finally pertinent research issues will be mentioned. The chapter will close with an annotated bibliography of pertinent books and papers dealing with these issues.

2. PRINCIPLES AND PURPOSES OF ASSESSMENT

Perhaps the only principle that should be applied is 'fitness for purpose'. To achieve this, assessment methods should be intimately related to the Aims and Objectives of the Module under consideration. And it should be born in mind that the assessment methods used will influence the learning behaviour of students to a considerable extent.

There are a number of purposes of assessment that should be considered:

1. to inform learners about their own learning.
2. to inform teachers of the strengths and weaknesses of the learners and of themselves so that appropriate teaching strategies can be adopted.
3. to inform other stakeholders – society, funders, employers including the next educational stage.
4. to encourage learners to take a critical-reflective approach to everything that they do, that is, to self assess before submitting.
5. to provide a summative evaluation of achievement.

3.   AIMS AND OBJECTIVES

Aims and objectives should be established both for a course and for each of the modules that comprise the course. The aims of a course are statements that identify the broad educational purposes of the course and may refer to the ways in which it addresses the needs of the stakeholders. Here are some examples; there are, of course, many more and each provider must write their own:

1. To provide a broad education in mathematics, statistics and computing for students who have demonstrated that they have the ability or who are considered to have the potential to benefit from the course.
2. To develop knowledge, understanding and experience of the theory, practice and application of selected areas of mathematics, statistics, operations research and computing so that graduates are able to use the skills and techniques of these areas to solve problems arising in industry, commerce and the public sector.
3. To develop key skills.
4. To provide students with an intellectual challenge and the practical skills to respond appropriately to further developments and situations in their careers.
5. To prepare students for the possibility of further study at post graduate level, including a PhD programme or a teacher training programme.

It would be necessary to indicate how each of the modules selected for a course helps to achieve the aims of the course. The aims of the individual modules should 'map' to the overall aims of the course. Objectives are statements of the intended learning outcomes that would demonstrate successful completion of the course or module, and that would warrant progression through the course and the eventual award of a degree. Module objectives should identify the knowledge, skills and attributes developed by a module, and course objectives should identify the knowledge, skills and attributes developed by the totality of modules selected for the course. Objectives may include reference to subject knowledge and understanding, cognitive skills, practical skills and key skills. They should be clearly relevant to fulfilling the aims and, above all, they should be assessable, that is, we should be able to devise assessment instruments that allow students to demonstrate that they have achieved the learning intended, and, if appropriate, to what extent. Here are some examples of course objectives: -

On completion of their studies graduates will have:

1. an understanding of the principles, techniques and applications of selected areas of mathematics, statistics, operations research and computing.
2. the ability and confidence to analyse and solve problems both of a routine and of a less obvious nature.
3. the ability and confidence to construct and use mathematical models of systems or situations, employing a creative and critical approach.
4. effective communication skills using a variety of media.
5. effective teamwork skills.

A course document should demonstrate how the aims and objectives of the constituent modules contribute to the overall course aims and objectives. Here is an example of the aims and objectives of a module, taken from an introductory module on mathematical modelling. (These aims and objectives are those of module MAT112J2, University of Ulster. Full details may be read under 'Syllabus Outline' at http://www.infj.ulst.ac.uk/cdmx23/mat112j2.html.) Note that an indication of the method of assessment of each objective is given.

Aims: The aims of this module are to:

1. enable students to understand the modelling process, to formulate appropriate mathematical models and to appreciate their limitations.
2. develop an understanding of mathematical methods and their role in modelling.
3. study a number of mathematical models.
4. develop in students a range of key skills.

It can be seen how these module aims help to meet the aims of the course listed above. Thus this module contributes to developing mathematical understanding, problem solving, and key skills.

Objectives: On completion of this module, students should be able to:

1. Formulate mathematical models and use them to solve problems of an appropriate level. (Assessed by coursework and written examination.)
2. Solve simple differential equations using calculus and computer algebra systems. (Assessed by written examination.)
3. Describe and criticise some mathematical models. (Assessed by coursework.)
4. Work in groups and report their work in a variety of media. (Assessed by coursework.)
5. Work both independently and in support of one another. (Assessed by coursework.)
6. Demonstrate other key skills. (Assessed by coursework.)

Again, it can be seen how these module objectives map to the course objectives listed above. There are references to the assessment of mathematical techniques, the construction and use of mathematical models, and key skills.

Of course, aims and objectives are not created in a vacuum. They evolve from the previous and present experiences of the lecturing staff who design the course and its constituent modules, and they are reviewed and modified from time to time as circumstances permit or demand. Nevertheless, the objectives for each module and for the course as a whole should be stated and essentially should be a form of contract between the lecturer and the students. Furthermore, detailed assessment criteria should be drawn up so that lectures have a well-defined framework in which to work and students have clear guidelines to what they have to do in order to succeed. This contractual arrangement does to some extent limit the power traditionally wielded by lecturers. This is a necessary and desirable consequence of the innovations described in this paper. It makes students more powerful in the right context, namely their own learning, in that it does require students to take more responsibility for themselves.

4. EXTERNAL ASSESSMENT OR EVALUATION

While this paper is primarily about the assessment of student learning, it may be appropriate to mention current developments in the evaluation of institutions, their courses and modules, and the teaching and other staff who deliver these. In the UK, for example, the government agency, The Quality Assurance Agency for Higher Education has a remit to review the quality of provision of education by institutions. Mathematics courses were reviewed between 1998 and 2000, along with several other subjects. The whole of university life is covered in approximately six-year cycles. Institutions are required to write an evidence-based self-assessment document (the SAD), which a visiting team of reviewers will scrutinise and make a judgement on. The SAD outlines the aims and objectives of the provision and provides details of the physical and human resources available. It then gives the institution's own assessment of its quality of provision under six headings:

1. Curriculum Design, Content and Organisation;
2. Teaching, Learning and Assessment;
3. Student Progression and Achievement;
4. Student Support and Guidance;
5. Learning Resources;
6. Quality Management and Enhancement.

Evidence to support the claims made in the SAD must be provided and may be found in documents and in observation of teaching.

This peer-review process (the reviewers are selected from other, similar institutions) devours a considerable amount of academic time and energy, and it remains to be seen if the improvements justify the cost. It is part of the general move in society to satisfy the public demand for public accountability of public funds. On the positive side, it has encouraged institutions to think about their course provision in a way that would be new to many.

5. METHODS OF ASSESSMENT

Once the learning outcomes or objectives have been articulated, suitable assessment methods have to be selected. (In practice, the articulation of objectives and the selection of assessment methods will proceed hand-in-hand.) This should be done in such a way as to ensure that the assessment methods are appropriate and allow students to demonstrate positive achievement. There should be transparent assessment criteria, which should be explained to students, if possible with examples of good work and not so good work harvested from previous cohorts, or descriptions of excellent, median and pass level performance. Ideally the assessment criteria should be drawn up in debate with the students, without sacrificing the lecturer's expertise and experience. Assessment should blend with the teaching and learning pattern. This section will now review some assessment practices that have been developed and used successfully.

5.1 Individual project work

Project work, both individual project work and group project work, is used widely. This has been a feature of many undergraduate mathematics courses for over twenty-five years, so it can hardly be described as innovative. Individual projects are often given to final year students and are substantial pieces of work. The very least the project is worth is about one sixth of final year studies, but it may be worth more than that. The topics set for investigation can be quite demanding and give scope for considerable initiative and independent work by students. Projects demand research and investigation and the production of a written report and some sort of presentation, such as a seminar, a poster or a *viva voce* examination. Students learn to conduct research and to organise information and present it cogently.

But now comes the hard part – assessing it reliably and validly – and ensuring that students know how it is to be assessed. If students know the assessment criteria and have some idea of what constitutes good or not so good work, then they are in a better position to assess their own work before submitting it, and in a better position to assess one another's work in a peer support activity. Project work like this is a good method for assessing many of the objectives outlined above that lead to the development of 'the way of life' of a working mathematician in whatever guise they may find themselves.

Experienced project assessors can usually come to an accurate judgement of a student's work fairly quickly and can defend that judgement to their peers. But there still is an element of subjectivity in this and, to remove as much of this as possible and to achieve consistent marking by several assessors, consultation and training are necessary. The team of assessors should develop assessment criteria, should trial their use and should analyse and reflect on their judgements. In this way hard markers and lenient markers will be identified, and all will learn how to apply the detail of the criteria. Inexperienced assessors need this sort of training exercise at the start of their careers.

5.2 Group project work

Group project work is often introduced at an earlier stage in a student's career. Again this gives opportunity for encouraging research, investigation and communication. But it also introduces students to group work and the problems associated with that. Often the internal working of the group can best be assessed by the members of the group themselves. This can be accomplished through confidential self and peer assessment. Sometimes it is more appropriate for the lecturer to observe the working process; this method has the added advantage that the lecturer can intervene in crisis situations. Assessors face dilemmas when assessing project work carried out by groups. If the same grade is given to each member of the group, then some may benefit and some may suffer from the work, or lack of it, of their peers, and this could be considered to be unfair, both by students and by society. Experience shows that this is a price worth paying. The dilemmas can be overcome by including an element of confidential, within-group, peer assessment, by observing group work, and by ensuring that students experience a good mix of group and individual assessment methods throughout their course. In working life, after all, group leaders often carry the blame for the poor performance of their group. By that stage, of course, they will be much more experienced and will have more control over their staff, but it is a useful lesson for students to discover the difficulties of working with other people, provided it is not disastrous to the overall outcome of their time at university.

5.3 Variations on written examinations

Variations on the theme of written examinations have been tried. These include open book examinations, seen examinations wherein students are given the questions some time in advance and they prepare their answers to them, examinations conducted in a computer laboratory with ready access to computer algebra systems and other mathematical software, and examinations which involve conceptual questions.

5.4 Comprehension tests

Some experiments on the use of Comprehension Tests have been carried out. This method of assessment is widely used in other subjects. Students are given an article or part of a book to read in advance. They study it very carefully and then take an unseen written paper, which is designed to explore the extent to which each student has comprehended the article. This can be useful for assessing a student's understanding of mathematical processes. Furthermore it encourages students to read critically and reflectively, to try to get into the mind of the author, and to think deeply about the topic of the article. It helps them to see that mathematics is alive and active in some contemporary context.

5.5 Journal writing

Student journal writing through the course can be used to help diagnose learning difficulties and to address these at an early stage. Students may be given time at the end of a teaching session to reflect on their learning during that session and to write down their thoughts and feelings, their worries and concerns, what they have learnt and what they are having difficulty with. Or they may be asked to do this overnight, thus giving them at least a little time to digest the day's work. The journals should be read frequently by the lecturer so that formative feedback can be given in good time and appropriate intervention strategies introduced if necessary.

Other strategies have been developed and used, such as brief, 'one-minute' quizzes or student-written summaries of key points learned (or not) at the end of class periods to provide feedback to instructors, particularly at early stages of modules. Student portfolios, student lectures and combined written-oral examinations are other strategies that have been used to good effect

6. DISSEMINATION OF INNOVATIVE IDEAS AND CONCLUSIONS OF RESEARCH STUDIES

The impetus for change and innovation usually comes from individuals who are dissatisfied with what they have been doing. They will have experimented, preferably with the approval of their head of department but sometimes covertly, and evaluated the effects of their ideas, and then adopted or scrapped them. The wider mathematics community can be informed about these developments in the same ways that research findings are disseminated, that is, by word of mouth at seminars and conferences, and by publication. It is helpful if papers relating to teaching and learning research are included in mainstream mathematics conferences and journals. Then lecturers who would never dream of attending a 'teaching' conference or reading a 'teaching' journal might just be exposed to these ideas and might be persuaded to accept them and adopt them.

Very occasionally a charitable foundation or a government agency will fund the production and dissemination of material relating to teaching innovation. Some examples are given in the annotated bibliography at the end of this chapter.

7. OBSTACLES TO CHANGE AND STRATEGIES FOR OVERCOMING THEM

Ignorance and prejudice are, perhaps, the greatest obstacles to change. Lack of resources is another. Many teachers in higher education will not have attended a course on teaching as part of their pre-service or in-service preparation for the job. (Most will have completed a PhD or equivalent and will be well versed in research methodology.) And so they will most probably teach as they themselves were taught. They are ignorant (in the nicest possible sense of the word!) of new ideas and new scholarship in student learning. Overcoming ignorance is relatively easy but requires an extensive programme of dissemination of ideas targeted particularly at new lecturers. One strategy being introduced in the UK is the requirement of many institutions for new lecturing staff to complete a post graduate certificate in university teaching. Usually this will be a two-year, part-time course delivered by the lecturer's own institution (or local consortium) and completing it successfully is a requirement of probation. Courses will usually include modules on generic and subject specific teaching and learning and the assessment instruments will include a portfolio of work relating to the lecturer's own teaching. Another strategy introduced in the UK is the recommendation that all lecturers join, and maintain their membership of, the newly constituted Institute for Learning and Teaching in Higher Education (ILT). Membership of the ILT gives public recognition that the member has had training as a teacher and continues to develop professionally. It will function as a professional association. (For two years there will be a special route to membership for experienced teachers, who will not be required to undergo an initial training programme but will, instead, submit a relatively short document outlining their experience as teachers and including a section of critical reflection on their work. See Mason, this volume, pp. 529-538.) It helps greatly if the head of department and other senior officers in the institution are sympathetic to the aims of such a programme, are knowledgeable about developments in teaching and learning, and support and encourage their colleagues to overcome their ignorance.

Prejudice is much harder to overcome. Prejudice is when a person, in full knowledge of developments, still rejects them unreasonably and out-of-hand, just because they are innovations or perhaps because once they encountered badly written arguments for change or are suspicious of research in education. This requires greater evangelistic effort. Research findings must be carefully presented and arguments for innovation persuasively written. Personal contacts, one-to-one over a meal or a drink, are good opportunities for this. Again, encouragement from heads of department and higher is very valuable.

Resource issues are important also. Recent years have seen resources diminish in universities all over the world. Classes are bigger and lecturers have conflicting and very strong demands on their time, particularly the demand to carry out high quality research. Money to pay for professional development is in short supply. There are no easy answers to the problem of lack of resources. It is a matter of commitment and priority, particularly on the part of heads of department and other resource managers in institutions. If the leader of a unit is committed to innovation and development in teaching and learning, then it is more likely to happen. One possible help might come from the bodies that determine the resources given to universities or departments to conduct research. If they were to allow research into the teaching and learning of a subject to have equal status with research into the subject itself, and if they were to allocate funds for this, then it is likely that more attention would be paid to teaching development. Also other carrots within institutions, such as promotion criteria which include good teaching, would help to stimulate the developments outlined in this paper.

Perhaps one of the more serious obstacles to be faced, as regards change and innovation in assessment at the university level, is that there may be genuine conflicts of interest between different stakeholders and parties. For instance, there may be a clash between, on the one hand, academics who tend to insist on the (in)formative purpose of assessment and on the ensuing necessity of multi-faceted and complex assessment instruments for validly capturing a fair range of knowledge and skills with their students, and, on the other hand, institution heads and administrators who tend to insist on the summative or ranking aspects of assessment in order for the institution to live up to common expectations in the environment or to make a positive appearance in a highly competitive 'university market'. However, even if this sort of clash is not present, clashes may arise if, as is often the case, the innovative assessment methods advocated or adopted by academics turn out to be considerably more time consuming or resource intensive than the traditional methods. At times when university funding is scarce heads and administrators may be inclined to counteract the use of such methods, not because of scepticism towards their relevance but simply because of the resources they consume. Such sorts of clashes are of an objective nature and cannot always be easily reconciled.

8. PERTINENT RESEARCH ISSUES

As mentioned above, there are sceptics in universities who are not yet convinced that teaching innovations are necessarily a good thing, but who may still be receptive to persuasive arguments and research findings. So research which seeks to evaluate innovative teaching methods and which demonstrates that aims and objectives are being met, is needed. Of course, the teaching developments themselves must have a rationale which is based on research into student learning. Another field of study is the robustness of the assessment methods themselves.

There are a number of internationally renowned teams who have published widely their research into student learning. Most of the work on assessment has been carried out by the Assessment Research Group (ARG) in the UK and some members of the International Community of Teachers of Mathematical Modelling and Applications (ICTMA). The work of the ARG is reported at length elsewhere in this volume (see Haines and Houston, this volume, pp. 431-442). Their main work has been to develop, test and evaluate robust methods for assessing several different forms of student project work and associated communication skills. They were also the nucleus of a group who received UK government funding to develop and disseminate resource material relating to innovative learning and assessment. Since some of this work related to teaching mathematical modelling, members of ARG are also active in the ICTMA, and some original research is published in the ICTMA series of books and conference abstracts.

But every teacher can be a researcher in their own classroom, picking up good ideas, developing them, evaluating them and then telling the world about them. This is actually quite an exciting thing to do and people who do it get a buzz from the experience which invigorates themselves, their teaching and their students.

ANNOTATED BIBLIOGRAPHY

American Association of University Professors (1990). Mandated Assessment of Education Outcomes. *Academe*, Nov./Dec. Discusses impact of mandated assessment on traditional arenas of professorial autonomy; focuses on five assessment issues (institutional diversity, skills, majors, value-added, and self-improvement). Concludes with recommendations for learning to live with mandated assessment.

Angelo, Thomas A. and Cross, K. Patricia (1993). *Classroom Assessment Techniques: A Handbook for College Teachers*, 2nd Ed, San Francisco: Jossey-Bass Publishers. This text focuses on formative classroom assessment. In addition to describing what classroom assessment is and how one might plan and implement classroom assessment tasks, the authors present 50 different classroom assessment techniques, many of which can be used in or modified for the mathematics classroom.

Astin, Alexander, Banta, Trudy W., Cross, Patricia K., et al. (1992). *Principles of Good Practice for Assessing Student Learning*. Washington, DC: American Association of Higher Education. Nine principles for assessing student learning developed by the long-standing annual Assessment Forum of the American Association of Higher Education (AAHE).

Ball, G, Stephenson, B, Smith, G, Wood, L, Coupland, M, and Crawford, K. (1998). Creating a Diversity of Mathematical Experiences for Tertiary Students. *Int J Math Edu Sci Technol*, 29, 827-841.

Bass, Hyman (1993). Let's Measure What's Worth Measuring. *Education Week*, October 27, 32. An 'op-ed' (opinion) column supporting Measuring What Counts from the Mathematical Sciences Education Board (MSEB). Stresses that assessments should (a) reflect the mathematics that is most important for students to learn; (b) support good instructional practice and enhance mathematics learning; and (c) support every student's opportunity to learn important mathematics.

Benjamin, Ernst (1990). The Movement to Assess Students' Learning will Institutionalize Mediocrity in Colleges. *Chronicle of Higher Education*, July 5. A brief op-ed column criticizing the indefensible consequences for higher education of rapidly spreading accountability systems that rely on narrow tests.

Benjamin, Ernst (1994). From Accreditation to Regulation: The Decline of Academic Autonomy in Higher Education. *Academe*, July/Aug., 34-36. A worried analysis by the retired general secretary of the American Association of University Professors (AAUP) concerning the impact of increased regulation based on accountability measures that, Benjamin believes, might distort traditional goals of the academy.

Berry, J. and Haines, C.R. (1991). *Criteria and Assessment Procedures for Projects in Mathematics*. Plymouth: University of Plymouth. The first of four reports written by the UK Assessment Group (ARG). It is a report of a workshop held in 1991, which aimed to review the assessment schemes and assessment criteria currently in use in UK universities and to begin to develop robust criteria-based assessment procedures for a wide range of topics. It describes the use of the FACETS data analysis package to consider the differing effects of the criteria themselves and the ways in which the assessors used them.

Berry, J. and Houston, K. (1995). Students using Posters as a means of Communication and Assessment. *Educational Studies in Mathematics*, 29, 21-27. Gives a thorough literature review of the use of posters by students in higher education generally, and suggests ways in which student posters could be used and could be assessed in mathematics classes.

Bok, Derek (1986). Toward Higher Learning: The Importance of Assessing Outcomes. *Change*, Nov./Dec., 18-27. A classic short essay by then Harvard President Derek Bok outlining the benefits to higher education of assessing the accomplishments and value of a college education.

Burton, L. and Izard, J. (1995). *Using FACETS to Investigate Innovative Mathematical Assessments*. Birmingham: University of Birmingham. The last of four reports written by ARG. It reports on the 1994 workshop and discusses continuing issues in assessing student project work. It deals with reliability of marking schemes, self, peer and tutor marked assignments, equity, and criteria for the assessment of students' posters.

Committee on the Undergraduate Program in Mathematics (CUPM) (1995). Assessment of Student Learning for Improving the Undergraduate Major in Mathematics. *Focus: The Newsletter of the Mathematical Association of America*, 15:3 (June), 24-28. Recommendations from the Mathematical Association of America (MAA) for departments of mathematics to develop a regular 'assessment cycle' in which they (1) set student goals and associated departmental objectives; (2) design instructional strategies to accomplish these objectives; (3) select aspects of learning and related assessments in which quality will be judged; (4) gather assessment data, summarize this information, and interpret results; and (5) make changes in goals, objectives, or strategies to ensure continual improvement.

Edgerton, Russell (1990). Assessment at Half Time. *Change*, Sept./Oct., 4-5. A brief summary of the political landscape of assessment in higher education by the president of the American Association of Higher Education (AAHE). Claims that state pressures for accountability will continue, but that if institutions define assessment in worthy terms the faculty will find the effort worthwhile.

Ewell, Peter T. (1997). Strengthening Assessment for Academic Quality Improvement. In Marvin W. Peterson, David D. Dill, and Lisa A. Mets (Eds.), *Planning and Management for a Changing Environment: A Handbook on Redesigning Postsecondary Institutions*, pp. 360-381. San Francisco: Jossey-Bass Publishers. Historical survey of assessment efforts in the U.S. during the last decade in the context of increased accountability requirements, decreased financial resources, and increased experience with assessment on college campuses. Discusses relation of assessment to academic planning and syndromes to avoid.

Ewell, Peter T. and Jones, Dennis P. (1996). *Indicators of Good Practice in Undergraduate Education: A Handbook for Development and Implementation*. Boulder, Co: National Center for Higher Education Management Systems (NCHEMS). Intended to provide colleges and universities with guidance in establishing an appropriate system of indicators of the effectiveness of undergraduate instruction, and to build on this foundation by cataloguing a range of exemplary indicators of 'good practice' that have proven useful across many collegiate settings.

Ferren, Ann (1993). *Faculty Resistance to Assessment: A Matter of Priorities and Perceptions*. Commissioned paper prepared for American Association of Higher Education. Analyzes faculty priorities to help understand why assessment is rarely valued by faculty. Argues that assessment must derive from widely agreed goals, must be connected to clear outcomes that the faculty see as beneficial, and must not be simply added to already overburdened faculty loads.

Frechtling, Joy A., 1995. *Footprints: Strategies for Non-Traditional Program Evaluation*. Washington, DC: National Science Foundation. A series of papers suggesting diverse strategies for assessing the impact of funded programmes both short- and long-term, both intended and unintended.

Gaither, Gerald H. (1995). *Assessing Performance in an Age of Accountability: Case Studies*. San Francisco: Jossey-Bass Publishers. Case studies from several states and public institutions about the shift from campus-based assessment in the 1980s to state-based accountability systems in the 1990s.

Glassick, Charles E., Huber, Mary T. and Maeroft, Gene I. (1997). *Scholarship Assessed: Evaluation of the Professoriate*. Carnegie Foundation for the Advancement of Teaching. San Francisco: Jossey-Bass Publishers. Companion to Ernest Boyer's widely-cited Scholarship Reconsidered: Priorities of the Professoriate, this report outlines standards for evaluating scholarship that transcend differences among disciplines: clear goals, adequate preparation, appropriate methods, significant results, effective presentation, and reflective critique.

Gold, Bonnie, Keith, Sandra and Marion, William (Eds.), (1999). *Assessment Practices in Undergraduate Mathematics*. Washington, DC: Mathematical Association of America. A collection of over fifty brief reports from dozens of different U.S. colleges and universities providing a wide variety of methods of assessing the major, teaching, classroom practice, the department's role, and calculus reform.

Haines, C.R. and Dunthorne, S. (Eds.) (1996). *Mathematics Teaching and Learning-Sharing Innovative Practices*. London: Arnold. This resource pack is a collection of articles describing innovative practices in teaching and assessment. It was written by mathematics lecturers from a consortium of UK universities, including members of ARG.

Haines, C.R. and Izard, J. (1995). Assessment in Context for Mathematical Modelling. In C. Sloyer, W. Blum and I. Huntley (Eds.). *Advances and Perspectives in the Teaching of Mathematical Modelling and Applications* pp. 131-149. Yorklyn, Delaware: Water Street Mathematics. Credible assessment schemes measure evidence of student achievement, as individuals or within a group, over a wide range of activities. This paper shows that item response modelling can be used to development rating scales for mathematical modelling. It draws on the work of ARG and work in Australia.

Haines, C.R., Izard, J. and Berry, J. (1993). *Awarding Student Achievement in Mathematics Projects*. London: City University. The second of four reports written by ARG. It investigates in depth the use of assessment criteria for judging oral presentations by students of their project work. It also proposes criteria for the assessment of written reports of different types of student project work.

Hilton, Peter (1993). The Tyranny of Tests. *American Mathematical Monthly*, April, 365-369. Several suggestions for 'reducing the distorting effect' which tests exert, principally on undergraduate mathematics.

Houston, S.K. (1993). *Developments in Curriculum and Assessment in Mathematics*. University of Ulster. This pamphlet contains papers presented at a one day symposium at the University of Ulster following the 1993 meeting of ARG.

Houston, S.K. (1993). Comprehension Tests in Mathematics (I and II), *Teaching Mathematics and its Applications*, 12, 60-73 and 113-120. These are the original papers describing the use of comprehension tests in mathematics.

Houston, SK. (1995). Assessing Mathematical Comprehension. In C. Sloyer, W. Blum, and I. Huntley (Eds.), *Advances and Perspectives in the Teaching of Mathematical Modelling and Applications*, pp. 151-162. Yorklyn, Delaware: Water Street Mathematics. This paper examines the rationale for comprehension tests in mathematics, and outlines possible aims and objectives. It describes the author's experiences in setting and using such tests and outlines the extent of their use in secondary and tertiary education in the UK and extends the work reported in Houston (1993).

Houston, S.K. (1997). Evaluating Rating Scales for the Assessment of Posters. In S.K. Houston, W. Blum, I. Huntley and N.T. Neill, (Eds.), *Teaching and Learning Mathematical Modelling*, pp. 135-148. Chichester: Albion Publishing (now Horwood Publishing). Deals with the use of posters by university students as a means of communication and as a vehicle for assessment. There is a summary of a literature review and a rationale for the activity. The author claims that it is an enjoyable activity, which is beneficial for students. The main purpose of the paper is to describe the development and evaluation of assessment criteria and rating scales.

Houston, S.K., Haines, C.R. and Kitchen, A. (1994). *Developing Rating Scales for Undergraduate Projects*. University of Ulster. The third of four reports written by ARG. It reports on the 1993 workshop, giving details of assessment criteria (or descriptors) for the assessment of written reports on projects in pure mathematics, mathematical modelling, statistical investigations and investigations of a more general nature. The report describes how the group members trialled the criteria and how the data analysis led to the development of robust assessment procedures. It also introduces the use of criteria for the assessment of student posters.

Izard, J. (1997). Assessment of Complex Behaviour as Expected in Mathematics Projects and Investigations. In S.K. Houston, W. Blum, I. Huntley and N.T Neill, (Eds.), *Teaching and Learning Mathematical Modelling*, pp. 109-124. Chichester: Albion Publishing (now Horwood Publishing). No single assessment method is capable of providing evidence about the full range of achievement. This paper reviews the problems faced in devising better assessments to monitor learning, and provides practical suggestions for meeting these problems. The methods presented are applicable to traditional examinations, project and investigation reports, presentations and posters, judgements of performance and constructed projects, observations of participation, collaborative group work and ingenuity. The paper concludes with advice on monitoring the quality of the assessment process.

Joint Policy Board for Mathematics (1994). *Recognition and Rewards in the Mathematical Sciences*. Providence, RI: American Mathematical Society. Discussion of faculty expectations in relation to institutional rewards. Findings include a general dissatisfaction with current methods of evaluating teaching as well as uncertainty about the weight of effective teaching in college expectations and rewards.

Katz, Stanley N. (1994). Defining Education Quality and Accountability. *Chronicle of Higher Education*, November 16, A56. An op-ed statement by the president of the American Council of Learned Societies (ACLS). Urges that colleges and universities heed the wake-up call of assessment from elementary and secondary schools and figure out how to define educational quality in terms that are worthy of higher education.

Linn, Robert L. and Herman, Joan L. (1997). *A Policymaker's Guide to Standards-Led Assessment*. Denver, CO: Education Commission of the States. Analysis of policy implications involved in shifting from norm-referenced assessments (which compare each students' performance to that of others) to standards-led assessments which incorporate pre-established performance goals, many of which are based on real-world rather than 'artificial' exercises.

Madison, Bernard (1992). Assessment of Undergraduate Mathematics. In Lynn A. Steen (Ed.), *Heeding the Call for Change: Suggestions for Curricular Action*, Washington, pp. 137-149. DC: Mathematical Association of America. Analysis of issues, benefits, worries, and pressures associated with the increasing demand for assessment of undergraduate mathematics. A background paper preceding release of the CUPM report on assessment.

Mathematical Sciences Education Board (1993). *Measuring What Counts: A Conceptual Guide for Mathematics Assessment.* Washington, DC: National Research Council. Intended primarily as advice for K-12 mathematics assessment, this report stresses the need for assessment to measure good mathematics, to enhance learning, and to promote access for all students to high quality mathematics.

National Council of Teachers of Mathematics (1995). *Assessment Standards for School Mathematics.* Reston, VA: National Council of Teachers of Mathematics. This third and final volume in NCTM's original set of standards for school mathematics focuses on six standards: effective assessment should reflect appropriate mathematics, enhance learning, promote equity, be based on an open process, promote valid inferences, and fit together coherently.

Niss, Mogens (Ed.) (1993). *Investigations into Assessment in Mathematics Education - An ICMI Study.* Dordrecht: Kluwer Academic Publishers. This book is one of two resulting from the ICMI Assessment Study. The book offers a variety of approaches to the conceptual, philosophical, historical, societal, and pedagogical investigation of assessment in mathematics education, by prominent mathematics educators from Europe, North America and Australia. Both survey chapters and specific empirical or theoretical studies are included in the book.

Open University Course Team (1998). *Assessment of Key Skills in the Open University Entrance Suite.* MU120, MST121, MS221. Open University, Milton Keynes.

Romer, Roy (1995). *Making Quality Count in Undergraduate Education.* Denver, CO: Education Commission of the States. Report by the then-Governor of Colorado on behalf of all U.S. state governors concerning what parents and students expect of higher education and what research says about the characteristics of high-quality undergraduate education. Concludes with recommendations for steps to make higher education more accountable to its public purposes.

Schilling, Karen Maitland and Schilling, Karl L. (1993). Professors Must Respond to Calls for Accountability. *Chronicle of Higher Education*, March 24, A40. An op-ed column arguing that faculty must take seriously the public's demand for evidence that students are learning, and learning the right things. Suggests portfolio assessment as an effective strategy.

Schoenfeld, Alan (1997). *Student Assessment in Calculus.* Washington, DC: Mathematical Association of America. Report of an NSF working group convened to support assessment of calculus reform projects by providing a conceptual framework together with extensive examples. Emphasizes the 'fundamental tenet' that, since tests are statements of what is valued, new curricula need new tests.

Seldin, Peter (1993). The Use and Abuse of Student Ratings of Professors. *Chronicle of Higher Education*, July 21, A40. An op-ed column lamenting the propensity of colleges to misuse student evaluations of faculty. Gives research-based advice for how to use such ratings intelligently and effectively.

Smith, G., Wood, L., Coupland, M., Stephenson, B., Crawford, K. and Ball, G. (1996). Constructing Mathematical Examinations to Assess a Range of Knowledge and Skills. *Int J Math Edu Sci Technol*, 27, 65-77.

Steen, Lynn Arthur (1999). Assessing Assessment. Preface to Bonnie Gold, Sanda Z. Keith and William A. Marion (Eds.), *Assessment Practices in College Mathematics*, pp. 1-6. Washington, DC: Mathematical Association of America. An exploration of issues, principles, and options available to address the wide variety of assessment challenges facing college mathematics departments.

Stevens, Floraline, Lawrenz, Frances and Sharp, Laure (1993). *User-Friendly Handbook for Project Evaluation.* Washington, DC: National Science Foundation. A 'how-to' guide to effective assessment for project directors who have neither experience in nor enthusiasm for evaluation.

Tucker, Alan C. and Leitzel, James R. C. (1995). *Assessing Calculus Reform Efforts.* Washington DC: Mathematical Association of America. A 'mid-term' review of the NSF-supported calculus reform movement in the United States, providing background on the motivation and goals of the movement, as well as evidence of changes in content, pedagogy, impact on students, faculty, departments, and institutions.

Wiggins, Grant (1989). A True Test: Toward More Authentic and Equitable Assessment. *Phi Delta Kappa*, May, 703-713. Argues that misunderstanding about the relation of tests to standards impedes progress in educational improvement. Suggests that only tests that require the 'performance of exemplary tasks' can truly monitor students' progress towards educational standards.

Wiggins, Grant (1990). The Truth May Make You Free, but the Test May Keep You Imprisoned: Toward Assessment Worthy of the Liberal Arts. In *Assessment 1990: Understanding the Implications*, pp.17-

31, Washington, DC: The American Association for Higher Education. (Reprinted in 1992, Lynn A. Steen (Ed.). *Heeding the Call for Change: Suggestions for Curricular Action*, pp. 150-162. Washington, DC: Mathematical Association of America.) Philosophical reflections on the purposes of education in the liberal arts or in basic science or mathematics. Focuses on ten principles of education that testing tends to destroy (e.g., justifying one's opinions; known, clear, public standards and criteria; self-assessment in terms of standards of rational inquiry; challenging authority and asking good questions).

Ken Houston
University of Ulster, Northern Ireland
sk.houston@ulst.ac.uk

JIM RIDGWAY, MALCOLM SWAN AND HUGH BURKHARDT

ASSESSING MATHEMATICAL THINKING VIA FLAG

1. ABSTRACT

Teachers of undergraduate mathematics face a range of problems which include an increasing number of less well qualified students, and increasing academic diversity in the student population. Students face courses which are radically different from mathematics courses encountered in school; they often face assessment systems which are ill-aligned to course goals, and which use assessment methods likely to encourage a surface rather than a deep approach to learning. The paper describes materials developed by the MARS group for the US National Institute for Science Education for use on their FLAG Web site. Collections of assessment materials can be downloaded free of charge which assess a range of valuable process skills in mathematics – proof, reasoning from evidence, estimation, creating measures, and fault finding and remediation. They are designed to offer a wide repertoire of assessment styles, as part of a process of encouraging a broadening of teaching and learning styles.

2. CHALLENGES FACING TEACHING

A number of challenges face teachers of undergraduate mathematics. Several authors have documented a downward drift of entry requirements for mathematics-based subjects such as engineering and physics (e.g. Hunt and Lawson, 1996; London Mathematical Society, 1995; Sutherland and Dewhurst, 1999; Sutherland and Pozzi, 1995). Serious conceptual problems have been documented in students who seem appropriately qualified (e.g. Armstrong and Croft, 1999).

A further difficulty for teachers in higher education is the increased demand for mathematics as a component of other courses; more and more disciplines use mathematical tools as part of their repertoire, notably the social sciences. Both of the above factors make teaching far more difficult, because of a rise in the heterogeneity of the population to be taught, and the corresponding challenges to teaching posed by a general lack of confidence in mathematics.

There are important differences in the nature of school mathematics and undergraduate mathematics. Mathematics as a discipline in its own right requires an intellectual approach which is quite distinct from school mathematics; this give rise to problems for students. A particular characteristic of university mathematics is its emphasis on proof and rigour – notions which may not figure large in school studies (e.g. Tall, 1992).

Derek Holton (Ed.), The Teaching and Learning of Mathematics at University Level: An ICMI Study, 423—430.
© 2001 Kluwer Academic Publishers. Printed in the Netherlands.

Brown, Bull and Pendlebury (1997) surveyed courses and assessment systems in higher education, and demonstrated a mismatch between stated course aims and the system of assessment in place. Assessment systems too often reflected traditional modes of assessment used within the organization, or in the discipline, and a response to the challenges of assessing large numbers of students. In particular, dependence on multiple choice and short answer questions encourages students to adopt a 'surface' approach to study (in contrast to a 'deep' approach where understanding is paramount) see Biggs (1999), and Marton and Saljo (1976).

Each of these aspects – reduced levels of student competence in mathematics on entry to university, students with increasingly diverse academic backgrounds, an intellectual transformation between school and university, and a poor match between curriculum intentions and assessment schemes – shows the need for serious attention to be paid to a growing problem in undergraduate mathematics. This paper describes some recent attempts to improve the quality of assessment systems, and thereby to improve the quality of undergraduate mathematics.

Some grounds for optimism are contained in evidence that educational attainment can be raised by better assessment systems (Black and William, 1998; Torrance and Pryor, 1998). Such assessment systems are characterized by: a shared understanding of assessment criteria; high expectations of performance; rich feedback; and effective use of self-assessment, peer assessment, and classroom questioning. The work reported here sets out to provide tools for academic staff to develop appropriate assessment schemes, and arises from a collaboration between the Mathematics Assessment Resource Service (MARS) and the College Level 1 group at the National Institute for Science Education (NISE), University of Wisconsin-Madison.

MARS is an international collaboration, based at Michigan State University, with groups at Berkeley, Nottingham, Durham and other centres. MARS supports programmes of improvement in mathematical education by offering assessment systems designed to exemplify and reward the full spectrum of mathematical performance that such programmes seek to develop.

NISE has developed FLAG – a Field-Tested Learning Assessment Guide, delivered via the Web. The overall ambition for FLAG is to enhance the first science, mathematics, engineering and technology learning experiences for all college students. Its purpose is to offer a readily accessible, free, up-to-date resource of classroom-tested assessment tools for instructors who have an interest in sharing and implementing new approaches to evaluating student learning. Each of the techniques and tools in this guide has been developed, tested and refined in classrooms and teaching laboratories at colleges and universities. MARS has created the mathematics component of FLAG.

The FLAG Web site has a number of components, which include: a basic primer on assessment; a tool based on Bloom's taxonomy (e.g. Bloom, Hastings and Madaus, 1971) to help instructors match their course goals with appropriate Classroom Assessment Techniques (CATs); CATs themselves, which are self-contained modules that introduce techniques for assessing conceptual, attitudinal and performance-based course goals, all of which can be downloaded in a form ready for use; links to related Web sites; and an annotated bibliography.

Traditional testing methods often have provided limited measures of student learning, and equally importantly, have proved to be of limited value for guiding student learning. The methods are often inconsistent with the increasing emphasis being placed on the ability of students to think analytically, to understand and communicate, or to connect different aspects of knowledge in mathematics. Because assessment exemplifies what is to be learned, all changes in educational ambition require associated changes in assessment.

3. THE CLUSTERS OF TASKS

The assessment tasks developed for FLAG are primarily concerned with sampling critical mathematical thinking rather than testing performances of technical skills or the acquisition of mathematical knowledge. They focus on the flexible and accurate use of relatively simple mathematics – an important capability which many undergraduates find difficult. They are best worked on by small groups in a collaborative, discursive atmosphere.

Five types of activity have been developed, 'Finding and fixing faults'; 'Creating measures', 'Making plausible estimates', 'Convincing and proving' and 'Reasoning from evidence'. These are described and illustrated more fully, below. These constitute valid, coherent activities which are commonly performed by mathematicians, scientists, politicians and critical observers of the media. They are less commonly found in mathematics classrooms.

3.1 Finding and fixing faults

Identifying and fixing faults is an important survival skill for everyday life, and an important metacognitive skill applied by every mathematician to their own work. Students are shown a number of mistakes which they are asked to diagnose and rectify. These require students to analyze mathematical statements and deduce from the context the part which is most likely to contain the error (there may be more than one possibility), explain the cause of the error and put it right.

Such tasks can be quite demanding for students. It is often more difficult to explain the cause of another's seductive error than to avoid making it oneself.

An elementary (genuine) example is shown below.

Shirts

The other day I was in a department store, when I saw a shirt that I liked in a sale. It was marked "20% off". I decided to buy two. The person behind the till did some mental calculations. He said, "20% off each shirt, that will be 40% off the total price."

(Incident in local supermarket.)

3.2 Creating Measures

We constantly construct measures for physical and social phenomena and use these to make decisions about our everyday lives. These can vary from measures of simple quantities such as 'speed' or 'steepness' to complex and subjective social ones such as 'quality of life'. All measures are constructions and are thus open to criticism and improvement. When is a measure appropriate for its purpose? While it is possible to define a range of alternative measures for any concept, they will differ in utility; some clearly will prove more useful than others.

A 'Creating Measures' task consists of a series of questions which prompt students to evaluate an existing plausible but partial or inadequate measure of an intuitive concept and then invent and evaluate their own improved measure of the concept.

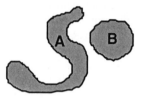

Figure 1. Which island is most compact and why?

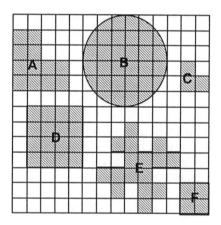

Figure 2. Compactness of island shapes

Examples we have used include defining measures for: the 'steepness' of a staircase; the 'compactness' of an island; the 'crowdedness' of a gathering of people; and the 'sharpness' of a bend in a road.

The first part of each task invites the student to rank-order examples, using their own intuitive understanding of the concept. For example, in one task, we offer students two 'islands', A and B (see Figure 1), and ask students to say *which* they think is most 'compact' and *why*.

The second part of the task provides students with an inadequate measure of the concept and asks them to order a set of 'islands', using this measure and explain why it might be inadequate. For the 'compactness' example, we ask students to use the measure "Compactness = Area ÷ Perimeter" on island shapes A to F (see Figure 2).

At first sight, this may seem a reasonable measure, but islands with similar shapes but different sizes show that this measure is not independent of scale and so a dimensionless measure needs to be sought.

The third part of the task invites students to devise and use their own measure on the given examples and the final part asks them to scale their measure so that it ranges from 0 to 1.

3.3 Making plausible estimates ('Fermi Problems')

Enrico Fermi posed problems which at first seemed impossible but, on reflection, can be solved by making assumptions based on common knowledge and following simple-but-long chains of reasoning. Fermi used such problems to make the point that problem solving is often not limited by incomplete information but rather by an inability to use information that is already available within the immediate community.

These tasks assess how well students can: decide what assumptions they need to make; make reasonable estimates of everyday quantities; develop chains of reasoning that enable them to estimate the desired quantity; ascertain the reasonableness of the assumptions upon which the estimate is based.

For example: The population of the USA is approximately 270 million. How many dentists are there in the USA?

3.4 Convincing and proving

These tasks are of two types. The first type presents a series of statements that students are asked to evaluate. These typically concern mathematical results or hypotheses such as 'the square of a number is greater than the number'. Students are invited to classify each one as 'always true', 'sometimes true' or 'never true' and offer reasons for their decision. The best responses contain convincing explanations and proofs; the weaker responses typically contain just a few examples and counter-examples. These tasks vary in difficulty, according to the statements being considered and the difficulty of providing a convincing or rigorous explanation. They also provide useful diagnostic devices as the statements are sometimes typical student 'misconceptions' which arise from over-generalizing from limited domains. Sample statements are:

- When you add two numbers, you get the same result as when you multiply them.
- $(a + b)^2 = a^2 + b^2$.
- The centre of a circle that circumscribes a triangle is inside the triangle.
- Quadrilaterals tessellate.

- A shape with a finite area has a finite perimeter.

Attempt 1:

    Assuming that

$$\frac{a+b}{2} \geq \sqrt{ab}$$

$$a+b \geq 2\sqrt{ab}$$
$$(a+b)^2 \geq 4ab$$
$$a^2 + b^b + 2ab \geq 4ab$$
$$a^2 + b^b - 2ab \geq 0$$
$$(a-b)^2 \geq 0$$

    Which is true for positive numbers, so the assumption was true.

Attempt 2:

    For all positive numbers

$$(\sqrt{a} - \sqrt{b})^2 \geq 0$$
$$a - 2\sqrt{a}\sqrt{b} + b \geq 0$$
$$a + b \geq 2\sqrt{a}\sqrt{b}$$
$$\frac{a+b}{2} \geq \sqrt{ab}$$

So the result is true.

Attempt 3:

    The area of the large square
$$= (a+b)^2 \ldots$$

    The unshaded area = $4ab$.
    Clearly

$$(a+b)^2 > 4ab$$
$$or \quad (a+b) > 2\sqrt{ab}$$
$$or \quad \frac{(a+b)}{2} > \sqrt{ab}$$

So the result is true.

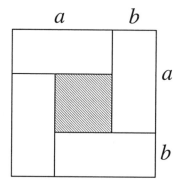

Figure 3. The relation between the arithmetic mean and the geometrical mean.

    The second collection involves the students in evaluating 'proofs' that have been constructed by others. Some of these are correct and some are flawed. (For example, in one question, three 'proofs' of the Pythagorean theorem are given). The flawed 'proofs' may be inductive rather than deductive arguments that only work with special cases, arguments that assume the result to be proved, or arguments which contain invalid assumptions. There are also some partially correct proofs that contain large unjustified 'jumps' in reasoning. In these tasks, students adopt an 'assessor' role and attempt to identify the most convincing proof and provide critiques for the remaining attempts. For example, one task provides three attempts to prove that the arithmetic mean is greater than or equal to the geometric mean for any two numbers (see Figure 3).

3.5 Reasoning From Evidence

These tasks ask students to organize and represent a collection of unsorted data and draw sensible conclusions.

For example, students are given a collection of data concerning male and female opinions of two deodorants. The experiment has been designed to test two variables; the deodorant name/packaging (Bouquet and Hunter) and the fragrance (A and B). Both forms of packaging are tested with both forms of fragrance. Thus 'Bouquet A' and 'Hunter A' both contain exactly the same fragrance, A.

The results are given on a five-point scale (from 'Love it' to 'Hate it'). The data are presented to students as a collection of 40 unsorted responses which they have to analyze.

Students are asked to present their findings in the form of a short report saying which fragrance and name are likely to be the most successful if they are to be sold to females and males.

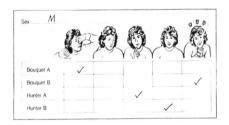

4. CONCLUDING REMARKS

Undergraduate teaching is problematic; shifts in student characteristics require shifts in teaching styles. These CATs are designed to offer a variety of assessment styles, as part of a process of encouraging a broadening teaching and learning styles. Each cluster of tasks requires students to demonstrate some core mathematical skills: proving; reasoning from evidence; estimating; creating measures; and fault finding and remediation. Mathematical challenges are presented at a variety of difficulty levels, and so these CATs can be used with groups of students with greatly differing experiences of mathematics.

REFERENCES

Armstrong, P., and Croft, A. (1999). Identifying the learning needs in mathematics of entrants to undergraduate engineering programmes. *European Journal of Engineering Education*, 14(1), 59-71.

Biggs, J. (1999). *Teaching for Quality Learning at University*. Buckingham: SRHE and OUP.

Black, P. and William, D. (1998). Assessment and classroom learning. *Assessment in Education*, 5(1), 7-73.

Bloom, B. S., Hastings, J.T. and Madaus, G.F. (1971). *Handbook on Formative and Summative Evaluation of Student Learning*. New York: McGraw Hill.

Brown,G., Bull,J. and Pendlebury,M (1997). *Assessing Student Learning in Higher Education*. London: Routledge.

Field-Tested Learning Assessment Guide (FLAG). URL: http://www.wcer.wisc.edu/nise/cl1/.

Hunt, D. and Lawson, D. (1996). Trends in mathematical competency of A level students on entry to university. *Teaching Mathematics and its Applications*, 15(4), 167-173.

London Mathematical Society (1995). *Tackling the Mathematics Problem*. London: London Mathematical Society.

Marton, F. and Saljo, R. (1976). On qualitative differences in learning: I - outcome and process. *British Journal of Educational Psychology*, 46, 4-11.

Mathematics Assessment Resource Service (MARS). URL: http://www.educ.msu.edu/mars/.

Sutherland,R., and Pozzi, R. (1995). *The Changing Mathematical Background of Undergraduate Engineers*. London: The Engineering Council.

Sutherland,R., and Dewhurst, H. (1999). *Mathematics Education Framework for progression from 16-19 to HE*. University of Bristol, Graduate School of Education.

Tall, D. (1992). The transition to advanced mathematical thinking: functions, limits, infinity and proof. In D.A.Grouws (ed.) *Handbook of Research on Mathematics Teaching and Learning*, pp. 495-511. New York: Macmillan.

Torrance, H. and Prior, J. (1998). Investigating Formative Assessment. Buckingham: Open University Press.

MARS is funded by NSF grant no. ESI 9726403; NISE is funded by NSF grant no. RED 9452971.

Jim Ridway
University of Durham, England
jim.ridgway@durham.ac.uk

Malcolm Swan
University of Nottingham, England
malcolm.swan@nottingham.ac.uk

Hugh Burkhardt
University of Nottingham, England
hugh.burkhardt@nottingham.ac.uk

CHRIS HAINES AND KEN HOUSTON

ASSESSING STUDENT PROJECT WORK

1. INTRODUCTION

This paper tells the story of the ARG. Along the way it describes a methodology for developing robust instruments for the assessment of many aspects of student project work, and it gives examples of assessment criteria. But it is written in narrative form - it tells a story - to emphasise that the assessment of student project work is a human, rather than a mechanical, activity, and to illustrate how the attitudes of the people involved, the baggage they bring with them and the ways in which they interact all contribute to the task in hand. Telling this as a story also allows us, the authors, to include personal reminiscences and reflections.

We, mathematicians who also teach mathematics, are accustomed to precision and exactness. We can set examination papers and devise marking schemes that account for every single mark awarded. Given this marking scheme we can mark many papers accurately and consistently and many different markers will agree on the mark of a particular paper to a highly significant level. Precision is part of our culture and it flows over into our teaching and writing. But we forget sometimes about the messiness of the research investigation that preceded the writing of the research paper, and we forget about the fact that sometimes our allocation of marks to different parts of a question can sometimes be arbitrary.

Student project work is different from examinations. It has different aims and assesses different objectives. It introduces messiness and subjectivity into mathematics, but it *is* part of our professional activity and should be part of student activity as well. Our colleagues in the Humanities subjects handle the assessment of prose writing and of performance very well, and we should learn how to do this also. It is possible to do it accurately and consistently, but not quite with the same precision we apply to the marking of examination papers. And by learning how to do this, we also learn how to teach our students to do their work better.

Not only is project work different from an assessment point of view, but within curricula, various definitions of project work lead to a variety of student outcomes. In this paper project work could range from short investigational exercises in class time lasting up to two hours, to extended investigations over a period of a few weeks and through to final year projects and dissertations lasting one or two semesters. The activity, engaged in on an individual or group basis, for example, might be termed modelling, investigations or a critical review of the literature on some mathematical topic.

Derek Holton (Ed.), The Teaching and Learning of Mathematics at University Level: An ICMI Study, 431—442.
© 2001 *Kluwer Academic Publishers. Printed in the Netherlands.*

and 1996, and has left a legacy of robust criteria for the assessment of many aspects of student project work. It has also contributed to the development of another major resource for university mathematics teachers, which will be described below. The members of the ARG met annually for three-day workshops in various venues in the UK from April 1991 to April 1994. It produced reports of these workshops shortly afterwards (in 1991, 1993, 1994 and 1995) and some of its members published papers in journals and conference proceedings. During 1995 and 1996, as the result of a funding initiative by the Higher Education Funding Councils in the UK, its members diverted into a larger research and dissemination activity covering aspects of teaching as well as research. The products of this activity were widely disseminated in the UK. It has not met as a group since 1996, but the people involved, if they have not yet retired, still use the assessment instruments developed and still conduct research into aspects of teaching, learning and assessment.

This paper tells the story of the ARG and in so doing brings its work to the attention of a wider audience.

2. BEGINNINGS (1991)

In 1991, John Berry (Plymouth University) and Chris Haines (City University, London) took the initiative of calling together twelve people from universities and sixth form colleges to start to examine the whole issue of assessing student project work. A three-day workshop was organised for April that year with the following aims, quoted from the report of the workshop (Berry and Haines, 1991): -

1. to review current project schemes and their assessment of project work across a wide range of institutions and levels;
2. to begin to develop criteria based assessment procedures for use on a wide range of topics within mathematics.

A range of nine projects from several institutions was available at the workshop, which was structured as follows: -

1. a presentation on "Insights into Communication Skills" by Ros Crouch and Alan Davies (University of Hertfordshire);
2. a presentation on "Preliminary Thoughts on Indicators and Descriptors" by Chris Haines;
3. a first group-work session wherein delegates in two smaller groups read and discussed the nine projects;
4. a second group-work session wherein possible Indicators were discussed;
5. a session wherein these Indicators were applied to six of the projects;
6. a session wherein an analysis by John Izard (Australian Council for Educational Research) of the previous activity was presented and discussed;
7. a concluding session wherein a research agenda for the following year was established.

Crouch and Davies described the course they gave at Hertfordshire, which combined mathematical modelling and communication skills development. Communication is a key skill, required of all graduates, and it is best developed in mathematics students when they communicate mathematics. Development of this skill involves working in teams, writing reports, giving presentations and giving and receiving feedback. Furthermore the act of communicating reinforces the learning of mathematics that has taken place. Engagement in modelling or investigation requires active learning on the part of students as opposed to passive receptance. It requires the integration of previously gained knowledge of several different areas of mathematics, and the development of new skills like conducting research, making value judgements and providing evidence to substantiate claims.

In his presentation, Haines made the point that, while many institutions gave guidance on the doing and assessing of projects, the ways in which this guidance was presented do not give insights into how to apportion marks and are often non-specific.

The two group-work sessions provided much food for thought. Given the time constraints of the workshop, it was agreed that each person would only skim read each project but would study at least two projects in detail. Each group then placed the projects in rank order of quality and also constructed a list of the assessment indicators that had been in peoples' minds as they did their work. The groups then met in plenary session to discuss a larger reservoir of indicators. These were honed down somewhat and a list of agreed indicators was then used by the whole workshop to judge six of the projects, with each person reading all six and assigning a score of 0 or 1 or 2 to each indictor for each project. A score of 0 implied that the assessor had seen no evidence of the behaviour suggested by the indicator; a score of 1 implied some evidence while a score of 2 implied that the assessor saw consistent evidence of the behaviour. Later in the process of developing assessment indicators, ARG were to use a scale from 0 to 4, to allow for greater differentiation.

The data were analysed by John Izard using the FACETS program developed by Mike Linacre at the University of Chicago (Linacre and Wright, 1994). This carries out a multifaceted Rasch analysis, which gives

- the overall score awarded to each project by the assessors;
- a score for each assessor measuring the severity or leniency with which each assessor awarded scores relative to the rest of the assessors; and
- a score for each indicator showing which were the most difficult and which were the least difficult to score on.

The programme also gives a consistency measure for each of the three facets - projects, assessors and indicators. Elsewhere Izard (1997) has described the application of this method of data analysis in some detail.

So we were able to identify whom the severe and lenient markers were and we could have taken this into account in awarding grades had we so wished. We were able to identify groups of indicators, which seemed to be measuring the same attribute, thus allowing us to hone down the list of indicators even further. We identified indicators that gave ambiguous results and, sure enough, on reflection we realised that they were badly phrased. Some described more than one measurable quantity. For example, we had the indicator "generalisations made and proofs attempted" which would be better de-coupled.

At this stage of the story, we will not give the lists of indicators we all took away with us to test in our own institutions because we were to do further work on them over the next two or three years. Our agenda was to use the agreed indicators with our own students to help inform our assessment of their project work, to submit the data to Chris Haines and John Izard for analysis (soon we were to learn how to use FACETS to carry out our own analysis), and to meet again in April 1992.

Our experiences in this workshop indicate that it is a non-trivial task for a group of assessors to devise robust assessment indicators and to learn to use them consistently and reliably. We recommend this learning process to whole departments who are engaged in project work with their students.

The activities of this our first workshop are recorded in Berry and Haines (1991).

3. AFTER ONE YEAR (1992)

We returned to Exeter in April 1992, all of those who attended the first meeting having had some experience of using the draft criteria in their own institutions. There were some new people in the Group, including some from the senior end of secondary schools. The organisers, again Berry and Haines, had prepared material for us to work on. In particular there was a video recording of groups of students giving oral presentations. We spent most of the time at the workshop judging these presentations, analysing our use of the draft criteria and refining them. We produced a set of criteria for the assessment of oral presentations of project work by individuals or small groups. Our development process leads us to believe that they are robust, and their use will lead to consistent and accurate judgements.

We also reflected on our use of the draft criteria for the assessment of written reports of modelling activities, and we produced second drafts of descriptors suitable for use with modelling projects, statistics projects and pure mathematics investigations, along with descriptors dealing with the writing of the reports. Our experience had guided us to use a five-point assessment scale for each descriptor (previously we had been using a three-point scale) and this allowed for greater discrimination in judging each of the aspects. But we recognised that further research would be necessary before we could use these with the same confidence as the oral descriptors. And so our agenda was set for the next twelve months.

4. AFTER TWO YEARS (1993)

We met next in Northern Ireland, in the village of Kells, in April 1993. At this workshop, using the same methodology as before, we finalised our proposals for assessment criteria for use with both mini-projects and major projects involving modelling, statistics, pure mathematics and projects of a more discursive nature such as those on the history of mathematics or mathematics education. These, together with a description of the workshop activities, are recorded in Houston, Haines and Kitchen (1994). The criteria may also be found in Crouch (1996), and one example is given in the Appendix to this paper. There was also a visit, on a very wet day, to the Giant's Causeway.

By this time, the activities of the ARG were being reported on the international stage. Several papers were presented, posters displayed and contributions made in workshops at ICME 7 (Quebec, 1992) and ICTMA 6 (Delaware, 1993). Research and development work from members of the ARG was presented and discussed at ICMEs 8 (Seville, 1996) and 9 (Tokyo, 2000) and at ICTMAs 7 (Jordanstown, 1995), 8 (Brisbane, 1997) and 9 (Lisbon, 1999). The distinctive ICTMA conferences were particularly useful platforms because they provided mathematicians and mathematical educators with more effective opportunities for discourse. See Sloyer et al (1995), Houston et al (1997), Galbraith et al (1998) and Matos et al (2001) for the papers from ICTMA 6, 7, 8 and 9, and Gaulin et al (1994) and Alsina et al (1998) for the papers from ICME 7 and 8. Berry and Houston (1995) published the first paper on the use of posters with mathematics students.

The Effective Teaching and Assessment (ETA) Programme (1993 to 1996). Perhaps the most significant development in the dissemination of the work of ARG started later in 1993. A consortium of academics from fifteen British Universities, with a nucleus of members of ARG, secured a large grant from the Higher Education Funding Councils in the UK for the Project "Mathematics Learning and Assessment - Sharing Innovative Practices" under the ETA Programme. During the three-year lifetime of this project, two national meetings and six regional workshops were held. Together these events involved over 200 academics from 69 universities. A major resource was also produced - a pack containing five booklets of exemplar material dealing with

- the context for innovation;
- projects and investigations;
- communication;
- teaching and learning;
- assessment.

There is also a video for use in a staff development context. (Haines and Dunthorne, 1996)

Burton and Haines (1997) describe in detail the experiences of this consortium project. They reflect upon the project implementation by focussing on four issues that are still pertinent to academic mathematicians as teachers. These are: -

1. the match between the key project concerns (Table 1) and the concerns of teachers of undergraduate students;
2. the willingness to address the question "Why change?"
3. the experience of promoting innovative practices;
4. the recognition of the need to promote and manage change.

Table 1. Key concerns of those teaching undergraduate mathematics (from Burton and Haines, 1997, p. 279)

Difficulties encountered at the interface between mathematics post-16 and HE
Students with variable mathematics backgrounds
Students' personal and professional skills
Mathematics in other disciplines
Teaching large classes
Group work
Devising appropriate experimental workshops
Use of projects and investigations as vehicles for teaching and learning
Production of learning materials to support the use of projects and investigations
Projects in statistics, pure mathematics, mechanics and other areas
Modelling in the first undergraduate year
The use of oral and written presentations for assessment
Communicating and assessing mathematics through written reports and posters
Learning diaries and comprehension exercises as assessment strategies
Appropriate tools for assessment for student work
Peer- and self-assessment appropriate to group activities
Home-based assignments

Burton and Haines (1997, pp. 279-281) looked particularly at the ETA project from the point of view of its impact on assessment. The focus of the project, exemplified by the key concerns expressed in Table 1, was the result of experimentation that the consortium members had been undertaking both collectively through ARG and individually within their own institutions. These concerns are to do with the student body, teaching, learning and, above all, assessment styles.

To progress the project, six regional workshops and two national meetings were offered to generate discussion in the mathematics community on a range of topics including the issues of concern with assessment. Bearing in mind the distinctive nature of project and investigational work in mathematics, the arguments raised questions about the appropriateness of traditional timed, unseen examinations which are said to allow students to demonstrate effective recall, selection, and use of mathematical facts, concepts and techniques. Continuing, Burton and Haines point out, that despite the premium placed on independence of thought and initiative by mathematicians in research and the professions and by employers generally, it is not clear that examinations in the above sense can possibly measure these qualities. That such qualities, often displayed in project and investigational work in mathematics, should be assessed is beyond question.

Our opening paragraphs commented upon the culture of mathematics and how within that culture commonly held beliefs develop. It is asserted by many colleagues in mathematics that an important strength of a traditional examination system lies in their belief that they are using a comparative scheme for assessment founded on fixed implicit standards and is therefore moderated to give exact measures on a fine scale of marks. Here we are concerned with the need to move towards standards or criterion referenced schemes, since for many of the innovative practices in the consortium, student achievement covers such a broad spectrum of behaviour that traditional methods are not appropriate. From the point of view of the practitioner, Izard (1991) discusses the detail of each of these forms of assessment, interpreting their interconnections through the work of Withers and Batten (1990). His analysis, founded on mathematics education in schools, provides valuable insights for the assessment of complex mathematical tasks in universities.

The regional workshops and those held at the two national meetings explored how to increase the ways in which evidence of student achievement might be obtained. Examples, most developed and considered within the ARG such as: essay-type questions within traditional mathematics examinations, poster, projects and 'comprehension' exercises were offered and their potential as *proper* assessment tools was discussed. Within them students are often required to construct mathematical arguments, formulate mathematical models and communicate mathematical ideas, and as well to show independence of thought and initiative.

It is clear that for mathematical projects and investigations in particular, a good assessment scheme will combine assessment objectives with appropriate schemes. Possible assessment objectives that emerged from the Mathematics Learning and Assessment project are listed in Table 2 (Burton and Haines, 1997, p. 281).

Table 2. Possible assessment objectives in mathematics (from Burton and Haines, 1997 p. 281)

Recall, select and use mathematical facts, concepts, techniques
Construct mathematical arguments
Formulate mathematical models
Evaluate mathematical models
Develop the skills of criticism
Organise mathematical information
Interpret mathematical information
Communicate mathematical ideas
Develop oral and written communication skills
Read and comprehend mathematics
Develop logical thinking
Provide students with vocational education
Encourage independence of thought and initiative
Develop groupworking skills

Suggested outline marking schemes, flexible enough to deal with a diversity of response from students, were available at the workshops and are contained in, for example, Crouch (1996). Robust assessment scales, as developed by the ARG, which can place students (and examiners) on an attainment continuum (or other continuum) can be constructed. The generic purposes here were to:

- ensure that assessment and learning support each other;
- offer a diversity of assessment styles;
- make the assessment strategies consistent with intentions of the curriculum.

Each of these is important for projects and investigations in mathematics.

5. AFTER THREE YEARS (1994)

After the first national meeting organised by the consortium in Birmingham in April 1994, the last of the ARG workshops to publish a report was held. We were, needless to say, quite buoyed up by the national meeting. But we soon settled to our work. In the previous twelve months, in between consortium meetings and writing, we had been using FACETS to investigate our innovative mathematical assessments. This workshop was a bringing together of this work, and the papers are contained in Burton and Izard (1995). The topics dealt with included: -

- reliability of marking schemes for assignments;
- self, Peer and Tutor marking;
- equity of assessment in a module-choice situation;
- using and analysing rating scales for the assessment of students' poster presentations.

6. AFTER FIVE YEARS (1996)

The members of ARG met for a day in London prior to the second national meeting organised under the ETA Programme, at which the resource pack produced by the consortium was launched. The ARG meeting reflected on what ARG had achieved over the last five years, what research had been undertaken and what was planned. Since then this work has been progressed by individuals or small groups and research findings published in the fora mentioned above. But this was to be the last formal meeting of the group. It had achieved its initial objective of looking at the assessment of student project work, and much more besides.

7. CONCLUSIONS

ARG produced the following major outcomes: -

- the bringing together of mathematicians, mathematics educators and communication specialists to think about the assessment of complex mathematical tasks;
- the establishment of a strong commitment to ensure that student achievement is properly rewarded;

- the production of rating scales which have the potential for robust assessment of these tasks;
- the "Mathematics Learning and Assessment" project.

Prior to 1991, assessment of complex tasks was emerging as an area of concern. This was evident at ICTMAs 4 (Roskilde, 1989) and 5 (Nordwijkerhout, 1991), and was the driving force behind the formation of ARG. See Niss et al. (1991) and de Lange et al. (1993). ARG personnel made a major contribution to ICTMA 6 at Delaware in 1993 by organising the "assessment" strand at that conference, bringing together a number of papers, plenaries and workshops (Sloyer et al., 1995). And the work is on-going, with researchers now looking at the factors that influence students' learning of modelling, and the role assessment has in that.

The value of the work of ARG was recognised when the consortium obtained the ETAP grant, and the work of the consortium was widely disseminated in the UK. Many universities have been influenced by this work. And the groundwork on the teaching and assessment of communication skills provided a platform for embedding the learning of Key Skills by undergraduate mathematics students in the curriculum. The teaching of Key Skills was a principal recommendation of the Dearing report of 1997 (Dearing, 1997). The implications of this are reverberating still through higher education in the UK. Similar issues are also being addressed in other countries.

So we have come to the end of our story, for the moment. The learning experiences we had through the ARG workshops and through the work of the consortium have, we believe, made us more thoughtful and better teachers. There is no substitute for this sort of experiential activity when it comes to personal development. We commend it to others, just as we commend the fruits of our work. You do not have to use our products, but you do have to think about the development of your own. Good luck, and may your students have a better experience as a consequence.

REFERENCES

Alsina, C., Alvarez, J.M., Niss, M., Perez, A., Rico, L. and Sfard, A., (Eds.), (1998). Proceedings of the 8th International Congress on Mathematics Education. Seville: SAEM 'Thales'.

Berry, J. and Haines, C.R. (1991). *Criteria and Assessment Procedures for Projects in Mathematics*. Plymouth: University of Plymouth.

Berry, J.S. and Houston, S.K. (1995). Students Using Posters as a Means of Communication and Assessment. *Educational Studies in Mathematics*, 29, 21-27.

Burton, L. and Haines, C.R. (1997). Innovation in Teaching and Assessing Mathematics at University Level. *Teaching in Higher Education*, 2, 3, 273-293.

Burton, L. and Izard, J. (1995). *Using FACETS to Investigate Innovative Mathematical Assessments*. Birmingham: University of Birmingham.

Crouch, R.M. (1996). Communication. In: C.R. Haines and S. Dunthorne (Eds.), *Mathematics Learning and Assessment: Sharing Innovative Practices*, pp. 3.01-3.30. London: Arnold.

Dearing, R. (1997). *Report of the National Committee of Inquiry into Higher Education*. London: HMSO.

Galbraith, P., Blum, W., Booker, G. and Huntley, I.D. (Eds.), (1998). *Mathematical Modelling - Teaching and Assesment in a Technology-Rich World*. Chichester: Horwood Publishing Ltd.

Gaulin, C., Hodgson, B.R., Wheeler, D.H. and Egsgard, J.C. (Eds.), (1994). Proceedings of the 7th ICME, Sainte-Foy: Les Presses de l'Université Laval.

Haines, C.R. and Dunthorne, S. (Eds.) (1996). *Mathematics Teaching and Learning-Sharing Innovative Practices*. London: Arnold.

Haines, C.R., Izard, J. and Berry, J. (1993). *Awarding Student Achievement in Mathematics Projects*. London: City University.

Houston, S.K., Blum, W., Huntley, I.D. and Neill, N.T. (Eds.), (1997). *Teaching and Learning Mathematical Modelling*. Chichester: Albion Publishing.

Houston, S.K., Haines, C.R. and Kitchen, A. (1994). *Developing Rating Scales for Undergraduate Projects*. University of Ulster.

Izard, J.F. (1991) *Assessment of Learning in the Classroom*. Geelong, Victoria: Deakin University Press.

Izard, J. (1997). Assessment of Complex Behaviour as Expected in Mathematics Projects and Investigations. In S.K. Houston, W. Blum, I. Huntley and N.T Neill, (Eds.), *Teaching and Learning Mathematical Modelling*, pp. 109-124. Chichester: Albion Publishing (now Horwood Publishing).

de Lange, J., Keitel, C., Huntley, I.D. and Niss, M. (Eds.), (1993). *Innovation in Mathematics Education by Modelling and Applications*. Chichester: Ellis Horwood Ltd.

Linacre, J.M. and Wright, B.D. (1994). *FACETS: Many-Facet Rasch Analysis*. Chicago Il: MESA Press.

Matos, JF, Carreiara, S.P., Blum, W. and Houston, S.K. (Eds.), (2001). *Modelling, Applications and Mathematics Education - Trends and Issues*. Chichester: Horwood Publications (to appear).

Niss, M., Blum, W. and Huntley, I.D., (Eds.), (1991). *Teaching of Mathematical Modelling and Applications*. Chichester: Ellis Horwood Ltd.

Sloyer, C., Blum, W. and Huntley, I.D., (Eds.), (1995). *Advances and Perspectives in the Teaching of Mathematical Modelling and Applications*. Newark DE: Water Street Mathematics.

Withers, C. and Batten, M. (1990). Defining types of assessment. In B. Low and G. Withers, (Eds.), *Developments in School and Public Assessment* (Australian Educational Review No 31). Camberwell, Victoria: Australian Council for Educational Research.

APPENDIX: COMMUNICATION SKILLS (WRITTEN)

		High			Low	Not shown	Not applicable
W1	**Gives a free standing abstract or summary** Includes a statement of the problem which may need redefinition, states the methodology used and gives specific conclusions. This section is free standing, brief and precedes the report itself.						
W2	**Gives an introduction to the report** States the problem, gives the background to the problem, sets it in context, explains the strategy.						
W3	**Structures the report logically** Connects main points logically, gives supporting evidence succinctly and concisely.						
W4	**Makes the structure of the report verbally explicit** Explains relative importance, emphasises important points (Can see what is there without digging). Makes clear the logical function of constituent parts, uses good and consistent internal referencing (labelling).						
W5	**Demonstrates a command of the appropriate written language** Uses good spelling and grammar. Uses an appropriate style (technical and linguistic). (This table is continued on the next page.)						

W6	**Complements logical structure with visual presentation and layout** Draws clear tables and diagrams, spaces report so it is easy to read. Makes it aesthetically pleasing, gives a contents list if needed.						
W7	**Makes appropriate use of references and appendices** Has a good external referencing system, lists references, uses clearly labelled appendices for secondary data, gives bibliography if appropriate.						
W8	**Gives a concluding section in the main report** Appears near the end of the report, makes a summary of the important points which have been logically derived, conclusions represent an adequate solution to the problem, (does not introduce any new information at this point).						
W9	**Gives a well reasoned evaluation** This could appear before the end. Considers limitations of the solution, discusses possible extensions and makes recommendations for appropriate action.						

Chris Haines
City University, England
c.r. haines@city.ac.uk

Ken Houston
University of Ulster, Northern Ireland
sk.houston@ulst.ac.uk

SECTION 7

TEACHER EDUCATION

Edited by Michèle Artigue and Derek Holton

HONOR WILLIAMS

PREPARATION OF PRIMARY AND SECONDARY MATHEMATICS TEACHERS:

A working group report

1. INTRODUCTION AND KEY ISSUES

Compiling a chapter which can adequately represent all the complexities associated with the preparation of teachers to teach mathematics throughout both primary and secondary level while at the same time incorporating different countries' views, would be rather foolish to try to attempt. What the working group endeavoured to accomplish was to identify commonalties and to agree on some of the issues which needed to be addressed if we are to train teachers to be better prepared to teach mathematics. We reiterated and reaffirmed that pre-service teacher education has to involve strong partnerships with schools and used as our focus, our role as mathematics teachers and teachers of mathematics in higher education. In other words, we attempted to clarify for ourselves an articulation of Judge, Lemosse, Paine, and Sedlak's statement (1994, p. 262). 'This may suggest that just as teachers teach best what they know best, so universities should concentrate on doing what they do best.' He then reminds us that this articulation 'will of course vary in definition and in balance from one country to another (and that these countries) will continue to be very different places'.

Views about the best preparation for intending (mathematics) teachers are as diverse as the teacher training models and programmes that have been adopted in order to produce an `effective practitioner'. In many countries, the preparation for teaching mathematics to pupils in primary schools is essentially different to the preparation for secondary teaching. Often the stereotypical view is that to teach at primary level you need to know about children and at secondary level to have a good understanding of your subject with an understanding of how pupils learn effectively, taking second place.

It is still the case, that although most teachers have undertaken some form of pre-service courses, much of mathematics teaching is seen to be inadequate. This gives rise to new structures being debated. The pressures for pre-service curriculum change do not necessarily (or mainly) come as a result of these debates among academic communities but often from governmental views on the nature of reform

445

Derek Holton (Ed.), The Teaching and Learning of Mathematics at University Level: An ICMI Study, 445—454.
© 2001 *Kluwer Academic Publishers. Printed in the Netherlands.*

required. In particular, the balance within training programmes of the following elements is under review:

- mathematics as a subject study;
- mathematics as it applies to classroom contexts;
- general pedagogical knowledge and understanding;
- actual teaching practice i.e. working with pupils in classrooms.
-

In many countries, the shortage of people who want to teach mathematics at secondary level and those with a subject expertise in mathematics at primary level is becoming critical. Different aspects of traditionally accepted views of being a teacher - as an expert in the subject, as a facilitator of learning, as a motivator and a source of inspiration and even an upholder of moral standards - are rapidly being replaced by more prescriptive descriptions of teachers' roles. This concerns being a deliverer of a prescribed curriculum necessitating the acquisition of particular skills and competencies by pupils. In an era when schools are also supposed to be responsible for developing children to face the needs of the next century by acquiring key skills such as co-operation, communication, flexibility and preparing them for the world of work, the above poses a dilemma.

Throughout the discussion, the 'level' of subject knowledge and understanding of intending teachers was a recurring theme. The group focussed, therefore, on principles which could be attributable to the most effective mathematics teacher education courses and views about what pre-service mathematics should be taught.

The main questions posed by the working group were:

- to what extent is it possible to have an agreed view about what constitutes appropriate pre-service training for both secondary and primary teachers?
- are there essential core/or key mathematical experiences which all students should acquire?
- what balance of pedagogy and mathematical knowledge should be offered at the pre-service level?
- with the increased emphasis on school-based teacher training, what is the role of higher education in the training of teachers?
- what does lifelong learning constitute for a teacher of mathematics?
- how could the wider availability of on-line technology influence the methodology for training teachers?

2. CURRENT ISSUES

Comparisons of structures across countries

Comparisons of models, structures and content of pre-service courses from other countries are not new. Bob Moon (1998) reminds us that, 'Political interest in how other countries organise their education systems goes back at least into the nineteenth century' (p. 3). What is of interest is the shifting location of where the responsibility for course models lie, whether in specialist teacher training colleges, universities or, more recently, solely within schools themselves. The key themes addressed by Moon in his paper are: Course Design; System building; Outcomes and standards; Partnerships; Pedagogies and didactics; and Values. All these provide a good basis for discourse, but what clearly emerges is that the strong cultural and political influences on models of training may well determine the nature of the curriculum (in our case, mathematics) that pre-service teachers encounter. Judge (1992) in his comparisons of France, USA and the UK, suggests that although there are sufficient salient features which permit valid discussions, the influences that need to be borne in mind are: the differing definitions of a university; the contrasting models of schooling and hence models of teacher training; and, the role of the state and public authority.

Teacher beliefs

Improvements to the quality of mathematics teaching are often accompanied by associations and assertions about teacher beliefs and how these are a major influence on pupils' learning. Teachers' conceptions of mathematics, and consequently how these relate to views on how mathematics is taught and learned, have been the subject of many research studies (Cobb, 1986; Cooney, 1985; Ernest, 1994; Thompson 1984, 1992;). A challenge for pre- and in-service courses and for curriculum development projects and initiatives, is in how to shape these strong convictions to orientate more closely with the aims and purposes of sound mathematics teaching and learning.

> '... we have to conclude that a good implementation of innovations implies preliminary work on teachers' beliefs, both in the field of subject matter knowledge and pedagogical knowledge. This does not mean that teachers' beliefs be conditioned or individual initiative and creativity abolished, but that teachers be made conscious of their beliefs and be able to discuss them in the light of innovations.' (Furinghetti, 1994)

This implies that however thoroughly the teacher training programme courses are conceived, unless they involve the trainees in articulating and discussing beliefs and practices associated with mathematics, they will often fail to grasp all that is being required of them in the classroom.

Teacher confidence and competence

Underlying discussions surrounding teacher confidence and competence we need to constantly ask, 'Knowledge and understanding of what? To what level?' It was agreed that pre-service teachers would need to gain confidence in using different approaches in the classroom in order to engage pupils in meaningful mathematics. The issue of what should take precedent - actions in the classroom or underpinning subject knowledge - somewhat polarises the complexities. Included within pre-service teachers' experiences has to be a realisation of how both teaching methods and appropriate subject knowledge can help to contribute to a continuing improvement in their craft of teaching.

In our discussions it was clear that many countries are moving towards more clearly centrally described teacher training models. In the UK, for instance, 'Teaching: High Status, High Standards', (DfEE, 1998) sets out criteria which all courses of pre-service teacher training must meet. 'The criteria set out the standard of knowledge, understanding and skills all trainees must demonstrate in order to successfully complete a course of initial teacher training and be eligible for Qualified Teacher Status.' Although describing threshold criteria can aid clarity and the development of shared meaning and purposes, this approach has the inherent danger of being seen as a set of hoops to jump through in order to become a 'confident and competent' teacher.

As Calderhead and Shorrock (1997), describe:

> 'What counts as 'good' may also be context-dependent: the 'good' teacher in one school with one particular class may not be 'good' in a different type of school teaching different children. By providing a list of competences we may contribute to an illusion that teaching is easy and that once a pre-determined knowledge base and set of skills have been learned, the task of teaching has been mastered. We may also be neglecting the creative element of teaching, and failing to acknowledge the individual ways in which teachers develop.' (p. 193)

The subtleties and complexities of maintaining effective mathematics teaching has to be considered alongside a continuing process of re-orientation by many teachers. Governments need to acknowledge the central role of professional development throughout a teacher's career in order to up-date both subject knowledge and teaching approaches so that they match changing climates, for example, use of technology, demand of the workplace, public expectations, changing emphases in mathematics curricula.

The role of theory, its application to mathematics education and hence to practice in the classroom

The debates will continue about the appropriateness of developing links within pre-service education of educational theory, application to mathematics teaching and learning and hence to actual practice in the classroom. A pre-service teacher's comments, "Don't explain to me why this is a good approach, just tell me what activities will work with my 7 year olds," resonates clearly with a pupil's comments in a mathematics lesson, "I don't want to know why it works in this and other situations, just give me the rule to follow."

Perhaps it is the extent to which pre-service teachers are exposed to underlying theories and principles which distinguishes individual country approaches to initial teacher education. Essentially we need to arrive at a situation where teachers will regularly ask questions such as,

- why did that activity work in the classroom?
- would it have worked with pupils of different ages, abilities and cultures? If not, why not?
- is there any research in this area to help me to understand and improve my practice?

Recently we can see a more profound shift of emphasis within in-service mathematics teacher education to a more central role for practitioner based research (Askew and Wiliam, 1995). This has helped us to view more carefully the nature and role of research in mathematics education (Balacheff, Howson, Sfard, Steinbring, Kilpatrick and Sierpinska, 1993); to relate to models of mathematics education (Wittmann, 1995) and to develop appropriate research protocols (Bishop, 1996). These elements provide a platform for us to research further mathematics classroom situations which will/can have a direct impact on the effective teaching and learning of mathematics in the classroom and hence be seen as being relevant and necessary to the pre-service teacher as part of their training.

Aims and intentions and Course models
As a group, we recognised that there needed to be a closer match between the aims and intentions through to the descriptions and content of pre-service courses. One example presented at the conference (Shoaf, 1998), provides a good précis of our discussions. The example described goals which relate to a pre-service secondary mathematics course, however, it can be seen that the main thrusts could be attributable to primary courses. The challenge is to adapt this agenda by varying the depth and balance across the individual elements.

The goals of this course (in following those stated by the NCTM 1989) are to help pre-service secondary mathematics teacher to:

- develop the mathematical ability to explore, conjecture, and reason logically thereby leading to an increase in students' mathematical knowledge and skills;
- solve non-routine problems and be exposed to the breadth of mathematics and many of its interesting problems and applications;
- communicate about and through mathematics while encouraging themselves and their future students to have fun with mathematics;
- connect ideas within mathematics and between mathematics and to exhibit the unity of diverse mathematical fields;
- develop personal self-confidence and a disposition to seek, evaluate, and use quantitative and spatial information in problem solving and making decisions to help promote their own creativity and that of their students;
- develop proficiency in:
  selecting mathematical tasks to engage students' interests and intellect;
  providing opportunities to deepen students' understanding of the mathematics being studied and its application;
  orchestrating classroom discourse in ways that promote the investigation and growth of mathematical ideas;
  using, and helping students to use, technology, in particular, graphing calculators and computer software, and other tools to pursue mathematical investigations;
  seeking, and helping students see, connections to previous developing knowledge;

    guiding individual, small-group, and whole-class work;
- identify and model strategies used for problem solving;
- gain increased competence with open-ended questions whose answers are not known, and with ill-posed questions;
- use oral and written discourse between teacher and students and among students to develop and extend students' mathematical understanding and creativity;

    be aware of the application of knowledge to current research in mathematics education;

    be aware of national, state, and local guidelines relating to mathematics education.

However, matching course aims to effective actions in mathematics classrooms requires the active participation and partnership of those closely involved - the schools, the pre-service teachers and teacher educators. "Aims can only act as forces for change and development when those who are to use and understand them are actively involved in either their evolution or in working at their implications as they effect their own situation" (Ahmed, 1987).

3. KEY MATHEMATICAL EXPERIENCES FOR PRE-SERVICE TEACHERS

As indicated earlier, notions of mathematical competence and levels of achievement for pre-service teachers at primary level taxed us severely. The mathematical entry requirements onto pre-service courses vary from country to country, from prescribed grades in national examinations to no specified qualifications. It was acknowledged that subject knowledge alone would not necessarily indicate an understanding and appreciation of the mathematics the pre-service teachers would find themselves teaching in school. In addition, many of the students found themselves taught within the mainstream undergraduate mathematics courses. By itself, this may not cause a problem, but the effectiveness of this route depends entirely on the mathematics courses student teachers are required to undertake by higher education institutions. It was recognised that if we are to increase the pool of students studying mathematics in higher education, and hence increase the number of students who would wish to teach mathematics in schools, then we would need to help both primary and secondary teachers to have a more thorough understanding of not only why mathematics is important but also how to teach mathematics in ways which can maintain interest and enjoyment as well as motivate pupils to study the subject beyond compulsory schooling. A lack of direct relevance or association with a student's chosen career could result in the student teacher being 'competent' in areas of, say, linear algebra or algorithms and programming, without having a fundamental understanding of place value. This possibility then leads to questions about what experiences are essential and what would seem to be an appropriate knowledge base for intending teachers. Some key mathematical and mathematical application experiences were identified. These are not intended to provide definitive lists but represent areas where there appears to be a shared understanding of meaning, although the forms and depth of treatment will vary in line with the models of pre-service training adopted.

Key mathematical experiences

- Developing mathematical thinking (reasoning, proving).

- Solving problems (unknown/open/non-routine).

- Using knowledge in new (non maths) situations/contexts.

- Modelling.

- Creating 'new' (for themselves) mathematical knowledge.

- Contact with historic and contemporary mathematics.

- Communicating mathematics - reading, writing, talking, listening.

- Connecting mathematical ideas.

- Recognising cross-curricular links.

These experiences imply a pro-active practical approach where pre-service teachers are able to both reflect on their own learning and understanding of mathematics as well as reflecting on the approaches used by their teachers to introduce and discuss topics. It is hoped that this would encourage the development of a strong facility to draw productive links between different areas of mathematics, show insight into the nature of complex problems, demonstrate an in-depth understanding of areas of mathematics, and enable effective communication of methods and arguments.

To parallel the processes of developing pre-service teachers' own mathematical thinking we would seek to develop an awareness of and a sensitivity towards children's mathematical thinking from an early stage. The following ingredients were identified as contexts for course planning and as a basis for pre-service teachers to abstract and incorporate insights gained in order to begin to apply these to lesson preparation and practice.

Key educational experiences of pre-service education for mathematics

- Gaining an understanding and insight into children's representations and ideas.

- Developing pre-service students' metacognitive skills.

- Making connections between professional knowledge and practices.

- Analysing and appraising pedagogy and didactics.

- Demonstrating and using technology to enhance the learning of mathematics.

- Encountering effective teaching to gain confidence in mathematical knowledge and understanding.

---

4. THE ROLE OF HIGHER EDUCATION

As indicated earlier, discussions will continue to take place on the role of higher education in the training of teachers. Perhaps what needs to be made more overt is an articulation of the distinctive contribution that mathematicians and mathematics educators can make to this training and the areas where schools are best placed to provide the appropriate environment and training.

One of the main points which we considered at some length was the areas of mathematics that pre-service teachers should study. It was important that the students engaged in-depth in some mathematics beyond the level of mathematics they proposed to teach. It was agreed that the best description for primary pre-service training would be for students *to revisit elementary mathematics topics from an advanced point of view*. In other words, putting an up-to-date interpretation of Klein's early thoughts into a practical proposition within pre-service education (Klein 1924). It was felt that this would help to generate confidence through a deeper appreciation of the mathematics that the students are intending to teach. Additional features which higher education is well placed to contribute are given below.

> ### *Contributions from higher education*
>
> - Helping students to revisit elementary topics from an advanced point of view.
>
> - Drawing links between theoretical education and beliefs in education.
>
> - Approaches and materials for learning mathematics used should be shown to be transferable to school settings.
>
> - Engender confidence in mathematics and teaching mathematics.
>
> - Using technology at a student's own level and relating it to work at pupil level.
>
> - Running problem solving 'seminars'.
>
> - Ensuring a match between teaching approaches used and approaches in the classroom.

Although the working group focussed mainly on pre-service education, it was recognised that in order to improve the quality of mathematics taught, that the continuing professional development of teachers is central to achieve this aim. It was agreed that the elements as outlined in this chapter are as pertinent for teachers who have been practising for some time as it was for those at the beginning of their careers.

In conclusion, the working group considers that pre-service courses for mathematics teachers need to be conceived in such a manner that they prepare teachers for lifelong learning. The principles underpinning the above key mathematical experiences, key educational experiences and the contributions from higher education should form a basis for rethinking pre-service mathematics courses for teachers at all levels.

REFERENCES

Ahmed, A G. (1987). *Better Mathematics.* London: HMSO.

Askew, M. and Wiliam, D. (1995). *Recent Research in Mathematics Education 5-16.* London: HMSO.

Balacheff, N. Howson, AG. Sfard, A. Steinbring, H. Kilpatrick, J. and Sierpinska, A. (1993). What is research in mathematics education, and what are its results? Discussion Document for an ICMI study. *International Reviews on Mathematics Education.* ZDM93/3 114-116.

Bishop, AJ. (1996). A researcher's code of practice? Paper presented ICME8 Seville, Spain, Working Group 24: Criteria for quality and relevance in mathematics education research.

Calderhead, J. and Shorrock, S. B. (1997). *Understanding Teacher Education.* London: The Falmer Press.

Cobb, P. (1986). Contexts, goals, beliefs and learning mathematics, *For the Learning of Mathematics,* 6(2), 2-9.

Cooney, T J. (1985). A beginning teacher's view of problem solving. *Journal for Research in Mathematics Education*, 16(5), 324-336.

DfEE (1998). *Teaching: High Status, High Standards Circular 4/98 Requirements for courses of initial teacher training*. London: Department of Education and Employment.

Ernest, P. (1994). *Mathematics Education and Philosophy: An International Perspective*. London: The Falmer Press.

Furinghetti, F. (1994). Mathematics teachers and changes in curricula: problems of reshaping beliefs and attitudes. In N.A. Malara and L. Rico (Eds.), Proceedings of the First Italian-Spanish Symposium in Mathematics Education, Modena, Italy, pp. 173-180.

Judge, H. (1992). Schools of education and teacher education. In D. Phillips (Ed.), Lessons of Cross-National Comparison in Education. *Oxford Studies in Comparative Education*, Vol 1, pp. 37-55. Wallingford: Triangle Books.

Judge, H., Lemosse, M., Paine, L., and Sedlak, M., (1994). The university and the teachers - France, the United States, England, *Oxford Studies in Comparative Education*, Vol 4,(1/2). Wallingford: Triangle Books.

Klein, F. (1924). *Elementarmathematik vom höheren Standpunkte aus*. Berlin: Springer Verlag. Translated into English by Hedrick, ER. and Noble, CA. (1924) *Elementary Mathematics from an Advanced Standpoint*. New York: Dover.

Moon, B. (1998). *The English exception? International perspectives on the initial education and training of teachers*. Occasional Paper 11, Universities' Council for the Education of Teachers.

Shoaf, M. (1998). A Capstone Course for pre-service secondary mathematics teachers' paper presented ICMI study conference on teaching university mathematics, Singapore, Working Group B4 preparation of Primary and Secondary mathematics teachers.

Thompson, A. G. (1984). The relationship of teachers' conceptions of mathematics and mathematics teaching to instructional practice, *Educational Studies in Mathematics*, 15, 105-127.

Thompson, A. G. (1992). Teachers' beliefs and conceptions: a synthesis of the research. In D. A. Grouws (Ed.), *Handbook of Research on Mathematics Learning and Teaching*, pp. 127-146. New York: Macmillan.

Wittmann, E. Ch. (1995).Mathematics education as a design science, *Educational Studies in Mathematics*, 15, 25-36.

Honor Williams
University College Chichester, England
h.williams@ucc.ac.uk

THOMAS J. COONEY

USING RESEARCH TO INFORM PRE-SERVICE TEACHER EDUCATION PROGRAMMES

Many forces at all levels of schooling contribute to the shaping of pre-service teacher education programmes. Prominent among them are the expectations of administrators and parents regarding mathematics and its teaching. Those expectations often are communicated in various ways to students, who are party to defining the didactical contracts of classrooms. But if teacher education is to be more than an activity driven by practical demands, as important and forceful as they may be, it must be grounded in a foundation of inquiry. In this paper, I will focus on what has been learned about both in-service and pre-service teacher education in an effort to provide footings for such a foundation.

1. LESSONS LEARNED FROM IN-SERVICE TEACHER EDUCATION PROGRAMMES

Of the many lessons to be learned from research involving in-service teachers, fundamental to them all is that teachers operate in a social-political arena that has its own defining characteristics. Nelson (1998), for example, found that school administrators have fairly well articulated views of mathematics and about its teaching. Some administrators' views reflect what might be called a transmission mode in that ideas tend to flow from the top down, much as when teachers teach by telling. When couched in this context, reform as defined by documents such as the NCTM Standards is difficult to achieve because the climate is not conducive to process-oriented types of teaching. Other administrators have more of an inquiry mentality in terms of promoting ideas about teaching and learning. In their schools, reform has a better chance of taking hold as support exists for teachers taking risks when they use innovative teaching methods. The point is that reform in the teaching of mathematics is not just a function of the individual teacher; rather, it is combination of circumstances in which that teaching occurs.

In light of this, care must be taken to think of teacher change in a pluralistic way. Reform can only make sense to the teacher when proposed changes are not antithetical to the core beliefs of the community in which that teaching occurs. Usually this leaves considerable room for change, for few would argue against the notion of such processes as problem solving and reasoning even though they are too seldom seen in most classrooms. To begin my analysis, I will entertain the notion of what teacher change might mean in the context of today's classrooms and teacher education programmes.

Derek Holton (Ed.), The Teaching and Learning of Mathematics at University Level: An ICMI Study, 455—466.
© 2001 *Kluwer Academic Publishers. Printed in the Netherlands.*

1.1 The Notion of Teacher Change

The construct **teacher change** has several different connotations. Historically, the notion of teacher change had to do with teachers changing certain aspects of their teaching behaviour. For example, certain teaching behaviours such as clarity and flexibility were associated with increased student learning. But, as the field moved away from deterministic research paradigms and toward those paradigms that took into account the contexts in which teaching occurred, a different view of teacher change emerged. Much of this shift was grounded in the notion of constructivism, social constructivism in particular. Concomitantly, research on teachers' beliefs became increasingly important. Thompson's (1992) review of this literature stands as significant testimony to the importance and contribution of this research to mathematics education. Primarily, the research reviewed by Thompson focused on what individual teachers believed about mathematics and its teaching and learning. Some of this research made use of Likert-type scales as teachers responded to various queries, thus revealing what they believed. For the most part, however, research on teachers' beliefs has been more qualitative in nature and is grounded in various philosophical or psychological perspectives. Perhaps spurred by the writing of Davis and Hersh (1981), the issue of what constituted mathematics has been called into question at least more extensively than had previously been the case in mathematics education. As Ernest (1991) pointed out, a fallibilistic view of mathematics led to the conclusion that its teaching was problematic as well and to a variety of questions, not the least of which was "What constitutes good teaching of mathematics?"

The work of Perry (1970) has gained credence among mathematics educators in terms of describing teachers' beliefs and orientations toward knowing. Although Perry's work has been refined and redefined by the works of Belenky, Clinchy, Goldberger, and Tarule (1986), and Baxter Magolda (1992), as well as inside the field of mathematics education, e.g., Cooney, Shealy, and Arvold (1998), and Wilson and Goldenberg (1998), it still serves as a framework for conceptualizing teacher change by determining whether a teacher can move away from absolutes and toward an appreciation of the context in which events occur. Related to this orientation is a considerable body of research on reflection to which Dewey (1916) was an early contributor and to which King and Kitchener (1994) are more recent contributors. Perhaps more than any other single construct, reflection is judged to be an essential ingredient of most teacher education programmes as teacher educators strive to produce what Schön (1983) has termed *reflective practitioners*.

The above mentioned research is important because of the current emphasis on process-oriented teaching. Seen from an absolutist perspective, mathematics is considered a body of knowledge subject to transmission from teachers to students; but from a fallibilistic perspective, the subject is seen as an evolving body of knowledge for the research mathematician as well as for the child. Teaching then becomes a matter of finding ways to facilitate that emerging body of knowledge in the child, not just in terms of knowing more facts and principles but as a way of thinking and approaching our world. Consequently, teacher change is less a matter of demonstrating specific behaviours and more a matter of demonstrating the kind of teaching in which process, context, and the development of one's own voice in creating a philosophy of teaching take precedence.

Wilson and Goldenberg's (1998) two-year study of Mr. Burt illustrates the difficulty associated with teachers making change. Mr. Burt was a middle school teacher who was interested in reforming his teaching of mathematics. He was successful in that his teaching became more concept oriented than procedure oriented. But his classroom instruction remained teacher centred where he was the determiner of what was correct and what was not. The authors reached the following conclusion regarding Mr. Burt's efforts to reform his teaching.

> Mr. Burt's approach generally portrayed mathematics as a rigid subject to be mastered and correctly applied, rather than as a way of thinking or as a subject to be explored. (p. 287)

It is worth noting that the authors concluded that Mr. Burt felt that he was making substantial changes in his teaching, albeit their observations of his teaching suggested otherwise. This circumstance reflects other efforts at reform and may be indicative of teachers who are on the verge of change but who have not yet transformed their teaching into what would likely be characterized as process-oriented teaching. This study also demonstrates the difficulty associated with teachers changing their beliefs to the extent that substantial changes in their teaching follow. Teachers, perhaps wisely so, are careful not to move too quickly from the land of certainty into uncharted waters.

Much of the research on teachers' beliefs focuses on the individual's beliefs and ways of knowing, sometimes at the expense of attending to the contexts that contributed to the formation of those beliefs. A particular theoretical perspective that attends to context and has gained considerable credence in mathematics education is that of social constructivism. For example, Cobb, Yackel, and Wood (1993) view the classroom as a place where the teacher establishes a mathematical community that honours shared mathematical meanings. This approach is constructivist not only in terms of realizing that students construct their own mathematical knowledge but also in the sense that interactions with others contribute to the construction of mathematical ideas.

In light of this orientation, Goldsmith and Shifter (1997) suggested that teachers and researchers alike need to reconceptualize fundamental notions of teaching, learning, and the nature of mathematics. These fundamental notions, rooted in social constructivism, require researchers and teachers to share their expertise in an effort to redefine the classroom environment, thus blurring the distinction between researcher and subject. Similarly, Simon (1997) presented models of teaching that serve as alternatives to the traditional lecture-demonstration mode of instruction that typify the teaching of mathematics around the world. Central to Simon's notion of teaching is the ability of the teacher to "harness the students' inherent ability to learn and to self-regulate" (p. 59). From Simon's perspective, the teacher is a facilitator who shapes mathematical discussions yet allows students to actively participate and shape those mathematical discussions.

Lave and Wenger's (1991) view that learning is a function of group participation in which group members share a sense of value and meaning provides one way of thinking about learning from a constructivist perspective. From this perspective, teacher change is a matter of the teacher learning to work in a facilitative way to engender and promote a mathematical community. Consistent with this perspective is that of Tharp and Gallimore (1988), who characterize change as an individual moving from a position in which he/she is dependent on others to one in which he/she is independent. The development of in-service programmes that promote these perspectives on change requires a model that is consistent with the kind of process teaching that is expected from participating teachers. That is, a community of teachers must be developed in which pedagogical problems are addressed in the same way that mathematical problems are addressed.

Other research on teacher change reveals the difficulty of establishing such communities. Lloyd (1999) studied two high school teachers' conceptions of the cooperative and exploration components of a reform-oriented mathematics curriculum. One teacher valued the open-ended nature of the curriculum's problems, whereas the other teacher saw the materials as too structured and consequently implemented the materials from a teacher-centred perspective. For both teachers, the curriculum remained fixed, as neither teacher seemed to appreciate the potential interactive effect of curricular materials and student thinking.

From one perspective, teacher change can be thought of as the propensity of the individual to manifest shifts in beliefs or in actual teaching practices. From another perspective, teacher change can be viewed as the ability of the teacher to participate in and contribute to professional communities that attempt to address problems of reform. From yet another perspective, teacher change can be conceived as the teacher's ability to formulate and nurture mathematical communities within his/her classroom. These perspectives are not mutually exclusive but rather compatible, each having its own merit. I will now examine various attempts to promote teacher change.

1.2   Strategies That Effect Change

Cooney and Krainer (2000) introduce the notion of networking to describe teachers' relationships with other professionals or with ideas they encounter in teacher education programmes. Simply put, horizontal networking refers to teachers forming relationships with their peers either in pre-service programmes or in in-service programmes. Vertical networking refers to the individual's commitment to ideas in a deep and meaningful way that leads to changes in his/her teaching practices. What seems apparent is that some sort of networking is fundamental to professional growth. Krainer (in press) tells the story of Gisela, who emerged as a 'lone fighter' early in her career and became a community leader later in her career. Gisela relished the opportunity to engage and network with other professionals. In her position as a school administrator and mathematics teacher she actively sought out ways for teachers in her school to engage in professional development programmes that promote interaction and mutual support.

The professional development programme described by Borasi, Fonzi, Smith, and Rose (1999) placed an emphasis on teachers interacting with inquiry-based materials that provided them with experiences in inquiry and concomitantly modelled inquiry for their own teaching. Teachers interacted with the materials in a summer programme and then were supported in their use of those same materials during the subsequent school year. This study points out the need for materials to be developed not only for teacher education programmes but for classroom teaching as well. Further, teachers need support in using materials if they are to realize reform as it is intended.

Even's (1999) approach to reforming the teaching of mathematics consisted of training in-service teachers to become in-service teacher educators. A critical factor in the programme's success was the continual contact the programme staff had with the emerging leaders. The focus of the different meetings included determining what constitutes a good problem, focusing on students' conceptions and ways of thinking, and deciding what it takes to become an effective teacher leader and provider of in-service programmes. Each of these foci were developed over a period of extensive teacher involvement.

Campbell and White (1997) described the project Increasing the Mathematical Power of All Children and Teachers (Project IMPACT) and its effect on teachers. The authors identified several characteristics of the programme that they considered critical to the project's success. First, the project explicitly addressed teachers' knowledge and their needs for understanding the rationale for reform. Second, in an effort to eliminate or reduce teacher isolation, entire mathematical faculties from schools were involved in the project. Finally, the project honoured the intellectual and pedagogical culture of the schools as they challenged, encouraged, and supported each school's attempt to define reform.

An approach that is attracting attention in teacher education is the use of cases to serve as focal points for engaging teachers in critical analyses of classroom events. Merseth (1996) described instances of using cases to increase teachers' mathematical and pedagogical content knowledge and promote teachers attending to student thinking in the process. In particular, Walen and Williams (2000) found that teachers' discussions of and reflections about cases were particularly helpful in acknowledging and addressing their concerns about the teaching of mathematics.

In summary, it is clear that there are two defining characteristics of successful professional development programmes: longevity and networking. Change is not likely to occur because of single experiences. Neither is change likely to occur when teachers are isolated and 'lone fighters' for reform. Perhaps the breaking down of the isolation that teachers face is the single most important element that leads to teacher change. Certainly, there are other aspects of this networking process that deserve attention, some of which will be addressed in the subsequent sections. But whatever they are, they must be grounded in the context of teachers sharing ideas and experiences regarding attempts to reform their teaching.

2. THE ISSUE OF REFLECTION

Most of the professional development programmes described above provide multiple opportunities for teachers to engage in reflective thought and to consider what aspects of their instructional programme are problematic. The notion of reflection is hardly a newcomer to teacher education programmes. The work of Dewey (1916) was, perhaps, the first significant effort to connect reflective thinking to the educational process. Schön's (1983) notion of reflective practitioner has considerable credence in today's teacher education programmes. Mathematics educators encourage teachers at all stages of their professional journey, including pre-service teachers, to engage in reflective analyses about what constitutes mathematics and what it means to be a good teacher of mathematics. For example, Jaworksi (1998) focused on what she calls the *development of teaching*, which entails the researcher and the teacher working together in a cyclical research process embedded in an atmosphere of reflective analysis. Through the act of reflection, problems are identified by the teacher and ways of dealing with those problems are addressed. The development of teaching is an evolutionary approach to teacher change fuelled by the act and art of reflecting thinking.

2.1 Theoretical Perspectives

There is good reason why reflection is such an important commodity in present-day teacher education programmes. If teacher change is to become a reality, there must be a sense that change can bring about learning in a way that the traditional means of teaching cannot. That is, there must be an element of doubt about existing instructional modes. To say it another way, if teachers are quite pleased with their current methods of teaching, they may see little need to change. If they do not see something problematic either about the nature of mathematics or about their students' understanding about mathematics, what rationale exists for embracing a different way of teaching?

The very essence of reflective thinking is to examine from a critical perspective what is and to consider what might be. Almost by definition, reflective thinking necessitates attention to context which, in general terms, Perry (1970) refers to as relativistic thinking, a precondition for true reform. This kind of thinking allows one to see mathematics and its teaching and learning as problematic. Instead of thinking of students' responses as right or wrong, reflective thinking leads one to the question of what kind of thinking resulted in a particular response. Reflective thinking prohibits one from thinking of mathematics as a fixed body of knowledge to be delivered in a fixed format for the simple reason that there is nothing problematic about this process. From both Dewey and Schön, we learn that being a professional involves moving beyond technical knowledge and toward seeing the act of a professional as a process of making judgements and decisions. The reflective practitioner makes those judgements based not only on technical knowledge but with respect to the broader context in which those questions arose in the first place.

This emphasis on reflective thinking has its roots in a variety of circumstances that have arisen more or less simultaneously. Primary among them is the current emphasis on constructivism and the notion that the individual constructs his or her own unique reality. In particular, Von Glasersfeld has been a significant player in this movement. His definition of reflection is one of the most cited passages in the literature.

> [Reflection is the capability] that allows us to step out of the stream of direct experience, to re-present a chunk of it, and to look at it as though it were direct experiences, while remaining aware of the fact that it is not. (Von Glasersfeld, 1991, p. 47)

Reflection so defined is a far cry from simply recalling events. It requires analysis and the ability to see an event contextually. Certainly the many decisions teachers make every day provide opportunities for reflection, although it would be nonsense to argue that every classroom decision is a product of explicit reflection. Nevertheless, the essence of reform is to consider an array of possibilities that contrast the typical. Reflection is the vehicle by which the process of reform moves forward. This perspective argues for teachers encountering opportunities to engage in reflective thinking and decision making, early in their professional training beginning with their pre-service teacher education programmes.

2.2 Reflection as a Basis for Educating Teachers

Most teacher education programmes today honour those research findings which demonstrate that teachers' beliefs about mathematics and its teaching are significant factors that influence what happens in classrooms. Consequently, they devote part, and often a significant part, of their programme to providing teachers the opportunity to reflect on a variety of issues. Mewborn (2000), for example, writes explicitly about the role of reflective thinking in her work with pre-service elementary teachers and how it influenced their professional development. The project Research and Development Initiatives Applied to Teacher Education (RADIATE) used reflective thinking as the means by which pre-service secondary teachers could examine and redefine their notions of mathematics and mathematics teaching (Cooney, Wilson, Albright, and Chauvot, 1998). Some teacher educators, e.g., Goffree and Oonk (in press) and Sullivan and Mousley (in press), use electronic media as a stimulus for teachers reflecting on their own practice and those of others. Edwards and Hensien (1999) use a reflective model for change as the basis for dialogues between researcher and teacher as they engage in action research.

The phenomena of reflection-based teacher education programmes are not unique to mathematics education. The theme of teacher socialization is recurrent in the literature and throughout teacher education programmes (see Zeichner and Gore, 1990). Teachers are shaped by the context in which they are educated and in which they educate and yet they can contribute to their own shaping through the process of reflective analysis. But it takes time and support from both colleagues and in-service personnel. This circumstance is not unique to mathematics teacher education, as subject matter alone is not the sole determinant of the socialization process or of reflective analyses that permit individuals to shape their own professional development. Nevertheless, mathematics is a unique subject with its own history, methods of generating knowledge, and expectations regarding what constitute mathematical thinking. Indeed, teacher socialization is, in part, influenced by communities that are identified with mathematics. The nature of the subject is relevant to reflective thinking about classroom events. As Dossey (1992) has argued, the nature of mathematics does and should play a significant role in shaping our views about what and how the subject should be taught.

3. THE ROLE OF MATHEMATICS

Many years ago Thom (1973) opined that "all mathematical pedagogy, even if scarcely coherent, rests on a philosophy of mathematics" (p. 204). I have come to appreciate this perspective more and more as I have worked with pre-service and in-service teachers. Teachers who see mathematics as essentially an exercise in problem solving tend to teach in that away, albeit their numbers are few. In contrast, those who see mathematics from a utilitarian perspective tend to emphasize procedures, often computational in nature, in their teaching. Evidence abounds that elementary teachers' knowledge of mathematics is often lacking in terms of supporting significant reform with respect to both curriculum and teaching methods. There is less evidence with respect to secondary teachers, albeit the evidence does point in the direction that both pre-service and in-service secondary mathematics teachers have a rather narrow view of mathematics (Thompson, 1992), which can be a limiting factor for reform. Given this set of circumstances, some projects have explicitly focused on teachers reconceptualizing mathematics.

3.1 The Reconceptualization of Mathematics

Schifter (1998) involved teachers in reconceptualizing the big ideas in mathematics by having them reflect on the kinds of understandings that their students had. She found that the two teachers whom she studied in detail "developed new conceptions of the nature of mathematics and a heightened sense of their own mathematical powers" (p. 83). Schifter went on to make the following claim.

> As learners of mathematics, they experienced a new kind of classroom, one in which the social character of doing mathematics was realized, and in which student thinking took centre stage. ... Finally, the fact that the mathematics content under study in this professional development setting is the very content they teach prepared them to engage their students more responsively and with greater mathematical fluency. (p. 83)

Although there are many features to Schifter's programme, one of the key elements was engaging teachers in mathematical experiences involving the same mathematics they were teaching. Similarly, Cooney, Brown, Dossey, Schrage, and Wittmann (1999) have approached the mathematics of the secondary school from a perspective that fuses mathematics and pedagogy. Their aim was to provide secondary teachers with mathematical experiences that mirror reform and to engage teachers in the doing of substantial mathematics that is grounded in the concepts of secondary school mathematics.

The evidence suggests that students' learning of a higher level of mathematics is not sufficient to ensure a relativistic view of the subject. The development of a broader and more process-oriented view of mathematics is more likely to occur when students study the mathematics they will be teaching but from a more sophisticated and broader perspective. That is, teachers need to study mathematics that is familiar to them, but from a more sophisticated perspective that embodies the kinds of connections called for in the NCTM Standards, for example, and that embraces a pedagogy of inquiry as suggested by Borasi, et al. (1999). This seems to be particularly true with respect to teaching mathematics using technology. Laborde (in press) makes the point that technology can enable teachers to see mathematics as a more process-oriented subject. The implications for pre-service teacher education programmes seem clear: Students preparing to teach mathematics need to experience the power of technology when investigating mathematics. My own experiences with pre-service teachers suggests that they are generally very adept at using various types of computer-based technology when solving their own mathematical problems. But these experiences alone do not translate into a reasonable comfort level when teaching with technology. Consequently, their students seldom realize the power of using technology and how it can transform mathematical investigations.

3.2   Attending to Students' Mathematics

The constructivist orientation places considerable emphasis on how and what the student defines mathematics to be. In some sense, it might be said that the mathematics curriculum *is* that which the student understands and defines mathematics to be. Although not every teacher education programme explicitly bases its programme on constructivist principles, it is still the case that attention to students' thinking is of paramount importance. That is, teacher education programmes strive to teach teachers how they can access and use their students' thinking about mathematics as a guide to designing their instructional programme.

Perhaps the most notable programme that is based on educating teachers to understand students' cognition is the Cognitively Guided Instruction (CGI) programme which was based on research on children's learning at the University of Wisconsin. As described by Carpenter, Fennema, and Franke (1996), CGI helps teachers understand young children's mathematical thinking in well-defined content domains and demonstrates how that knowledge can be used to shape their instructional approach. The authors are quick to point out that the research-based knowledge of children's learning is central to the project but that how teachers interpret and use that knowledge in their classrooms is problematic. Nevertheless, the evidence suggests that teachers so informed did become more adept at listening to their students' interpretations of mathematical ideas and changed their beliefs about how young children construct mathematical concepts and procedures.

Evidence of other cognitively guided instructional approaches with teachers of older children is sparse, perhaps because research on student learning at higher levels of schooling lacks the sophistication and conclusiveness that generally characterize research on the conceptual development of younger children. Yet, unquestionably teacher education programmes make considerable effort to educate teachers as to how they can access their students' thinking. This orientation towards teaching fits well with the notion of reflective thinking.

4. SOME CONCLUDING THOUGHTS

The field of mathematics teacher education has come a long way. Gone are the days when researchers saw teaching as an isolated behaviour that can be dissected into minuscule parts ready for predetermined analyses. Teaching occurs in a context. Teachers are socialized into their profession. These circumstances alone suggest that mathematics teacher education must attend to teachers' needs, to the contexts in which they teach, and to strategies for enabling them to monitor and promote their own professional growth.

So what have we learned from research that can guide our thinking about the education of pre-service teachers? Given that teaching occurs in a context that shapes teachers' beliefs and ways of teaching, it behoves us to educate teachers to appreciate that context and to understand what constraints they face, and which of those constraints are subject to their own personal influence. Enter the notion of reflective thinking. If teacher education is to be more than a matter of giving teachers the 'tricks' we have learned, then we must enable them to be analytical about the circumstances in which they find themselves, for otherwise circumstances will control them. Furthermore, it must be appreciated that this kind of reflective analysis should be nurtured and supported over a long period of time and in a context in which ideas can be shared with critical friends. The development of such an ability need not and should not wait until a teacher has been socialized into the profession. It should begin in the pre-service years where questions such as "What is mathematics?"; "What do you believe about mathematics?"; and "How do you think mathematics is best taught and learned?" are examined critically. We should never forget the unique heritage of our beloved subject. But neither should that heritage blind us to the ways that students, and our teachers as well, acquire mathematical ideas.

I can think of no greater force for improving the teaching of mathematics than to instil in teachers, at all levels of schooling and experience, an attitude of inquiry about the teaching of mathematics. In some sense, we should train researchers, not teachers, in that we want teachers to hypothesize why some students learn and others do not and why some mathematical ideas are difficult to conceptualize and others less so; and to challenge their own teaching methods in the interest of improving those methods. Finally, we need to help them see the value in challenging themselves by asking the question, "How should my students think about mathematics?" It is not that we want teachers to suffer paralysis from their analysis. Rather, we want them to be empowered with their own philosophy of teaching that stands the test of a thoughtful rationale. Such empowerment would enable them to judge for themselves what aspects of reform works for them, rather than blindly accepting or rejecting whatever proclamations are in the wind at any particular point in time. Research suggests that such education is not easy to come by. But research also has provided us with some important clues as to how we can address this problem of education. What greater challenge can we face professionally than to engage in a reflective analysis of our own programmes, much as we would like our teachers to challenge their programmes? Knowledge, acquired in a relativistic perspective, is power. That power should be the focal point of our teacher education programmes.

REFERENCES

Baxter Magolda, M. (1992). *Knowing and Reasoning in College.* San Francisco: Jossey-Bass, Inc.

Belenky, M., Clinchy, B., Goldberger, N. and Tarule, J. (1986). *Women's Ways of Knowing: The Development of Self, Voice, and Mind.* New York: Basic Books.

Borasi, R, Fonzi, J., Smith, C. and Rose, B. J. (1999). Beginning the process of rethinking mathematics instruction: A professional development programme. *Journal of Mathematics Teacher Education, 2,* 49-78.

Campbell, P. and White, D. (1997). Project IMPACT: Influencing and supporting teacher change in predominantly minority schools. In E. Fennema and B. Nelson (Eds.), *Mathematics Teachers in Transition,* pp. 309-355. Mahwah, NJ: Lawrence Erlbaum Associates.

Carpenter, T., Fennema, E. and Franke, M. (1996). Cognitively guided instruction: A knowledge base for reform in primary mathematics instruction. *Elementary School Journal, 97,* 3-20.

Cobb, P., Yackel, E. and Wood, T. (1993). Learning mathematics: Multiple perspectives, theoretical orientations. In T. Wood, P. Cobb, E. Yackel, and D. Dillon (Eds.), *Rethinking Elementary School Mathematics: Insights and Issues.* Journal for Research in Mathematics Education Monograph Series, 6, pp. 21-32. Reston, VA: National Council of Teachers of Mathematics.

Cooney, T., Brown, S., Dossey, J., Schrage, G. and Wittmann, E. (1999). *Mathematics, Pedagogy, and Secondary Teacher Education.* Portsmouth, NH: Heinemann.

Cooney, T. and Krainer, K. (2000). The notion of networking: The fusion of the pubic and private components of teachers' professional development. In M. Fernández (Ed.), *Proceedings of the Twenty-Second Annual Meeting of the North American of the International Group for the Psychology of Mathematics Education,* V2. pp. 531-537. Columbus, OH: ERIC.

Cooney, T., Shealy, B. and Arvold, B. (1998). Conceptualizing belief structures of preservice secondary mathematics teachers. *Journal for Research in Mathematics Education, 29,* 306-333.

Cooney, T., Wilson, P., Albright, M. and Chauvot, J. (1998). Conceptualizing the professional development of secondary preservice mathematics teachers. Paper presented at the American Educational Research Association annual meeting. San Diego, CA.

Davis, P. and Hersh, R. (1981). *The Mathematical Experience.* Boston: Birkhäuser.

Dewey, J. (1916). *Democracy and Education.* New York: The Free Press.

Dossey, J. (1992). The nature of mathematics: Its role and influence. In D. Grouws (Ed.), *Handbook Of Research On Mathematics Teaching And Learning,* pp. 39-48. New York: Macmillan Publishing Company.

Edwards, T. and Hensien, S. (1999). Changing instructional practice through action research. *Journal of Mathematics Teacher Education, 2,* 187-206.

Ernest, P (1991). *The Philosophy Of Mathematics Education.* London: Falmer Press.

Even, R. (1999). The development of teacher leaders and inservice teacher educators. *Journal of Mathematics Teacher Education, 2,* 3-24.

Goffree, F. and Oonk, W. (in press). Digitizing real teaching practices for teacher education programmemes: The MILE approach. In F.L. Lin and T. Cooney (Eds.), *Making Sense of Mathematics Teacher Education.* Dordrecht: Kluwer Academic Publishers.

Goldsmith, L. and Schifter, D. (1997). Understanding teachers in transition: Characteristics of a model of developing teachers. In E. Fennema and B. Nelson (Eds.), *Mathematics Teachers In Transition,* pp. 19-54. Mahwah, NJ: Lawrence Erlbaum Associates.

Jaworski, B. (1998). Mathematics teacher research: Process, practice, and the development of teaching. *Journal of Mathematics Teacher Education,* 1, 3-31.

King, P. and Kitchener, K. (1994). *Developing Reflective Judgment.* San Francisco: Jossey-Bass, Inc.

Krainer, K. (in press). Teachers' growth is more than the growth of individual teachers: The case of Gisela. In F.L. Lin and T. Cooney (Eds.), *Making Sense Of Mathematics Teacher Education*. Dordrecht: Kluwer Academic Publishers

Laborde, C. (in press). The use of new technologies as a vehicle for restructuring teachers' mathematics. In F.L. Lin and T. Cooney (Eds.), *Making Sense of Mathematics Teacher Education*. Dordrecht: Kluwer Academic Publishers

Lave, J. and Wenger, E. (1991). *Situated Learning: Legitimate Peripheral Participation*. Cambridge, UK: Cambridge University Press.

Lloyd, G (1999). Two teachers' conceptions of a reform-oriented curriculum: Implications for mathematics teacher development. *Journal of Mathematics Teacher Education*, 2, 227-252.

Merseth, K. (1996). Cases and case methods in teacher education. In J. Sikula, T. Butery and E. Guyton (Eds.), *Handbook of Research on Teacher Education* (2nd ed.), pp. 722-744. New York: Macmillan Publishing Company.

Mewborn, D. (2000). Learning to teach elementary mathematics: Ecological elements of a field experience. *Journal of Mathematics Teacher Education*, 3, 27-46.

Nelson, B. (1998). Lenses on learning: Administrators' Views on reform and the professional development of teachers. *Journal of Mathematics Teacher Education*, 1, 191-215.

Perry, W. (1970). *Forms Of Intellectual And Ethical Development In The College Years*. New York: Holt, Rinehart, and Winston.

Schifter, D. (1998). Learning mathematics for teaching: From a teachers' seminar to the classroom. *Journal of Mathematics Teacher Education*, 1, 55-87.

Schön, D. (1983). *The Reflective Practitioner: How Professionals Think In Action*. New York: Basic Books, Inc.

Simon, M. (1997). Developing new models of mathematics teaching: An imperative for research on mathematics teacher development. In E. Fennema and B. Nelson (Eds.), *Mathematics Teachers In Transition*, pp. 55-86. Mahwah, NJ: Lawrence Erlbaum Associates.

Sullivan, P. and Mousley, J. (in press). Thinking teaching: Seeing mathematics teachers as active decision makers. In F.L. Lin and T. Cooney (Eds.), *Making Sense Of Mathematics Teacher Education*. Dordrecht: Kluwer Academic Publishers.

Tharp, R. and Gallimore, R. (1988). *Rousing Minds to Life: Teaching, Learning, and Schooling in a Social Context*. Cambridge, UK: Cambridge University Press.

Thom, R. (1973). Modern mathematics: Does it exist? In G. Howon (Ed.), *Developments In Mathematical Education*, pp. 194-212. Cambridge, UK: Cambridge University Press.

Thompson, A. (1992). Teachers' beliefs and conceptions: A synthesis of the research. In D. Grouws (Ed.), *Handbook Of Research On Mathematics Teaching And Learning*, pp. 127-146. New York: Macmillan Publishing Company.

Von Glasersfeld, E. (1991). Abstraction, re-presentation, and reflection: An interpretation of experience and Piaget's approach. In L. Steffe (Ed.), *Epistemological Foundations Of Mathematical Experience*, pp. 45-67. New York: Springer-Verlag.

Walen, S. and Williams, S. (2000). Validating classroom issues: Case method in support of teacher change. Journal of Mathematics Teacher Education, 3, 3-26.

Wilson, M. R. and Goldenberg, M. P. (1998). Some conceptions are difficult to change: One middle school mathematics teacher's struggle. *Journal of Mathematics Teacher Education*, 1, 269-293.

Zeichner, K. and Gore, J. (1990). Teacher socialization. In R. Houston (Ed.), *Handbook of Research on Teacher Education*, pp. 329-348. New York: Macmillan Publishing Company.

Thomas Cooney
University of Georgia, Georgia, USA
tcooney@coe.uga.edu

CATHY KESSEL AND LIPING MA

MATHEMATICIANS AND THE PREPARATION OF ELEMENTARY TEACHERS

1. INTRODUCTION

What should university mathematicians do to prepare prospective elementary teachers to teach? Often this question is answered by a list of topics that outline university mathematics courses for teachers. We contend that selection of topics is extremely important, but there is more that must be considered, particularly in countries that are attempting to change the way mathematics is taught in primary school. So, rather than provide a list of topics, we consider the relationship of teacher preparation with the structure of schooling and teaching. We claim that this relationship should be considered not only in how topics for teacher preparation courses are selected, but also with regard to the aims and teaching of these courses.

In many countries, part of teacher preparation takes place in universities and mathematicians are involved in teaching some of the courses. But teacher learning doesn't necessarily begin or end with teacher preparation. The life of a teacher typically includes schooling, teacher preparation, and teaching as depicted in Figure 1 (taken from Ma, 1999, p. 145). That teaching is a part of the schooling of later generations. In this sense, education, and mathematics education in particular, is part of a cycle in which student learning is affected by teacher knowledge.[1]

From that perspective, teacher preparation is a time to help prospective teachers to acquire the mathematical knowledge they did not receive in their schooling and will need for teaching. Research on the relationship between teacher knowledge and teaching suggests this is important: what a teacher knows about mathematics affects the way he or she teaches (see, e.g., Borko and Putnam, 1996).

But knowledge, or the lack of it, is not the only outcome of schooling. Students (including prospective teachers) develop beliefs about mathematics, what it means to learn mathematics, and what it means to teach mathematics — as mathematicians sometimes discover with dismay. Left unchanged, these beliefs can act as a filter in later learning and later in life. Prospective teachers' beliefs about mathematics, and its teaching and learning may influence their learning in teacher preparation courses, and later their teaching (see, e.g., Borko and Putnam, 1996; Zakkis, 2000).

Outcomes associated with the schooling-preparation-teaching cycle may vary with country. In particular, some countries do not appear to have a body of

---

[1]One might think of the cycle as a projection of a spiral. The last stage of the cycle for one person (teaching) overlaps with the first stage (learning) for later generations and closes the loop of Figure 1.

467

Derek Holton (Ed.), The Teaching and Learning of Mathematics at University Level: An ICMI Study, 467—480.
© 2001 *Kluwer Academic Publishers. Printed in the Netherlands.*

mathematical knowledge that is peculiar to the profession of teaching. In other countries, teachers have developed knowledge for teaching that includes knowledge of student strategies, how to connect those strategies with fundamental mathematical ideas, and ways of representing and formulating topics to make them comprehensible to students.

The examples from different countries that follow are not meant as a contribution to 'the horse race' — the competition to see which country (or students, or teachers) is the 'best', judged under absolutely 'fair' conditions. This is difficult to determine, partly because the definition of 'fair' and 'best' varies. More importantly, knowing the outcome of the horse race is not necessary to this article. Instead, we wish to understand what kind of knowledge teachers need for teaching and how to prepare prospective teachers to acquire it. Our strategy is not to determine the 'best' method and to say that other countries should copy it. It is rather to understand how teaching and teachers' knowledge are related with an eye to changes that have the potential, according to what we know about teaching and learning, to improve matters. Descriptions of what occurs in one country can help to illuminate knowledge and practices that are taken for granted within another. It is for this purpose that we use international comparisons.

We begin with some examples of the mathematical knowledge that teachers may acquire — the end product of the schooling-preparation-teaching cycle in various countries. We examine some descriptions of classroom teaching — which influence two aspects of this cycle: students' schooling and teachers' experiences of teaching. We then consider relationships of teaching, teacher preparation, and the profession of teaching in different countries. (Note that information available for one country may not be available for another. Because of that the countries in the following sections vary.) Finally, we return to the question of the mathematical preparation of teachers, and how courses may be designed for them.

2. SCHOOLING

2.1 Examples of mathematical knowledge for teaching

Example 1
[To compute] 21 – 9, [students] would need to know that you cannot subtract 9 from 1, then in turn you have to borrow a 10 from the tens space, and when you borrow that 1, it equals 10, you cross out the 2 that you had, you turn it into a 10, you now have 11 – 9, you do that subtraction problem then you have 1 left and you bring it down. (Ma, 1999, p. 2.)

Example 2
[An experienced teacher describes what happens when she asks her class to compute 34 – 6.] I put the problem on the board and ask students to solve the problem on their own. ... After a few minutes they finish. I have them report to the class what they did. They might report a variety of ways. One student might say, "34 – 6, 4 is not enough to subtract 6. But I can take off 4 first, get 30. Then I still need to take 2 off because 6 = 4 + 2. I subtract 2 from 30 and get 28. So, my way is

$$34 - 6 = 34 - 4 - 2 = 30 - 2 = 28.$$"

Another student who worked with sticks might say, "When I saw that I did not have enough separate sticks, I broke one bundle. I got 10 sticks and I put 6 of them away. There were 4 left. I put the 4 sticks with the original 4 sticks together and got 8. I still have another two bundles of tens. Putting the sticks left all together, I had 28." Some students might report. ... "We have learned how to compute $14 - 8$, $14 - 9$, why don't we use that knowledge? So, in my mind I computed the problem in a simple way. I regrouped 34 into 20 and 14. Then I subtracted 6 from 14 and got 8. Of course I did not forget the 20, so I got 28." (Ma, 1999, pp. 13–14.)

Example 3
[There are five different ways that second-grade students at a particular point in the curriculum might arrive at a solution for $12 - 7$. These ways can be put into three categories: Counting-subtraction, Subtraction-addition, Subtraction-subtraction. The second two methods] do not involve counting. Since the counting method is not involved, the students [who use those methods] have a better sense of composing and decomposing numbers, particularly the number 10 that is important for understanding decimal numbers. (Yoshida, 1999, p. 117.)

These are just a few examples of the very different kinds of knowledge a teacher might use in teaching. The first is an example of what is sometimes called procedural knowledge. It gives a clear description of how to subtract 9 from 21, but does not give a rationale for the procedure. Unlike the first example, the second and third examples involve not only a mathematical topic for a particular grade, but also how students in that grade often understand it. The second example describes possible student responses to a problem. Implicit in this description is the teacher's understanding that all the students' methods for computation are correct. The third example describes possible student responses using technical terms for students' methods and evaluates them with respect to future learning of a particular topic.

These examples are meant to suggest the *kinds* of knowledge that appear to be typical of various countries and meant to typify trends rather than individual teachers. The first example comes from the United States, the second from China, and the third from Japan. Research on elementary teachers' mathematical knowledge (e.g., Borko and Putnam, 1996; Ma, 1999) suggests that the U.S. and Chinese examples are representative of how mathematical knowledge for teaching is construed in those countries. Classroom research suggests one reason why teachers in different countries might develop different forms of knowledge for teaching.

2.2 Lesson structure in three countries

'Teaching' often means 'what teachers do in the classroom.' Its possible meanings are illustrated by the findings of the TIMSS Videotape Classroom Study. For this study, videotapes of 231 randomly selected eighth grade mathematics lessons in Germany, Japan, and the United States were collected and analysed (Stigler and Hiebert, 1999). Table 1 outlines the structures typical of lessons in the three countries.

Table 1. Eighth grade lesson structure

Germany	Japan[a]	United States
Review previous material: Review homework or remind of previous work.	Review the previous lesson: Brief teacher lecture, discussion, or students recite main points.	Review previous material: Check homework or engage in warm-up activity.
Teacher presents the topic and problems for the day.	Teacher presents the problem for the day.	Teacher demonstrates how to solve problems for the day.
Teacher develops solution procedures.	Students work individually or in groups.	Students practice.
Students practice.	Class discusses student or teacher solution methods. Teacher summary.	Correct seatwork and assign homework.

Source: Stigler and Hiebert, 1999, pp. 78–81.

Notes: Lessons in the three countries were generally about the same length (45 to 50 minutes). In each country, the mathematics to be learned is made explicit at different parts of the lesson (indicated by boxes).

[a]Accounts of teaching in Japan suggest this lesson structure may not be typical of all later grades.

Although this article concerns elementary rather than eighth grade mathematics, these findings are relevant in two ways. Prospective teachers are very likely to have attended eighth grade lessons as students. And the structure of eighth grade lessons is likely to be consistent with students' expectations for lessons after their first seven grades of school. Comparative studies (e.g., Stevenson and Stigler, 1992) show differences among U.S. and East Asian elementary classrooms consistent with those of the TIMSS eighth grade study. Earlier studies of U.S. classrooms (e.g., Schoenfeld, 1988) show teaching practices consistent with the TIMSS findings for the U.S.

Table 2. Structure of two fifth grade lessons in Japan and the USA

Japan	United States
Teacher asks students to name kinds of triangles they have studied.	Teacher reviews concept of perimeter.
Presentation of the problem: Students are given different paper triangles and asked: What would be the best way to find the area of this triangle?	Area of a rectangle: Teacher asks students for area of various rectangles, gives formula.
	Students do two practice problems using the formula.
Students attempt to solve the problem on their own.	Area of a triangle: Teacher gives explanation and formula.
Class discussion about students' solutions, leading to the general formula for the area of a triangle.	Students do three practice problems using the formula.
Students do further problems from the textbook.	Students begin homework assignment: Teacher walks round and helps.

Source: Stigler, Fernandez and Yoshida, 1996, pp. 221–222.

In their descriptions of lessons, Stigler and Hiebert distinguish between 'stated' and 'developed' mathematics. Stated mathematics is correct statements without mathematical justification or examples (e.g., Example 1 of the previous section). In contrast, developed mathematics helps to provide a rationale for *why* mathematical statements hold (e.g., the possible student responses in Example 2). In the lessons analysed for the TIMSS video study, stated mathematics occurred as teacher statements. In contrast, developed mathematics occurred in various forms, for example, as a lecture in which a teacher goes through the proof of a theorem or develops a formula from student solutions. Table 2 illustrates the stated–developed distinction and provides an example of mathematics developed from student solutions.

2.2.1 Outcomes for students

Cross-national studies suggest that differences in teaching may be associated with differences in student learning. In the 1980s, Stevenson, Stigler, and their colleagues studied first and fifth graders in Japan, Taiwan, and the United States using tests and individual interviews. On average, Japanese students outperformed Taiwanese students and Taiwanese students outperformed U.S. students (Stevenson and Stigler, 1992). Less detailed studies, such as the three international mathematics studies, have had similar findings.

Differences in teaching mathematics are also associated with differences in beliefs about what constitutes teaching, learning, and knowing mathematics. In the United States, the following beliefs have been documented:

- Doing mathematics means following the rules laid down by the teacher.
- Knowing mathematics means remembering and applying the correct rule when the teacher asks a question.
- Mathematical truth is determined when the answer is ratified by the teacher. (Lampert, 1990)
- Students who understand the subject matter can solve assigned mathematics problems in five minutes or less. Corollary: Students stop working on a problem after just a few minutes because, if they haven't solved it, they didn't understand the material (and therefore will not solve it). (Schoenfeld, 1988, p. 151.)

Such beliefs may affect prospective teachers in two ways: They are likely to teach students who hold these beliefs, or at the very least, to teach students with parents who hold these beliefs. And, they are likely to hold these beliefs themselves and interpret preparatory courses through the lenses of these beliefs. For example, if they believe that all mathematics problems can be solved in five minutes, they may (unless further intervention is made) feel uncomfortable working on problems that require more time. In contrast, Stevenson and Stigler (1992, p. 105) report that Japanese elementary students will continue to work on impossible problems until someone stops them.

This contrast suggests an association between beliefs and lesson structure. Elementary mathematics lessons in Japan often focus on one problem which students attempt (and sometimes solve) before solutions are presented and discussed by the class. In contrast, mathematics lessons in the United States seem to focus on many short exercises which students practice after seeing a teacher demonstration of a solution method. It is perhaps unsurprising that Japanese students are willing to persevere in trying to solve a problem and U.S. students are not.

This correlation between beliefs and lesson structure suggests a conjecture for further investigation as well as a heuristic for use when designing courses:

- the structure of mathematics lessons exerts a strong influence on beliefs about mathematics and how it's taught and learned.

2.2.2 Outcomes for teachers

Teachers in the three countries may acquire different sorts of mathematical knowledge to teach according to these lesson structures, and in turn, lesson structures may be dependent on the knowledge that teachers bring to teaching. In the United States, to teach according to the lesson structure described above, teachers must be able to state procedures correctly and to be able to correct homework — to know how. In Germany, teachers need to understand how procedures (and at eighth grade, theorems) are developed — to know why as well as know how. In Japan, teachers need to understand students' solutions to the problems and be able to connect those solutions to the mathematics to be learned — to know how students know and know why that student knowledge is connected to the mathematics to be learned. The second and third examples that began this article illustrate this kind of knowledge.

The differences in lesson structure may suggest why Japanese and Chinese teachers have developed a specialized mathematical knowledge for *teaching* that focuses on connecting student understandings to the mathematics to be learned and why U.S. teachers do not seem to have developed such knowledge (Borko and Putnam, 1996; Ma, 1999; Yoshida, 1999). Comparisons of the structure of teaching and teacher preparation suggest some clues about how individual teachers acquire such knowledge and the conditions that foster the maintenance of a body of knowledge unique to teachers. This suggests a conjecture for further investigation:

- the structure of mathematics lessons exerts a strong influence on the kind of knowledge of mathematics that teachers develop.

3. TEACHING

3.1 The profession of teaching in four countries

The structure of teaching careers in European countries, China, Japan, and the United States, differs considerably (Comiti and Ball, 1996; Ma, 1999; Stevenson and

Nerison-Low, 1999). As with all cross-national comparisons, categories need to be interpreted with care. Activities with the same name — in particular, 'teaching' (as discussed earlier) — may be quite different in different countries. The interested reader is encouraged to consult the references for further details.

Cross-national studies suggest some important differences in teacher professional development. Information imparted by non-teachers to teachers appears to be the emphasis in U.S. and German professional development, and is sometimes required for continued certification or promotion. In contrast, communication among teachers about teaching is a large part of teachers' professional lives in China and Japan. Such communication occurs in study groups both in and out of school, via lesson study, and in written publications. Teachers in China and Japan, like university faculty in other countries, publish articles and books for an audience of their peers.

Chinese and Japanese elementary schools are organised for teacher learning as well as for student learning. The physical organisation supports teacher communication, for example, teachers have desks together in a single 'teachers' room' rather than separately in their classrooms as is the practice in the United States (Ma, 1999; Yoshida, 1999). The practice of teaching in 'rounds' (teaching the same group of students for two or more years in succession) offers an opportunity for teachers to understand how student knowledge changes from grade to grade and to understand the elementary mathematics curriculum as a whole.

Given the superior performance of Japanese students on international mathematics comparisons, the following may come as something of a surprise. When Japanese teachers were asked "What is a good teacher in Japan?" many responded "of course, they also have to know the subject matter well," but most of their responses focused on personal qualities: "motivation for teaching," kindness, and liking children.

Teachers in China and Japan are expected to continue learning throughout their careers and the structure of teaching affords them opportunities for that learning in their daily interactions as well as during more formally scheduled events such as lesson study. It may be for this reason that Japanese teachers stress the importance of personal qualities — a teacher who is motivated to improve has opportunities for improvement, but a teacher who is not motivated is unlikely to change. Interest in children and their ways of thinking may also be connected with the kind of knowledge for teaching that is generated in Japan.

In addition to learning throughout their careers, teachers in China and Japan are also researchers who produce new knowledge about teaching and curriculum that benefits other teachers. They have a specialised language for discussing teaching and learning (illustrated in Example 3 of Section 1).

In the United States, researchers and teachers are separate groups. Researchers have a specialised language for discussing student strategies that teachers do not use, for example, the terms 'counting on' and 'counting up.' Rather than learning from each other, teachers are expected to learn from experts by participating in workshops or taking university courses. This association suggests a conjecture:

• the structure of communication among teachers exerts a strong influence on the knowledge of mathematics associated with the profession of teaching.

3.2   The curriculum[2]

Accounts of teachers' knowledge (e.g., Ma, 1999) suggest that how topics are construed and how they are sequenced in the curriculum may play an important role in the development of teachers' knowledge. Topics may be sequenced in a way that promotes their learning as well as their logical development. For example, Chinese teachers construe 15 – 9 as involving 'decomposing a higher value unit' (in this case, the 1 in 15 is considered to represent 1 ten and ten is a higher value unit that can be decomposed as 10 ones). Subtraction problems like 15 – 9 are the first stage (subtraction with decomposing within 20) of a curriculum sequence that continues with subtraction with regrouping involving minuends between 20 and 100. In Chinese elementary mathematics textbooks, these topics are preceded by a section on the composition and decomposition of 10 (i.e., pairs of numbers that sum to 10). This sequencing may contribute to teachers' learning of connections among topics as well as to students' learning of the topics themselves.

Few (if any) fine-grained comparisons of curricula have been done that examine which topics are taught, the order in which they are taught, and how previous topics are connected with later ones. However, examinations of standards documents, textbooks, and programmes of study (e.g., Baldwin and Kessel, 1999) together with accounts of teachers' knowledge such as the one above suggest this conjecture:

• the structure of the curriculum exerts a strong influence on the knowledge of mathematics associated with the profession of teaching, and on student learning.

4.   SOME SUGGESTIONS FOR TEACHER PREPARATION

We return to our original question: What should university mathematicians do to prepare prospective elementary teachers for teaching? Teaching and learning practices in different countries suggest that this question has different answers in different countries. For example, a brief description of changes that occur at different stages of a teacher's career in China is given in Table 3. (See Ma, 1999 for more detail.) In elementary school, prospective teachers appear to acquire a conceptual knowledge of mathematics; for example some are able to create story problems for a given division by fractions problem. During normal school, they acquire a concern for teaching and learning. When given a scenario about students who make a mistake in a multi-digit multiplication, prospective teachers reacted by providing explanations of why it was wrong and what they would do if they encountered such a mistake. Some of the teachers in Ma's sample, after many years of teaching, had developed what she called Profound Understanding of Fundamental

---

[2] We use this term rather loosely. Education researchers distinguish between intended and enacted curriculum. The differences among countries that we discuss occur in both.

Mathematics (PUFM): an organisation of elementary mathematics for *teaching* that allows a teacher to connect a student's current understanding with fundamental principles of mathematics as well as with the student's future learning. Example 2 of Section 1 illustrates one aspect of PUFM: knowledge of possible student strategies.

Table 3. Outcomes associated with schooling and teaching in China and the USA

	Schooling	Teacher preparation	Teaching
China	Conceptual knowledge	Conceptual knowledge Concern for teaching and learning	PUFM
USA	Procedural knowledge	Procedural knowledge Concern for teaching and learning	Procedural knowledge

The picture is very different for the U.S. (see Table 3), and perhaps also for other countries that are trying to make major changes in mathematics education. In the U.S., several specialised groups are concerned about mathematics education: teachers, mathematics education researchers, curriculum developers, and mathematicians. Members of these groups do not share a common vision of mathematics teaching and learning. Although individual teachers may have conceptual knowledge of mathematics and knowledge for teaching, there appears to be no body of mathematical knowledge specific to the profession of teaching.

Evidence from China and Japan suggests that developing such knowledge may require changing the structure of teaching, lessons, and curricula. How this may be done in the United States and other countries is outside the scope of this article, except as it concerns teacher preparation. For some suggestions, see Ma (1999).

4.1 Teacher preparation courses

We now arrive at the topic readers might have expected at the beginning of the article, courses for teacher preparation. But we don't arrive empty-handed. The previous discussion suggests four heuristics that should be considered in designing courses for prospective teachers:

- lesson structure influences beliefs about mathematics, teaching, and learning;
- lesson structure influences teachers' knowledge of mathematics;
- the structure of communication among teachers exerts a strong influence on the knowledge of mathematics associated with the profession of teaching;
- curriculum structure influences the knowledge of mathematics associated with the profession of teaching.

We sketch how these heuristics might be used in designing courses for teachers with respect to two main themes: lesson structure and curriculum.

4.1.1 Lesson structure

Some mathematics education researchers have thought carefully about designing mathematics courses that address students' beliefs about knowing and doing mathematics. Borko and Putnam (1996) have summarised descriptions of such courses for prospective teachers and some studies of their effectiveness. Consistent with the heuristics above, the lesson structure in these courses is very different from those typical of the United States. Instead:

- Lessons tend to concern one or two problems.
- Solutions tend to be presented by the students.
- Correctness tends to be judged by students.

Such a structure violates all of the U.S. beliefs listed earlier; for example, one doesn't follow procedures given by the teacher, the teacher is not the sole mathematical authority, and one doesn't stop working on problems after five minutes. This is far from meaning that the teacher has nothing to do, but it does mean that the role of the teacher is quite different from that in the U.S. lesson structure described above.

The first days of a non-standard course. Instantiating a lesson structure that violates students' initial expectations is non-trivial and should probably begin on the first day of class. When they are simply asked to follow a different lesson structure without further explanation, students sometimes resist. Reformed calculus courses, for example, have received reactions like "I know calculus and this isn't it!," "How can we tell if we have the correct answers?," "I wish I had to memorize more" (Culotta, 1992; Brown, Shure, Megginson, Shaw, and Black, 1994). Such reactions have brought calculus reform to a halt at some universities. And such reactions are not exclusive to reformed calculus courses, but occur in other university courses that differ from the expectations that students have developed during their schooling.

Solutions to this problem exist, but they are often not well documented. The case of a problem solving course for students with at least one semester of calculus has been studied (Arcavi, Kessel, Meira and Smith, 1998; Schoenfeld, 1998), and the structure of its first days of class may be adapted to a course for elementary teachers.

The problem-solving course has many overall goals (e.g., students experience mathematics as sense-making), and these influence the design of the entire course. We focus on three goals particular to the beginning of the course:

- convince students that the course is interesting and valuable;
- introduce students to daily routines;
- establish a climate of trust.

The first two goals are addressed in several ways. Students are told about the course in detail: how it was created and its successes; shown that they have something to learn and that they can learn it in the course; and daily routines are enacted. Showing students that they have something to learn is done in a way that allows them to reflect on their previous classroom experiences and why those previous experiences did not result in satisfactory learning. Students are given carefully chosen problems to work on; they present solutions at the board, and other students judge the correctness of solutions, scaffolded by questions from the teacher.

Although a similar structure could be used for a course for prospective teachers, its specifics should be different. The prospective teachers are likely to lack confidence in doing mathematics and their experiences of mathematics are likely to differ from those of the students in the problem solving course. Also, they are likely to take the course in order to fulfil a requirement rather to learn more mathematics. Ways of convincing them that the course is of value should probably draw on their past experiences and interest in teaching children.

Videotapes of U.S. and Asian classrooms may be of use in getting students to reflect on their previous classroom experiences, as well as providing them with examples of other lesson structures and different ways in which a class may approach a mathematical topic. Helping students to reflect on their previous learning is one aspect of convincing them that it was unsatisfactory, both in terms of the mathematics they learned and in the way they learned it. However, they also need to be convinced that the course in which they are enrolled will help them to better understand mathematics and to better teach their students. One possibility is for the instructor to choose a videotape of an elementary classroom that involves a problem that prospective teachers will find difficult to solve on their own, but straightforward after questions and suggestions from the instructor. (A similar type of activity, without videotape and geared to different students, is used in the problem-solving course on the first day of class.) The problem should be something of importance to the prospective teachers, for example, understanding whether or not a child's strategy for solving an arithmetic problem is correct or relationships between two strategies.

Instructors should expect the design of a course (or of a teacher preparation programme) to require trial and refinement, and the first few days are no exception. For a description of such trial and refinement in the case of a high school mathematics course, see Magidson (1997 April).

4.1.2 Curriculum

Topics that might be included in a course for elementary teachers have been discussed in the recent draft report from the Conference Board of the Mathematical Sciences (2000 March). This report does not subscribe to the 'trickle-down theory' of mathematics education in which prospective teachers are taught abstract versions of ideas and expected to apply them in their teaching. For example, teachers may be taught the distributive law in a college algebra course and expected to use it in teaching arithmetic. 'Non–trickle-down' teaching may be difficult to imagine for instructors used to discussing concepts and properties in abstract terms. Part of the solution may be to represent concepts and properties in terms of the meanings of operations — as the prospective teachers may do later in their teaching.

Doing this creates another consideration for curriculum planning, both for teacher preparation programmes and for elementary curricula. Meanings and representations for operations need to be introduced with the operations; and these change when the domains of the operation change. All need to be sequenced with forethought for future learning.

The trickle-down theory is related to the U.S. lack of "a solid and substantial school mathematics" (Ma, 1999, p. 149). Current school mathematics does not support the development of a knowledge of mathematics for teaching. A step toward this is to connect topics with respect to meaning and representation. One example is relationships between operations with positive integers and with fractions. Meanings and properties of the operations change depending on their domain.

School mathematics in China supports such connections. The Chinese teachers interviewed by Ma pointed out that interpretations of multiplication and division are different for whole numbers and fractions. For example, a very experienced teacher said,

> With integers students have learned the partitive model of division. ... *finding the value of a unit when the value of several units is known.* In division by fractions, however, the condition has been changed. Now what is known is not the value of several units, rather, the value of a part of the unit. For example, given that we paid 1 3/4 Yuan to buy 1/2 of a cake, how much would a whole cake cost? Since we know that 1/2 of the whole price is 1 3/4 Yuan, to know the whole price we divide 1 3/4 by 1/2 and get 3 1/2 Yuan. In other words, *the fractional version of the partitive model is to find a number when a part of it is known.* (Ma, 1999, p. 75.)

This is one example of knowledge for teaching. Preparation that might allow prospective teachers' later development of this knowledge might begin with helping them to understand and make interpretations of arithmetic operations with positive integers, then to understand how the operations can take on additional meanings as the domain of the operations is expanded to include fractions.

Viewing arithmetic operations in this way helps to make sense of a common U.S. student difficulty. 'Multiplication makes larger' is a stumbling block for U.S. students when they learn about multiplication with fractions. 'Multiplication makes larger' is true for positive integers (with the exception of multiplication by 1), but false for fractions and for integers. One might simply give a counterexample to this statement in these two cases. However, for students (and prospective teachers), understanding why it is false for these cases may be tied to an interpretation for multiplication. Graeber and Campbell (1993) suggest that part of the problem is that U.S. students understand multiplication as repeated addition, which is hard to interpret when both factors are fractions. Understanding multiplication of positive integers in terms of an area model provides an interpretation of multiplication that makes sense for fractions and which can be used in later learning. Interpreting multiplication of positive integers in terms of a bar model can also help students make sense of multiplication with fractions and serve as a foundation for later learning (see, e.g., NCTM, 2000, p. 282). Such models may be taught in methods courses for prospective teachers (see, e.g., Van de Walle, 1998). But pre-service and in-service teachers may not have the chance to understand how these models may be related to students' current and future learning, from methods courses or from school curricula in the United States.

Another consideration is prospective teachers' prior knowledge and their beliefs about that knowledge. Some mathematics (e.g., statistics) will often be new to them and some won't (e.g., operations with numbers). Whether or not a topic is old or new may affect the teachers' beliefs about it and hence be an important consideration for course design. For example, Zazkis (2000) interviewed prospective teachers after a 3-week college course unit on factors, multiples, and divisors. The teachers tended to use the meanings for multiples and divisors learned in their primary and secondary schooling. Zazkis says: "What often occurs in a content course for pre-service elementary school teachers is not construction of new meanings or concepts, but reconstruction of previously constructed meanings." This suggests that prospective teachers' prior learning should be considered in the choice and sequencing of topics in a course and in a teacher preparation programme.

5. CONCLUDING REMARKS

What prospective teachers need to learn and how they need to learn it is not a simple story. It depends on their previous schooling and the teaching they will do in the future. And, it includes mathematical attitudes as well as mathematical topics.

We have not attempted to give a comprehensive prescription for preparing teachers. Instead, we have endeavoured to give a framework to think about this process, illustrated by examples. What lies ahead is not a matter of throwing out the old and replacing it with the new, but of understanding the old and weaving it together with new knowledge and practices from several fields and countries.

ACKNOWLEDGMENTS

Our thanks to Rudy Apffel, Abraham Arcavi, Judith Epstein, Derek Holton, Susan Magidson, Ann Ryu, Alan Schoenfeld, and Natasha Speer for helpful comments of all kinds on earlier versions of this article.

REFERENCES

Arcavi, A., Kessel, C., Meira, L. and Smith, J. P. (1998). Teaching mathematical problem solving: A microanalysis of an emergent classroom community. In A. H. Schoenfeld, E. Dubinsky, and J. Kaput (Eds.), *Research in Collegiate Mathematics Education III*, pp. 1–70. Providence, RI: AMS.

Baldwin, J. and Kessel, C. (1999). Concept and computation: The role of curriculum. *MER Newsletter* *12*(1), 1, 4–5, 10. URL: http://www.math.uic.edu/~jbaldwin/mathed.html.

Borko, H. and Putnam, R. (1996). Learning to teach. In D. Berliner and R. Calfee (Eds.), *Handbook of Educational Psychology*, pp. 673–708. New York: Macmillan.

Brown, M., Shure, P., Megginson, R., Shaw, D., and Black, B. (1994). *Math 115 Calculus: Instructor's Guide*. Ann Arbor, MI: University of Michigan.

Comiti, C. and Ball, D. (1998). Preparing teachers to teach mathematics. In A. Sierpinska and J. Kilpatrick (Eds.), *Mathematics Education As A Research Domain*, pp. 1123–1153. Dordrecht: Kluwer.

Conference Board of the Mathematical Sciences (2000 March). *CBMS Mathematical Education of Teachers Project Draft Report*. Washington, DC: CBMS. URL: www.maa.org/cbms.

Culotta, E. (1992). The calculus of education reform. *Science*, 255, 1060-1062.

Graeber, A. and Campbell, P. F. (1993). Misconceptions about multiplication and division. *Arithmetic Teacher*, 40(7), 408–411.

Lampert, M. (1990). When the problem is not the question and the solution is not the answer: Mathematical knowing and teaching. *American Educational Research Journal*, 27, 29–63.

Ma, L. (1999). *Knowing and Teaching Elementary Mathematics*. Mahwah, NJ: Erlbaum.

Magidson, S. (1997 April). *On the Complexities of Research-based Instructional Design*. Paper presented at the annual meeting of the American Educational Research Association, Chicago.

National Council of Teachers of Mathematics (2000). *Principles and Standards for School Mathematics*. Reston, VA: NCTM.

Schoenfeld, A. H. (1988). When good teaching leads to bad results: The disasters of 'well-taught' mathematics classes. *Educational Psychologist*, 23(2), 145–166.

Schoenfeld, A. H. (1998). Reflections on a problem solving course. In A. Schoenfeld, E. Dubinsky, and J. Kaput (Eds.), *Research in Collegiate Mathematics Education III*, pp. 81–113. Providence, RI: AMS.

Stevenson, H. W. and Nerison-Low, R. (1999). *To Sum It Up: Case Studies of Education in Germany, Japan, and the United States*. URL: http://www.ed.gov/offices/OERI/SAI.

Stevenson, H. and Stigler, J. (1992). *The Learning Gap*. New York: Simon and Schuster.

Stigler, J., Fernandez, C. and Yoshida, M. (1996). Cultures of mathematics instruction in Japanese and American elementary classrooms. In T. Rohlen and G. LeTendre (Eds.), *Teaching and Learning in Japan*, pp. 213–247. Cambridge: Cambridge University Press.

Stigler, J. and Hiebert, J. (1999). *The Teaching Gap*. New York: Free Press.

Van de Walle, J. A. (1998). *Elementary and Middle School Mathematics: Teaching Developmentally* (3rd ed.). New York: Addison Wesley Longman, Inc.

Yoshida, M. (1999). *Lesson Study*. Unpublished Ph.D. dissertation, University of Chicago.

Zazkis, R. (2000). Factors, divisors, and multiples. In E. Dubinsky, A. Schoenfeld, and J. Kaput (Eds.), *Research in Collegiate Mathematics Education IV*, pp. 210–238. Providence, RI: AMS.

Cathy Kessel
Berkeley, USA
kessel@soe.berkeley.edu

Liping Ma,
Palo Alto, USA
maliping@juno.com

MICHEL HENRY AND BERNARD CORNU

MATHEMATICS TEACHERS' EDUCATION IN FRANCE: FROM ACADEMIC TRAINING TO PROFESSIONALIZATION

1. TEACHERS' ACADEMIC TRAINING IN MATHEMATICS: SOME CHARACTERISTICS.

In France, teacher training for secondary mathematics teachers presents some unique features. Unlike in some other countries there is no tradition of an integrated curriculum and future teachers have first to graduate in mathematics. It is essentially on the basis of their academic mathematics knowledge, through highly competitive national examinations, that students are selected to become teachers. After being recruited, they are then trained in the professional dimensions of teaching. Nevertheless, in the present context of teaching, the Ministry of Education has become more and more sensitive to the limits of the traditional system of training. This resulted ten years ago in the creation of the IUFMs[1]. Their objective was both to preserve the quality of the academic training of future teachers, and, at the same time, to efficiently provide new teachers with professional knowledge and competencies. In this paper, we present the training system which was then organised, the way it has functioned up to now, and discuss what we can learn from this as regards difficult and crucial issues such as the following: how to develop professional competencies in pre-service training and how to built efficient links between professional and more academic (including mathematics) forms of knowledge ?

In France, the minimal university level required for becoming a teacher (both secondary and primary) is that of a 'Licence' (three years after the 'Baccalauréat', examination which ends secondary education and allows access to university). With the Licence, students can take the CAPES[2], and if they have a 'Maitrise' degree (one year more), they can take the Agrégation, a competition which is more selective and gives them higher status. CAPES and Agrégation are national competitions, through which the state recruits teachers. The number of positions offered each year is fixed by taking into account the national needs and the demography of the teacher population. They are now very selective. In 1999, for instance, 8 950 candidates

---

[1] IUFMs: Instituts universitaires de formation des maitres – university institutes for teacher education. They are university professional institutions, for pre-and in-service training for all teachers (primary and secondary). There are 29 IUFMs in France, one in each administrative region for education; they were created in 1990 and 1991, and they are linked with universities through contracts.

[2] CAPES : Certificat d'Aptitude Professionnelle pour l'Enseignement Secondaire.

Derek Holton (Ed.), The Teaching and Learning of Mathematics at University Level: An ICMI Study, 481—499.
© 2001 Kluwer Academic Publishers. Printed in the Netherlands.

population. They are now very selective. In 1999, for instance, 8 950 candidates applied for the CAPES in mathematics (respectively 3 045 for the Agrégation) while only 945 positions were offered (respectively 368 for the Agrégation).

The general route to becoming a teacher is thus the following:

- Students first take three years at university to get their Licence degree in a given subject (for instance mathematics);
- Then they go to an IUFM for two years: the first year is mainly devoted to preparation for the CAPES (recruiting competition), and mainly focussed on subject content. Teachers are recruited by the state, not by schools or local authorities as in some other countries. The state provides posts to the schools, and nominates teachers to these posts.
- If they succeed, the students enter the second year at IUFM, with the status of a civil servant (they then get a salary). The second year is focused on professional preparation, and includes practice periods in schools, where the students have the full responsibility of teaching for a third of a normal duty.
- After two years at an IUFM, graduates are given a teaching post.

Students do not have to attend an IUFM in order to take the CAPES. If they are successful, they then enter directly into the second year of the IUFM. The Agrégation is much more selective and reserved for students who took one more year at university studying for the Maitrise degree. Students prepare for the Agrégation at university; if they are successful, they enter directly into the second year at IUFM. It is interesting to notice that primary teachers have the same scheme for training: 3 years at university, plus 2 years at IUFM, with a recruiting competition including all the subjects to be taught in primary schools. As a consequence, in France since 1990, primary teachers have received the same salaries as secondary teachers. This was a great change in the relative status of both kinds of teachers!

In what follows, we briefly describe the academic training corresponding to a Licence degree in mathematics in France and the way its characteristics tend to shape the mathematical culture and views of prospective teachers. The secondary schools mathematics curriculum is an integrated curriculum which focuses on elementary algebra, geometry (mainly synthetic geometry with strong emphasis on transformations) and elementary calculus, but includes also combinatorics, probability and statistics. Students are used to working with vectors and groups of transformations in geometry but algebraic structures are not introduced. Calculus is introduced in an intuitive way, without formal definitions, by relying on the graphical and numerical explorations allowed by graphic calculators (which are now compulsory). Then it is developed, algebraically, in order to allow students to solve simple but interesting problems of variation and optimisation.

During the first two years at university level, solving techniques become more complex, and they are applied to new objects. At the same time, the mathematics field is progressively structured:

- by introducing topics such as algebraic structures;
- by introducing formal definitions for the basic concepts of calculus such as the concept of limit;
- by formally proving the previously assumed fundamental theorems of calculus; and
- by developing theories such as the Riemann integral theory.

The curriculum focuses on algebra and analysis and, in most universities, geometry, which played an important role at secondary level, nearly disappears, if one excepts some analytical geometry linked with linear algebra. The third year at university corresponds to a qualitative jump in the level of abstraction of mathematical objects students deal with: general topology, differential calculus in normed vector spaces, integration, functional analysis, set theory, abstract algebraic structures, and, but often not compulsory, probability, affine, Euclidean and even projective geometry. Courses in computer technology are part of the two first years' curriculum and, more and more, students are also introduced to CAS software such as Maple, Mathematica or Mathlab.

This quick overview shows that university teaching in France remains influenced by the values of the 'Bourbakist' era and has difficulty taking into account the evolution of the mathematical sciences. Very soon, the large majority of mathematics students lose contact with other scientific disciplines and even the vision that they develop of the mathematical field is inclined to be a restricted one. Discrete mathematics, probability and statistics, and numerical analysis are still only given a marginal role up to the Licence level, and the same applies to applications of mathematics and modelling, if one excepts some rare interesting experimental projects.

Thus, most university students have a very abstract view of mathematics and their mathematical culture makes it difficult to relate what they have been taught with the contemporary issues and methods of the scientific, economic or technological world. They are used to formal discourse, symbolic writing, and to the traditional way of presenting mathematics knowledge that they have been mainly exposed to at university level. Changes have been introduced in recent years but, up to now, there has been little success in substantially transforming the whole picture.

According to some research, these conditions seem to give way to two extreme profiles among prospective mathematics teachers.

- Some of them, after completing three year courses at university, still rely on the mathematical values they were exposed to at secondary level: doing good mathematics is being able to fluently use the computing techniques and the properties of the elementary objects they encountered in geometry, in algebra or in calculus, in order to solve school problems. They have a functionalist vision of mathematics, they have experienced pleasure doing mathematics in a certain way and want to share it, and this reinforces their desire to become a mathematics teacher.

- For some others, the role of mathematics teaching, both at school and at university, is to provide students with a well built construction of mathematics, enhancing unities of organisation and language. They would like to transpose to secondary level, the vision of mathematics they have been exposed to at university level.

Even if quite different, these two profiles share one common characteristic: for different reasons, they are insensitive to the epistemological richness of mathematics, to the importance of its connections with other scientific fields, and to the limits of functionalist or usefulness visions, as they are generally transposed in school contexts.

2. NEW EXPECTED COMPETENCIES FROM MATHEMATICS TEACHERS

Much more competencies are nowadays expected from mathematics teachers than was the case some decades ago, both from the mathematical and didactical points of view. Hence the inadequacy of the mathematics culture offered by universities becomes more critical.

From a mathematical point of view, teachers have to face a strong evolution of mathematical balances between domains. Mathematics is more and more present in the surrounding social, scientific and technological worlds, providing these with an increasing diversity of conceptual and technical tools. Thus mathematics teaching, even at secondary level, cannot be restricted to its traditional fundamentals. New balances have to be found. Topics that were marginal, such as probability, statistics, and discrete mathematics, are likely to play increasing roles in the curriculum. Links have to be made with other scientific domains. All of these require new mathematical expertise from teachers and, most often, they are not prepared for these by their university training. At the same time, mathematics practices are quickly evolving, due to modern technology, and teachers are expected to adapt to this evolution, in real time, both for doing mathematics, and for finding and exploiting professional resources.

From a didactical point of view, teachers have to adapt to the fact that today, in France as in many countries, secondary mathematics teaching is mass teaching. 92% of pupils complete the compulsory part of education, while 68% of an age class take the Baccalaureat, in a general (34%), technical (21%) or vocational (13%)[3] stream. Within the last decade, the secondary student population has evolved quickly. It is now much more heterogeneous and it doesn't necessarily share the values associated in the traditional culture to school and to mathematics teaching. Moreover, students live today in a society where traditional forms of authority: parents, family, and so on, are being strongly challenged. They want to know why such or such a topic is taught and no longer accept the answer "because it is part of the syllabus".

---

[3] 1998 figures, from DPD (Direction de la Programmation et du Développement, Ministry of National Education, France).

Moreover, in these changing conditions, mathematics teachers are no longer asked to select and teach an elite as was the case up to the 1980s. They are expected to teach all students and to find ways to achieve success for the great majority of their students. They cannot expect their students to be able to learn in an autonomous way. They have to be aware of the difficulties these students may encounter, and of the ways these difficulties can be overcome. They must be able to analyse students' behaviour and products, in order to evaluate their level of understanding of complicated notions. They have to be aware that an efficient treatment of errors is not achieved by simply correcting them, but must take into account the conditions in which they were produced and use them as a lever for learning. They have to be able to design classroom situations that address the diversity of students' needs and potentials, in classes which are more and more heterogeneous from a cultural and social point of view, but also be able to create a common classroom culture.

These conditions make the teaching profession undoubtedly more exciting, but much more difficult. As we pointed out previously, their university academic training doesn't prepare teachers very efficiently from a mathematical point of view. As regards the didactical part, students are offered during the three first years at university short professionalization modules which allow them to have a first contact with the work of teachers and the functioning of secondary institutions, but in most universities, that is all. Didactics is rarely offered as an optional course at Licence level; it is mainly reserved for Doctorate programmes.

The Ministry of Education began to be sensitive to the increasing inadequacy of the training offered to prospective teachers more than ten years ago. This resulted in the creation of IUFMs in the early nineties. IUFMs had the goal of providing future teachers with a consistent professional background during their initial training and of integrating a minimal theoretical basis in didactics, psychology and educational sciences and classroom in a coherent way practice. Professional activity and in-service training would then allow the integration of this background into a real professional competency.

Regarding the competencies that teachers should acquire, the 'Bancel Report'[4] says:

> Each of these competencies puts in action three pillars of knowledge, which delineate a global professionality:
>
> - the first pillar consists in the knowledge related to the subjects (knowledge to be taught, history, epistemology and social impacts of the different disciplines);
> - the second pillar consists in the knowledge related to the management of the learning processes (didactical and pedagogical);
> - the third pillar consists in the knowledge related to the educational system (national educational policy, structures and functioning of the educational institution, understanding of the dynamic of school projects, etc.).

---

[4] "Creating a new dynamic for teachers training", a report to the Minister of Education, by Daniel Bancel, October 1989; this report is the result of the work of a commission; the two authors of this paper were members of that commission.

In the next two paragraphs, we will present in a detailed way the training system that was then defined.

3. THE TEACHERS' RECRUITING COMPETITION

We focus in this paragraph on the CAPES preparation, which is the main training stream for secondary mathematics teachers. The CAPES competition is open to any candidate with a Licence degree, but IUFMs jointly with universities offer a specific training. IUFMs also provide their first year students with some professionalization in terms of specific units[5] and training sessions in secondary schools (about fifteen days) but training during this first year is mainly a mathematics training which aims at preparation for the CAPES. Theoretically, mathematical knowledge required for the CAPES doesn't go much beyond what is taught during the first two years at university. But what is required in order to pass this selective competition is much more than what is learnt by students during these first two years.

The competition has both a written and an oral part. The written part consists of two problems: the first one deals more with algebra and geometry, the second one more with calculus, but interactions between domains are frequent. These are long problems with many parts and questions (the text is generally about 5 to 6 pages) and students are given 5 hours to solve each of them. The oral examination consists of two parts: in the first, students are asked to synthetically present a mathematical theme as it could be done at the secondary level; in the second, they are asked to choose and present a set of activities and exercises on a given theme, and to justify and comment on the pertinence of their choices. Lists of themes (about 80) are known in advance but each year new themes are added and others dropped out.

The preparation for CAPES is certainly the first time for students in their long school career when, theoretically, nothing new is added in terms of mathematical content to what they have been taught before. One year is given for reorganizing, connecting, deepening previously learnt knowledge, and for developing more efficient solving skills and attitudes. In fact, the situation is not so simple and, generally, students have to learn new content, for instance in geometry, probability and statistics. But, after three years at university with many different and separated courses, the preparation of CAPES certainly offers an opportunity for developing more synthetic and integrated views about mathematics and for developing some reflexive thinking about mathematical objects.

Preparing the oral part of the competition leads to a global consideration of the curricula of secondary schools, and thus can introduce students to some questions

---

[5] For example, the IUFM of Grenoble has proposed a 'continuum', consisting in considering the continuity of mathematical knowledge and concepts from primary to secondary education, and the continuity of the pupil in his or her mathematical learning all along the educational continuum. This gives a new vision of mathematics. It provides a vision of the articulation between scientific knowledge and knowledge to be taught, and gives an opportunity to study some notions, concepts and methods from their epistemological or historical approach to their didactical transposition into education.

related to the didactical transposition of knowledge[6]. It, of course, requires a good mastery of these elementary topics, but it also needs more. Selecting exercises in the huge amount of existing textbooks and literature, analysing their potential for the teaching of such or such a theme, their relevance with respect to the present curriculum, finding different ways of solving these exercises and comparing them, and correcting the texts proposed in textbooks if not adequate or adding new questions to a given exercise, is a very demanding task, even if students can use a wide diversity of documents freely during their two hour preparation. All the year-long training for this oral test, is certainly a first step towards professionalization. We nevertheless have to confess that, in recent years, the higher selectivity of the competition tends to favour those students with a strong academic background in the classical mathematics of the written part of the competition, at the expense of those who have been especially sensitive to the pre-professional part of their training. We will come back to this point later when noting some changes in the system that are presently being widely debated.

Even with the limitations pointed out above, CAPES preparation has certainly had positive effects. The goal of the reorganization of mathematical knowledge is at least partially achieved and the oral part of their preparation leads students to question the didactical dimension of the content they will be asked to teach. Surveys were made which show that, when entering their second year at IUFM, students feel ready to enter the teaching profession and aware of its multiple dimensions. This creates a strong will for practical training for most of them.

4. THE SECOND YEAR AT IUFM

The training curriculum for secondary teachers in the second year at IUFM is organised into four main parts (Perrenoud, 1994 and Henry, 2000). We will examine them successively:

1. practice period with full responsibility: pedagogical initiation through both imitation and analysis of personal practices;
2. training for mathematics teaching: study of secondary school curricula, didactics and epistemology;
3. general professional training: a variety of objectives, particularly cultural ones;
4. the professional dissertation: analysing some professional issue, linking theory and practice in one particular study.

4.1 Predominance of the practical training

Students entering the second year at IUFM, in order to become mathematics teachers, have succeeded in one of the recruiting competitions for secondary

---

[6] Didactical transposition is the complex set of processes which transform scholarly knowledge, as it appears in professional and in research institutions, into 'taught knowledge', as it appears in the classroom. Cf. *La Transposition Didactique* (Chevallard, 1991).

education: either CAPES, or Agrégation[7]. They are civil servants and they get a salary; if they are approved at the end of this second year, they will have a permanent position in a school.

From the beginning of the year, these student-teachers have full responsibility ('responsibility practice') of one or two mathematics classes in a secondary school (they teach 6 hours weekly, which is one third of the normal duty). Each student-teacher is assisted by a tutor (pedagogical advisor) in the same institution (and if possible with class(es) at the same grade(s)). With such assistance, student-teachers progressively begin to build some pedagogical practice: organising lessons and differentiating different phases in these, designing and following up the annual progression of the curriculum, establishing fruitful relationships with the pupils, preparing and marking assessment tasks, participating in the institutional life of their secondary school, etc. (Borreani and Tavignot, 2000).

After some weeks for adaptation, student-teachers are offered a second practice period, called 'accompanied practice', which may last up to 6 to 7 weeks. This takes place in another secondary school (upper secondary if their responsibility practice takes place in lower secondary, and conversely). This enables the student-teacher to compare different levels of education and different pedagogical practices in different environments. Thus the student-teachers, quite autonomous in their school practice, can quickly build a professional identity, which allows them to develop specific expectations and a critical attitude towards the training they get at the IUFM.

The practice period in full responsibility happens to be a key element of this initial pedagogical training. Student-teachers enjoy it, they feel excited as beginning teachers, leaving behind them their student status. But starting from the very beginning as teachers, without enough pedagogical preparation, is not totally satisfactory. Student-teachers sometimes complain about this lack of preparation, about the difficulties they have at creating a calm, quiet atmosphere allowing real mathematical work in their classes. Due to social evolution and crisis, these difficulties have increased a lot in recent years. In spite of this, adjusting their practice step by step, most student-teachers progressively overcome these difficulties and succeed in creating an acceptable atmosphere in their classroom. However, once this point has been reached, they are not necessarily ready for further and deeper didactical reflection about the quality of the mathematical work that takes place in their classes and the ways in which this quality could be improved.

One must admit that, up to now, didactical research does not provide immediate answers to the student-teachers' needs, as they analyse them. As stressed by many authors (see Margolinas and Perrin-Glorian, 1998, for instance), didactical research in its first stages focuses on the students' conceptions and learning processes and doesn't consider the teacher as a real object of enquiry. Even if in recent years promising research has been developed about teachers, exploring their conceptions and beliefs (Cooney, this volume, pp. 455-466) and the effects of these, analysing

---

[7] CAPES and Agrégation allow graduates to teach mathematics in general and technological secondary schools. Mathematics teachers in vocational high school are also sciences teachers and recruited by a specific competition. After passing this competition they have a similar professional training at an IUFM, but their training also includes a six-week period in the workplace.

their professional work and trying to model their actions and behaviour (Chevallard, 1998, 1999; Schoenfeld, 2000), research does not provide the student-teachers with the answers and knowledge they immediately need in order to make the right decisions for classroom management.

The articulation between theory and practice, which is a main aim of the training at IUFM, is thus not easy to manage. For student-teachers, practice seems more important than more reflexive and theoretical work and such a focus on practice is reinforced by the full responsibility they have in their classes. The feeling that they progressively get good results with their pupils, despite the fact that they have had no previous didactical preparation, may make some of them doubt the necessity for such a preparation. And this doubt may even suggest to some of them that no preparation is needed for the teaching profession other than imitation and progressive adjustments with the assistance of more experienced teachers such as that offered by their pedagogical tutor. Most student-teachers are not so radical but they also question the training they get at the IUFM, facing the existing discrepancy between their immediate needs and the long-term objectives their trainers aim at. These objectives are very diverse, in terms of basic didactical knowledge, epistemological views, and new pedagogical abilities to be developed by the student-teachers, who have a strong tendency to reproduce the conditions they experienced when they were students themselves. Such a questioning has certainly positive effects as it leads the trainers themselves to regularly question the didactical designs they carry out at the IUFM, in order to better link local and particular demands coming from practice to more reflexive and theoretical work.

4.2   Training to mathematics teaching

This epistemological and didactical part of the preparation consists of a variety of topics dealing with mathematics teaching, which are implemented in very diverse ways, by the different IUFMs, in a reduced amount of time (about 100 hours). These include:

- refreshing some parts of mathematical knowledge taught in secondary schools (such as for instance spatial geometry, probability and statistics),
- introducing basic terminology and concepts in the didactics of mathematics, developing an epistemological and didactical analysis of the four main areas of mathematics secondary education (geometry, arithmetic, algebra and calculus),
- reflecting on the role that computer technologies can play in mathematics teaching and learning,
- reflecting about proof and the development of proving competencies,
- designing and managing problem solving situations,
- reflecting about assessment issues.

In some cases, some elements of the history of mathematics are taught.

Whatever the quality of this theoretical part may be, it is difficult to adapt it to the short-term needs of the student-teachers. This is shown and confirmed by some surveys made among former students (Robert, 1996).

The wide diversity of the training content and the limited number of hours available, tend to give the training delivered by IUFM an eclectic and abstract aspect. Relationships with students' individual practices are difficult to set up in a way that doesn't systematically go from theory to practice, in the framework of standard course designs. This is why new devices have been developed and are taking more and more importance such as: small collaborative groups for exchange and discussion about practice, analysis of pedagogical innovations, analysis of class progressions and class preparations, specific sessions for stimulating and addressing student-teacher questions, classroom video analysis, classroom observation analysis, etc. (Hébert and Tavignot, 1998).

Such a training, based more on classroom situations actually managed by the student-teachers, needs some individualisation and has an important cost. It is actually done, here and there, enhanced by the visits the trainers make to the student-teachers in their classes, and by the interviews which follow these visits. But there are too few visits (two or three) in order to allow, by way of the sole visits, an adequate balance between theory and practice in the didactical training. Designs such as those mentioned above are thus likely to take an increasing role in the training.

The link between mathematics teaching and the teaching of other subjects is difficult to establish in such conditions. It nevertheless appears to be more and more necessary so that the students can develop adequate views of the way mathematics interacts with disciplines outside of their field. Mastering an inter-disciplinarily oriented training, is one more open question for the IUFMs.

These remarks show that it is not possible to conceive an efficient initial training for teachers if it only consists of one year of school practice. They show the necessity for earlier pre-professional didactical and epistemological study (Henry, 1996), integrated in university courses. But they also show that a deep didactical reflection can only take place in the perspective of a lifelong in-service training, based on progressive experience and on the questions class practice provides. IUFMs, which when created were only for the pre-service training of primary and secondary teachers, now also undertake their in-service training. This important change should promote a better connection between pre-service and in-service teacher training in the future. One can hope that it will give a new opportunity for developing individualised three or four year programmes for training which will allow each teacher, taking into account his or her previous preparation, to build consistent professional knowledge and adapt their initial training to the necessary evolution of mathematics teaching.

4.3 General professional training.

This part of the training (generally about 100 hours) aims at creating a professional culture, which is common for all levels of education (primary and secondary). It addresses many themes, linked with the competencies expected today for a beginning teacher: history, philosophy and sociology of education, cognitive psychology, pedagogical practices, teaching in difficult situations and specific areas, institutional functioning of schools, legal and ethical aspects of the profession, international comparisons and perspectives, etc. But it also aims at developing individual abilities, such as using word processors, spreadsheets and multimedia tools, or mastering one's body and voice in the classroom. Part is compulsory, part is optional and various possibilities are offered to students, allowing a real individualization of their training. This general training takes several forms: lectures, workshops, seminars and group work.

General professional training modules are received in different ways by different students. Some of them, based on recent research developments, are not easily connected with the real conditions of mathematical learning. Those modules that address what happens in the classroom, the pedagogical relationship and its instrumentation, are the most popular. This general training is also too much separated from the didactical one. For instance, the student-teachers find it difficult to understand the failures of some pupils in mathematics and to help them efficiently. They do not find adequate answers either in the general professional training or in analysis limited to the didactical field. Simultaneously taking into account the pedagogical, psychological and sociological parts of the difficulties pupils meet, enlightening this approach by a didactical analysis based on a good understanding of the epistemological dimension of the knowledge at stake, would be necessary. However, the training at the IUFM is not really an integrated training and all these different dimensions tend to be addressed separately, even if there is a clear institutional potential for more coordinated views. This gives new challenges for training but also for didactical and educational research to develop in the IUFMs.

4.4 The professional dissertation: linking theory and practice.

Writing a professional dissertation is an opportunity for working on the connection between theory and practice, and to put into action some methodologies taken from research. The dissertation has to focus on a question linked with mathematics teaching, but this opens up a great diversity of issues: analysis of students' difficulties in a specific domain, design and experimentation of sessions with specific mathematics goals, use of specific tools such as calculators or mathematical software, comparison of modes of assessment, introduction of new pedagogical practices, deep analysis of didactical situations, analysis of taught content, and so on. Students are expected to articulate some questions clearly and then to work on these by taking into account the relevant literature.

IUFMs have now accumulated lots of dissertations, diverse in quality, but including some very interesting work. These documents reveal the social impression those beginning teachers have about mathematics, about mathematics teaching and mathematics learning, about their teacher role, etc., and the present difficulties of the profession. Some IUFM have made a specific effort for their diffusion to disseminate this material and have published structured synthesies of the best dissertations (Amiot, Barbin and Mariller-Bonnot, 1998; Artigue, 1999). But this very wide sample, which could be of great interest for many researchers, remains poorly exploited.

Trainers also accumulate rich experience when directing such dissertations. Various devices have been developed by the IUFMs in order to organise the tutoring of these dissertations. These choices and their impact on student-teachers' products and training, as well as on the trainers could be productively analysed (Comiti, Nadot and Saltiel, 1999). It is indeed a new role for most of the trainers who, even if they have been involved in teacher training for a long time, do not necessarily have a solid experience in educational research.

When the IUFMs were designed, at a time when the training objectives became more explicit, many studies showed the value of what was called 'training through research'. Even if it was not possible to have the student-teachers work in university laboratories, it seemed necessary to introduce future teachers to research, since a part of their professional activity would need some research ability. Such things as following a scientific development, being aware of the main results of research, doing bibliographical research on a given topic, investigating open questions, working in teams, writing texts, and communicating their ideas, would be of value to them. The professional dissertation was then considered to be a device to provide an introduction to the research world. In some IUFMs this resulted in adopting very strict criteria with respect to the professional dissertation. Today, the opinion is generally more balanced and it has been acknowledged that good and efficient work in the framework of the professional dissertation is something different from good academic research. The context, the time available for reflection, the theoretical and methodological background available to student-teachers, appropriate questions well adapted to such a short investigative and reflective work, the essential ambition of finding ways for linking theoretical and practical approaches, all make a clear difference. Material and methodological means, the bibliographical tools, the theoretical data, are not comparable. This does not reduce the value of the professional dissertation. When we do not set inaccessible aims we get excellent work from students showing that they have actually been engaged in research. It is fascinating to see the efforts students make to reflect on education, even if their academic studies did not prepare them to do so. The dissertation helps them to approach learning as a complex dynamical process and overcome their initial dependence on textbooks. More than any theoretical training, the dissertation makes them aware of the existence on nearly every topic, of previous reflection and research work, including previous dissertations, on which they can rely and to which they can add their personal contribution. They often provide trainers with material they can use in the didactical sessions, material that is especially accessible to student-teachers.

So focusing more clearly on the role of the dissertation in the training, one can better attain the objective of situating the students in what we call a 'research attitude'. One can then hope they will feel close to their academic colleagues, and interested in the research they produce.

5. FURTHER COMMENTS AND REFLECTIONS

The professional preparation of teachers needs a wide variety of content and devices linked to one another. The second year at the IUFM is not long enough. It is the core moment for establishing relationships between pedagogical practices and theoretical knowledge about learning and didactics, and it should rely on a mathematical culture that, beyond academic knowledge, includes some epistemological background. Far from aiming at a complete training, this training year plans to enable the teachers to start their profession with enough knowledge that they do not initiate a pedagogical disaster in their first teaching position. It also aims to sow the seeds for professional expertise and hopes that these seeds will then grow through in-service training, personal work and experience.

The French system gives full teaching responsibility to student-teachers immediately after they have succeeded in the recruiting competition, which is mainly an academic one, and concentrates professional training into one year. As we have tried to show, this certainly produces difficulties for the professional training. But, at the same time, one has to take into account the strong desire most students have, after at least four years of academic studies and often much more, to leave their student's position and take such a responsibility. More integrated teacher training at university certainly allows a better balance between the development of mathematics knowledge and professional competencies. It is not part of the French tradition and mathematicians clearly fear that changing the tradition would lead to a reduction of the mathematical expertise that they think is an essential component of professional expertise, without necessarily producing substantial gains on the professional side.

In the present situation, nevertheless, introducing teachers, during their short initial professional training, to the epistemological, historical and didactical components of such a professional expertise, in a coherent and efficient way, is not an easy task. As regards the historical and epistemological component, even if academic mathematical studies integrate a first approach (but it is hardly the case), epistemological issues take a particular dimension when addressed from the point of view of teaching. For instance, from this point of view, one is not only interested in the identification of the epistemological obstacles which have been associated to the historical development of such or such notion, (s)he wants to understand how such epistemological obstacles reflect in present students' learning, and how they can be overcome in the present context of teaching. As regards the didactic dimension, the context of the second year at the IUFM does not allow the presentation of didactical theories, such as the theory of didactical situations (Brousseau, 1997) or the theory of didactical transposition (Chevallard, 1991) because such a theoretical knowledge is too far away from the students' expectations and knowledge. There is no doubt that beginning teachers mainly need to know about the applications of these theories to the teaching of the content of secondary school curricula, which they actually encounter, that it is the level at which studying didactics must essentially take place in the second year at IUFM. Nevertheless, introducing some didactic basic concepts and methods is quite necessary for fulfilling these practical aims as well as for analysing teaching practices. Didactic concepts and methods enable trainers and students to communicate in a more precise way. They enable teaching and learning questions to be stated more deeply, emphasising the problematics for study, and accessing the bibliographical references for articles and books about these questions. Higher education textbooks (Joshua and Dupin, 1993), books especially aimed at teacher training (Briand and Chevalier, 1995), (Legrand, 1997), make available and accessible the results of the last thirty years in the didactics of mathematics. Journals such as *Reperes-IREM*, *Petit x* or the Bulletin of the APMEP[8], provide the student-teachers with professional papers which they can use in their daily practice and in the elaboration of their professional dissertation. But, in spite of these resources, finding an appropriate way for introducing students to didactic issues and perspectives in a so short amount of time is not easy at all.

There are some university courses about didactics at the Licence or Maitrise levels. They might seem too abstract, not linked to the classroom reality, but it does not appear to be an obstacle for mathematics students, since they are used to abstraction. But, since there is no evaluation, one can ask what long-term impact such courses have on the professional practice of future teachers. It seems that the university courses on didactics prepare the IUFM students in a certain way to better state the questions they actually have to solve in their classes.

After several years of practice, some teachers may be interested in a new access to didactical theories, if they are able to provide a framework for more advanced studies, such as for instance the study of the questions raised by new technologies in education. In this respect, one can observe that the present structure for teachers' initial training adapts too slowly to introducing computer tools and more generally information and communication technology in classes as a support for teaching. Information and communication technologies change the knowledge itself, the way one can do mathematics, and makes mathematics a more experimental science. It changes the way one can access knowledge, and new resources are now available for students and teachers at a distance. Communication technologies allow different types of work, combining distance and interactive activities. Information and communication technologies deeply change the way teachers work, they change the teaching profession. Technical training limited to the use of software is not enough, since the didactical effects of powerful software deeply transform the management of learning, and even its objectives in terms of knowledge. Very soon, new competencies will thus be expected from teachers, and it is only a significant investment in in-service training, based on specific research, which will enable teachers to cope with the new competencies. Here is a new great challenge for the IUFMs.

The diversity of training delivered in the different IUFMs, the diversity of contexts in which new teachers must practice their profession, make it difficult to measure precisely what IUFMs have achieved, ten years after their creation. Such a state of the art would be important, but perhaps too schematic and reducing. The National Committee of Evaluation, in charge of universities, was asked to evaluate the different IUFMs over the last four years. Their reports, addressed to the President of the Republic, are public and accessible on Internet (the CNE web site

---

[8] APMEP: Association of Mathematics Teachers.

URL is http://www-cne.evaluation.fr). What are their results? The practical part of training seems globally satisfactory but it remains too faintly connected with the more theoretical parts of the training. Reports stress the negative effect of time constraints and the fact that student-teachers are faced with too many different trainers. They suggest that, in several cases, ambitions should be reduced and better rationalised. A major difficulty for IUFMs is to develop their own academic and professional competency, and to work in close cooperation with universities and with schools. The cultures of higher education, school education, and teacher education have been too separated in the past, and IUFMs now try to bring more coherence into the system. Their relations with universities, and with schools, are not always easy. The major difficulty is to articulate theory and practice in teacher education properly, not as juxtaposed components, but with permanent links, and with a real reflexive analysis of their practice. This is why often students feel that the theoretical input is too far from the concrete necessity of practice!

On the other side, there are strong debates in France about the "contradiction between content knowledge and pedagogy". Some people think that being in favour of developing teachers' pedagogical competencies means denying the importance of their content knowledge, and they think that IUFMs will contribute to a decrease in the mathematical level of teachers, since IUFMs want to increase their professional ability. Such a debate sounds strange, but is still present 10 years after the creation of the IUFMs. IUFMs aim at preparing teachers with both a high subject level, and a strong professional training. But they are sometimes accused of being the 'temples of pedagogism' – places of dogmatism or false science.

The difficulty of assessing students' academic competencies is well known. Evaluating the professional competencies acquired by student-teachers in the IUFMs raises new and difficult problems. The literature, especially in comparative studies, is scarce. We have the feeling though, that newly trained teachers start their professional life well prepared for their work. This can pragmatically be seen through the increasing role they are playing in local pedagogical teams, through their investment in the tutoring of new student-teachers, through their participation in the conferences offered by IREMs[9] or APMEP[10], as well as in summer schools for in-service training which the Ministry of Education funds every year, and through their integration into IREM teams.

What remains from initial training after some years of professional practice? And how to identify among the successes or the difficulties that new teachers encounter in their practice, what comes from their training and what comes from their environment, behaviour or psychology?

Teaching is a global act, even if it needs specific competencies that in our opinion can be learnt. It is from this point of view that the question of evaluating the

---

[9] IREM: Institut de Recherche sur l'Enseignement des Mathematiques - Research Institute for Mathematics Education. There are 26 such institutes in France, created in universities at the beginning of the 70s. They develop research on mathematics education, provide in-service training for teachers, and publish documents and books for mathematics teachers.

[10] APMEP: Association des Professeurs de Mathematiques de l'Enseignement Public - Public Education Mathematics Teachers Association. This association is particularly dynamic, and a huge number of mathematics teachers, among the more active, are members of it.

training at IUFMs and its impact on practices can be raised. It therefore seems important to us that some specific aspects could be identified and compared to training experiences carried out in other countries, taking the opportunity of international cooperation such as the ICMI studies.

It has now been ten years since teacher education was reformed in France with the creation of IUFMs, training primary and secondary teachers in a university context, and aiming at articulating subject training and professional preparation. The French Minister of Education wants now to evaluate the present situation of teacher education, and to produce improvements in teacher education. He has asked two academics (Bernard CORNU, director of the IUFM of Grenoble from 1990 to 2000, one of the authors of this paper, and Jean BRIHAULT, rector of the University of Rennes) to produce a report proposing recommendations for a renovation of teacher education. This report will be available soon in its definitive form. The two main aims of such a renovation are to change the competitive recruiting examinations and to create the opportunity of an articulation between pre-service and in-service training, by designing a training for all new teachers during their first year of teaching (a kind of 'third year' under the responsibility of the IUFMs). But there is strong opposition to change in the recruiting examinations: some think that increasing the professional aspects will necessarily decrease the academic aspects.

The following are some of the measures proposed in the report:

- Reinforce the professional dimension of the recruiting examinations. A practice period in a school should be compulsory for everyone who wants to apply for the recruiting examinations.
- Develop pre-professional modules at university for students prior to their entering an IUFM.
- Make the way applicants are admitted to IUFMs through the country more equitable: due to local contexts, the procedure may differ from one IUFM another, and the level of selectivity varies, according to the number of applicants.
- Improve collaboration between IUFMs and universities, especially for the first year at IUFMs, where the universities contribute to the subject content.
- Improve the quality of the training in the second year at IUFMs, through national recommendations about the content and methods of training. The training should be better integrated and not limited to the juxtaposition of a wide set of modules, as it often is at present.
- Develop and reinforce the integration of information and communication technologies in teacher education, for all new teachers.
- Make more precise and accurate the way teachers are evaluated and assessed at the end of their two years at IUFMs.
- Create a 'third year' for teacher education. During their first year of teaching, new teachers should be given a part ($1/6^{th}$?) of their time for compulsory training activities. It would articulate pre- and in-service training; it is the right time for some parts of the training, which need a first professional experience and the encountering of concrete problems in schools.

- Confirm and develop the role of IUFMs in teachers' in-service training, so that pre- and in-service training can be strongly articulated, leading to a 'life long training' for teachers, and so that the content and methods of in-service training can be designed and operated by university institutions.
- Improve the balance between the different categories of trainers composing the staff of IUFMs: trainers coming from university, from secondary schools, from primary schools; permanent staff and staff recruited for a fixed term (3 or 4 years); part time trainers; improve the role of tutors in teacher education.
- Develop and reinforce the activity of IUFMs in the field of research, so that the training can be strongly linked with research, and that the staff in the IUFMs are strongly involved in research.

6. SOME CONCLUDING REMARKS

The creation of IUFMs in France introduced a break in the conceptions that previously determined secondary teachers' training. These were that first purely academic studies in mathematics at university were taken. This was followed by professional training with the assistance of tutors involving direct contact with classes where teacher trainees imitated what they saw others do and pragmatically adapted to what happened in the classroom.

In 1989, a Commission at the Ministry of Education, chaired by Rector Bancel[11], developed a project stressing the need for new theoretical and practical competencies (not only mastering the subjects), and considering them as objects of training.

IUFMs and their training schemes were born from a double objective of developing competencies in psychology, sociology, philosophy, pedagogy, epistemology and didactics as core components of the training, and articulating theory and practice. This needs an initiation into learning theories, psycho-pedagogical theories and didactical theories, and teacher confrontation with the actual practical management of the classroom, the management of actual learning, and the management of a progression in knowledge acquisition. Practice analysis, exchange groups, and other reflective devices, play a role in the need for theoretical references. The professional dissertation, subject training, transverse training, and visits to classes followed by interviews, then play the role of sampling and applying theoretical data.

Based on such an articulation, IUFMs are in an intermediate position between practical training schemes 'at the chalk face', and higher education colleges or universities in educational sciences. In that sense, they have a particular and original place, which is worth considering. Even if their first years were ones of 'adjustment', they now reveal new paths, particularly regarding professional dissertations and the analysis of practices. They were able to progressively emphasise where strides forward in the preparation of future teachers could be made.

---

[11] Rector Bancel produced a report in 1989, recommending the creation of new higher education institutions (IUFM) in order to train future teachers (primary and secondary), give them professional competencies, and having as a core objective the "articulation between theory and practice".

They also showed what new features could be expected from the trainers, whose specific training and preparation is still a key question for the future of IUFMs. The diversity of training situations and content requires the contribution of different categories of trainers. These range over pedagogical advisors or tutors in the secondary institutions, university teachers from different disciplines, and secondary teachers appointed by the IUFMs for didactical training. IUFMs also need to have permanent trainers, together with trainers who keep a strong link with 'the chalk face' by acting part time at IUFMs and part time in schools, or by working at IUFMs for a fixed term. Coordinating all these contributions is not at all easy and certainly requires a strong institutional impulse. It certainly also requires the development of mixed research and development teams. Developing coherent and consistent programmes of research about teacher training is today a major issue for the IUFMs. It needs also the development of specific training for trainers and opportunities for them to meet and discuss the problems they encounter at different levels, from local seminars to national meetings and summer schools. Professional acknowledgement of the knowledge and expertise they develop remains today an open problem: trainers do not have any official status.

Such potentialities are very promising for all decision-makers and those engaged in teacher education in our country.

REFERENCES

Amiot, M., Barbin, E. and Mariller-Bonnot, M.C. (1998). *Enseigner les mathématiques en collège et en lycée.* Le Perreux/Marne: C.R.D.P. de l'académie de Créteil.

Artigue, M. (Ed.), (1999). *Entre théorie et pratique, le mémoire professionnel de Mathématiques à l'IUFM.* Reims: I.U.F.M. de Reims et C.R.D.P. de Champagne-Ardenne/C.N.D.P.

Borreani, J. and Tavignot, P. (2000). *Pratiques d'enseignement des mathématiques observées en classe de sixième.* Rouen: I.U.F.M. et C.R.D.P. de Rouen/CNDP.

Briand, J. and Chevalier, M.C. (1995).*Les enjeux didactiques dans l'enseignement desmathématiques.* Paris: Hatier.

Brousseau, G. (1997). *Theory of didactical situations in mathematics.* Dordrecht: Kluwer Academic Publishers.

Chevallard, Y. (1991). *La transposition didactique, du savoir savant au savoir enseigné.* Grenoble: La Pensée Sauvage Editions.

Chevallard, Y. (1997). Familière et problématique, la figure du professeur. *Recherches en Didactique des Mathématiques,* vol. 17.3, 17-54.

Comité National d'Evaluation. *Rapports d'évaluation des IUFM des académies: Caen, Grenoble, Lyon*(1996). *Amiens, Bourgogne, Nord-Pas-de-Calais, Reims, Rouen* (1998). *Besançon, Créteil, Paris, Versailles, Orléans-Tours* (1999). *Toulouse* (2000). Official reports available on the Internet: http://www-cne.evaluation.fr.

Comiti, C., Nadot, S. and Saltiel, E. (1999). *Le Mémoire Professionnel - enquête sur un outil de formation des enseignants.* Grenoble: I.U.F.M.

Cooney, T. (2001). Using research to inform pre-service teacher education programmes, this volume, pp. 455-466.

Cornu, B. (1999). Training today the Teacher of tomorrow. In C. Hoyles, C. Morgan and G. Woodhouse (Eds.), *Rethinking the Mathematics Curriculum,* pp. 195-202. London: Falmer Press.

Cornu, B. (1999). Teacher Training and Technology Training Systems. In Donald P. Ely, Linda E. Odenthal and Tjeerd Plomp (Eds.), *Educational Science and Technology,* pp.93-103. Enschede: Twente University Press.

Cornu, B. (sous la direction de) (2000). *Le nouveau métier d'enseignant,* Commission Nationale Française pour l'UNESCO. Paris: La Documentation Française.

Davisse, A. and Rochex, J.Y.(1997). *"Pourvu qu'ils m'écoutent": autorité et discipline dans la classe.* Le Perreux/Marne: C.R.D.P. de l'académie de Créteil.

Davisse, A. and Rochex, J.Y. (1998)."*Pourvu qu'ils apprennent": face à la diversité des élèves.* Le Perreux/Marne: C.R.D.P. de l'académie de Créteil.

Hébert, E. and Tavignot, P. (Eds.), (1998). *Entrez dans nos classes.* Rouen: I.U.F.M. et C.R.D.P. de Rouen/CNDP.

Henry, M. (1996). IUFM: quelle formation? De la théorie à la pratique, témoignage d'un acteur engagé. *Repères-IREM* n° 23, 59-82.

Henry, M. (2000). Evolution and prospects of the pre-service education of mathematical teachers in France. *Journal of Mathematics Teacher Education*, 00: 1-9.

Joshua, S. and Dupin, J.J. (1993). *Introduction à la didactique des sciences et des mathématiques.* Paris: Presses Universitaires de France (PUF).

Legrand, M. (Ed.), (1997). *Profession enseignant, les maths en collège et en lycée.* Paris: Hachette Education.

Perrenoud, P. (1994). *La formation des enseignants en IUFM. Comparaison de 18 IUFM, théorie et pratique.* Paris: l'Harmattan.

Robert, A. (1996). IUFM: réflexion sur la formation professionnelle initiale des professeurs de mathématiques des lycées et collèges. *Repères-IREM* n° 23, 83-108.

Robert, A., Lattuati, M. and Penninckx, J. (1999). *L'Enseignement des Mathématiques au Lycée, un point de vue didactique.* Paris: Ellipses.

Michel Henry
University and IUFM of Franche-Comté, France
henry@math.univ-fcomte.fr

Bernard Cornu
IUFM of Grenoble, France
bernard.cornu@grenoble.iufm.fr

BERNARD HODGSON

THE MATHEMATICAL EDUCATION OF SCHOOL TEACHERS: ROLE AND RESPONSIBILITIES OF UNIVERSITY MATHEMATICIANS

1. INTRODUCTION

The preparation of mathematics teachers for the primary and secondary school is a multifaceted task. It spans various periods of the teachers' life, encompassing their experiences first as pupils, then as undergraduate students and finally as professionals learning from their own action and from in-service activities. Although relatively short, the formal part of this preparation. Besides the need to include actual teaching practice as early as possible (i.e., working with pupils in real classrooms), three components of this preparation can be identified, which reflect the belonging and interests of the university educators responsible for the formal education of teachers: mathematics itself, didactics of mathematics and psychology of learning. The focus of this paper is on the first of these components.

Mathematicians have a major and unique role to play in the education of teachers — they are neither the sole nor the main contributors to this complex process, but their participation is essential. Maybe this will be seen as a truism, at least in connection with the preparation of secondary school mathematics teacher. But I wish nonetheless to present here some comments about the context in which this role can and should be played. I also want to support the view that mathematicians should take part in the education of primary school teachers. I see such an involvement as important because of the perspective on mathematics itself mathematicians can bring to student teachers. Moreover, I believe this involvement can be a source of gratifying and stimulating mathematical moments for the mathematicians themselves.

In the final part of this paper, I will briefly suggest a few examples of mathematical topics which, from my experience, nicely illustrates the richness of the mathematical content pertaining to student teachers, both of the primary and secondary level. But first I want to examine some aspects of the role and responsibilities of mathematicians in the preparation of schoolteachers, in particular from an historical perspective.

Derek Holton (Ed.), The Teaching and Learning of Mathematics at University Level: An ICMI Study, 501—518.
© 2001 *Kluwer Academic Publishers. Printed in the Netherlands.*

social and cultural factors as well as local traditions, that generalisations are almost impossible. Even within a single country the models are sometimes extremely varied.

In this paper, I want to concentrate on the problem of pre-service mathematical preparation of primary and secondary school teachers, which is typically done in colleges, universities or institutes for teacher education. The mathematicians who are at the centre of my discussion may thus belong to various units within a university. While some of them are attached to educational divisions, either directly or through cross-appointments, the majority of the mathematicians I have in mind are members of departments of mathematics. A key issue is then to what extent their department is really supportive of their involvement, actual or potential, in teacher education.

It does not make much sense to hope for every individual in every department of mathematics to develop interest or expertise in teacher education. For instance, it would be unreasonable to expect a young Ph.D. in pure mathematics with a high proficiency in research, who has been hired in a top-level department world-renowned for research, to devote much energy to the education of teachers. Even in more typical departments, the duties to be fulfilled year after year are in most cases extremely varied and each faculty member tends to develop some speciality. The question, I believe, is whether the task of teaching teachers is regarded as *bona fide* work in mathematics, on the same level, say, as teaching to engineers or to students specialising in mathematics. Looking at various indicators (for instance, peer recognition, promotions, grants, etc.), it is quite tempting to conclude that teaching mathematics to prospective teachers, and especially to those of the primary school, is regarded by many as low-level work in comparison, for example, with teaching to undergraduate mathematics students or still more to graduate students — not to speak of mathematical research *per se*. However, such a view denies the fact that the quality of the students in university classrooms is conditioned to a large extent by the quality of the teachers they have had in school, even at the primary level.

There is, within the mathematics community, a very long tradition of mathematicians getting seriously involved not only in general pedagogical issues, but also more particularly in the problems related to the preparation of schoolteachers. I will now recall briefly a few historical facts related to this. I will refrain in the present text from commenting on the mathematicians' role in education from an epistemological perspective; such a view, presented in connection with the nature of mathematics itself and its contribution to society, can be found in de Guzmán (1993). See also the influential paper of Thurston (1990).

2.1 Some historical milestones

It might be an interesting task to trace back to past centuries the influence exerted on educational matters by mathematicians, either directly or indirectly. But I will restrict my comments mainly to a few events that took place in the last century.

The history of the International Commission on Mathematical Instruction (ICMI) is a rich source of information on the implication of mathematicians in education (see Howson, 1984, for a description of the origins, history, work and aims of ICMI). In fact, one could even see ICMI as having been formed on the very assumption that university mathematicians should have an influence on school mathematics — at least at the secondary level. This is suggested by the following resolution adopted during the fourth International Congress of Mathematicians, held in Rome in 1908, and which marks the inception of ICMI:

> The Congress, recognizing the importance of a comparative study on the methods and plans of teaching mathematics at secondary schools, charges Professors F. Klein, G. Greenhill, and Henri Fehr to constitute an International Commission to study these questions and to present a report to the next Congress. (Lehto, 1998, p. 13)

This resolution was submitted on the initiative of David Eugene Smith, who had suggested the idea of such a commission three years earlier in the international journal *L'Enseignement Mathématique* (see Smith, 1905), in response to a survey proposed by the editors of the journal, Charles-A. Laisant and Henri Fehr, on reforms needed in mathematics education.

Since the establishment of ICMI, many distinguished mathematicians have played a major role in it. The list of officers of the Commission is quite eloquent in this respect — to give just a few names (see Howson, 1984): Felix Klein, Jacques Hadamard, Marshall H. Stone, André Lichnerowicz, Hans Freudenthal, Hassler Whitney, Peter Hilton, Jean-Pierre Kahane; or the current ICMI President, Hyman Bass.

The birth of ICMI took place at a time when many events with an educational flavour were happening in various countries. For instance, the journal *L'Enseignement Mathématique* had been founded in Switzerland a few years earlier. In France, an extensive reform of national curricula, concerning especially the teaching of geometry, had been set forth in 1905 by the government. Many mathematicians, such as Émile Borel, were then involved in discussions about the teaching of elementary mathematics (see Borel, 1914). But the most striking example is surely that of Felix Klein in Germany.

Already, during the last decade of the nineteenth century, Klein was very active in making accessible to secondary school teachers some of the most recent mathematical developments of his time. For example, he presented in his 1894 summer course for teachers (*Vorträge über ausgewählte Fragen der Elementargeometrie*, published in 1895 and soon to be translated into French, Italian, English and Russian), a proof of the impossibility of the three Greek geometrical constructions, Cantor's proof of the existence of transcendental numbers, as well as proofs of the transcendence of e and π. These lectures, he wrote, were due to his desire "to bring the study of mathematics in the university into closer touch with the needs of the secondary schools" (Klein, 1897, p. 1) and they are still considered, a century later, as a masterpiece of exposition and an excellent entrance to modern science (see Kahane, 1997, p. ii).

Some ten years later, in his lectures entitled *Elementarmathematik vom höheren Standpunkte aus*, Klein intended to provide teachers with a comprehensive view of basic mathematics (Arithmetic, Algebra, Analysis, Geometry) of such a range, he wrote in the Preface, "as I should wish every teacher in a higher school to have" (Klein, 1939, p. v). His concern was not "with the different ways in which the problem of instruction can be presented to the mathematician", but rather

> with developments in the subject matter of instruction. I shall endeavor to put before the teacher, as well as the maturing student, from the view-point of modern science, but in a manner as simple, stimulating, and convincing as possible, both the content and the foundations of the topics of instruction, with due regard for the current methods of teaching. (Klein, 1932, p. iii)

He concluded his introductory remarks with the wish that his exposition of elementary mathematics

> may prove useful by inducing many of the teachers of our higher schools to renewed use of independent thought in determining the best way of presenting the material of instruction. This book is designed solely as such a mental spur, not as a detailed handbook. The preparation of the latter I leave to those actively engaged in the schools. (Klein, 1932, p. iv)

It was Klein's hope that this "renewed use of independent thought" may help teachers overcome the "double discontinuity" which they meet when going from secondary school to university, and then back to school as a teacher (see Klein, 1932, p. 1). As commented by Furinghetti (2000), teachers, as a consequence of this double discontinuity,

> tend to reproduce in teaching what was taught them in secondary school. This is one of the causes of the conservative style characterizing school systems around the world. It is because of this conservative style that changes in the curricula or in the method of teaching are so difficult to make.

The idea initiated by Klein of looking at elementary mathematical topics from an advanced standpoint has proved to be extremely fruitful and can be seen, in many respects, to constitute the core of the mathematical preparation of schoolteachers.

The list of prestigious mathematicians significantly involved in matters related to education could be extended much further. Leonhard Euler, for instance, while not a teacher in the usual sense of that word today, surely possessed an acute sense of the importance of expository quality, as can be judged from such gems as his *Letters to a Princess of Germany*, his *Elements of Algebra*, or his *Foundations of Differential Calculus*, to name three of his masterly written elementary works. As an example of more recent times, Henri Poincaré, one of the most distinguished mathematicians at the turn of the century, enjoyed confronting pedagogical issues, as is testified to by some of his papers published in *L'Enseignement Mathématique* during its first decade and devoted to topics such as the links between differential notation and teaching (vol. 1, 1899), logic and intuition in mathematics and in teaching (vols. 1,

1899 and 10, 1908), or the role of definitions in mathematics (vol. 6, 1904). (This perception of pedagogically inclined distinguished mathematicians could be challenged with statements such as "I have a real aversion to teaching" (Gauss, 1802) or even "I hate 'teaching'" (Hardy, 1967, p. 149). But this in turn could be contrasted with a famous assertion from Poisson (see Hauchecorne and Suratteau, 1996, p. 287) claiming that life is good for only two things: discovering mathematics and teaching mathematics.)

I would now like to turn to a major mathematician, George Pólya, who exerted an exceptional influence in the educational realm during the last half-century (the first edition of the celebrated 'How to Solve It' appeared in 1945).

2.2 Pólya on the education of teachers

The impact of Pólya's works in general mathematics education is remarkable and has been presented in various places (see, for instance, the ICMI Presidential Address of Kahane (1988), which also includes comments on his mathematical works). His views on the teaching of mathematics, and on the consequential question of the mathematical preparation of mathematics teachers, have been examined in depth by Kilpatrick (1987). I would like to point here to just a few aspects of these.

The basis of Pólya's ideas about the education of mathematics teachers is probably to be found under the item *Rules of teaching* of his 'Short dictionary of heuristic' (Pólya, 1957, p. 173):

> The first rule of teaching is to know what you are supposed to teach. The second rule of teaching is to know a little more than what you are supposed to teach. ... [I]t should not be forgotten that a teacher of mathematics should know some mathematics, and that a teacher wishing to impart the right attitude of mind toward problems to his students should have acquired that attitude himself.

The same ideas come back in his 'Ten commandments for teachers' (see Pólya, 1959, and Pólya, 1981, p. 116) under items 1, 2 and 5: (1) Be interested in your subject; (2) Know your subject; and (5) Give your students not only information, but *know-how*, attitudes of the mind, the habit of methodical work.

Knowledge, in Pólya's view, consists partly of *information* and partly of *know-how*, and teachers need to impart both. Mathematical know-how has been described by Pólya as "the ability to solve problems, to construct demonstrations, and to examine critically solutions and demonstrations". And he imposes on these two components of knowledge a very sharp hierarchy: "know-how is much more important than the mere possession of information". Hence his somewhat provocative conclusion: "it may be more important in the mathematics class how you teach than what you teach". (Quotations from Pólya, 1959, p. 528, and Pólya, 1981, p. 118.)

Pólya cannot help observe that secondary school mathematics teachers are very badly equipped when they begin their teaching career. Having themselves left "the high school, more often than not, with no knowledge or with a wobbly knowledge of high school mathematics" (Pólya, 1959, p. 531), the question arises: where and when should they acquire the knowledge of the mathematics necessary for their teaching? And more precisely: where and when should they not only learn basic facts or improve their skills in secondary school mathematics, but, more to the point, develop know-how, i.e., their ability to reason and think creatively?

As commented by Kilpatrick (1987, p. 87), Pólya's appreciation of the teacher preparation offered by colleges or universities is that it does very little to improve this situation: "In Pólya's view, departments of mathematics stress abstruse information at the expense of developing mathematical know-how, whereas schools of education dwell too heavily on content-free teaching methods."

The remedy that Pólya advocated was to give prospective teachers an opportunity to pursue mathematical investigations of their own at a level appropriate to their interest and expertise: this was the idea behind his seminar in problem-solving for teachers, where the requisite knowledge was at the high school level and the difficulty of the problems to solve, just a little above high school level (Pólya, 1959, p. 532). Pólya has presented in great detail the philosophy underlying this seminar and the types of problems he worked with teachers (see Pólya, 1981).

The creativity involved in such an approach can help the teachers develop a dynamic relation to mathematics and eventually bring some original contribution to it. Cassidy and Hodgson (1982) have suggested that finding new solutions to given problems, and in particular 'elementary' solutions requiring a minimal level of prior knowledge or even of mathematical maturity, could be seen as a type of research activity fully appropriate for teachers. A parallel can be made here with the merit, in traditional mathematics, of finding a new proof of an already established theorem. On this matter, Wittgenstein has maintained the following:

> It might be said: ' — that every proof, even of a proposition which has already been proved, is a contribution to mathematics'. But why is it a contribution if its only point was to prove the proposition? Well, one can say: 'the new proof shews (or makes) a new connexion'. (Wittgenstein, 1978, p. 191)

More on Pólya's views about the mathematical preparation of secondary school mathematics teachers can be found in Chapter 14 of Pólya (1981) entitled 'On learning, teaching, and learning teaching'. (Although many of Pólya's ideas concerning the education of teachers could obviously be extended to primary school teachers, Pólya did not, as far as I know, write explicitly about them.)

2.3 Towards a conceptual understanding of mathematics

It is quite common to find among the general public a distorted image of mathematics, a widespread view reducing it to rote calculations. Prospective teachers may also share such biases to a certain extent, even those at the secondary school level. The formal preparation offered to teachers at the university should thus be seen as a unique occasion to help them develop a better appreciation of what mathematics really is. While furthering the development of know-how advocated by Pólya, this preparation should put them in contact with and see in action some of the 'deep ideas' which form the heart of mathematics (see Steen, 1990): *structures* (numbers, shapes, algorithms, functions, ...) ; *attributes* (linear, periodic, symmetric, random, ...) ; *actions* (experiment, model, classify, prove, ...) ; *abstractions* (symbols, infinity, similarity, recursion, ...) ; *attitudes* (wonder, beauty, ...) ; *behaviours* (motion, convergence, stability, chaos, ...) ; *dichotomies* (discrete–continuous, finite–infinite, ...).

Reports have been published recently, which provide a survey of several researches confirming the importance of teachers' knowledge of mathematical content (see Brown and Borko, 1992, Fennema and Franke, 1992, Grossman, Wilson and Shulman, 1989; Swafford, 1995). For instance, a contrast is made in Grossman et al. (1989) between one teacher with a weak knowledge, who drilled pupils in algorithms to be memorised and applied to predictable problem sets, and another one with a strong mathematical background, who emphasised the 'whys' of mathematics and led pupils to think through problems. Moreover, not only do teachers with a shallow knowledge constitute inadequate models for their pupils of the attitude of mind to develop, but also, as they need to fill holes in their content knowledge, they are bound to have less time to devote to properly pedagogical issues:

> Without adequate content knowledge, student teachers spend much of their limited planning time learning content, rather than planning how to present the content to facilitate the [pupils'] understanding. Student teachers with strong content preparation are more likely to be flexible in their teaching and responsive to [pupils'] needs, and to provide conceptual explanations, instead of purely procedural ones. They also tend to place greater emphasis on the organization and connectedness of knowledge within the discipline and less on the provision of specific information. Student teachers without adequate content knowledge are likely to lack confidence in their ability to teach well.
> (Brown and Borko, 1992, p. 220)

Teachers with weak competencies in mathematics may be unable to reach the level of autonomy enabling them to appraise critically the adequacy and accuracy of textbooks or to distinguish between more or less legitimate claims. They lack the expertise to identify 'good ideas', as they can only tell the difference between a familiar and an unfamiliar situation. As commented by Moise (1984), "[w]hen unfamiliar insights are ignored as if they were worthless, or actually rejected as wrong, the effect is to discourage students, or corrupt their mentality, or both."

The expression *conceptual understanding* is often used to describe the type of knowledge of mathematics a teacher needs to develop in order to fully play the role of 'facilitator' between pupils and mathematical knowledge. Going much beyond mere factual information, conceptual understanding of mathematics stresses organising principles and central concepts. It allows teachers to perceive mathematics not as a set of facts to be memorised, but as a co-ordinated system of ideas. It makes explicit the paradigms proper to mathematics and shows proofs and reasoning, in various forms, playing a crucial role in certifying facts. It gives a thread linking key mathematical ideas. Such a notion of conceptual understanding is at the core of the famous study of Ma (1999), comparing teachers from China and the USA with respect to a "profound understanding of fundamental mathematics".

I would like to illustrate this point with an elementary example. What most adults remember about the notion of area they learned at school is a list of a few formulas: formula for the area of a square or a rectangle with respect to the sides, for the area of a triangle with respect to a side and the corresponding altitude, etc. The *Tangram* is a familiar geometrical puzzle made of seven geometrical shapes which provides a very nice setting for allowing prospective primary school teachers to better understand the concept of area. Many of them are at a loss when asked, for instance, to find the area of the large triangle in terms of the parallelogram (these are two of the seven pieces): where are the formulas? This indicates a confusion between the concept of *measuring* an area (i.e., comparing a certain surface with a given unit-surface, eventually via intermediate units), and that of *calculating* an area (i.e., replacing the measurement of an area by measurements of lengths followed by arithmetical manipulations on these measures, according to some appropriate formula). A teacher must be aware of the fact that, while reducing an area problem to an arithmetical calculation is often a useful device, it does not reflect the quintessence of measurement of area *per se*.

As can be expected, the more teachers are comfortable with the concepts of mathematics in a deep way, the better they are equipped to work these concepts with their pupils: "The evidence is beginning to accumulate to support the idea that when a teacher has a conceptual understanding of mathematics, it influences classroom instruction in a positive way." (See Fennema and Franke, 1992, p. 151.) However, research also suggests that prospective mathematics teachers are often lack adequate content knowledge, especially those of the primary level who in addition frequently show high levels of 'math anxiety' — see Brown and Borko (1992, p. 220) and Fennema and Franke (1992, p. 148). This leads us to the difficult question of identifying the type of mathematical experiences prospective teachers should meet during their formal education in universities or colleges.

2.4 Programmes for prospective teachers

Programmes for prospective teachers of mathematics must address the many dimensions of the teachers' task, preparing them for the decision-making process in which they will be involved daily in their classrooms. Some of the teachers' major roles, as identified in the NCTM report 'Professional Standards for Teaching Mathematics' (see NCTM, 1991, p. 5) are: setting goals and selecting or creating mathematical tasks to help students achieve these goals; stimulating and managing classroom discourse so that both the students and the teacher are clearer about what is being learned; creating a classroom environment to support teaching and learning mathematics; and analysing students' learning, the mathematical tasks, and the environment in order to make ongoing instructional decisions.

A central aim of teacher education programmes is to impart the ability of transforming disciplinary knowledge into a form of knowledge appropriate for pupils and specific to the task of teaching. These programmes hence consist of various components dealing with the multiple aspects of the knowledge teachers need to acquire to that effect, including: knowledge of mathematics, knowledge of instructional representations of mathematics suitable for pupils, knowledge about learners' cognition, knowledge of curricula, general pedagogical knowledge, and knowledge of didactics of mathematics. Underlying teacher programmes and calling for their specificity is thus a vision of teachers as professionals, the 'professionals of the pedagogical act'.

I would now like to comment on the first of these components, the acquisition of mathematical content by prospective teachers and to suggest some means of furthering the development of a conceptual understanding of mathematics in such a context. My purpose here is not to enter into a detailed description of possible courses, but rather to suggest some general guidelines — in the final part of this paper, I will indicate a few examples of mathematical topics suitable in such a context. (In addition to reports produced in various countries about mathematics teacher education, the studies of Morris (1984), although more than a decade old, still form a precious source on teacher education as seen from an international perspective.)

In many colleges or universities, the mathematical preparation offered to prospective schoolteachers can be roughly described as follows. Primary school teachers often have no mathematics courses at all to take after having completed high school. Their only post-secondary mathematical activities take place in pedagogy courses. Hence they have no occasion to develop adult-level insights into the mathematics they will be teaching to children. Secondary school teachers will quite obviously take several mathematics courses, typically in the context of a major programme in mathematics. However, one faces here the criticism made by Pólya: these courses are not explicitly intended for prospective teachers. Some will be offered to various categories of students (scientists, engineers, etc.) while others will be advanced courses for mathematics specialists. It is only in the rare case that prospective secondary school teachers will be given mathematics courses specifically designed for them. In spite of the fact that they do acquire some high-level mathematical knowledge, they have no explicit occasion for making connections with the mathematical topics for which they will be responsible in school, nor of looking at those topics from an advanced point of view *à la* Klein. This situation clearly paves the way to the 'double discontinuity' paradigm discussed by Klein (1932), as mentioned above.

A consensus has emerged to the effect that it is not by simply having primary and secondary school teachers take a greater number of traditional mathematics courses that the situation will improve. There is an urgent need for courses specifically intended for teachers and in which a different spirit is conveyed through the choice of topics, the way they are treated, the teaching methods used. As Swafford (1995), p. 161, puts it, "it is not enough to know more mathematics. Teachers must also know more about the mathematics they will teach." Such courses should thus provide prospective teachers with the opportunity to analyse school mathematics from a deeper perspective than the one which prevailed in their own schooling, giving them the ability to explicate the meanings underlying elementary mathematical topics. The 'O-script/A-script method' of Wittmann (this volume, pp. 539-551), can be seen as a particular instance of this teaching principle. Based on the general philosophy that the mathematical education of any category of students should reflect in a fundamental and systematic way their professional context, the approach of Wittmann allows, for instance, primary student teachers, through learning by discovery, to develop a better appreciation of reasoning as an objective in mathematics education. Wittmann also puts forward a strong plea for mathematicians not to look down on primary mathematics education.

Although it does not seem possible to speak here of a universal consensus, the idea that mathematicians should see the mathematical education of primary school teachers as part of their responsibilities clearly appears as an emerging trend. This is for instance the standpoint taken recently in the USA by the American Mathematical Society — basically a research society! — in its 'National Policy Statement' (AMS, 1994). Involvement of mathematics departments in primary teachers education is also one of the recommendations found in a report from the Mathematical Association of America on the mathematical preparation of teachers, 'A Call for Change' (see Leitzel, 1991, p. 11): "The mathematical experiences recommended for teachers at the K–4 level require that mathematics departments offer courses specifically designed for this audience."

In order to succinctly suggest general objectives that can be attached to mathematics courses for schoolteachers education, I would like to state some standards proposed in the report 'A Call for Change' (Leitzel, 1991, p. 1): primary and secondary school teachers should come to view mathematics as a system of interrelated principles; to communicate mathematics accurately, both orally and in writing; to understand the elements of mathematical modelling; to understand and use calculators and computers appropriately in the teaching and learning of mathematics; and to appreciate the development of mathematics both historically and culturally.

The same report also proposes a set of standards pertaining to particular mathematical topics and related either to primary or secondary school teaching. These proposals are presented not as descriptions of specific courses, but rather as indications of content areas in very general terms. The philosophy can probably be best perceived through the following comment — which reminds one of Pólya's statement quoted above: "The specific topics covered are not as important as *how* those topics are taught." (Leitzel, 1991, p. 27)

Putting in practice recommendations such as those appearing in 'A Call for Change' may require a substantial change of mentality among many mathematicians. The design of a specific disciplinary content using as framework the needs of future teachers is far from being part of the general tradition of departments of mathematics, especially in the case of primary school teacher education. But a number of examples can be exhibited of successful accomplishments of such objectives, some having even been going on for many years.

2.5 Contexts for the implication of mathematicians in education

I would like to conclude this part of the paper by commenting on possible frameworks through which mathematicians can become involved both in school mathematics and teacher education issues. I will distinguish three different levels at which such an implication can happen: the level of the mathematician as an individual; the level of an academic unit like a mathematics department; and the level of a professional body such as a national mathematical society.

There are numerous instances of mathematicians being involved in education as a result of a personal initiative. In some cases, this educational interest will be only occasional, while in others it will be much more sustained. As indicated above, one can even easily think of quite a few famous mathematicians in this connection. In addition to Felix Klein and other mathematicians mentioned earlier, a famous example is Hans Freudenthal who, after a long and substantial career as a research mathematician, became a leader in mathematics education as a research field. He had a long-lasting influence through such accomplishments as the founding of a research institute in his country, the promotion of ICMI activities during his . presidency (among which were the inception of the quadrennial International Congresses on Mathematical Education), or the creation of the research journal *Educational Studies in Mathematics*.

But such individual initiatives, however prestigious their proponents, are always at risk of having limited influence if they are not fully supported by the 'system'. A crucial aspect is thus the recognition and encouragement by the academic unit to which the individual professionally belongs (typically, in my discussion, a mathematics department). This includes, among other aspects, career issues such as promotions, tenure, etc. Sometimes the educational activities of mathematicians will even be officially presented as part of the agenda of the academic unit. A number of examples of varied depth could be given here and I will briefly present a rather unusual 'success story', the Community Teaching Fellowship Program of the University of California.

The CTF programme is an original example of a possible context for mathematicians to get seriously involved in matters of education. As the CTF programme started more than three decades ago and as it covers the eight campuses of the University of California, there have been variations in its functioning. But the basic idea at its origin was to involve mathematics graduate students (most of them in the Ph.D. programme, but also some in the M.A. programme) in providing mathematics enrichment to primary school children in minority and poverty neighbourhoods — this is a way for these students to receive financial support from the university, instead of through the regular teaching assistantship. Departing from usual remedial activities, the CTF programme concentrates on developing children's potential and building their confidence by the introduction of higher-level mathematics topics in simple and appealing ways. A graduate student participating in the programme goes to a local classroom three to five times a week to present activities in which children are encouraged to 'discover' mathematics. A premise of the programme is that the ability to use a discovery teaching method requires an ease and familiarity with mathematics that comes only through years of study, so that the average primary school teacher should not be expected to have the mathematical background necessary to teach by discovery the mathematics covered in a typical CTF class.

The CTF programme appears rather successful from various perspectives: schoolchildren seem to benefit from it, according to testimony from advisors, teachers and parents as well as attitude questionnaires; graduate students report a change in their vision of the teaching of mathematics — in a few cases, the CTF fellowship was even the beginning of a long-term career interest in mathematics education within a department of mathematics; and finally there is an impact on the mathematics departments, both from the point of view of the perception of the importance of educational matters and with respect to new collaborations established on some campuses between the department of mathematics and the school of education.

A third possible context for the involvement of mathematicians in educational issues is the general professional environment surrounding their work, in particular in connection with mathematical associations to which they may adhere. It is a most encouraging sign that in many countries — this is the case at least in the two countries I am most familiar with, Canada and the USA — professional societies of mathematicians clearly identify education as part of their responsibilities, including the education of teachers of all levels. For instance, the following can be found in a recent policy statement of the American Mathematical Society (see AMS, 1994): "The AMS will support increased participation of mathematicians in programmes for the professional development of teachers of mathematics." Numerous education sessions are now regularly organised at the annual meetings of societies of mathematicians, something almost unheard of a few decades ago. The strong opinion expressed on these matters by such a distinguished mathematician as Thurston (1990) is definitely influential in fostering evolution of mentalities.

3. MATHEMATICAL TOPICS FOR TEACHERS

In order to help prospective teachers develop the necessary deep conceptual understanding of the school mathematics content which falls under their responsibility, they should encounter in their university courses mathematical topics linked to the school curricula and presenting a good potential for instilling strong insights. I will now briefly sketch a few examples which I have successfully used with my primary and secondary school student teachers. I make no claim for high originality here, as most of these problems will probably be already familiar to many of the readers. The point I wish to bring out is the richness of the mathematical phenomena which are accessible in this context: working with prospective teachers can be fully gratifying, even from the strict point of view of the mathematical problems to be dealt with. I also want to show how some themes are appropriate for both the primary and secondary school teachers, as they can be subjected to an increasing depth of treatment. While many of the following examples could be adapted so to be presented to pupils of the appropriate level, the target audience I have in mind here is the teachers themselves.

3.1 Dilemmas of decimal developments

A **decimal fraction** is a fraction equivalent to one whose denominator is a power of 10. It thus corresponds to a terminating decimal development, i.e., an expansion with a finite number of non-zero digits following the decimal mark. It is natural to ask for a characterization of the decimal fractions. And the answer is remarkably simple: a fraction (in its lowest terms) is decimal if and only if the sole prime factors of its denominator are 2 and 5. Moreover the length of the expansion can be easily deduced from the prime factorisation of this denominator. The proof of these basic facts involves only the notion of prime factorisation and is readily accessible to primary school teachers.

If a fraction is not decimal, then its decimal expansion is non-terminating, but periodic: a certain sequence of digits is bound to repeat indefinitely. The expansion is then made of two parts: the **period**, comprising the block of digits which reappear *ad infinitum*, eventually preceded by a non-repeating part, the **pre-period**. A problem suitable for secondary school teachers is the following: given a non-decimal fraction a/b (in lowest terms), can we predict, by simple inspection of the numerator a and denominator b, the lengths of the pre-period and of the period of its decimal expansion? The answer to this question is again a simple application of elementary number theory and can be easily found in various textbooks (see for instance Rosen, 1988, Chapter 10, where the discussion is made for expansions in any base).

There are many ways of pursuing this study of decimal expansions. I indicate one, which brings into the picture more advanced results from the theory of numbers. It is striking that the periods of $1/7$ and $1/17$ are optimal, in the sense that all possible remainders are used before repetition — the former has period of length 6 and the latter, 16; such is not the case for $1/3$, $1/11$ or $1/13$, which have 'short' periods. How can this phenomenon be explained? It is not difficult to show (see Rademacher and Toeplitz, 1957, Chapter 23) that in general the length of the period of a/b is a divisor of $\varphi(b)$, the Euler phi-function. The period of $1/p$ is thus optimal when it is equal to $p-1$, which happens only in the exceptional cases where 10 is a *primitive root modulo p*. For instance, it can be checked that the only primes below 100 generating optimal periods are 7, 17, 19, 23, 29, 47, 59, 61 and 97. Very little is known in general about such primes (see Hardy and Wright, 1979, Section 9.6).

3.2 A night at the hotel

The following problem is nearly a classic in the mathematics literature for primary school student teachers — see among others the version in Bell *et al.* (1976), p. 620.

> At Long Hotel, there are n rooms all located along a very long corridor and numbered consecutively from 1 to n. One night, after dinner, the n guests play the following game. The first guest runs down the corridor and opens all the doors. Then the second guest runs down the corridor and closes every second door beginning with door 2. Afterwards, the third guest changes the position of every third door beginning with door 3 (that is, the guest opens the doors that are closed and closes those that are open). In a

similar way, the fourth guest changes the position of doors 4, 8, 12, ... This process continues until the nth guest runs down to the end of the corridor to change the position of door n. Which doors are left open and which ones are left closed at the end of the game?

Although complex at first glance, this problem obeys a remarkably simple rule: *all doors end up closed except those whose numbers are perfect squares.* This stems from the fact that at the end of the process, door d will be open or closed depending only on the parity of the number of divisors of d. The fact that the only numbers with an odd number of divisors are the perfect squares can readily be justified by primary school student teachers. This problem also provides a nice setting for exploring notions playing an important role in primary school teaching, such as (common) divisor, multiple, gcd or lcm.

Based on the 'what-if-not?' strategy, the following variant of the preceding problem was proposed by Cassidy and Hodgson (1993) for secondary school student teachers.

The rooms of Circle Hotel are built around a circular courtyard and are numbered consecutively from 1 to n. One night, after dinner, the n guests want to play the same game as at Long Hotel. But since the hotel has the form of a circle, each of the guests could go endlessly round the corridor. So it is agreed that a guest should stop as soon as he or she changes the position of door n — which is bound to happen for every guest. Which doors are left open and which ones are left closed at the end of the game?

This problem is of the sort that responds to an inductive approach. Experimenting with a few concrete cases leads to the following observation, which is easily proved using basic notions of modular arithmetic: in a hotel with n rooms, if guest k $(k \neq n)$ changes the position of a door, then so does guest $n - k$. The problem thus boils down to the single action of guest n, when n is odd, and to the combined action of guests n and $n/2$, when n is even. This directly leads to the conclusion that *for any n, all the doors are left closed except a single one; and this exceptional door is door n when n is odd, and door $n/2$ when n is even.*

The Circle Hotel problem can be extended in various directions. For instance, from the point of view of a given guest, one can ask how many times this guest goes round the corridor before stopping, and how many doors have been touched at the end of this action. And from the point of view of a given door, one can look for a characterization of the guests changing the position of a certain door during the process, and for the number of such guests. These questions directly relate to notions such as gcd and lcm. It can thus be shown, using Bachet-Bézout's relation expressing the gcd of two integers as a linear combination of these, that in a circular hotel with n rooms, guest k touches door d exactly when gcd (k,n) is a divisor of d.

3.3 *Kaleidoscopic visions*

My final example is from geometry and concerns the images that can be observed in a kaleidoscope. Ever since its invention by the Scottish physicist Sir

David Brewster in the early 19th century, the kaleidoscope has fascinated people of all ages through the richness and beauty of the pictures created by the interplay of mirrors. Using such an 'attention-catcher', teachers can bring their pupils into discovery activities about mathematical topics closely connected to central themes of the school geometry curriculum.

The understanding of the mathematical principles underlying the kaleidoscope is a challenge fully appropriate for primary school student teachers. The mastery of such a mathematical 'micro-theory' can have a positive impact on their perception of mathematics and their personal relation to it. The study of the kaleidoscopic phenomenon should go through various phases before a formal description is obtained: first looking through genuine kaleidoscopes — I always bring my small collection of kaleidoscopes into the classroom when I work this topic; then manipulation of real mirrors in a free setting (instead of fixed mirrors in a tube), using as a motif small objects or a figure drawn on a sheet of paper; and finally paper-and-pencil activities involving abstractly defined transformations, eventually performed with ruler and compass. (A further step could be the implementation of the kaleidoscopic phenomenon on the computer.) A guided exploration of elementary 'kaleidoscope geometry' can be found in Hodgson (1987).

With secondary school student teachers, the rosaces (or rose-patterns) produced by kaleidoscopes could be studied from the point of view of group theory, namely with the help of symmetry groups. A further extension would be to modify the type of mirrors 'inside' the kaleidoscope — some of the resulting kaleidoscopes have no real physical counterparts. For instance (standard) axial reflection could be replaced by central reflection (i.e., reflection in a point instead of a straight line). Or one could consider kaleidoscopes based on the geometric transformation of circular inversion. The resulting patterns can be interpreted as a new type of tessellation enjoying remarkable geometrical properties (Hodgson and Graf, 2000). Here again the computer is an exceptional tool in exploring these fictitious kaleidoscopes (see Graf and Hodgson (1998) for a discussion of such pedagogical environments).

4. CONCLUSION

While I do believe that mathematicians have a crucial contribution to bring to the mathematics education of both primary and secondary school teachers, I do not pretend that this could or should be done to the exclusion of their colleagues from schools of education. The contribution of mathematicians is different from that of mathematics educators, each having its specificity, its own aims. My point in this paper was to stress that the presence of mathematicians in the complex process of teacher preparation is not only desirable, but in fact essential.

An interesting challenge for mathematicians involved in teacher education is to identify topics and problems that can or should be worked with their students. One possible approach is to concentrate on the idea of seeing *elementary mathematics from an advanced standpoint* which I discussed in the first part of the paper: it can be truly profitable for students teachers to review from an adult perspective basic notions of mathematics they have learned in school, trying to gain a global and unifying understanding and to develop new insights. I have argued above that this constitutes the crux of the mathematical education of teachers. Another approach is to identify topics from more advanced mathematics which lend themselves to an elementary discussion; to borrow the title from a remarkable book by Rademacher (1983), one now looks at *higher mathematics from an elementary point of view*. The point here is not to trivialise mathematics, but to convey, when possible, the core ideas of a topic in elementary terms, avoiding the use of non-essential sophisticated tools. The above examples illustrate, I hope, both these approaches.

The implication of mathematicians in teacher education raises many difficult issues. One is the professional recognition of their educational work by universities or by their mathematical colleagues, as I discussed above. Another one is the reception of their efforts by professional mathematics educators. Still another is the acquisition by active mathematicians of an expertise in matters pertaining to teacher education, as they have had no formal training in mathematics education during their own graduate studies.

A major problem that arises is thus the education of mathematicians serving as teachers' educators. It is recognised that one of the main factors influencing prospective teachers is the way they are taught themselves: *what* they learn is fundamentally connected with *how* they learn it. Hence college and university mathematicians must use with student teachers the same methods these persons will be expected to use in their own classrooms: more student interaction, less lecturing and memorisation, open-ended problem-solving, etc. This might require substantial changes in the teaching habits of many faculty members.

For the involvement of the mathematics community in educational issues to be fully productive, there is a great need for a better collaboration between mathematicians in departments of mathematics, mathematics educators in schools of education and experienced mathematics teachers in the schools. It is important to tear down the 'iron curtains' which exist in many countries between the various educational levels, and to invigorate the communication channels between them. For instance the question of the 'dialogue' of mathematics education researchers "with other scientific communities, in particular the mathematics research community" (Sierpinska and Kilpatrick, 1998, p. x) was one of the issues raised at the outset of the recent ICMI Study on research in mathematics education. While it is important for educators to keep informed on the evolution of contemporary mathematics, there is also an urgent need for mathematicians to become better aware of the researches of their colleagues in education and of their potential impact both on their own teaching and on school teaching.

ACKNOWLEDGEMENTS

I am grateful to the following colleagues for the support they provided me in the preparation of this paper: Claude Gaulin, Frédéric Gourdeau, Miguel de Guzmán, Leon Henkin, Jean-Pierre Kahane, Jeremy Kilpatrick, Mogens Niss, Lynn A. Steen.

REFERENCES

AMS (1994). National policy statement 1994–1995. *Notices of the American Mathematical Society*, 41, 435–441.

Bell, M.S., Fuson, K.C. and Lesh, R.A. (1976). *Algebraic and Arithmetic Structures: A Concrete Approach for Elementary School Teachers*. New York: The Free Press.

Borel, É. (1914). L'adaptation de l'enseignement secondaire aux progrès de la science. *L'Enseignement Mathématique*, 16, 198–210.

Brown, C.A. and Borko, H. (1992). Becoming a mathematics teacher. In D.A. Grouws (Ed.), *Handbook of Research on Mathematics Teaching and Learning*, pp. 209–239. New York: Macmillan.

Cassidy, C. and Hodgson, B.R. (1982). Résolution élémentaire de problèmes. *For the Learning of Mathematics*, 2(3), 24–28.

Cassidy, C. and Hodgson, B.R. (1993). Because a door has to be open or closed: an intriguing problem solved by some inductive exploration. In S.I. Brown and M.I. Walter (Eds.), *Problem Posing: Reflections and Applications*, pp. 222–228. Hillsdale: Lawrence Erlbaum Associates. (Reprinted from *Mathematics Teacher*, 75 (1982) 155–158.)

Fennema, E. and Franke, M.L. (1992). Teachers' knowledge and its impact. In: D.A. Grouws (Ed.), *Handbook of Research on Mathematics Teaching and Learning*, pp. 147–164. New York: Macmillan.

Furinghetti, F. (2000). The history of mathematics as a coupling link between secondary and university teaching. *International Journal of Mathematical Education in Science and Technology*, 31, 43–51.

Gauss, C.F. (1802). Excerpt from a letter to Wilhelm Olbers. Quoted in M. Kline (1977), *Why the Professor Can't Teach*, p. 87. New York: St. Martin's Press.

Graf, K.-D. and Hodgson, B.R. (1998). The computer as a context for new possible geometrical activities. In C. Mammana and V. Villani (Eds.), *Perspectives on the Teaching of Geometry for the 21st Century: An ICMI Study*, pp. 144–158. Dordrecht: Kluwer Academic Publishers. (New ICMI Studies Series, Volume 5.)

Grossman, P.L., Wilson, S.M. and Shulman, L.S. (1989). Teachers of substance: subject matter knowledge for teaching. In M.C. Reynolds (Ed.), *Knowledge Base for the Beginning Teacher*, pp. 23–36. Oxford: Pergamon.

Guzmán, M. de (1993). El papel del matemático frente a los problemas de la educación matemática. *X Semana de Metodología de las Matemáticas*, pp. 9–20. Madrid: Facultad de Matemáticas, Universidad Complutense de Madrid.

Hardy, G.H. (1967). *A Mathematician's Apology*. Cambridge: Cambridge University Press.

Hardy, G.H. and Wright, E.M. (1979). *An Introduction to the Theory of Numbers*. (5th edition) Oxford: Oxford University Press.

Hauchecorne, B. and Suratteau, D. (1996). *Des mathématiciens de A à Z*. Paris: Ellipses.

Hodgson, B.R. (1987). La géométrie du kaléidoscope. *Bulletin de l'Association mathématique du Québec*, 27(2), 12–24. (Reprinted in (1988) *Plot (Supplément: Symétrie — dossier pédagogique)* 42, 25–34.)

Hodgson, B.R. and Graf, K.-D. (2000). Visions kaléidoscopiques. In R. Pallascio et G. Labelle (Eds.), *Mathématiques d'hier et d'aujourd'hui*, pp. 130–145. Montréal: Modulo.

Howson, A.G. (1984). Seventy-five years of the International Commission on Mathematical Instruction. *Educational Studies in Mathematics*, 15, 75–93.

Kahane, J.-P. (1988). La grande figure de Georges Pólya. In A. Hirst and K. Hirst (Eds.), *Proceedings of the Sixth International Congress on Mathematical Education*, pp. 79–97. Budapest: János Bolyai Mathematical Society.

Kahane, J.-P. (1997). Préface. In F. Klein, *Leçons sur certaines questions de Géométrie Élémentaire*, pp. i–iii. Paris: Diderot.

Kilpatrick, J. (1987). Is teaching teachable? George Pólya's views on the training of mathematics teachers. In F.R. Curcio (Ed.), *Teaching and Learning: A Problem-solving Focus*, pp. 85–97. Reston, VA: National Council of Teachers of Mathematics.

Klein, F. (1897). *Famous Problems of Elementary Geometry*. Boston and London: Ginn.

Klein, F. (1932). *Elementary Mathematics from an Advanced Standpoint: Arithmetic, Algebra, Analysis*. New York: Macmillan.

Klein, F. (1939). *Elementary Mathematics from an Advanced Standpoint: Geometry.* New York: Macmillan.

Lehto, O. (1998). *Mathematics without Borders: A History of the International Mathematical Union.* Berlin: Springer-Verlag.

Leitzel, J.R.C. (Ed.), (1991). *A Call for Change: Recommendations for the Mathematical Preparation of Teachers of Mathematics.* Washington: Mathematical Association of America, Committee on the Mathematical Education of Teachers (COMET).

Ma, L. (1999). *Knowing and Teaching Elementary Mathematics: Teachers' Understanding of Fundamental Mathematics in China and in the United States.* Mahwah: Lawrence Erlbaum Associates.

Moise, E.E. (1984). Mathematics, computation, and psychic intelligence. In V.P. Hansen and M.J. Zweng (Eds.), *Computers in Mathematics Education* (1984 Yearbook), pp. 35–42. Reston, VA: National Council of Teachers of Mathematics.

Morris, R. (Ed.), (1984). *Studies in Mathematics Education.* Volume 3: *The Mathematical Education of Primary School Teachers.* Volume 4: *The Mathematical Education of Secondary School Teachers.* Paris: UNESCO.

NCTM (1991). *Professional Standards for Teaching Mathematics.* Reston, VA: National Council of Teachers of Mathematics.

Pólya, G. (1957). *How to Solve It: A New Aspect of Mathematical Method.* (2nd edition) Princeton: Princeton University Press.

Pólya, G. (1959). Ten commandments for teachers. Reprinted in: G.-C. Rota (Ed.), *George Pólya: Collected Papers*, (Volume 4, pp. 525–533). Cambridge: MIT Press.

Pólya, G. (1981). *Mathematical Discovery: On Understanding, Learning and Teaching Problem Solving.* (Combined edition) New York: Wiley.

Rademacher, H. (1983). *Higher Mathematics from an Elementary Point of View.* Boston: Birkhäuser.

Rademacher, H. and Toeplitz, O. (1957). *The Enjoyment of Mathematics: Selections from Mathematics for the Amateur.* Princeton: Princeton University Press.

Rosen, K.H. (1988). *Elementary Number Theory and Its Applications.* (2nd edition) Reading: Addison-Wesley.

Sierpinska, A. and Kilpatrick, J. (1998). *Mathematics Education as a Research Domain: A Search for Identity: An ICMI Study* (2 volumes). Dordrecht: Kluwer Academic Publishers. (New ICMI Studies Series, Volume 4.)

Smith, D.E. (1905). Réformes à accomplir dans l'enseignement des mathématiques — Opinion. *L'Enseignement Mathématique*, 7, 469–471.

Steen, L.A. (1990). Pattern. In L.A. Steen (Ed.), *On the Shoulders of Giants: New Approaches to Numeracy*, pp. 1–10. Washington: National Academy Press.

Swafford, J.O. (1995). Teacher preparation. In I.M. Carl (Ed.), *Prospects for School Mathematics*, pp. 157–174. Reston, VA: National Council of Teachers of Mathematics.

Thurston, W.P. (1990). Mathematical education. *Notices of the American Mathematical Society*, 37, 844–850.

Wittgenstein, L. (1978). *Remarks on the Foundations of Mathematics.* (3rd edition) Oxford: Basil Blackwell.

Wittmann, E.C. (2001). The alpha and omega of teacher education: Organizing mathematical activities, this volume, pp. 539-551.

Bernard R. Hodgson
Université Laval, Québec Canada
bhodgson@mat.ulaval.ca

MARC LEGRAND

ON THE TRAINING OF FRENCH PROSPECTIVE
UNIVERSITY TEACHERS

1. PEDAGOGICAL TRAINING OF UNIVERSITY TEACHERS

The decision in 1989 to create Centres d'Initiation à l'Enseignement Supérieur (CIES) in France has certainly begun a small revolution. It has meant the end of the acceptance of the institutional myth that specialists in a given field are automatically qualified to teach their discipline at university level. At an institutional level, they have been a success, since the myth above, and the aggressiveness of a lot of academics towards this training, have almost disappeared. Furthermore, most of the graduates of the scheme have had no difficulty in finding a university position after completing their PhDs. University lecturers are still recruited mainly on the basis of their research but, when two candidates are otherwise equally matched, to have been involved in a CIES programme turns out to be an advantage.

In France, doctoral studies begin with a year of essentially academic courses. During this year, students prepare for what is called the Diplome d'études approfondies (the DEA). At the end of this year, students can prepare for the PhD, ostensibly a three year programme. Some of them are selected, on the basis of their academic ability and their research potential, to become *moniteurs*. Moniteurs receive a grant to work on their PhD and simultaneously are expected both to teach at university level for 64 hours a year and also to undertake specific training for tertiary teaching in a CIES.

2. THE RISK THAT SUCH TRAINING PLAYS ONLY A SYMBOLIC ROLE

Very quickly it has become clear that if the decision to set up moniteur training was not considered globally, if it was not reconsidered in terms of issues for the university, of issues for knowledge, of issues for the democratic organization of society, then the setting up of the training could be counter-productive. It could reinforce the negative image that the majority of academics have about anything aiming to structure and support pedagogical views.

Why should this be the case? In their first period of instruction moniteurs are given general information on the tertiary system. The most common feeling shared by the students is that listening to formal descriptions they cannot relate to, is a waste of time. They feel that it would be more efficient for them to stay in their research laboratory and for the material to be provided as a hand out.

Derek Holton (Ed.), The Teaching and Learning of Mathematics at University Level: An ICMI Study, 519—528.
© 2001 *Kluwer Academic Publishers. Printed in the Netherlands.*

In subsequent training sessions, researchers in didactics and in the science of education give specialized talks. Most of the time, these talks are unrelated to each other and unrelated to the problems of the students. Indeed, either these new young teachers have very good rapport with their students and hence are convinced that everything is running smoothly or they are faced with serious difficulties. In both cases, the moniteurs are not ready to engage in a theoretical study of specific points that they feel are unimportant. It turns out that very often the moniteurs disdain these training sessions. At best they try to grasp a few pedagogical tricks that they very quickly discover to be of limited use.

Other sessions deal with communication techniques intended to introduce the students to the social and theatrical sides of teaching. When a good group dynamic is taking place during these sessions, students like them very much. This is mainly because they have just discovered, through being in charge of tutorials, the difficulties of leadership in front of a group of students, the difficulties of communicating with them and the difficulties of allowing fruitful discussions to take place between students. The issue raised by the success of these sessions, when separated from the rest of the moniteur's training, is that they introduce a kind of epistemological misinterpretation. Whereas the cross-disciplinary sessions deal with general difficulties about communicating within a group, they are not based on the specific difficulty of didactical communication - a communication aimed at teaching specific knowledge.

The moniteur may then have the feeling that these communication techniques are of universal value. He is led to think that mastering a few professional tricks will exempt him from carrying out an epistemological analysis of what has to be taught. For instance, constructivist teaching strategies, whose aim is to help students construct meaningful mathematical knowledge, generally rely on situations that students are expected to find problematic and where solutions cannot be easily obtained. Both the choice of adequate problems and their management are difficult. (See Artigue, this volume, pp. 207-220 and Legrand, pp. 127-136, for some illustrative examples.) It is not enough to set up a rich and open problem and to organise a lively discussion between students and teacher to ensure that students will develop the targeted mathematical activity. Questions may be too simple or too difficult, too abruptly set up or, on the contrary, discussed too much by the teacher. Without a deep analysis of the mathematical knowledge at stake and of the way it can be accessed by students, the teacher can radically misinterpret both their positive and negative reactions.

Let us give an elementary illustration. We have often asked different groups of first year students, the same simple question in the following two very similar forms:

Form A: *What is the area of the following parallelogram whose sides are 3m and 4m long?*

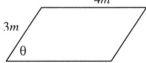

In that case, a great majority of the students give the correct answer:12.sin θ. These students are showing that they know something, but exactly what?

Form B: The same question and the same drawing except that θ is not mentioned. In that case, more than 50% of first year students regularly answer that the area is 12 m^2!

Some of them even, give a spontaneous proof for this, drawing the figure below and treating 3m as the width of the rectangle.

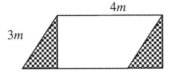

This simple experience provides evidence of the complexity of the situation. Answers to Form A prove that students are able to correctly solve the problem when the pertinent variable θ is labelled. Answers to Form B prove that students are not able to find the pertinent variables for this type of problem by themselves. This cannot be seen if they are only faced with Form A. Passing from B to A, the real difficulty has been suppressed.

A superficial analysis of the situation would not lead one to suspect that such similar questions (with just a detail varying in the associated representation) are in fact very different in terms of the mathematical competencies needed to solve them.

In our opinion, the most important and the most difficult thing to teach is the ability to identify the pertinent variables in a mathematical problem. Hence we believe that it is essential that future university teachers overcome the didactically naïve idea that two such questions are similar. That is why we give great importance in the training of the moniteurs, to the development of competencies in the epistemological analysis of the knowledge at stake in teaching situations.

Finally, because of the increasing unemployment of highly qualified people, a session on "how to find a job" has been introduced. During this session, it is confirmed for the moniteur that teaching criteria will not play a major role in selection.

In short, whenever these sessions are disconnected, whenever they are not considered as part of the lecturer's profession, they tend to reinforce the negative image of pedagogical and didactical reflection. This makes the moniteurs feel they are wasting their time and that specialists in pedagogy and didactics spend their time quibbling about marginal details unrelated to real teaching issues. Under these conditions, it is unfortunately likely that the main knowledge the moniteurs will gain from experienced teachers is standard communication techniques and some useful tricks.

Therefore this kind of introduction to pedagogy and didactics does not convince them of the necessity to study the issue more thoroughly. It doesn't allow them to see that finding the proper way to teach a given subject to a given population includes previously defining the outcome of the teaching as a scientific project, to analyse different epistemological ways to realise that project, and finally to compare this epistemological analysis with a social and ethical range of questioning. In other words, to ask what kind of relationship with knowledge, with the world and society will be induced through the way I teach such and such a topic?

A last disadvantage of this lack of global thinking is that a moniteur, prepared this way, doesn't understand that without referring to a theory, without a precise vocabulary, it will be very difficult for him to discuss his work seriously with his counterparts. He will not be able to engage in fruitful exchanges on pedagogical questions with his colleagues, whether or not they have different points of view on knowledge and the world.

3. THE MAIN POINTS AT STAKE FOR A GLOBAL ORGANISATION OF THE TRAINING OF MONITEURS

When considering the moniteurs' situation, we believe that that they are in fact in a good position to look at university teaching in a new way.

First they are just coming out of the system. So, presumably, they are aware of its advantages and shortcomings. In fact, we need to moderate this point of view. They are half aware that something must be changed. On the one hand, as ex-students, they like this system. After all, it has acknowledged them as successful. They believe that they have a good understanding of what they have studied for their degrees. Most of them have not realised that to go through the system successfully it is more important to be very skilled in reproducing mathematical techniques in standard situations than to be creative or to have a deep understanding of the mathematics used.

Second, as regards research, the moniteurs' integration into their departments does not usually raise any difficulty but moniteurs are often destabilised by the behaviour of their elders (lecturers and professors) when it comes to teaching matters. Hence by contrast, they are led to feel the need to change teaching practices. Indeed, the moniteurs are sometimes shocked by the way their superiors consider teaching, by their lack of interest if not disdain. Sometimes the moniteurs take no part in decisions regarding teaching. Their 'job' is reduced to giving, in seminars, solutions of exercises they have not chosen. They are not asked to give their opinion on the progression of the course or on the examination papers. They are not part of a pedagogical team. Sometimes, on the contrary, they may be considered as "the maid of all work" and responsible for everything. In that case, they have to conceive a course on their own. They are in charge of the assessments and so on. They may, at first, feel flattered by this trust put in them. But soon they realise that all of this is not considered to be very important by their superiors and so they and the students are in fact being abused.

Because of all this, moniteurs are suddenly faced with serious teaching problems. They often suffer from the injustices produced by the casualness of lecturers and professors. That is why moniteurs may be encouraged to think and act differently from their elder colleagues.

Consequently, in such a situation of crisis, if what is proposed concerning their teaching training is inadequate (just the description of the university system), is not closely linked to their daily difficulties (the specialists' talks on theoretical matters), is too technical (the professionals' tricks), is too general (the communication techniques), and is too much aimed at getting a position at university, it is very likely that the main thing the moniteurs will learn from these sessions is how to behave so that they can integrate smoothly into the system once they are recruited. That is to say, after having been scandalised by their elder colleagues, after having criticised them, the moniteurs will more or less consciously subject themselves to the rules that allow them to take their place in the system. Then, after a few years, they will reproduce what they violently criticised. This attitude can be explained by the fact that, during their formative years, the moniteurs did not acquire any knowledge that could help them to anticipate what would cause them to change their mind. They will soon get used to and collaborate with the mediocrity and the unfairness they previously rejected.

The risk is enhanced because, despite their political and ethical thinking about social injustice occurring first in schools then in universities, most of the moniteurs don't have any idea about the powerful humanising and socialising effects of knowledge. They are not aware of how different methods of learning and ways of succeeding at school affect people and society.

For example, the incentive to discuss points of view as opposed to the refusal of any arguing, the elaboration of a rigorous intellectual way of thinking as opposed to the adoption of systematic intellectual practices, are fundamental to school and to university. These factors are strongly linked to the reinforcement or not of the democratic life of a country.

For the great majority of these student-teachers, knowledge is the syllabus, knowledge is what one must be able to recite. Moniteurs do not reflect on the nature of their discipline and its specific way to view the world.

These students are either mathematicians, or physicists, or biologists, or lawyers. Consequently they do not question the methods or approaches of their different professions. They do not suspect that dogmatism in some professional practices may arise from former dogmatic training. Finally, because during their studies, they met only one sort of didactics, they do not realise that some didactics can be better than others depending on the complexity of the knowledge involved. They do not have the faintest hint of how the understanding and construction of concepts by students can be improved by didactic means. For most of the moniteurs, there is only one way to learn and to teach - the way they have learnt and were taught. The only qualitative differences they make concern individuals: there are 'good' and 'bad' teachers, 'good' and 'bad' students, students who work, and students who do not work. Without willing it, all this thinking takes place in a deep misunderstanding. It may even happen that moniteurs end up by considering contemptuously the students they teach who, as slow learners, are not like them.

4. EXAMPLES FROM GRENOBLE'S CIES AND SOME ULTIMATE CHOICES

At the Grenoble CIES it took us several years to become aware that the unease in most of the CIES is due to the fact that the moniteurs were not able to resolve all the contradictions mentioned above. They felt as if they had no other choice than to close their eyes and avoid any change or, on the contrary, to open them wide to work for a deep and global transformation of the system. But, because of their position, this latter course is out of their reach.

If the answers provided by the CIES are superficial and do not take into account their deep questioning, the moniteurs will reject their training except when it can give them useful tricks.

Then the moniteurs work against any attempts made to raise the level of debate. They oppose the different disciplines. The 'strong' scientific subjects (maths, physics, etc.) do not want to consider problems where definite answers cannot be given; there is a strong conscious or unconscious difference between scientists and non-scientists.

In short, these young teachers/old students tend to systematically denigrate the training sessions, their organisation and their content. The generalisation of such behaviour means that the goals that the trainers had for them cannot be achieved. It also becomes more and more difficult to find people willing to give seminars to such an audience.

Having become gradually aware of the situation, we decided to tackle the basic problems at the very first session. We strongly felt the necessity for such action. As we were afraid to fail, we started by organising a one day session, 50 students at a time, on the theme 'the lecturers' job'. Then two years later, because of the importance of the questions that arose during this session, we extended the duration of this session to two days. This year we organised a residential two and a half day session with one hundred moniteurs. We chose this form because it allowed the moniteurs to get on more friendly terms, with meals and evenings giving the opportunity to engage in further debate.

From the start of this first session we invited the moniteurs to consider the following fundamental questions:

- What could be the nature and the function of our job and what role does it play in society?
- What are our epistemological responsibilities (nature of the knowledge taught)?
- What are our social responsibilities (relationships between, on the one hand truth socially constructed, truth imposed and, on the other hand, the possibility of democratic life)?
- We are researchers. In what ways does this compel us to bring something more to our students?
- Which ethical values can the lecturer hold?

Should not university teaching face the following epistemological obstacle: the more a model seems without any failure and mathematically perfect, the more it is by essence far from the reality of the world it aims at representing? Should not university teachers who are themselves researchers have the responsibility of addressing such issues, even if it requires them to break with the common culture, even if it looks difficult? These are the questions we discuss during this first phase of the training.

It seemed to us that the best way to work was the following one: to gather in one session all the first year moniteurs from all disciplines and alternate the work in small groups with plenary sessions:

- Short statements during the plenary sessions to expose personal convictions or experiences or to raise a specific problem.
- Two- or three-hour workshops running in parallel on a common theme with 15 moniteurs.
- Reports of the workshops and debates during the evening.

This session has continually been much more successful than we had expected and has matched the unexpressed wishes of many of the moniteurs. The moniteurs participate actively and are very satisfied with this session. One can say (with very few exceptions that can be managed) that all the negative behaviours described in the first pages of this paper have practically vanished in the training sessions at Grenoble.

This session for beginners is now coupled with two other sessions at the start of the second year.

Session 1: Pedagogy and general didactics for all disciplines

In this session, we introduce some elements from cognitive theories, as well as the notion of epistemological obstacle. We also introduce some didactic notions, with specific emphasis on the notion of 'didactic contract' (see Brousseau, 1997) and the distinction between the 'game of the teacher' and the 'game of the learner' in any didactic. In another words, we try to make the moniteurs discover that, when analyzing students' behaviour, they must not suppose that these students react as they could do in another situation. In lessons or tutorials, students play a role: the role they think is expected from students in the university game, and they also expect that teachers will play their specific role too. The notion of didactic contract models these respective expectations which, although they have a strong effect on the learning and teaching processes, for the most part remain unstated. The didactic contract induces specific behaviours that can be totally paradoxical. For instance, we make the moniteurs reflect on the following amazing example. In an elementary school, the teachers asked Paul who was 8 years old the following question: "You have ten pens in each of your pockets, how old are you?" Paul answered, "I am 20".

At first sight, one is tempted to say that Paul is a stupid boy, but one can also consider that Paul's behaviour, far from being stupid, is logical and results from the characteristics of the didactic contracts to which he has been exposed. Perhaps Paul is thinking: this teacher taught me the addition of numbers; in order to check that I have correctly learnt, he gives me a problem where addition is hidden behind a stupid problem of pens and age. I add the numbers and give the result as my age, of course it isn't my age but this isn't the point here.

Such an interpretation was corroborated by the following fact: the teacher reacted to Paul's answer by saying: "But Paul you know perfectly well that you are not 20." Paul replied: "It is your fault, you didn't give me the right numbers." Through this reply, Paul clearly expresses that the pertinence of what is said in class is not his business. Pertinence, in a form of didactic contract that we would qualify as a 'students' scientific irresponsibility', is the exclusive responsibility of the teacher.

In our opinion, it is very important to discuss with the moniteurs problems such as this one, or the parallelogram one, in order to help them to avoid a simplistic analysis of their students' responses or behaviour in terms of success-failure, truth–falsity, meaningfulness–stupidity, with scientific interest-without scientific interest, etc. All these binary classifications tend to deprive teachers of their responsibility by inducing them to definitively categorize students as good or bad.

We thus invite them to postulate the existence of didactic contracts that have been progressively established over the years and may constrain their own teaching. They have to try to identify the main characteristics of these didactic contracts when beginning a course, as these are not necessarily adapted to what they want to teach. Such a point of view restores the teachers' responsibilities. They have to act explicitly in order to change what they believe to be the inadequacies of the implicit didactic contract.

The notion of epistemological obstacle (Artigue, this volume, pp. 207-220) has to be introduced carefully with the moniteurs as it goes against the strong belief they have of the efficiency of good explanations. We have to make these young teachers (who, very often, haven't yet discovered that they don't understand very well what they feel they know perfectly), become aware of the following fact: frequently the most important thing to understand is also nearly impossible to teach directly in the form that we want it to be understood. When one faces an epistemological obstacle, explanations, 'mise en gardes' are not sufficient. All students have to come to grips with their own contradictions, have to experience the necessity of changing their mind. All of this, in order to be effective, requires time, errors and errors, and the opportunity of analyzing these.

In order to concretely address this kind of issue, we use paradoxical situations such as the 'jeans problem' (see Legrand, this volume, pp. 127-136), where a situation from daily life that looks very simple, can reveal the hidden complexity of vectorial sums and the interest of such complexity.

Session 2: Pedagogy and practice in didactics

Moniteurs in disciplines close to each other work together here. For example, when we can put together specialists in mathematics, physics, mechanics and computing for this session, it is possible, within three days, to organize a scenario such as the following:

Part 1 (one and a half days) Preparation of a course:
- the moniteurs, in groups of five, choose a topic which is difficult to teach (limits, infinity, forces, potential, proof, etc.);
- the moniteurs epistemologically analyse the theme;
- the moniteurs choose an approach (constructivist or not; although we invite them to take the advantage of the special conditions they are in to choose a constructivist approach) and organize a sequence of classes to be presented to the other moniteurs who will act as students.

Part 2 (one and a half days) Teaching and analysis:

- teach the sequences that were prepared in Part 1. The teaching lasts half an hour. The audience is the other moniteurs;
- analyse the teaching session at three levels:

> First level: As a student: "What have I understood? What have I not understood? What was boring? What was interesting? etc." As a teacher, "I had a good or a bad experience when succeeding to answer (or not) such or such a question."

> Second level: As a colleague, "I agree or not with such a point of view". Usually the moniteurs who prepared the lesson question the differences between what was planned and what happened, recall the discussions they had and the choices made during the preparations, and make it clear if they had chosen to hide or not hide such or such a difficulty.

> Third level: As a didactician "Which didactic mechanisms are at stake, what worked, what was not mastered."

It seems that the best we can offer to moniteurs is when such sessions take place with seriousness, kindness and humour, and personal attacks are not allowed. We meet them at the core of their own practice and we strive to accompany them as far as we can, in a utopian but reasonable way. On the one hand we do not stick to first order trivialities and on the other we do not give them the illusion of great achievements they could not realize by themselves. We give them the opportunity to live the fundamental experience "it is possible to teach and learn in another way". The distance from the teaching act given by the analysis of the lesson shows how 'this other way' is not a mere gadget. The analysis shows how didactic choices induce in the students different attitudes towards knowing. We think that teaching at university should be based on such ideas.

It is clear that during their formative years and even later on, the moniteurs will have to decide to carry on in this direction or not. But since they have touched on a truly different way of teaching, they will be able to make a real choice. They will have experienced the cost and the benefits of this other way of teaching, costs and benefits both for the students then engaged in constructing meaning, and for the lecturer who, while 'problematizing' his knowledge to share it with the students, rediscovers what he thought he knew well.

5. CONCLUSION

The main point I would like to emphasise here is that the success of the pedagogy and practice in the didactics sessions, the most important sessions to offer to prospective university teachers, is due to several closely related factors.

First, the multidisciplinarity of the moniteurs who come from different fields. It is not possible for specialists of the same discipline to show that they 'do not know'. It is not possible for them to make mistakes, to not understand what the other is saying. But, if these sessions are to be of any value, it is absolutely necessary to make mistakes, to be in a position where it is possible to say that we do not understand.

Second, a shared vocabulary, the shared knowledge of a theoretical framework which enables the moniteurs to move away from their own actions and their own behaviour and hence give them the possibility of talking about themselves and their own actions. This is why we set up a session on general didactics; this session must be followed by the practice session described above. Without the practice, the general didactics session remains empty talk. But without the general didactics session, the practice session runs the risk of taking place without the theoretical tools. The practice didactic session would then end up by expressing personal judgements and in personal criticism.

Third, the possibility given to the moniteurs to question the meaning and the effects of our teaching. To get used to the idea that teaching is more than delivering mathematical knowledge and that, through teaching, one can transmit more global values. One can choose, by the way one handles the teaching situation, to minimize the gap between 'talking' and 'doing'!

Once more, we can say that the first session in the first year, the session where we question the meaning of the profession, must be followed by a practice session. Otherwise, it will have no effect. At best it will be an elaboration of a list of 'good intentions'. But conversely, we noticed that, without a previous deep questioning, it turns out to be very difficult to propose common didactic actions to 'problematize' knowledge. Indeed, if the reasons (very costly reasons) of this 'problematization' have not been discussed previously, it becomes very quickly unreasonable to try to teach differently. From an academic point of view, it is not worth it!

For this training to be fruitful for both the moniteur and university teaching, it must be conceived globally in a coherent way. Our many years' experience has shown that, when the trainers' team is courageous enough to face the issues of the meaning of our profession, to face what is, in teaching, at stake for society, then the moniteurs greatly benefit from these sessions. My observation of moniteurs in tutorials has convinced me that a true change in practice can take place in the first years.

I'll be honest in saying that moniteurs who do wish it can offer their students ways of working akin to the researcher-teacher position. It seems to me that those who resist the 'pressure of the environment' and carry on teaching this way, improve their pedagogical skills and efficiency. In particular, their epistemological knowledge allows them to take the risks to open up the situations to force the students to be more scientifically responsible.

REFERENCES

Artigue, M. (2001). What can we learn from educational research at the university level? this volume, pp. 207-220.
Brousseau, G. (1997). *The Theory of Didactic Situations*. Dordrecht: Kluwer Academic Publishers.
Legrand, M. (2001). Scientific Debate in Mathematics Courses, this volume, pp. 127-136.

Marc Legrand
IREM de Grenoble, France
marc.legrand@ujf-grenoble.fr

JOHN MASON

PROFESSIONALISATION OF TEACHING IN HIGHER EDUCATION IN THE UNITED KINGDOM

1. INTRODUCTION

Professional development for staff in higher education has suddenly become a hot topic in the U.K. Universities not already possessing Learning and Teaching Support Groups are setting them up. These are running courses for new staff, but encouraging experienced lecturers to attend as well. They are finding a willing ear amongst those with little experience, and also among those with considerable experience who are dissatisfied in some way with their teaching (or with the students' learning), or who are teaching in novel ways and seeking approval and support for their experiments. Furthermore, there is a new institution in the UK which aims to become the professional body for higher education staff with the aim of establishing a culture of continuing professional development. The driving force is governmental: transparency and close inspection of all public expenditure. The effect could be far reaching, for with transparency comes close attention to details such as distinguishing between time spent on research, on teaching, on links with industry, and on administration, but also requiring evidence that teaching is effective, and raising questions about that effectiveness (e.g. why do so many students with good university entrance qualifications end up with a third class degree in mathematics?).

2. BACKGROUND

In 1996, the United Kingdom Government set up the National Committee of Enquiry into Higher Education, under the chairmanship of Sir Ron Dearing. The Committee was appointed in May 1996 with bipartisan support, to make recommendations on how the purposes, shape, structure, size and funding of higher education, including support for students, should develop to meet the needs of the United Kingdom over the next 20 years. They were asked to recognise that higher education embraces teaching, learning, scholarship and research. In their report (Dearing 1997a) which is wide ranging, they make 9 points concerning "A vision for the 21st century" (Dearing 1997b), of which the following seem the most pertinent:

2   Our title, `Higher Education in the Learning Society', reflects the vision that informs this report. Over the next 20 years, the United Kingdom must create a society

Derek Holton (Ed.), The Teaching and Learning of Mathematics at University Level: An ICMI Study, 529—538.
© 2001 *Kluwer Academic Publishers. Printed in the Netherlands.*

life enriching and desirable in its own right. It is fundamental to the achievement of an improved quality of life in the UK.

That future will require higher education in the UK to:

encourage and enable all students – whether they demonstrate the highest intellectual potential or whether they have struggled to reach the threshold of higher education – to achieve beyond their expectations;

safeguard the rigour of its awards, ensuring that UK qualifications meet the needs of UK students and have standing throughout the world;

be at the leading edge of world practice in effective learning and teaching; undertake research that matches the best in the world, and make its benefits available to the nation;

ensure that its support for regional and local communities is at least comparable to that provided by higher education in competitor nations;

sustain a culture which demands disciplined thinking, encourages curiosity, challenges existing ideas and generates new ones;

be part of the conscience of a democratic society, founded on respect for the rights of the individual and the responsibilities of the individual to society as a whole;

be explicit and clear in how it goes about its business, be accountable to students and to society, and seek continuously to improve its own performance.

6   To achieve this, higher education will depend on:

professional, committed members of staff who are appropriately trained, respected and rewarded;

a diverse range of autonomous, well-managed institutions with a commitment to excellence in the achievement of their distinctive missions.

7 The higher education sector will comprise a community of free-standing institutions dedicated to the creation of a learning society and the pursuit of excellence in their diverse missions. It will include institutions of world renown and it must be a conscious objective of national policy that the U.K. should continue to have such institutions. Other institutions will see their role as supporting regional or local needs. Some will see themselves as essentially research oriented; others will be predominantly engaged in teaching. But all will be committed to scholarship and to excellence in the management of learning and teaching.

8 Higher education is fundamental to the social, economic and cultural health of the nation. It will contribute not only through the intellectual development of students and by equipping them for work, but also by adding to the world's store of knowledge and understanding, fostering culture for its own sake, and promoting the values that characterise higher education: respect for evidence; respect for individuals and their views; and the search for truth. Equally, part of its task will be to accept a duty of care for the wellbeing of our democratic civilisation, based on respect for the individual and respect by the individual for the conventions and laws which provide the basis of a civilised society.

9 There is growing interdependence between students, institutions, the economy, employers and the state. We believe that this bond needs to be more clearly recognised by each party, as a contract which makes clear what each contributes and what each gains.

A significant component of the compact Dearing identified is the professionalisation of teachers in higher education. To this end, *the Institute for Learning and Teaching* (ILT) was created in April 1999, based on the following three paragraphs taken from the full Dearing report:

Recommendation 13

We recommend that institutions of higher education begin immediately to develop or seek access to programmes for teacher training of their staff, if they do not have them, and that all institutions seek national accreditation of such programmes from the Institute for Learning and Teaching in Higher Education.

Recommendation 14

We recommend that the representative bodies, in consultation with the Funding Bodies, should immediately establish a professional Institute for Learning and Teaching in Higher Education. The functions of the Institute would be to accredit programmes of training for higher education teachers; to commission research and development in learning and teaching practices; and to stimulate innovation.

Recommendation 48

We recommend to institutions that, over the medium term, it should become the normal requirement that all new full-time academic staff with teaching responsibilities are required to achieve at least associate membership of the Institute for Learning and Teaching in Higher Education, for the successful completion of probation.

The principle aims of the ILT are to:
- enhance the status of teaching in higher education;
- maintain and improve the quality of learning and teaching in higher
- education;
- set standards of good professional practice that its members, and in
- due course all those with learning and teaching responsibilities,
- might follow.

The Institute is a membership organisation open to all those engaged in teaching and in the support of learning in higher education, who share their objective of improving the status of teaching and learning support as professional activities. It aims to be an innovative organisation that encourages diversity in approaches to learning and teaching, and is responsive to the needs of the whole sector. It intends to be shaped primarily by the concerns and experiences of the practitioners. This nationally initiated organisation is in addition to attempts by members of various university teaching-support staff to set up a similar organisation.

3. THE STAFF EDUCATIONAL DEVELOPMENT ASSOCIATION

Various fledgling groups with similar aims to the ILT made suggestions as to the nature of the ILT programme and membership qualifications. One of these was the Staff Educational Development Association (SEDA) which had already anticipated Dearing's recommendations and had begun to set up its own process of qualification. These included workshops developed from a wide-ranging programme developed by Graham Gibbs and colleagues in The Oxford Centre for Staff and Learning Development at the University of Oxford-Brookes. See their Web site (http://www.brookes.ac.uk/services/ocsd/home.html/) for books about 'lecturing to more students', 'assessing more students', 'supporting more students', and 'improving student learning' through various means, etc. SEDA has some 300 individual and institutional members, mostly in UK higher education. The SEDA programme consists of a range of workshops and conferences, electronic discussions, and the construction of a teaching portfolio.

> Teaching portfolios consist of a collection of evidence of teaching and related skills associated with the role of a teacher. They are now very widely used, in the UK and elsewhere, for the assessment of a teacher's competence in a formal programme of training in teaching and learning in higher education. Other main uses of the teaching portfolio are to provide evidence of teaching and related skills for tenure and/or promotion decisions and to provide the basis for continuing professional development. Debate is moving on from descriptions of how to develop and present a portfolio of evidence about one's teaching to how to use portfolios as part of wider educational development practice, involving such practices as mentoring and continuing professional development.

> (See SEDA Web site: http://www.mailbase.ac.uk/lists/seda/.)

The SEDA Web site has a particularly well-structured discussion structure, called Deliberations (see SEDA Web site), including a Deliberations Forum. There one can find time-limited but archived discussions on a wide range of issues including student feedback in the evaluation of teaching, group work, communications, information technology and learning, encouraging continuing professional development, can peer observation improve teaching and if so, how can this be done?, how do students learn study skills most effectively?, and the use of multiple choice questions. The SEDA site is hosted by the Educational Development Unit of London Guildhall University, and is one of a number of institutions running professional development courses for people teaching in higher education.

4. THE ILT PROGRAMME

The ILT decided to accept programmes with existing SEDA recognition for ILT accreditation, which is reasonable because they need to integrate themselves into the scene, and also because SEDA requirements were rather more rigorous than those decided upon by ILT. A balance had to be struck between overly demanding criteria for entry, and criteria which would encourage a significant number of teachers in higher education to join and participate.

To date, ILT energy has been directed toward attracting membership. Although at one time it was hoped that there would be financial recognition in salaries for members, this is not only overly ambitious, but redirects attention away from the principal concern of the institute, namely to foster and support professional development. Any programme of forced participation, such as takes place in some states in the US for school teachers, where registration as a licensed teacher depends on attendance at a stated number of professional development courses, naturally engenders a redirection of attention from professional development to form-filling hoops and obstacles. The ILT recognises that staff in higher education are not likely to respond well to compulsion, which is why it is trying to attract people to engage in professional development.

Candidates for membership who are already experienced teachers have to address the following substantive questions, and get two referees to comment on and confirm their responses. Each question is intended to have a notional 500 words maximum.

Teaching and the support of learning

Please indicate the range of teaching and learning support activities in which you are involved. Give examples of successful activities or techniques you use and comment on how you came to use them and why you think they are successful. Include proportion of time spent teaching/supporting learning, subjects and level taught, methods adopted and any special responsibilities you carry (such as Course, Programme or Subject Leadership or Head of a Learning Support Unit).

Contribution to the design and planning of learning activities.

Please identify the different ways in which you contribute to the design and planning of learning activities. These might include both involvement in the design or re-design of courses and programmes and identifying and planning different kinds of interaction with students in various contexts. Equally it might include indirect involvement through participation in validation panels, feedback to colleagues on team teaching or contribution to the creation of learning resource packs and computer-based or open learning materials

Assessment and giving feedback to students

Please indicate how you give feedback to students (e.g. in writing, orally, as part of the supervision of research students). Please also describe the types of assessment you use with students, both formal and informal, including the variety of methods used (e.g. essays, unseen time constrained exams, portfolios, live critiques etc.) and which approaches you use (e.g. tutor assessment, self assessment, computer-based assessment, group assessment). Also indicate how and why you choose the approaches and methods you use, in so far as this is your own decision, and to what extent, if any, you are involved in designing assessments. Describe how you try to ensure that the feedback you give to students helps them to improve and remedy deficiencies in performance

Developing effective learning environments and student learning support systems

Please comment on the range of ways in which you contribute to making the students' learning environment effective through such means as placement or project supervision, personal and academic tutoring, one-to-one advice, counselling, learning support in IT laboratories, laboratory work supervision, studio critiques, placement supervision, support of work-based learning, and information retrieval and management.

Reflective practice and personal development

Please include here a brief description of the means by which you evaluate your learning/teaching support activities, and how you build on what you learn about your working practices. You should also refer here to any activities you have undertaken to update yourself on aspects of teaching and learning, including staff development activities or conferences on learning and teaching. Also include participation in projects to develop learning methods (e.g. FDTL, UMI, TLTP, or in-house activities in your own university) and reading, research and/or publication specifically related to your teaching.

5. THE ROLE OF THE OPEN UNIVERSITY

The Open University saw that, even though many institutions were likely to set up courses for their own staff, not all would have access to expertise. So a Centre For Higher Education Practices (see CeHEP Web site: http://www2.open.ac.uk/) was set up within the Institute of Educational Technology at the Open University, and courses were written, designed to support the construction of a personal portfolio while engaging in personal and professional development. There are three courses: Teaching in Higher Education, Course Design in Higher Education and the two combined in one as Teaching and Course Design in Higher Education. They are generic, spanning all disciplines, but intensely practical. The first is aimed primarily at people beginning their career as teachers, either as graduate students or as newly appointed lecturers. The second is intended for people with more responsibility in course design. The materials for the first course comprise

- practice guides on leading discussion, lecturing, demonstrating, marking and giving feedback, supervising, organising, reviewing and improving teaching;
- reader: with chapters on leading discussion, lecturing, demonstrating, marking and giving feedback, supervising, how students learn, how students differ as learners, how students develop as learners, how lecturers learn;
- assessment guides, including building a portfolio, how to find and work with a mentor, as well as access to email-discussions, tutor, etc.

In developing these materials, and in common with institutions running face-to-face sessions, choices have to be made and issues confronted. The generic nature of professional development courses is a major issue. On the one hand, once you spend some time working with academics from different disciplines, you can see all sorts of similarities and in any case, few institutions can afford to mount specialist courses for specific disciplines. On the other hand, each discipline is convinced that their problems are special and unique to them.

To counter the customary dismissal of generic materials about teaching and learning in higher education, subject specific units are being developed to augment the first of the Open University courses, including a Guide to Mathematical

Practices (Mason 1999). The aim of the guide is to appeal to mathematicians and to anyone teaching mathematics to undergraduates in disciplines such as engineering or science. It provides very specific suggestions about ways to interact with students so as to support them in taking initiative and in making sense of the mathematics they are encountering.

However, you cannot discuss practices without discussing philosophies or perspectives on pedagogy, and you cannot discuss either of these significantly without a vocabulary of technical terms. One of the tell-tale features of absence of attention to details of teaching in higher education, at least in mathematics, is the lack of vocabulary for talking about what we do when we are engaging with students, or preparing for such engagement. For example, we can usually distinguish between lecturing and tutoring (running 'examples' classes etc.), though sometimes in tutorials the distinction is blurred. Exposition (in which the teacher tries to draw students into their world, and uses the presence of the students to make fresh contact with the ideas) often displaces explanation (when the tutor tries to enter and stay in the word of the students' experience of the content). But there is no recognised and accepted word to refer to acts of teaching. What do we call particular acts of teaching such as pausing for effect (when we expect students to be rehearsing what was just said, in their minds), or using different sides of the room or different boards or screens for the intuitive-informal and for formal expression?

In the Guide, I used the term *gambit*, building on resonance with chess in which a short sequence of moves appears at first to be a diversion, but later yields a stronger position. In teaching, getting students to take initiative often seems to take up valuable time or even result in temporary loss of control of the class. But often these gambits result in everybody learning more effectively and more efficiently, so that time can be made up later. In an expanded version of the guide, which is in preparation as a free standing book (Mason, in press), I use the term {\it tactic} as an acronym for Teaching Act Contributing To Initiative Control. This builds on the notion that the teacher uses tact to act in ways that provoke students to take initiative. For example, getting students to repeat to each other in pairs, some complex sentence just uttered by the lecturer, or their answer to some question, can encourage many more students to participate in thinking about the topic, not just taking notes about it. The notion of a tactic as an act of teaching also reflects my own view that 'teaching takes place in time (through specific acts) while learning takes place over time (as an essentially invisible maturation process)'. The book extends the gambits included in the guide to over 40 different tactics, as well as suggestions on how to advise students to develop good mathematical study habits, ways of constructing tasks for students to explore, to exercise on, and to be assessed by, all based on a collection of inescapable teaching tensions and fundamental frameworks for informing acts of teaching (a vocabulary). A different collection of tactics can be found in Hubbard (1990) and more generic ones in Gibbs (see Technical and Educational Services Web site: http://www.Web-direct.co.uk/tes/). A compendium of techniques for assessing students in mathematics can be found in Gold, Keith and Marion (1999).

6. DIFFICULTIES IN SUPPORTING COLLEAGUES

There are a number of obstacles that must be overcome if mathematicians are to be attracted to working on their teaching.

First, teaching must be acknowledged as a fundamental component of being expert in any subject, and certainly in mathematics. It is not enough just to prove theorems. It is vital, as one's career develops, to become more aware of historical roots, of the conceptual stumbling blocks or obstacles that made development of the topic bumpy and irregular rather than smooth and logical. The reason is that, unless circumvented, some of these may be reflected in student experience. It is also a major contribution to simplify the presentation of complex ideas. Indeed, this is vital if a domain is to remain alive and fresh, and if it is to attract younger researchers to work on the current unsolved problems.

Second, and at the same time, evidence-based quality teaching must be acknowledged in promotion cases. Of course, this requires development of acceptable and meaningful evidence. Student comments have been used extensively in North America, but they provide little evidence of working on one's teaching and are subject to the whims of students who see themselves as having no further stake in the outcome.

Third, work on teaching can actually improve your research, for there is a mutuality between research and teaching. As you become more sensitive to the needs and experiences of your students, that awareness can feed into awareness about how you are engaging in research, thereby enabling you to make more informed choices and so to improve your research. But it also works the other way round. As you become more aware of your own thinking processes, you are likely to become more sensitive to student needs. Of course, as an expert in mathematics who probably had few difficulties in the early stages, you may find entering the experience of some students difficult. They seem not to know very much. But if you make use of analogy, noting where you yourself are struggling and how you make progress, and then seeking analogous behaviour (or lack of it) among your students, you can develop sensitivity to students as well.

Fourth, mathematicians need assistance in recognising that mathematical epistemology does not work well in human sciences. Mathematicians are used to definitions, theorems, and proofs, and there are precious few if any of these in education. Almost any assertion that has some informative power or use in teaching has the property that its converse is sometimes true as well. It is exceedingly difficult in education to state both what might be invariant and what the allowable range of conditions might be to preserve that invariant, if indeed it is ever possible, whereas this is the very essence of mathematics. This tension between mathematical and educational thinking can be a major stumbling block for many who are used to the relative confidence and certainty of mathematics. Education is, for me at least, about sensitivity to people, about becoming aware of my own experience so as to be able to recognise the experiences of others and thus be able to be of assistance to them. Decisions made in the moment are not logical or even rational, though they may have a rational basis. Rather, they are intuitive, sensitive, made because an idea or possibility comes to mind just then.

Fifth, the vast and growing literature on adult education is not attractive to someone who wants to get on with their teaching and their research. Yet many of the principal issues encountered by most lecturers have been studied and there is a growing vocabulary for them. The more this expands, the more it becomes a hurdle for the subject expert who wants a few pointers to enable them to tinker with their teaching and get on with their research.

The ILT is trying to address the reasonable complaint that teaching is not sufficiently valued in higher education especially when it comes to promotion. In the past the only evidence available to include in the teaching component of a case for promotion has been student feedback and quotations from colleagues. Yet student feedback is notoriously unrobust. The ILT has initiated a programme which is intended to provide more concrete evidence, and this is essential if teaching is to form a serious component of promotion cases, and hence to be seen as a significant part of every academic's professional life.

7. FURTHER DEVELOPMENTS

The ILT have begun publishing their own journal: Active Learning in Higher Education (ILT Web site: http://www.ilt.ac.uk). They will also publish two series of books, one a series of monographs on key issues associated with teaching and learning, the other practitioner-orientated edited collections with a subject-teaching focus. Their Web site also provides access to a database of communication and information technology teaching materials. Whether the benefits of membership will carry the Institute to its goal of 25% or more of the teachers in higher education in a few years time, and whether it will be able to establish an ethos in which it is accepted as a norm that teaching involves working on one's teaching not just showing up for lectures and tutorials, remains to be seen. As one eminent mathematician remarked, having read through an early draft of my book, "This makes me feel guilty that I do not put as much time into teaching as I should". I suggested that perhaps "could" would be more helpful than "should".

For mathematicians, mathematical symbol-processing software coupled with on-screen manipulation such as in dynamic geometry software suggests a powerful medium waiting to be exploited for the presentation of mathematics. While there are many fascinating mathematical Web sites, the most useful seem to be the encyclopaedias and glossaries, whether of curves, of number sequences, of technical terms, or of historical mathematicians. We are witnessing a renewal of that tremendous force in the 18th century to capture everything in a compendium of knowledge. When you look at Web sites devoted to mathematical topics, you tend to find exposition. There are, it is true, sites featuring dynamic images and interactivity, and as Java programming becomes easier, there will be many more. But mostly what people seem to be doing is using the Web to tell others what they have found out about some topic. The novelty of expressing one-self in the new media may blind us to the need for students to take initiative and work on ideas themselves. Replacing static textbooks with clickable hypertext and diagrams may give the illusion of interactivity, but may not lead to significant changes in student learning. Or perhaps this is old educational thinking. Perhaps the whole structure of what it means to know is in flux. Whereas we used to understand when we could reconstruct for ourselves, in future we may be considered to understand if we can quickly find an explanation on the Web which we can make sense of and use locally. Perhaps what students need in the future is the confidence to make use of brief exposition in order to make use of some idea found through a search engine. I am not convinced, but I detect movement in this direction. Whatever happens, it is vital that we work on our teaching, in order that students are stimulated to work at learning mathematics. It is essential that we approach teaching in a professional manner.

Postscript: The Open University courses have been withdrawn because they were unable to attract the necessary 400+ participants to make them viable.

REFERENCES

CeHEP site: URL: http://www2.open.ac.uk/cehep/.
Dearing, R., (1997a). *Higher Education in the Learning Society.* London: HMSO. URL: http://www.ncl.ac.uk/ncihe/index.htm
Dearing, R. (1997b). *Higher Education in the Learning Society: Introduction.* London: HMSO. URL: http://www.ncl.ac.uk/ncihe/sr_008.htm\#Introduction.
Deliberations Site: URL: http://www.lgu.ac.uk/deliberations/forum/index.cgi.
Gold, B., Keith, S. and Marion, W. (Eds.) (1999). *Assessment Practices in Undergraduate Mathematics.* Washington: MAA.
Hubbard, R. (1990). *53 Interesting Ways to Teach Mathematics.* Bristol: Technical and Educational Services.
ILT Web site: URL: http://www.ilt.ac.uk/.
Mason, J. (1999). *Mathematics Practice Guide,* (HE851). Milton Keynes: Open University.
Mason, J. (in press). *Teaching Mathematics at College Level.* Oxford-Brookes (The Oxford Centre for Staff and Learning Development: OCSLD). URL: http:// www.brookes.ac.uk/services/ocsd/home.html;
Course description: URL: http://www.brookes.ac.uk/services/ocsd/fwtitle.htm
SEDA Site: URL: http://www.mailbase.ac.uk/lists/seda/.
Technical and Educational Services Web site: URL: http://www.Web-direct.co.uk/tes/.

John Mason
The Open University, England
j.h.mason@open.ac.uk

ERICH CH. WITTMANN

THE ALPHA AND OMEGA OF TEACHER EDUCATION: ORGANIZING MATHEMATICAL ACTIVITIES

In future not guidance and receptivity, but organisation and activity will be the special mark of the teaching/learning process.

Johannes Kühnel (1869-1928)

1. INTRODUCTION

The aim of this paper is to describe an introductory mathematics course for primary student teachers and to explain the philosophy behind it.

The paper is structured as follows: It starts with a general plea for placing the mathematical training of any category of students into their professional context. Then the context of primary education in Germany, with its strong emphasis on the principle of learning by discovery, is sketched. The third and main section of the paper presents the 'O-script/A-script method', a special teaching/learning format for stimulating student teachers' mathematical activities along the principle of learning by discovery. In section 4 special attention is given to the notion of proof in the context of primary teacher education. The paper concludes with some observations of how student teachers evaluate this approach.

2. MATHEMATICS IN CONTEXTS

It is a most remarkable phenomenon that the teaching and learning of mathematics at the university level which was hardly a subject of public discussion in the past is now attracting world wide attention. The Discussion Document for the ICMI Study on this topic (ICMI 1997) lists five external reasons for this changing attitude:

1. the increase in the number of students who are attending tertiary institutions;
2. pedagogical and curriculum changes that have taken place at the pre-university level;
3. the increasing differences between secondary and tertiary mathematics;
4. the rapid development of technology;
5. demands on universities to be accountable.

Derek Holton (Ed.), The Teaching and Learning of Mathematics at University Level: An ICMI Study, 539—552.
© 2001 Kluwer Academic Publishers. Printed in the Netherlands.

steady rise of formalism and structuralism culminating in Bourbaki's monolithic architecture of mathematics. However, by the end of the seventies this programme despite its success in some fields of mathematics turned out as a failure as a universal programme, as did similar structuralistic programmes in other areas, for example linguistics and architecture. At that time it was widely recognized that in no field of study could semantics be replaced by syntax. Postmodern philosophy rediscovered the meaningful context as an indispensable aspect of all human activity, including mathematical activity. As far as details of the changing views of mathematics are concerned I refer to Davis and Hersh (1981) and Ernest (1998).

As a consequence, we have to conceive of 'mathematics' not solely as an academic field of study but as a broad societal phenomenon. Its diversity of uses and modes of expression is only in part reflected by the kind of specialized mathematics which we typically find in university departments. I suggest a use of capital letters to describe MATHEMATICS as mathematical work in the broad sense including mathematics in science, engineering, economics, industry, commerce, craft, art, education, daily life, and so forth, and including the customs and requirements specific to these contexts. Of course, specialized mathematics is a central part of MATHEMATICS. But mathematicians cannot and must not claim a monopoly for the whole. It is unjustified to assume that any piece of mathematics would form an absolute body of knowledge carrying its potential applications in itself. In his paper "The pernicious influence of mathematics on science" J.T. Schwartz used drastic words to warn mathematical specialists of applying mathematics to other fields without paying proper attention to the context (Schwartz 1986).

The consequences for the teaching and learning of mathematics at the university should be clear: In teaching mathematics to non-specialists the professional context of the addressees has to be taken fundamentally and systematically into account. The context of mathematical specialists is appropriate for the training of specialists, not for the training of non-specialists.

In the present paper the professional context to be considered is teaching mathematics at the primary level. There are mathematicians who look down on this task. In my view this is a fundamental mistake. The importance of primary mathematics within MATHEMATICS can hardly be overestimated. After all, it is at this level where the systematic encounter of children with mathematics begins and where the points for their whole mathematical education are set. I would like to refer here to the wisdom of the Tao-te-ching:

> Plan difficult things at the very beginning when they are still easy.
> Care for big things as long as they are still small.

Although many elements of the context of primary teacher education are specific the general approach adopted in this paper might be interesting for developing mathematical courses for other professional fields, too.

3. THE CONTEXT OF TEACHER EDUCATION

Since the beginning of the eighties the development of primary education in the State of Northrhine-Westfalia has exerted a great influence on the other German States[1]. The boundary conditions for primary mathematics education in Northrhine-Westfalia are special in two respects:

1. In the first phase[2] of their education at the university all primary student teachers have to study three subjects: German language, mathematics and a third subject (for example, environmental education, physical education, art, etc.). One of the three subjects has to be chosen as a major subject (45 credit hours out of the 120 credit hours of the whole 3-year programme). Two other (minor) subjects cover 25 credit hours[3]. As a consequence mathematics is compulsory for all primary student teachers. Roughly 90% of them choose mathematics as a minor subject (25 credit hours).

2. The syllabus for primary schools (grades 1 to 4) adopted in 1985 marked an important turning point in the history of public education in Germany. For the first time the principle of learning by discovery was explicitly prescribed as the basic principle of teaching and learning (Kultusminister des Landes Nordrhein-Westfalen 1985, §3):

> The tasks and objectives of mathematics teaching are best served by a conception in which learning mathematics is considered as a constructive and investigative process. Therefore teaching has to be organized such that children are offered as many opportunities as possible for self-reliant learning in all phases of the learning process:
>
> 1. starting from challenging situations; stimulating children to observe , to ask questions, to guess;
> 2. exposing a problem or a complex of problems for investigation; encouraging individual approaches; offering help for individual solutions;
> 3. relating new results to known facts in a diversity of ways; presenting results in a more and more concise way; assisting to memory storage; stimulating individual practice of skills;
> 4. talking about the value of new knowledge and about the process of acquiring it; suggesting the transfer to new, analogous situations.
>
> The task of the teacher is to find and to offer challenging situations, to provide children with substantial materials and productive ways of practising skills, and, above all, to build up and sustain a form of communication which serves the learning processes of all children.

This emphasis on mathematical processes instead of ready-made subject matter is visible in other parts of the syllabus, too. For example, the first section 'Tasks and objectives' lists the following four 'general objectives' of mathematics teaching: Mathematizing, Exploring, Reasoning and Communicating. Obviously, these

---

[1] With 17 million people Northrhine-Westfalia is the largest German State.

[2] The first phase (3 years) is followed by the second phase (2 years) which is spent at special institutions in close proximity of schools.

[3] 25 credit hours are for general education (pedagogy, psychology, ...).

objectives reflect basic components of doing mathematics at all levels. The fourth section of the syllabus describes in some detail why mathematical structures on the one hand and applications of mathematics on the other hand are two sides of one coin and how these two aspects can be interlocked in teaching. The explicit statement of this complementarity is also novel for German primary schools.

The development of this new syllabus was certainly very much influenced by similar developments in other European countries, in particular, the Netherlands. However, there has also been a strong trend towards active learning within German mathematics education. At the beginning of this century, Johannes Kühnel, one of the leading figures of progressive education in Germany, wrote his famous book 'Neubau des Rechenunterrichts' ('Reconstructing the Teaching of Arithmetic') in which he described the 'teaching/learning method of the future' as follows (Kühnel 1954, 70):

> The learner will no longer be expected to receive knowledge, but to acquire it. In future not guidance and receptivity, but organisation and activity will be the special mark of the teaching/learning process.

Since the late eighties considerable progress has been made in developing practical approaches and materials for this new conception of primary mathematics teaching including innovative textbooks (cf. Winter, 1987, Wittmann and Müller, 1994-97, and Becker and Selter, 1996). The project 'mathe 2000'[4] has played a leading role in this development. Of course the implementation of these materials depends crucially on the teachers' ability to abandon the deeply rooted guidance/receptivity model of teaching and learning in favour of the organisation/activity model. However, as experience shows, it is not enough just to describe new ways of teaching in general terms. The natural way to stimulate and to support the necessary change within the school system is to restructure teacher education according to the organisation/activity model. Only teachers with first hand experiences in mathematical activity can be expected to apply active methods in their own teaching as something natural and not as something imposed from outside. Therefore all efforts in pre-service and in-service teacher education have to be concentrated on reviving student teachers' and teachers' mathematical activity.

Interestingly, the new emphasis on student activity is not restricted to teacher education, it is a general phenomenon of the present discussion about teaching mathematics at the university level (cf. the section 'Student Activity' in ICMI 1997).

---

[4]'mathe 2000', founded in 1987, is a 'curriculum' project at the University of Dortmund (Directors: Gerhard N. Müller, Heinz Steinbring and Erich Ch. Wittmann). In the past the project has been concerned with developing theoretical concepts and practical materials for the teaching of mathematics at the primary level, including an innovative textbook series. However, the project is based on a comprehensive perspective of mathematics teaching and will be extended to the secondary level in the near future. As a special feature of the project, instructional design, empirical research, pre-service teacher training, in-service teacher training, counselling, and public relations are closely linked and are pursued simultaneously. Essential for this 'systemic-evolutionary' approach is the establishment of a 'Theory-Practice-Network' bringing together all partners of the educational system in the very process of development. Here the 'Handbook for Practising Skills in Arithmetic' (2 vols.) has played a fundamental role as a basic reference. For an overview see Müller, Steinbring and Wittmann 1997.

More and more mathematicians are taking special care of stimulating student activities. Bill Jacob's 'Linear Functions and Matrix Theory' (Jacob 1995) is a good example.

4. THE O-SCRIPT/A-SCRIPT METHOD

The traditional pattern of introductory mathematics courses at German universities is a combination of a 2 to 4 hours per week lecture ('Vorlesung') on the one hand and 2 hours of practice ('Übungen') which take place in groups of about 30 students on the other hand. I am well aware that expository teaching can be very stimulating and that work in groups based on good problems can arouse students' thinking and communication as well. Nevertheless I contend that *grosso modo* the lecture/practice pattern has a strong inherent tendency towards guidance and receptivity: Often the tasks and exercises offered to students for elaboration require mainly or even merely a reproduction of the conceptual and technical tools introduced in the lecture. So more or less students' individual work and work in groups tends to be subordinated to the lecture. Frequently, work in groups degenerates into a continuation of the lecture: The graduate student responsible for the group just presents the correct solutions of the tasks and exercises.

The lecture/practice format is particularly common in courses for large groups of students. In fact if you are confronted with numbers of students as large as 400 to 600, as we are in our primary teacher education programme, there is a strong pressure towards guidance/receptivity, and it is hard to think of alternatives.

However, the more I got involved in developmental research along the lines of learning by discovery the more I felt the contradiction between the teaching/learning model which I followed in my mathematical courses and the teaching/learning model which I recommended in my courses in mathematics education.

The O-script/A-script method has been developed as an attempt to mitigate this cognitive conflict. The basic idea, the **Alpha** and **Omega**, of this method is very simple: Just take Johannes Kühnel literally in teacher education and replace 'guidance and receptivity' by '**O**rganisation and **A**ctivity', that is, use both the lecture and the group work for organizing student activities.

An essential ingredient of this new teaching/learning format is a clear distinction between the text written down by the lecturer on the blackboard or the overhead projector and the text elaborated by the individual student. As the lecturer's main task is to organize students' learning her or his text is called the 'O-script'. It is not a closed text, but it contains many fragments, leaves gaps, and often gives only hints. Therefore it is a torso to be worked on. As the elaborated text expresses the student's individual activity it is called the 'personal A-script'.

The regulations of our teacher education programme do not allow for making the A-script obligatory. However, the A-script can be used as an additional qualification by students who fail the final test. Experience shows that the majority of student teachers is willing to write an A-script.

How to organize students' activity in a lecture? In trying to find an answer to this question I got inspired by two quotations:

> We should teach more along problems than along theories. A theory should be developed only to the extent that is necessary to frame a certain class of problems. (Giovanni Prodi)

> The main goal of all science is first to observe, then to explain phenomena. In mathematics the explanation is the proof. (David Gale)

Accordingly, I divided the course in two parts: The first part was devoted to introducing and clarifying a list of 50 carefully selected generic problems which should be elaborated in the A-scripts. The second systematic part should present a theoretical framework for these problems, however, based on students' experiences in writing the A-scripts. The second part did not differ from ordinary lectures. I think this format absolutely appropriate at this place of the learning process. Actually, I don't see a substitute for it.

The following areas which are closely related to the contents of the primary curriculum were covered in the course: (1) Place Value Systems, (2) Elementary Combinatorics, (3) Arithmetic Progressions, (4) Sequences, (5) Elementary Number Theory.

These areas are rich playgrounds for genuine mathematical activities. By using the opportunities offered in the course student teachers acquire not only the appropriate background knowledge which enables them to look at the primary curriculum from a higher level. They also acquire first-hand experiences in mathematizing, exploring, reasoning, and communicating.

The 10 problems selected for the area 'Arithmetic Progressions' are as follows:

1. (From Butts, 1973.) Try to decompose the set $\{1, 2, 3,..., n\}$ of the first n natural numbers into two subsets such that the sum of the numbers in one subset is equal to the sum of the numbers in the other subset. For which n is this possible? For which n not?
2. Investigate the analogous problem for the set $\{2, 4, ..., 2n\}$ of the first n even numbers.
3. Investigate the analogous problem for the set $\{1, 3, ..., 2n\text{-}1\}$ of the first n odd numbers.
4. Which numbers can be represented as sums of consecutive numbers?
5. Which numbers can be represented as sums of 2 (or 3, 4, ...) consecutive numbers?
6. In how many ways can 1000 be represented as a sum of consecutive numbers?
7. In how many ways can 1000 be represented as a sum of consecutive odd numbers?
8. From Monday to Friday 60 little lambs were born on a pasture: on Tuesday 3 more than on Monday, on Wednesday 3 more than on Tuesday, on Thursday 3 more than on Wednesday, and on Friday 3 more than on Thursday. How many lambs were born on each day?
9. (From Steinbring, 1997.) In the following scheme (see Figure 1) the number in the circle (the 'addition number') and the number in the first box ('the starting number') can be chosen arbitrarily. The numbers in the other four boxes are calculated inductively according to the following rule (see the example in Figure 1): The number in a box is the number in the preceding box plus the addition number. The numbers in all five boxes are added to give the final result (the 'target'). How to choose the starting number and the addition number in order to get the target 50? How many solutions do exist? Which numbers can be obtained as targets? (In this problem and the next one natural numbers and the number 0 are admitted.)
10. Investigate the same problem for 6 boxes instead of 5.

The list of these 10 problems has been constructed by employing the 'method of generating problems' (Wittmann, 1971). So the use of heuristic strategies is ensured. Problem 8 is taken from a textbook for grade 4, problem 9 from a paper on the findings of a teaching experiment based on this problem. Therefore student teachers can see explicit connections with the primary curriculum[5]. As these connections are reinforced in the subsequent maths education course the maths courses become meaningful for student teachers within their professional context.

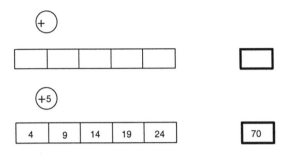

Figure 1

In the first part of the course each weekly lecture introduced 5 problems to the student teachers for investigation. The problems were explained in full detail and it was indicated how these problems could be attacked in different ways by using various 'enactive', 'iconic' and 'symbolic' representations. The main heuristic strategies as described, for example, in Polya (1981), Mason (1982) or Schoenfeld (1985), were explained by referring to the problems of the course. However, no solutions were given.

The student teachers had less problems with developing ideas. The real challenge was how to formulate a coherent text. "What should an A-script look like?" was a frequent question. So parts of the lecture as well as of the group work had to address this difficulty. Referring to some examples I indicated in my lectures how the gaps of the O-script can be filled to get an A-script. In addition, student teachers were allowed to submit drafts of their A-scripts for critical reading, and could revise them according to the comments they received.

At the end of the first part of the course the students (at least the brave ones) had intensively worked on 50 selected problems. Even when they hadn't solved all problems properly, they had experienced a variety of mathematical phenomena. This was a good basis for the theoretical framework developed in the subsequent second part of the course.

For example, the problems on arithmetic progressions were theoretically framed by proofs of the sum formula and of the following remarkable theorem by J.J. Sylvester: The number of representations of a number n as a sum of consecutive numbers is equal to the number of odd divisors of n.

Both proofs were based on ideas that had been developed by students before.

Interestingly, the extended work on problems in the first part paid off in the second part: the course 'covered' the same mathematical content as courses in the ordinary format usually do.

---

[5] After their own work on problem 9 student teachers were shown a video on a teaching experiment in which a group of 12 fourth graders had found all solutions within 30 minutes.

5. OPERATIVE PROOFS

As stated at the beginning, the basic tenet of the present paper is that the mathematical training of student teachers should reflect their professional context. This requirement is particularly critical when it comes to proofs.

It should be obvious that the notion of formal proof related to deductively structured theories is inappropriate or even counterproductive as a background for appreciating 'Reasoning' as an objective of primary mathematics. That is not to say, however, that the notion of proof is irrelevant for primary mathematics. On the contrary. Fortunately, contemporary views of proof allow for an intellectually honest incorporation of proof into both primary teacher education and primary teaching. Studies in the history and philosophy of mathematics have destroyed the long held formalistic doctrine that the only rigorous form of proof is a formal proof. It has turned out that the notion of formal proof has its clear limitations, particularly from the point of view of the practising mathematician (cf., for example, Branford, 1913, Hardy, 1929, Thom, 1973, Davis and Hersh, 1981, Atiyah, 1984, Long, 1986 and Thurston 1994). In a letter submitted to the working group on proof at ICME 7, Québec 1992, Yuri I. Manin expressed his broader understanding of 'proof as a journey' very nicely:

> Many working mathematicians feel that their occupation is discovery rather than invention. My mental eye sees something like a landscape; let me call it a 'mathscape'. I can place myself at various vantage points and change the scale of my vision; when I start looking into a new domain, I first try a bird's eye view, then strive to see more details with better clarity. I try to adjust my perception to guess at a grand design in the chaos of small details and afterwards plunge again into lovely tiny chaotic bits and pieces.
>
> Any written text is a description of a part of the mathscape, blurred by the combined imperfections of vision and expression. Every period has its own social conventions, and the aesthetics of the mathematical text belong to this domain. The building blocks of a modern paper (ever since Euclid) are basically axioms, definitions, theorems and proofs, plus whatever informal explanations the author can think of.
>
> Axioms, definitions and theorems are spots in a mathscape, local attractions and crossroads. Proofs are the roads themselves, the paths and the highways. Every itinerary has its own sightseeing qualities, which may be more important than the fact that it leads from A to B.
>
> With this metaphor, the perception of the basic goal of a proof, which is purportedly that of establishing 'truth', is shifted. A proof becomes just one of many ways to increase the awareness of a mathscape.
>
> Any chain of argument is a one-dimensional path in a mathscape of infinite dimensions. Sometimes it leads to the discovery of its end-point, but as often as not we have already perceived this end-point, with all the surrounding terrain, and just did not know how to get there.
>
> We are lucky if our route leads us through a fertile land, and if we can lure other travellers to follow us.

In mathematics education this new view of proof has been reflected in many papers (cf., for example, de Villiers, 1997). Based on Semadeni's and Kirsch's proposals of 'pre-mathematical' or 'pre-formal' proofs (Semadeni 1974, Kirsch 1979), the concept of 'operative proof' has been developed (Wittmann 1997). An operative proof is a proof which is embedded in the exploration of a mathematical problem context and which is based on the effects of operations exerted thereby on meaningfully represented mathematical objects.

For this reason operative proofs explain phenomena which were observed before (cf. Gale's statement quoted above) and thus they contribute to understanding mathematics.

As also non-symbolic representations can be used operative proofs are particularly useful for the early grades and for primary teacher education. I would like to demonstrate this by giving two examples from my introductory course on arithmetic.

Example 1: Infinity of primes.

The formal proof of the infinity of prime numbers runs as follows: Let us assume that the set of prime numbers is finite: $p_1, p_2, ..., p_r$. The number $n = p_1 p_2 ... p_r + 1$ has a divisor p that is a prime number. Therefore n is divisible by one of the numbers $p_1, ... , p_r$. From $p|n$ and $p|p_1 p_2 ... p_r$ we conclude that p also divides the difference $n - p_1 p_2 ... p_r = 1$. However, $p|1$ is a contradiction of the fact that 1 is not divisible by a prime number. Therefore the assumption was wrong.

The following operative proof of the infinity of primes is based on the representation of natural numbers on the number line. One of the problems that the student teachers had to investigate was the determination of primes by means of the sieve of Eratosthenes. Therefore they knew from their own experience how the sieve works. Using this knowledge the infinity of primes can be proved just by explaining why the iterative sieve procedure does not stop: Assume that in finding primes we have arrived at a prime number p. Then p is encircled and all multiples of p are cancelled. The product

$$n = 2\times3\times5\times7\times11\times ... \times p$$

is a common multiple of all primes sieved out so far. So it was cancelled at every previous step of the procedure. As no cancellation process following the selection of a prime can hit adjacent numbers the successor of n has not been cancelled yet. Therefore after every step there are numbers left and the smallest of them is a new prime number.

Example 2: Sylvester's theorem.

In the first part of the course the student teachers worked with arithmetic progressions and investigated the representation of natural numbers as sums of consecutive numbers. Based on their experiences the following operative proof of Sylvester's theorem emerged in a natural way: Sums of consecutive numbers are represented as staircases. Depending on the parity of the number of stairs, each staircase can be transformed into a rectangular shape that represents a product. If the parity is odd, there is a middle stair and the upper part of the staircase can be cut off and added to the lower part (Figure 2).

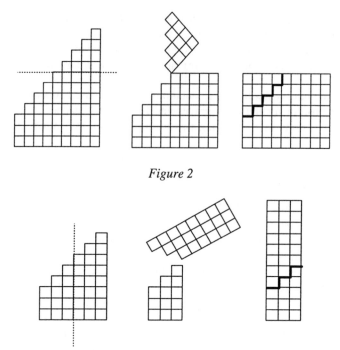

Figure 2

Figure 3

If the parity of stairs is even then the staircase can be divided vertically in the middle and the two parts fit together to make a rectangular shape (Figure 3).

A careful study of the effects of these two operations shows that in both cases an odd divisor of the represented number arises: either the number of stairs or the sum of the heights of the first and last stair (which must be odd for an even number of stairs). As a consequence any staircase representation of a number gives rise to an odd factor of n. But the converse is also true: A rectangle with an odd side can be transformed into a staircase of one of the two types depending on the relative size of the odd factor. A closer inspection reveals that this relationship between staircase representations and rectangular representations of n is bijective.

Again this operative proof explains phenomena which are well known from previous work on problems.

The advantage of operative proofs in the context of teacher education is obvious: These proofs are not separated from this context but closely related to it. In becoming acquainted with operative proofs student teachers learn to appreciate the use of informal means of representations for doing mathematics at early levels. Often, elements of such activities in teacher education can immediately be implanted into primary teaching. Consider, for example, the following exercise from a textbook for the second grade:

$$1 + 2 + 3 =$$

$$2 + 3 + 4 =$$
$$3 + 4 + 5 =$$
$$4 + 5 + 6 =$$
$$\cdots \quad \cdots$$

Looking at the results children discover the times 3-row. If the sums are represented by three columns of counters, the displacement of one counter to make a rectangle is obvious. This work with counters is a good – and in my view also a necessary – preparation for algebra where the same exercise can be resumed as follows:

$$(a - 1) + a + (a + 1) = 3a.$$

6.   EXPERIENCES WITH THE COURSE

Feedback from student teachers collected by means of a questionnaire after the introductory course on elementary geometry showed that the 'O-script/A-script' method was accepted by 75% of the population. The writing of the A-script was experienced as a very time-consuming, but effective exercise. In the same vein 70% affirmed that their understanding of the principle of learning by discovery had been improved.

However, only 59% of the students indicated that the course had had a more or less positive influence on their view of mathematics. 41% expressed their concerns about the openness of the first part. This result is not surprising as at school many students are programmed as receivers of knowledge. The adopted definitely mechanistic and formalistic attitude towards mathematics gives them a feeling of security and helps them 'to survive'. Feeling comfortable with mechanistic routines in the system of school and university (!) they do not want to be confronted with uncertainty.

The unfavourable influence of mathematical experiences from school is particularly apparent in student teachers' preconceptions of operative proofs. An instructive example was reported in Wittmann and Müller (1990). In a seminar student teachers studied figurate numbers[6]. In particular trapezoid numbers were introduced as a composition of square and triangular numbers (see Figure 4).

In looking for patterns the students guessed that for all n the trapezoid number T_n and n leave the same remainder modulo 3. For this relationship an operative proof (at that time called 'iconic proof') was given which was based on the corresponding pattern.

---

[6] In history figurate numbers played a fundamental role as a cradle of number theory. We are convinced that these numbers are also a wonderful context for stimulating mathematical activities in children. As a consequence figurate numbers play an important role in 'mathe 2000'.

Figure 4. Trapezoid Number $T_n = n^2 + n(n-1)/2 = (3n^2 - n)/2$

Right after this demonstration some students expressed their doubt on its validity. The teacher didn't intervene and quickly the whole group agreed that the demonstration could only claim the status of an illustration, not the status of a proof. The teacher then offered a formal proof and confronted it with the operative proof. The student teachers were invited to think about these two types of proof and to write down their opinions. The papers showed very clearly how the student teachers' appreciation of operative proofs was inhibited by the understanding of proof that they had acquired at school. For illustration I quote from some papers:

> The symbolic proof is to be preferred because it is more mathematical.

> The iconic proof is much more intuitive for me and explains much better what the problem is. For me the inferences drawn from patterns of dots are convincing and sufficient as a proof. Unfortunately we have not been made familiar with this type of proof at school. Only symbolic proofs have been taught.

> The iconic proof is very intuitive. One understands the connections from which the statement flows. I can't imagine how a counterexample could be found, because it does not matter how many 3-columns can be constructed. In my opinion it is nevertheless not a proof, but only a demonstration, which, however, holds for all n. At school I learned that only a symbolic proof is a proof.

> The symbolic proof is more mathematical. This proof is more demanding, as some formulae are involved which you have to know and to recall. The iconic proof can be followed step by step, and each step is immediately clear. However, I wonder if an iconic proof would be accepted in examinations.

Cognitive conflicts in accepting operative proofs as valid proofs have to be understood as natural symptoms of a metamorphosis lifting student teachers to higher professional levels. Experience shows that in retrospect student teachers consciously appreciate teacher education programmes which are embedded in the professional context. In a recent survey by the centre of teacher education at the

University of Dortmund 2700 student teachers in Northrhine-Westfalia in their second phase of training were asked to evaluate the courses in mathematics and mathematics education they had received in the first phase of their training at the university (Zentrum für Lehrerbildung 1997). The results are very encouraging (Figure 5). The evaluations of the programmes at the universities Paderborn and Dortmund which share the same philosophy were much higher than those of the six other universities in Northrhine-Westfalia which offer courses in primary teacher education.

Mathematics

(1 = very little; 2 = fairly little; 3 = fairly much; 4 = very much)

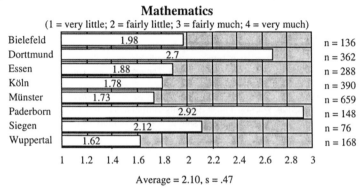

Bielefeld	1.98	n = 136
Dorttmund	2.7	n = 362
Essen	1.88	n = 288
Köln	1.78	n = 390
Münster	1.73	n = 659
Paderborn	2.92	n = 148
Siegen	2.12	n = 76
Wuppertal	1.62	n = 168

1   1.2   1.4   1.6   1.8   2   2.2   2.4   2.6   2.8   3

Average = 2.10, s = .47

Figure 5

A team of 16 authors has just written a book 'Arithmetic as a Process' (Müller, Steinbring and Wittmann, 2001) which is based on the O/A approach to teacher education described in this paper. This book is a truly mathematical book, but unlike other books it consciously puts mathematics in the context of teacher education – neither by sacrificing education to mathematics nor mathematics to education.

REFERENCES

Atiyah, M. (1984). Interview with Michael Atiyah. *Mathematical Intelligencer*. 6, 9-19.

Becker, J. and Selter, Ch. (1996). Elementary School Practices. In A. Bishop, C. Keitel, C. Laborde, K. Clements and J. Kilpatrick (Eds.), *International Handbook of Mathematics Education*, Part 1, pp. 511-564. Dordrecht: Kluwer.

Branford, B. (1913). *Betrachtungen über mathematische Erziehung vom Kindergarten bis zur Universität*. Leipzig: Teubner.

Butts, Th. (1973). *Problem Solving in Mathematics. Elementary Number Theory and Arithmetic*. Glenview, Ill.: Scotts Foresman.

Davis, Ph. and Hersh, R. (1981). *The Mathematical Experience*. Boston: Birkhauser.

Ernest, P. (1998). A Postmodern Perspective on Research in Mathematics Education. In A. Sierpinska and J. Kilpatrick (Eds.), *Mathematics Education as a Research Domain: A Search for Identity, An ICMI Study*, Book 1, pp. 71-85. Dordrecht: Kluwer.

Hanna, G. (1983). *Rigorous Proof in Mathematics Education*. Toronto: Ontario Institute for Studies in Education.

Hardy, G.H. (1929). Mathematical Proof. *Mind* 38, No. 149, 1-25.

ICMI 1997: On the Teaching and Learning of Mathematics at the University Level. Discussion Document. *ICMI Bulletin*, No. 43, 3-13.

Jacob, B. (1995). *Linear Functions and Matrix Theory*. New York: Springer.

Kirsch, A. (1979). Beispiele für prämathematische Beweise. In W. Dörfler, and R. Fischer (Eds.), *Beweisen im Mathematikunterricht*, pp. 261-274. Wien: Hölder-Pichler-Tempsky/Teubner.

Kühnel, J. (1954). *Neubau des Rechenunterrichrs*. Bad Heilbrunn:Klinkhardt.

Kultusminister des Landes Nordrhein-Westfalen (1985). *Richtlinien und Lehrpläne für die Grundschule in Nordrhein-Westfalen, Mathematik*, Köln.

Long, R. (1986). Remarks on the History and Philosophy of Mathematics. *American Mathematical Monthly*,93, 609-619.

Mason, J. (1982) *Thinking Mathematically*. London: Addison Wesley.

Müller, G.N., Steinbring, H. and Wittmann, E.Ch. (Eds.) (1997). *10 Jahre 'mathe 2000', Bilanz und Perspektiven*. Leipzig: Klett.

Müller, G.N., Steinbring, H. and Wittmann, E.Ch. (2001). *Arithmetik als Prozess*. Leipzig: Klett.

Polya, G. (1981). *Mathematical Discovery. On Understanding, Learning and Teaching Problem Solving*. Combined Edition. New York: Wiley.

Schoenfeld, A. (1985). *Mathematical Problem Solving*. New York: Academic Press.

Schwartz, J.T. (1986). The Pernicious Influence of Mathematics on Science. In: Marc Kac, Gian Carlo Rota and Jacob T. Schwartz (Eds.), *Discrete Thoughts. Essays on Mathematics, Science and Philosophy*, pp. 19-25. Boston: Birkhauser.

Semadeni, Z. (1974). *The Concept of Premathematics as a Theoretical Background For Primary Mathematics Teaching*. Warsaw: Polish Academy of Mathematical Sciences.

Steinbring, H. (1997). Epistemological investigation of classroom interaction in elementary mathematics teaching. *Educational Studies in Mathematics*, 32, 49-72.

Thom, R. (1973). Modern mathematics - Does it exist? In: G. Howson (Ed.), *Developments in Mathematical Education*, pp. 194-212. Cambridge: CUP.

Thurston W. P. (1998). On Proof and Progress in Mathematics. *Bulletin of the American Mathematical Society*, 30(2), 161–177.

de Villiers, M. (Ed.) (1997). Proceedings of Topic Group 8 'Proofs and Proving: Why, when, and how?', ICME 8 Seville, Spain, 1996. Centrahill: Ass.Math. Ed. South Africa.

Winter, H. (1987). *Mathematik entdecken. Neue Ansätze für den Mathematikunterricht in der Grundschule*. Frankfurt a.M.: Scriptor.

Wittmann, E.Ch. (1971). Complementary Attitudes in Problem Solving. *Educational Studies in Mathematics*, 4, 241-253.

Wittmann, E.Ch. (1997). Operative Proofs. In de Villiers, M. (Ed.) Proceedings of Topic Group 8 'Proofs and Proving: Why, when, and how?', ICME 8 Seville, Spain, 1996, pp. 15-22. Centrahill: Ass.Math. Ed. South Africa.

Wittmann, E.Ch. and Müller, G.N. (1990). When is a proof a proof? *Bull. Soc. Math. Belg.* 42, I, Ser. A, 15-42.

Wittmann, E.Ch. and Müller, G.N. (1994-97). *Das Zahlenbuch. Mathematik für die Grundschule* (4 vols.). Leipzig: Klett.

Zentrum für Lehrerbildung (1997). Grundschullehrer(innen)-Ausbildung zwischen Fachwissenschaft, Fachdidaktik und Praxis. *Universität Dortmund, ZfL-Info*, 2, 7-12.

Erich Ch. Wittmann
University of Dortmund, Germany
ewittmann@mathematik.uni-dortmund.de

INDEX

New ICMI Studies Series

KLUWER ACADEMIC PUBLISHERS – DORDRECHT / BOSTON / LONDON